# Mathematical Control Theory and Finance

Andrey Sarychev · Albert Shiryaev
Manuel Guerra · Maria do Rosário Grossinho
Editors

# Mathematical Control Theory and Finance

 Springer

Prof. Andrey Sarychev
DiMaD
University of Florence
via Cesare Lombroso, 6/17
50134 - Florence
Italy
asarychev@unifi.it

Prof. Albert Shiryaev
Steklov Mathematical Institute of the
Russian Academy of Sciences
Gubkin str. 8
119991 Moscow
Russia
albertsh@mi.ras.ru

Dr. Manuel Guerra
ISEG-TU Lisbon
Rua do Quelhas, 6
1200-781 Lisbon
Portugal
mguerra@iseg.utl.pt

Dr. Maria do Rosário Grossinho
ISEG-TU Lisbon
Rua do Quelhas, 6
1200-781 Lisbon
Portugal
mrg@iseg.utl.pt

ISBN 978-3-540-69531-8     e-ISBN 978-3-540-69532-5

Library of Congress Control Number: 2008929628

*Cover design:* WMXDesign GmbH, Heidelberg, Germany

Printed on acid-free paper

9 8 7 6 5 4 3 2 1

springer.com

# Preface

Control theory provides a large set of theoretical and computational tools with applications in a wide range of fields, running from "pure" branches of mathematics, like geometry, to more applied areas where the objective is to find solutions to "real life" problems, as is the case in robotics, control of industrial processes or finance.

The "high tech" character of modern business has increased the need for advanced methods. These rely heavily on mathematical techniques and seem indispensable for competitiveness of modern enterprises. It became essential for the financial analyst to possess a high level of mathematical skills. Conversely, the complex challenges posed by the problems and models relevant to finance have, for a long time, been an important source of new research topics for mathematicians.

The use of techniques from stochastic optimal control constitutes a well established and important branch of mathematical finance. Up to now, other branches of control theory have found comparatively less application in financial problems.

To some extent, deterministic and stochastic control theories developed as different branches of mathematics. However, there are many points of contact between them and in recent years the exchange of ideas between these fields has intensified. Some concepts from stochastic calculus (e.g., rough paths) have drawn the attention of the deterministic control theory community. Also, some ideas and tools usual in deterministic control (e.g., geometric, algebraic or functional-analytic methods) can be successfully applied to stochastic control.

We strongly believe in the possibility of a fruitful collaboration between specialists of deterministic and stochastic control theory and specialists in finance, both from academic and business backgrounds. It is this kind of collaboration that the organizers of the *Workshop on Mathematical Control Theory and Finance* wished to foster.

This volume collects a set of original papers based on plenary lectures and selected contributed talks presented at the *Workshop*. They cover a wide range of current research topics on the mathematics of control systems and applications to finance. They should appeal to all those who are interested in research at the junction of these three important fields as well as those who seek special topics within this scope.

The editors of these proceedings express their deep gratitude to those who contributed with their work to this volume and those who kindly helped us in peer-reviewing them.

We are thankful to the scientific and organizing committees of the Workshop on Mathematical Control Theory and Finance as well as the plenary lecturers and all the participants; their presence and their work formed the main contribution for the success of the event.

We thank the event financial supporters:

FCT – Fundação para a Ciência e a Tecnologia,
Banco de Portugal,
Fundação Calouste Gulbenkian,
REN – Rede Eléctrica Nacional,
UECE – Unidade de estudos sobre a Complexidade na Economia),
Luso–American Foundation,
Caixa Geral de Depsitos,
Fundação Oriente,
Delta Cafés.

Our gratitude goes to the academic institutions who jointly organized the Workshop on mathematical Control Theory and Finance:

CEMAPRE – Centro de Matemática Aplicada à Previsão e Decisão Económica,
CEOC – Centro de Estudos em Optimização e Controlo,
CIM – Centro Internacional de Matemática,
ISR – Institute of Systems and Robotics.

It goes equally to Instituto Superior de Economia e Gestão, Technical University of Lisbon which hosted the event and provided unreserved support on facilities, staff and logistics.

A special word of thanks to Maria Rosário Pato and Rita Silva from CEMAPRE for their secretary work and constant support.

Lisbon, Florence & Moscow,                      *Maria do Rosário Grossinho*
                                                *Manuel Guerra*
                                                *Andrey Sarychev*
                                                *Albert Shiryaev*

# Contents

**Extremals Flows and Infinite Horizon Optimization**
Andrei A. Agrachev, Francesca C. Chittaro .......................... 1

**Laplace Transforms and the American Call Option**
Ghada Alobaidi, Roland Mallier ................................. 15

**Time Change, Volatility, and Turbulence**
Ole E. Barndorff-Nielsen, Jürgen Schmiegel ....................... 29

**External Dynamical Equivalence of Analytic Control Systems**
Zbigniew Bartosiewicz, Ewa Pawłuszewicz ......................... 55

**On Option-Valuation in Illiquid Markets: Invariant Solutions to a Nonlinear Model**
Ljudmila A. Bordag ............................................ 71

**Predicting the Time of the Ultimate Maximum for Brownian Motion with Drift**
Jacques du Toit, Goran Peskir .................................. 95

**A Stochastic Demand Model for Optimal Pricing of Non-Life Insurance Policies**
Paul Emms ................................................... 113

**Optimality of Deterministic Policies for Certain Stochastic Control Problems with Multiple Criteria and Constraints**
Eugene A. Feinberg ............................................ 137

**Higher-Order Calculus of Variations on Time Scales**
Rui A. C. Ferreira, Delfim F. M. Torres ......................... 149

Finding Invariants of Group Actions on Function Spaces, a
General Methodology from Non-Abelian Harmonic Analysis
*Jean-Paul Gauthier, Fethi Smach, Cedric Lemaître, Johel Miteran* ..... 161

Nonholonomic Interpolation for Kinematic Problems, Entropy
and Complexity
*Jean-Paul Gauthier, Vladimir Zakalyukin* ........................... 187

Instalment Options: A Closed-Form Solution and the Limiting
Case
*Susanne Griebsch, Christoph Kühn, Uwe Wystup* ................... 211

Existence and Lipschitzian Regularity for Relaxed Minimizers
*Manuel Guerra, Andrey Sarychev* .................................. 231

Pricing of Defaultable Securities under Stochastic Interest
*Nino Kordzakhia, Alexander Novikov* .............................. 251

Spline Cubatures for Expectations of Diffusion Processes and
Optimal Stopping in Higher Dimensions (with Computational
Finance in View)
*Andrew Lyasoff* .................................................. 265

An Approximate Solution for Optimal Portfolio in Incomplete
Markets
*Francesco Menoncin* .............................................. 293

Carleman Linearization of Linearly Observable Polynomial
Systems
*Dorota Mozyrska, Zbigniew Bartosiewicz* .......................... 311

Observability of Nonlinear Control Systems on Time Scales -
Sufficient Conditions
*Ewa Pawłuszewicz* ................................................ 325

Sufficient Optimality Conditions for a Bang-bang Trajectory
in a Bolza Problem
*Laura Poggiolini, Marco Spadini* .................................. 337

Modelling Energy Markets with Extreme Spikes
*Thorsten Schmidt* ................................................ 359

Generalized Bayesian Nonlinear Quickest Detection Problems:
On Markov Family of Sufficient Statistics
*Albert N. Shiryaev* ............................................... 377

Necessary Optimality Condition for a Discrete Dead Oil
Isotherm Optimal Control Problem
*Moulay Rchid Sidi Ammi, Delfim F. M. Torres* ..................... 387

**Managing Operational Risk: Methodology and Prospects**
*Grigory Temnov* ..................................................397

# List of Contributors

**Agrachev, Andrei A.**
SISSA-ISAS,
via Beirut 2-4, 34014 Trieste, Italia.
agrachev@sissa.it

**Alobaidi, Ghada**
Department of Mathematics,
American University of Sharjah,
Sharjah, United Arab Emirates.
galobaidi@aus.edu

**Barndorff-Nielsen, Ole E.**
The T.N. Thiele Centre for Applied
Mathematics in Natural Science,
Department of Mathematical Sciences, University of Aarhus,
Ny Munkegade, DK-8000 Aarhus C,
Denmark. oebn@imf.au.dk

**Bartosiewicz, Zbigniew**
Białystok Technical University,
Faculty of Computer Sciences,
Wiejska 45A, Białystok, Poland.
bartos@pb.bialystok.pl

**Bordag, Ljudmila A.**
Halmstad University,
Box 823, 301 18 Halmstad, Sweden.
Ljudmila.Bordag@ide.hh.se

**Chittaro, Francesca C.**
SISSA-ISAS,
via Beirut 2-4, 34014 Trieste, Italia.
chittaro@sissa.it

**du Toit, Jacques**
School of Mathematics,
The University of Manchester,
Oxford Road, Manchester M13 9PL,
United Kingdom. Jacques.Du-Toit@
postgrad.manchester.ac.uk

**Emms, Paul**
Faculty of Actuarial Science and
Insurance, Cass Business School,
City University,
106 Bunhill Row, London
EC1Y 8TZ, United Kingdom.
p.emms@city.ac.uk

**Feinberg, Eugene A.**
State University of New York at
Stony Brook,
Stony Brook, NY 11794-3600, USA.
Eugene.Feinberg@sunysb.edu

**Ferreira, Rui A. C.**
Department of Mathematics,
University of Aveiro,
3810-193 Aveiro, Portugal.
ruiacferreira@yahoo.com

**Gauthier, Jean-Paul**
LE2I, UMR CNRS 5158,
Université de Bourgogne,
Bat. Mirande, BP 47870,
21078, Dijon CEDEX France.
gauthier@u-bourgogne.fr

**Griebsch, Susanne**
Frankfurt School of Finance &
Management,
Sonnemannstrasse 9-11, 60314
Frankfurt am Main, Germany
s.griebsch@frankfurt-school.de

**Guerra, Manuel**
CEOC and ISEG–T.U.Lisbon,
R. do Quelhas 6, 1200-781 Lisboa,
Portugal. mguerra@iseg.utl.pt

**Kordzakhia, Nino**
Macquarie University,
NSW 2109, Australia.
nino.kordzakhia@mq.edu.au

**Kühn, Christoph**
Frankfurt MathFinance Institute,
Goethe-University Frankfurt,
Robert-Mayer-Straße 10, 60054
Frankfurt am Main, Germany,
ckuehn@math.uni-frankfurt.de

**Lemaître, Cedric**
LE2I, UMR CNRS 5158,
Université de Bourgogne,
Bat. Mirande, BP 47870,
21078, Dijon CEDEX France.
Cedric.Lemaitre@u-bourgogne.fr

**Lyasoff, Andrew**
Mathematical Finance Program,
Boston University, USA.
alyasoff@bu.edu

**Mallier, Roland**
Department of Applied Mathematics,
University of Western Ontario,
London ON Canada, Canada.
rolandmallier@hotmail.com

**Menoncin, Francesco**
Dipartimento di Scienze Economiche,
Via S. Faustino, 74/B – 25122 Brescia
– Italia. menoncin@eco.unibs.it

**Miteran, Johel**
LE2I, UMR CNRS 5158,
Université de Bourgogne,
Bat. Mirande, BP 47870,
21078, Dijon CEDEX France.
miteranj@u-bourgogne.fr

**Mozyrska, Dorota**
Białystok Technical University,
Faculty of Computer Sciences,
Wiejska 45A, Białystok, Poland.
admoz@w.tkb.pl

**Novikov, Alexander**
University of Technology,
Sydney, NSW 2007, Australia.
alex.novikov@uts.edu.au

**Pawłuszewicz, Ewa**
Białystok Technical University,
Faculty of Computer Sciences,
Wiejska 45A, Białystok, Poland.
epaw@pb.edu.pl

**Peskir, Goran**
School of Mathematics,
The University of Manchester,
Oxford Road, Manchester
M13 9PL, United Kingdom.
goran@maths.man.ac.uk

**Poggiolini, Laura**
Dipartimento di Matematica
Applicata "G. Sansone",
Università di Firenze,
Via S. Marta, 3,
I–50139 Firenze, Italia.
laura.poggiolini@math.unifi.it

**Sarychev, Andrey**
DiMaD, University of Florence,
via Cesare Lombroso 6/17, 50134–
Firenze, Italia. asarychev@unifi.it

**Schmidt, Thorsten**
Department of Mathematics,
University of Leipzig,
D-04081 Leipzig, Germany.
thorsten.schmidt
@math.uni-leipzig.de

**Schmiegel, Jürgen**
The T.N. Thiele Centre for Applied
Mathematics in Natural Science,
Department of Mathematical Sci-
ences, University of Aarhus,
Ny Munkegade, DK-8000 Aarhus C,
Denmark. schmiegl@imf.au.dk

**Shiryaev, Albert N.**
Steklov Mathematical Institute of
the Russian Academy of Sciences,
Moscow, Russia.
albertsh@mi.ras.ru

**Sidi Ammi, Moulay Rchid**
Department of Mathematics,
University of Aveiro,
3810-193 Aveiro, Portugal.
sidiammi@ua.pt

**Smach, Fethi**
LE2I, UMR CNRS 5158,
Université de Bourgogne,
Bat. Mirande, BP 47870,
21078, Dijon CEDEX France.
smach_fethi@yahoo.fr

**Spadini, Marco**
Dipartimento di Matematica
Applicata "G. Sansone",
Università di Firenze,
Via S. Marta, 3,
I–50139 Firenze, Italia.
marco.spadini@math.unifi.it

**Temnov, Grigory**
Vienna University of Technology,
Institute for Mathematical Methods
in Economics, Financial and Actuar-
ial Mathematics,
Wiedner Hauptstrasse 8-10/105-1,
A-1040 Vienna, Austria.
gtemnov@fam.tuwien.ac.at

**Torres, Delfim F. M.**
Department of Mathematics,
University of Aveiro,
3810-193 Aveiro, Portugal.
delfim@ua.pt

**Wystup, Uwe**
Frankfurt School of Finance &
Management,
Sonnemannstrasse 9-11, 60314
Frankfurt am Main, Germany.
uwe.wystup@mathfinance.com

**Zakalyukin, Vladimir**
Moscow State University,
119 993, Leninski Gori, 1, Moscow,
Russia. zakalyu@mail.ru

# Extremals Flows and Infinite Horizon Optimization

Andrei A. Agrachev and Francesca C. Chittaro

SISSA-ISAS, via Beirut 2-4, 34014 Trieste, Italia.
`agrachev@sissa.it`, `chittaro@sissa.it`

**Summary.** We study the existence and the structure of smooth optimal synthesis for regular variational problems with infinite horizon. To do that we investigate the asymptotic behavior of the flows generated by the extremals (of finite horizon problems) using curvature–type invariants of the flows and some methods of hyperbolic dynamics.

## 1 Introduction

Given a smooth function $\varphi : \mathbb{R}^n \times \mathbb{R}^n \to \mathbb{R}$ we would like to minimize the functional

$$J(\gamma) = \int_0^\infty \varphi(\gamma(t), \dot\gamma(t)) \, dt$$

defined on the Lipschitzian curves $\gamma : [0, +\infty) \to \mathbb{R}^n$ such that the integral (1) converges. More precisely, we assume that $\varphi(0,0) = 0$, $\frac{\partial\varphi}{\partial q}(0,0) = 0$ and that $\varphi(q,0), \frac{\partial\varphi}{\partial q}(q,0)$ do not vanish simultaneously for $q$ different from 0; then we take $q \in \mathbb{R}^n$ and try to find

$$\min\{\int_0^\infty \varphi(\gamma(t), \dot\gamma(t)) \, dt : \gamma(0) = q, \ \lim_{t\to\infty} \gamma(t) = 0\}.$$

Such a problem has no solutions in too many interesting cases and it is natural to modify the functional introducing the discount or "forgetting" factor $\alpha > 0$. Namely, we set

$$J^\alpha(\gamma) = \int_0^\infty e^{-\alpha t} \varphi(\gamma(t), \dot\gamma(t)) \, dt. \tag{1}$$

Integral (1) may converge even in the case of an unbounded function $t \mapsto \varphi(\gamma(t), \dot\gamma(t))$ and we try to minimize $J^\alpha(\gamma)$ on the curves with the initial condition $\gamma(0) = q$ and certain asymptotic conditions as $t \to +\infty$ to be specified later.

The restriction of a minimizing curve $t \mapsto q(t)$ to the segment $[0, T]$ is automatically a minimizer for the finite horizon functional $\int_0^T \varphi(\gamma(t), \dot{\gamma}(t))\, dt$ with boundary conditions $\gamma(0) = q(0)$, $\gamma(T) = q(t)$. Extremal paths of the finite horizon problems are described either by the Euler–Lagrange second order ordinary differential equation on $\mathbb{R}^n$ or by the Hamiltonian system on $\mathbb{R}^n \times \mathbb{R}^n$. We use the Hamiltonian approach, in the spirit of the Pontryagin Maximum Principle.

Namely, we consider the Hamiltonian

$$H(p, q) = \max_{u \in \mathbb{R}^n}(\langle p, u \rangle - \varphi(q, u)).$$

If $(p, q) \mapsto H(p, q)$ is a well-defined smooth function on $\mathbb{R}^n \times \mathbb{R}^n$ then for any solution $q(t)$, $t \in [0, T]$, of the finite horizon problem (without discount) there exists a Lipschitzian curve $p(t)$, $t \in [0, T]$, such that the pair $(p(t), q(t))$ satisfies the Hamiltonian system

$$\begin{cases} \dot{p} = -\frac{\partial H}{\partial q}(p, q) \\ \dot{q} = \frac{\partial H}{\partial p}(p, q) \end{cases}. \tag{2}$$

In the case of the discount factor $\alpha$ system (2) has to be substituted by the nonautonomous Hamiltonian system

$$\begin{cases} \dot{p} = -e^{-\alpha t}\frac{\partial H}{\partial q}(e^{\alpha t}p, q) \\ \dot{q} = \frac{\partial H}{\partial p}(e^{\alpha t}p, q) \end{cases}. \tag{3}$$

It is convenient to rescale the auxiliary variable $p$ by setting $\xi = e^{\alpha t}p$; then we arrive to the autonomous "rescaled" system

$$\begin{cases} \dot{\xi} = \alpha\xi - \frac{\partial H}{\partial q}(\xi, q) \\ \dot{q} = \frac{\partial H}{\partial \xi}(\xi, q) \end{cases}. \tag{4}$$

In order to see what kind of solution we can expect, let us look at the elementary one-dimensional model.

**Example.** Let $n = 1$, $\varphi(q, \dot{q}) = \frac{1}{2}(\dot{q}^2 - rq^2))$, where $r$ is a constant. Then $H(p, q) = \frac{1}{2}(p^2 + rq^2)$ and Hamiltonian system takes the form

$$\begin{cases} \dot{p} = -rq \\ \dot{q} = p \end{cases}. \tag{5}$$

The phase portraits are saddles for $r < 0$ and centers for $r > 0$ (Fig. 1 and Fig. 2).

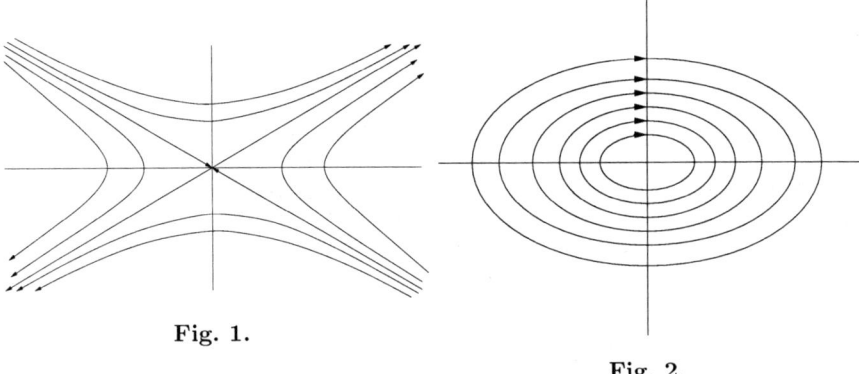

**Fig. 1.**

**Fig. 2.**

Let $r < 0$, then for any $q_0 \in \mathbb{R}$ there exists a unique $p_0$ such that $(p_0, q_0)$ belongs to the stable submanifold of the equilibrium $(0,0)$. The solution $(p(t), q(t))$ of system (5) with initial conditions $(p(0), q(0)) = (p_0, q_0)$ tends to $(0,0)$ with the exponential rate as $t \to +\infty$. It is easy to show that $q(\cdot)$ is optimal trajectory.

If $r > 0$, then no one solution of system (5) tends to the equilibrium and optimal trajectories simply do not exist. Moreover, even the finite horizon problems do not have solutions if the horizon $T$ is greater than one half of the period of the trajectories of system (5) and the infimum of the functional equals $-\infty$. What happens if we introduce the discount factor $\alpha$?

The rescaled system (4) takes the form:

$$\begin{cases} \dot{\xi} = \alpha\xi - rq \\ \dot{q} = p \end{cases} . \tag{6}$$

The phase portrait of system (6) is an unstable focus for $\alpha < 2\sqrt{r}$ and an unstable node for $\alpha > 2\sqrt{r}$. Optimal solutions of the finite horizon problems do not exist for sufficiently big $T$ while the phase portrait is a focus and do exist for all $T > 0$ when the phase portrait is a node (Fig. 3). Optimal solutions of the horizon $T$ problem with boundary points $\gamma(0) = q_0$, $\gamma(T) = 0$ is the projection to the axis $\{q\}$ (horizontal axis on Fig. 3) of the trajectory of system (6) which starts at the line $\{(\xi, q_0) : \xi \in \mathbb{R}\}$ and arrives to the line $\{(\xi, 0) : \xi \in \mathbb{R}\}$ at the time moment $T$.

Let $t \mapsto z_T(t; q_0)$ be such a trajectory of (6), $z_T(t; q_0) = (\xi_T(t; q_0), q_T(t; q_0))$. Now fix $q_0$ and send $T$ to $+\infty$. It is easy to see that there exists the limit $\lim\limits_{T \to +\infty} z_T(t; q_0) = z_\infty(t; q_0)$ and, moreover, $t \mapsto z_\infty(t; q_0)$ is a trajectory of system (6) which belongs to a proper invariant subspace of this system. System (6) has two proper invariant subspaces corresponding to the eigenvalues $\frac{\alpha}{2} \pm \sqrt{\frac{\alpha^2}{4} - r}$ of the matrix $\begin{pmatrix} \alpha & -r \\ 1 & 0 \end{pmatrix}$. Trajectories $z_\infty(\cdot; q_0)$ belong to the invariant subspace related to the lower eigenvalue. This sub-

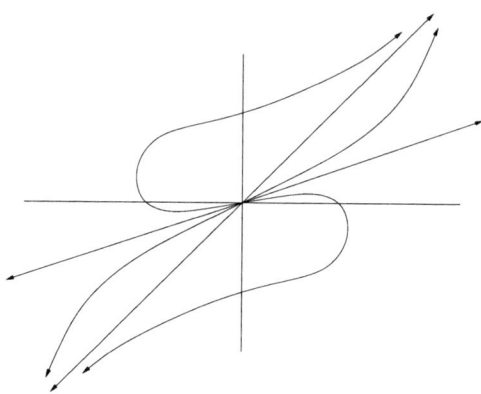

**Fig. 3.**

space can also be characterized as the locus of solutions to system (6) whose norm grows slower than $e^{\frac{\alpha t}{2}}$ as $t \to \infty$.

Clearly, the projection $q_\infty(\cdot; q_0)$ of the curve $z_T(\cdot; q_0)$ can be treated as optimal trajectory for the infinite horizon problem with discount factor $\alpha$. It is indeed optimal if we accept the following general definition.

**Definition 1.** *We say that a locally Lipschitzian curve* $\tilde{\gamma} : [0, +\infty) \to \mathbb{R}^n$ *is an optimal trajectory for the infinite horizon problem with the discount factor* $\alpha$ *if*

$$J^\alpha(\tilde{\gamma}) = \{\min J^\alpha(\gamma) : \gamma(0) = \tilde{\gamma}(0), \lim_{t \to +\infty} e^{-\alpha t}|\dot{\gamma}(t)|^2 = 0\}.$$

Optimality of $q_\infty(\cdot; q_0)$ in this sense is not immediately seen but easily follows from a general result (see Theorem 2 below).

Another observation derived from our model is that optimal trajectories corresponding to different initial conditions all together form a smooth optimal synthesis according to the following definition.

**Definition 2.** *We call smooth optimal synthesis a smooth complete vector field* $X$ *on* $\mathbb{R}^n$ *such that all solutions of the equation* $\dot{q} = X(q)$ *are optimal trajectories of the infinite horizon variational problem under consideration (the problem may be with or without discount).*

In what concerns our example, $X(q)$ is a unique number such that $(X(q), q)$ belongs to the stable invariant subspace of system (5) in the case of negative $r$ and to the "less unstable" invariant subspace of system (6) in the case of positive $r$ and discount factor $\alpha > 2\sqrt{r}$.

The goal of this paper is to characterize a broad class of nonlinear multi-dimensional problems which possess a smooth optimal synthesis similarly to the just considered elementary 1-dimensional example. The role of the parameter $r$ will be played by the "curvature" of the Hamiltonian system.

## 2 Curvature

Given a smooth Hamiltonian $H(p, q)$, we denote by symbol $\mathbf{H}$ the corresponding Hamiltonian vector field, i.e.

$$\mathbf{H} = \frac{\partial H}{\partial p} \frac{\partial}{\partial q} - \frac{\partial H}{\partial q} \frac{\partial}{\partial p}.$$

In particular, system (2) is shortly described as

$$\dot{z} = \mathbf{H}(z), \tag{7}$$

where $z = (p, q) \in \mathbb{R}^n \times \mathbb{R}^n$.

In this paper we deal only with complete vector fields and use exponential notations for the flows generated by systems of ordinary differential equations. In particular,

$$e^{t\mathbf{H}} : \mathbb{R}^{2n} \to \mathbb{R}^{2n}, \quad t \in \mathbb{R},$$

is the Hamiltonian flow generated by system (7).

Given $z \in (\mathbb{R}^n \times \mathbb{R}^n) = \mathbb{R}^{2n}$, we set

$$V_z^t = D_{e^{t\mathbf{H}}(z)} \left( e^{-t\mathbf{H}} \right) (\mathbb{R}^n \times 0),$$

a family of $n$-dimensional subspaces of $\mathbb{R}^{2n}$. Here $D_z \Phi$ is the differential of the diffeomorphism $\Phi$ at the point $z$. Note that $V_z^t$ are Lagrangian subspaces of the space $\mathbb{R}^n \times \mathbb{R}^n$ endowed with the standard symplectic structure $\sigma = dp \wedge dq$. Indeed, the "vertical subspace" $\mathbb{R}^n \times 0$ is Lagrangian and the Hamiltonian flow preserves symplectic structure.

The family of subspaces $t \mapsto V_z^t$ is thus a curve in the Lagrange Grassmannian $L(n)$ and $V_z^0 = \mathbb{R}^n \times 0$. We call it the *Jacobi curve* at $z$ associated to the Hamiltonian field $\mathbf{H}$.

Recall that a tangent vector to the Grassmannian $\xi \in T_V L(n)$ is naturally identified with a linear mapping $\underline{\xi} : V \to \mathbb{R}^{2n}/V$, where $\underline{\xi}$ is defined as follows: take a curve $V(t)$ such that $\xi = \dot{V}(0)$ and set $\underline{\xi}(v(0)) = \dot{v}(0) + V$, for any smooth curve $v(t) \in V(t)$ coordinate computations show that the result depends only on $\xi$ and $v(0)$.

Now assume that we are given a splitting $\mathbb{R}^{2n} = V_z^0 \oplus W_z^0$, $z \in \mathbb{R}^n$, where $W_z^0$ is a smooth family of Lagrangian subspaces. We set

$$W_z^t = D_{e^{t\mathbf{H}}(z)} \left( e^{-t\mathbf{H}} \right) W_z^0.$$

Obvious identifications $\mathbb{R}^{2n}/V_z^0 \cong W_z^0$, $\mathbb{R}^{2n}/W_z^0 \cong V_z^0$ give us the linear mappings

$$\underline{\dot{V}_z^0} : V_z^0 \to W_z^0, \quad \underline{\dot{W}_z^0} : W_z^0 \to V_z^0.$$

Composition of these mappings is a linear operator

$$R_z = -\underline{\dot{W}_z^0} \circ \underline{\dot{V}_z^0}$$

on $V_z^0$. Operator $R_z$ is called the curvature of the Hamiltonian field $\mathbf{H}$ at $z$ with respect to the given splitting.

In the next sections we deal only with Hamiltonians that are strongly convex with respect to $p$, i.e. $\frac{\partial^2 H}{\partial p^2}(z)$ is a positive definite quadratic form for any $z \in \mathbb{R}^{2n}$. In this case, the curvature $R_z$ is a self-adjoint operator on $V_z^0 = \mathbb{R}^n \times 0$ with respect to the Euclidean structure on $V_z^0$ defined by the quadratic form $\frac{\partial^2 H}{\partial p^2}(z)$. In particular, $R_z$ is a diagonalizable operator and all its eigenvalues are real. We write $R_z > 0 \, (< 0)$ if all eigenvalues of $R_z$ are positive (negative). Similarly, we write $R_z > cI \, (< cI)$ for some constant $c$, if $R_z - cI > 0 \, (R_z - cI < 0)$.

Any vector field $Y$ on $\mathbb{R}^{2n}$ is splitted $Y = Y_V + Y_W$, where $Y_V(z) \in V_z^0$, $V_W(z) \in W_z^0$, $\forall z \in \mathbb{R}^{2n}$. For any vector field $X$ which takes values in the distribution $V_z^0$, $z \in \mathbb{R}^{2n}$ we have:

$$R_z X(z) = -[\mathbf{H}, [\mathbf{H}, X]_W]_V(z),$$

where $[\cdot, \cdot]$ is the Lie bracket of vector fields, $[Y_1, Y_2] = \frac{dY_2}{dz}Y_1 - \frac{dY_1}{dz}Y_2$; see [1] for details.

The curvature, according to its definition, depends not only on the Hamiltonian field $\mathbf{H}$ but also on the choice of the Lagrangian distribution $W_z^0$, $z \in \mathbb{R}^{2n}$, that is transversal to the "vertical" distribution $V_z^0$. Is there any canonical choice of $W_z^0$? Yes, it is, under very modest regularity conditions on the Hamiltonian $H$.

The construction is described in [3], [1]. Briefly, a germ of a curve $t \mapsto \Lambda(t)$ in the Lagrange Grassmannian $L(n)$ is called *tame* if $\Lambda(t) \cap \Lambda(0) = 0$ for small $t \neq 0$ and this fact can be recognized from a finite jet of $\Lambda(\cdot)$ at 0. To any tame germ one can associate (in an intrinsic way) a *derivative element* $\Lambda^\circ \in L(n)$ such that $\Lambda(0) \cap \Lambda^\circ = 0$. This construction, applied to the germs at 0 of the curves $t \mapsto V_z^t$, defines a canonical splitting: $W_z^0 = (V_z)^\circ$.

In the case of a nondegenerate $\frac{\partial^2 H}{\partial p^2}(z)$ (the only interesting for us case in this paper) canonical $W_z^0$ is computed as follows: $W_z^0 = \{(Cq, q) : q \in \mathbb{R}^n\}$, where

$$2\frac{\partial^2 H}{\partial p^2} C \frac{\partial^2 H}{\partial p^2} = \left\{H, \frac{\partial^2 H}{\partial p^2}\right\} - \frac{\partial^2 H}{\partial p \partial q}\frac{\partial^2 H}{\partial p^2} - \frac{\partial^2 H}{\partial p^2}\frac{\partial^2 H}{\partial q \partial p}$$

and $\{\cdot, \}$ is the poisson brackets. In other words,

$$2\left(\frac{\partial^2 H}{\partial p^2} C \frac{\partial^2 H}{\partial p^2}\right)_{ij} = \sum_{k=1}^{n}\left(\frac{\partial H}{\partial p_k}\frac{\partial^3 H}{\partial p_i \partial p_j \partial q_k} - \frac{\partial H}{\partial q_k}\frac{\partial^3 H}{\partial p_i \partial p_j \partial p_k} - \right.$$

$$\left. \frac{\partial^2 H}{\partial p_i \partial q_k}\frac{\partial^2 H}{\partial p_k \partial p_j} - \frac{\partial^2 H}{\partial p_i \partial p_k}\frac{\partial^2 H}{\partial q_k \partial p_j}\right), \quad i, j = 1, \ldots, n.$$

The formulas are drastically simplified in the case of the Hamiltonian of a natural mechanical system: $H(p, q) = \frac{1}{2}|p|^2 + U(q)$. Then $W_z^0 = 0 \times \mathbb{R}^n$ and $R_z = \frac{\partial^2 U}{\partial q^2}$, the Hessian of the potential energy.

Now for any $\alpha \in \mathbb{R}$ we set:

$$H^\alpha = \mathbf{H} + \alpha p \frac{\partial}{\partial p},$$

the vector field which appears in the study of the variational problem with discount factor $\alpha$ (see (4)). This vector field is not Hamiltonian; it does not preserve symplectic form $\sigma$ but satisfies the identity:

$$\left(e^{tH^\alpha}\right)^* \sigma = e^{\alpha t} \sigma.$$

In particular, linear mappings $D_z \left(e^{tH^\alpha}\right)$ transform Lagrangian subspaces in the Lagrangian ones. This allows to define the curvature operator $R_z^\alpha$ of the field $H^\alpha$ in the same way as the curvature $R_z$ of the field $\mathbf{H}$: we simply substitute the Jacobi curves $V_z^t$ by the curves

$$V_z^{\alpha\, t} = D_{e^{tH^\alpha}(z)}\left(e^{-tH^\alpha}\right)(\mathbb{R}^n \times 0)$$

in the construction. Obviously, $V_z^{\alpha\, 0} = V_z^0$.

As we mentioned, the curvature is sensitive to the splitting $\mathbb{R}^{2n} = V_z^0 \oplus W_z^0$.

A simple calculation gives the following formula in the case of the canonical splitting. Let us consider Hamiltonians of the form

$$H(p, q) = g_q(p, p) + U(q),$$

where for any $q \in \mathbb{R}^n$ $g_q(p, p)$ is a positive definite quadratic form on $\mathbb{R}^n$. Let $R_z$ be defined by the canonical splitting $\mathbb{R}^{2n} = V_z^0 \oplus (V_z)^\circ$ associated to the curves $V_z^t$ and $R_z^\alpha$ defined by the canonical splitting $\mathbb{R}^{2n} = V_z^0 \oplus (V_z^\alpha)^\circ$; then we have:

$$R_z^\alpha = R_z - \frac{\alpha^2}{4} I. \tag{8}$$

# 3 Results

In what follows, we always assume (without additional mentioning) that function $u \mapsto \varphi(q, u)$, $u \in \mathbb{R}^n$, is strongly convex for any $q \in \mathbb{R}^n$, i.e. $\frac{\partial^2 \varphi}{\partial u^2} > 0$. As stated in the introduction, we also assume that $\varphi(0, 0) = 0$, $\frac{\partial \varphi}{\partial q}(0, 0) = 0$, and that $(0, 0)$ is the only point where $\varphi(0, 0)$ and $\frac{\partial \varphi}{\partial q}(0, 0)$ vanish simultaneously. This implies smoothness of the Hamiltonian

$$H(p, q) = \max_{u \in \mathbb{R}^n}(\langle p, u \rangle - \varphi(q, u))$$

on the domain of definition. Assumptions of any of Theorems 1, 2 below guarantee that $H(p, q)$ is defined for all $(p, q)$ and generates a complete Hamiltonian field.

**Theorem 1.** *Assume that there exist constants $a, b, c > 0$ such that*

- $\varphi(q, u) + c > 0$;
- $\dfrac{|u|}{\varphi(q,u)+c} \to 0$ *as* $|u| \to \infty$;
- $\left|\dfrac{\partial \varphi}{\partial q}(q, u)\right| \leq a(\varphi(q, u) + |u|) + b, \quad \forall q, u \in \mathbb{R}^n$.

*If the curvature $R_z$ of the Hamiltonian field $\mathbf{H}$ with respect to some Lagrangian splitting is negative for any $z \in \mathbb{R}^{2n}$, then the infinite horizon variational problem (without discount) admits a smooth optimal synthesis.*

**Theorem 2.** *Let $\varphi(q, u) = g_q(u, u) + f(q)$, where for any $q \in \mathbb{R}^n$ $g_q(\cdot, \cdot)$ is a positive definite quadratic form. Assume that there exists a nondegenerate quadratic form $\phi$ on $\mathbb{R}^n \times \mathbb{R}^n$ such that*

$$\left|\frac{\partial^i}{\partial q^i}(\varphi - \phi)(q, u)\right| \to 0 \quad as \quad (|q| + |u|) \to \infty \qquad (9)$$

*for any multi-index $i$ such that $|i| \leq 2$ (i.e. $\varphi$ is asymptotically quadratic at infinity). If the curvature $R_z$ of the Hamiltonian field $\mathbf{H}$ with respect to the canonical splitting satisfies inequalities $0 < R_z \leq C$ for some constant $C$ and any $z \in \mathbb{R}^{2n}$, then the infinite horizon variational problem without discount does not admit optimal trajectories, while the problem with any discount factor $\alpha > 2\sqrt{C}$ admits an optimal synthesis of class $C^1$.*

## 4 Sketch of proofs

We give only main ideas of the proofs. See [5] for the complete proof of Theorem 1; the complete proof of Theorem 2 will appear elsewhere.

The construction of the optimal synthesis is similar to one in the elementary examples described in the introduction. We find a stable Lagrangian submanifold of system $\dot{z} = \mathbf{H}(z)$ in the case of negative curvature and a "less unstable" invariant submanifold of system $\dot{z} = H^\alpha(z)$. Moreover, we show that these submanifolds have a form

$$\{(p, q) : p = \Phi(q), \ q \in \mathbb{R}^n\},$$

where $\Phi$ is a vector-function. In the case of the stable manifold, $\Phi$ has the same class of smoothness as $\mathbf{H}$, while in the "less unstable" case we can guarantee only $C^1$-smoothness. The equation

$$\dot{q} = \frac{\partial H}{\partial p}(\Phi(q), q)$$

provides optimal synthesis. Optimality is proved by a straightforward generalization to the infinite horizon of a standard sufficient optimality condition (see[4, Th. 17.1]).

Let us now explain the construction of the invariant submanifolds. To this end we need a better understanding of the geometric meaning of the curvature operator. Consider a splitting $\mathbb{R}^{2n} = \Lambda \oplus \Delta$, where $\Delta$ and $\Lambda$ are Lagrangian subspace. Symplectic form $\sigma$ defines a nondegenerate pairing $(\lambda, \delta) \mapsto \sigma(\lambda, \delta)$ of the subspaces $\Lambda$ and $\Delta$. We thus obtain a canonical isomorphism $\Delta \cong \Lambda^*$. Any transversal to $\Delta$ $n$-dimensional subspace $V$ is the graph of a linear mapping $S_V : \Lambda \to \Delta = \Lambda^*$. Subspace $V$ is Lagrangian if and only if $S_V$ is self-adjoint, i.e. $S_V^* = S_V$.

In other words, we may associate to $V$ a quadratic form $\bar{S}_V : v \mapsto \sigma(v, S_V v)$ on $V$ and the relation $V \mapsto \bar{S}_V$ is a one-to-one correspondence between transversal to $\Delta$ Lagrangian subspaces and quadratic forms on $\Lambda$. It is easy to see that $\dim(V \cap W) = \dim \ker(\bar{S}_V - \bar{S}_W)$. Of course, the type of the quadratic form (i.e. its signature and rank) depends on the splitting.

In fact, for any symmetric $(n \times n)$-matrix $A$ there exists a splitting $\mathbb{R}^{2n} = \Lambda \oplus \Delta$ and coordinates on $\Lambda$ such that $A$ is the matrix of the quadratic form $\bar{S}_V$. Moreover, given a pair of Lagrangian subspaces $V, W$ and a pair of symmetric $(n \times n)$-matrices $A, B$ with the only restriction $\dim \ker(A - B) = \dim(V \cap W)$, there exists a Lagrangian splitting $\mathbb{R}^{2n} = \Lambda \oplus \Delta$ and coordinates on $\Lambda$ such that matrices of the quadratic forms $\bar{S}_V$ and $\bar{S}_W$ are $A$ and $B$.

On the other hand, given a curve $t \mapsto V(t)$ the type of the form $\bar{S}'_{V(t)} = \frac{d}{dt} \bar{S}_{V(t)}$ does not depend on the splitting and is the same for all curves obtained from $V(\cdot)$ by linear symplectic transformations of $\mathbb{R}^{2n}$. In particular, quadratic forms $\bar{S}'_{V_z^\alpha t}$ and $\bar{S}'_{W_z^\alpha t}$ are equivalent to the forms $\bar{S}'_{V_{e^t H^\alpha}^{\alpha 0}(z)}$ and $\bar{S}'_{W_{e^t H^\alpha}^{\alpha 0}(z)}$ due to the identities

$$V_z^{\alpha t + \tau} = \left(D_{e^{tH^\alpha}(z)} e^{-tH^\alpha}\right) V_{e^{tH^\alpha}(z)}^{\alpha \tau}, \quad W_z^{\alpha t + \tau} = \left(D_{e^{tH^\alpha}(z)} e^{-tH^\alpha}\right) W_{e^{tH^\alpha}(z)}^{\alpha \tau}.$$

Finally, the curvature $R_z^\alpha$ is negative (positive) if and only if quadratic forms $\bar{S}'_{V_z^{\alpha 0}}$ and $\bar{S}'_{V_z^{\alpha 0}}$ are both sign-definite and have opposite (equal) signs.

The proofs and other information on geometry of Lagrangian Grassmannians can be found in [2], [1]. Anyway, all this implies the following

**Lemma 1.** *If $R_z^\alpha < 0$ $\forall z \in \mathbb{R}^{2n}$, then there exist limits $\lim_{t \to \pm\infty} V_z^{\alpha t} = V_z^{\alpha \pm}$.*

**Proof.** Quadratic form $\bar{S}'_{V_z^{\alpha 0}}$ is equivalent to the form $-\frac{\partial^2 H}{\partial p^2}$ and is thus negative. Hence the form $\bar{S}'_{W_z^{\alpha 0}}$ is positive, $\forall z \in \mathbb{R}^{2n}$. Moreover, the form $\bar{S}'_{V_z^{\alpha t}}$ is negative and $\bar{S}'_{W_z^{\alpha t}}$ positive for any $t \in \mathbb{R}$.

Now fix $z \in \mathbb{R}^{2n}$ and a splitting $\mathbb{R}^{2n} = \Lambda \oplus \Delta$ such that $\bar{S}_{V_z^{\alpha 0}} - \bar{S}_{W_z^{\alpha 0}} > 0$. Then the form $\bar{S}_{V_z^{\alpha t}}$ monotonically decreases with $t$ and the form $\bar{S}_{W_z^{\alpha 0}}$ monotonically increases, while their difference is never degenerate (and thus remains positive) since $V_z^{\alpha t}$ and $W_z^{\alpha t}$ are transversal. Hence there exists $\lim_{t \to +\infty} \bar{S}_{V_z^{\alpha t}} = \bar{S}_{V_z^{\alpha +}}$. If we take another splitting such that $\bar{S}_{V_z^{\alpha 0}} - \bar{S}_{W_z^{\alpha 0}} < 0$, then we catch $\lim_{t \to -\infty} \bar{S}_{V_z^{\alpha t}} = \bar{S}_{V_z^{\alpha -}}$. $\square$

It is easy to see that the Lagrangian splitting $\mathbb{R}^{2n} = V_z^{\alpha+} \oplus V_z^{\alpha-}$, $z \in \mathbb{R}^{2n}$, is invariant with respect to the flow $e^{tH^\alpha}$, i.e.

$$\left( D_z e^{tH^\alpha} \right) V_z^{\alpha\pm} = V_{e^{tH^\alpha}(z)}^{\alpha\pm}, \quad \forall z \in \mathbb{R}^{2n}, \ t \in \mathbb{R}.$$

Another important property of $V_z^{\alpha\pm}$ which follows directly from the construction:

$$V_z^0 \cap V_z^{\alpha-} = V_z^0 \cap V_z^{\alpha+} = 0, \quad z \in \mathbb{R}^{2n}. \tag{10}$$

Note that $H^0 = \mathbf{H}$, $V_z^{0\,t} = V_z^t$, and $W_z^{0\,t} = W_z^t$, so all written can be used not only in the setting of Theorem 2 but also of Theorem 1, just put $\alpha = 0$; we also set $V_z^\pm = V_z^{0\pm}$.

A more careful study gives the following estimates.

**Lemma 2.** *There exists an associated to* $\mathbf{H}$ *and the splitting* $\mathbb{R}^{2n} = V_z^0 \oplus W_z^0$ *Riemannian structure* $(\cdot \mid \cdot)_z$, $z \in \mathbb{R}^{2n}$, $\|\xi\|_z = \sqrt{(\xi \mid \xi)_z}$ *and constants* $\rho > 0$, $\varepsilon \in (0, \frac{\alpha}{4})$ *such that*

- *if* $R_z < 0$ *for any* $z \in \mathbb{R}^{2n}$, *then*

$$\left\| \left( D_z e^{t\mathbf{H}} \right) v_- \right\|_{e^{t\mathbf{H}}(z)} \le \rho e^{-\varepsilon t} \|v_-\|_z, \quad \left\| \left( D_z e^{t\mathbf{H}} \right) v_+ \right\|_{e^{t\mathbf{H}}(z)} \ge \frac{1}{\rho} e^{\varepsilon t} \|v_+\|_z,$$

$\forall v_\pm \in V_z^\pm$, $t > 0$;

- *if* $R_z > 0$, $R_z^\alpha < 0$ *w. r. t. canonical splittings for any* $z \in \mathbb{R}^{2n}$, *then*

$$\frac{1}{\rho} e^{\varepsilon t} \|v_-\|_z \le \left\| \left( D_z e^{tH^\alpha} \right) v_- \right\|_{e^{tH^\alpha}(z)} \le \rho e^{(\frac{\alpha}{2} - \varepsilon)t} \|v_-\|_z,$$

$$\left\| \left( D_z e^{tH^\alpha} \right) v_+ \right\|_{e^{tH^\alpha}(z)} \ge \frac{1}{\rho} e^{(\frac{\alpha}{2} + \varepsilon)t} \|v_+\|_z,$$

$\forall v_\pm \in V_z^\pm$, $t > 0$.

The proof is based on the *structural equations* from [1] and the hyperbolicity test from [8].

**Corollary 1.** .

- *The case of negative* $R_{\cdot}$. *Given* $z \in \mathbb{R}^{2n}$, *if* $\mathbf{H}(z) \in V_z^-$ *then*

$$\left\| \mathbf{H}(e^{t\mathbf{H}}(z)) \right\|_{e^{t\mathbf{H}}(z)} \to 0 \quad (t \to +\infty)$$

*with the exponential rate, otherwise*

$$\left\| \mathbf{H}(e^{t\mathbf{H}}(z)) \right\|_{e^{t\mathbf{H}}(z)} \to \infty \quad (t \to +\infty)$$

*with the exponential rate.*

- *The case of positive $R$ and negative $R^\alpha$ w.r.t. canonical splittings. We have*

$$\left\| H^\alpha(e^{tH^\alpha}(z)) \right\|_{e^{tH^\alpha}(z)} \to \infty \quad (t \to +\infty)$$

*with the exponential rate for any $z \in \mathbb{R}^{2n}$. Moreover, if $H^\alpha(z) \in V_z^{\alpha-}$ then*

$$e^{-\frac{\alpha}{2}t}\left\| H^\alpha(e^{tH^\alpha}(z)) \right\|_{e^{tH^\alpha}(z)} \to 0 \quad (t \to +\infty),$$

*with the exponential rate, otherwise*

$$e^{-\frac{\alpha}{2}t}\left\| H^\alpha(e^{tH^\alpha}(z)) \right\|_{e^{tH^\alpha}(z)} \to \infty \quad (t \to +\infty)$$

*again with the exponential rate.*

**Proof.** This is an immediate corollary of Lemma 2 and the fact that the flow $e^{tH^\alpha}$ preserves vector field $H^\alpha$. □

Another corollary of Lemma 2 in the case of negative $R$ is the existence of a stable invariant Lagrangian submanifold of the flow $e^{tH}$ that is an integral submanifold of the distribution $V^-$. This follows from the Hadamard–Perron theorem (see [6], [7]).

Unfortunately, standard Hadamard–Perron theorem is not applicable in the case of positive $R$ and negative $R^\alpha$ w.r.t. canonical splittings. Indeed, the statement of Lemma 2 in this case brings us in the "partially hyperbolic" framework with a "central distribution" $V^{\alpha-}$. Central distributions are not always integrable, but asymptotic condition (9) allows to reduce our study to the case where stable and unstable foliations are *quasi-isometric* that guarantees the integrability of the central distribution (see [7]). This implies the existence of a "less unstable" invariant Lagrangian submanifold of the flow $e^{tH^\alpha}$ that is an integral submanifold of the distribution $V^{\alpha-}$. For any $z$ from this submanifold, the quantity $\left|e^{tH^\alpha}(z)\right|$ tends to $\infty$ slower than $e^{\frac{\alpha}{2}t}$ as $t \to +\infty$.

Note that stable submanifold has the same smoothness class as $\mathbf{H}$, while for the "less unstable" submanifold only $C^1$-smoothness is guaranteed.

Relation (10) implies that both submanifolds are transversal to the distribution $V_{(p,q)}^0 = \mathbb{R}^n \times q$, $(p,q) \in \mathbb{R}^{2n}$. Hence the projection $(p,q) \mapsto q$ is a local diffeomorphism of any of these two submanifolds into $\mathbb{R}^n$. It remains to prove that this is a global diffeomorphism.

The rest of the proof runs differently for Theorems 1 and 2.

We start from Theorem 1. Let $\mathcal{U}^s$ be the stable submanifold. First, corollary 1 implies the following characterization of the points of this submanifold: $z \in \mathcal{U}^s$ if and only if $\mathbf{H}(z) \in V_z^-$. In particular, $\mathcal{U}^s$ is a closed subset of $\mathbb{R}^{2n}$. Moreover, $\mathcal{U}^s \subset H^{-1}(0)$ since $H$ is a first integral of the Hamiltonian flow. The growth condition on $\varphi$ implies that the projection $(p,q) \mapsto q$ restricted to $H^{-1}(0)$ is a proper mapping. Hence this projection restricted to $\mathcal{U}^s$ is also a proper mapping. Combining this with the fact that the projection

$(p, q) \mapsto q$, $(p, q) \in \mathcal{U}^s$ is a local diffeomorphism we obtain that this projection is a covering of $\mathbb{R}^n$ and hence a global diffeomorphism.

Now turn to Theorem 2. Identity (8) implies that $R_z^\alpha < 0$, $\forall z$. For any sufficiently big $N > 0$ we substitute $H^\alpha$ by a vector field $H_N^\alpha$ which is equal to $H^\alpha$ inside the ball of radius $N$ and is equal to the linear vector field outside the ball of radius $N + 1$, where the linear field is obtained from the quadratic form $\phi$ in the same way as $H^\alpha$ is obtained from $\phi$. We do this modification of $H^\alpha$ using a cut-off function; our asymptotic conditions imply that the modified field has a positive, close to $R_z$, curvature w. r. t. the canonical splitting. Now take a local "less unstable" submanifold $U^{lu}$ of $H^\alpha$, i. e. an integral manifold of the distribution $V^{\alpha-}$ which contains $0 \in \mathbb{R}^{2n}$. Then $\mathcal{U}_N^{lu} = \bigcup_{t \geq 0} e^{tH_N^\alpha}(U^{lu})$ is the global "less unstable" invariant submanifold of $H_N^\alpha$. Note that the contained in the radius $N$ ball part of $\mathcal{U}_N^{lu}$ is actually a part of the global "less unstable" invariant submanifold of $e^{tH^\alpha}$, while the part of $\mathcal{U}_N^{lu}$ out of the radius $N + 1$ ball is a part of the "less unstable" invariant submanifold of the linear flow, that is a fixed transversal to $\mathbb{R}^n \times 0$ $n$-dimensional vector subspace $E^{lu} \subset \mathbb{R}^{2n}$.

Any started from $U^{lu} \setminus 0$ trajectory of the system $\dot{z} = H_N^\alpha(z)$ leaves the radius $N + 1$ ball and thus arrives to $E^{lu}$. We obtain that the mapping $(p, q) \mapsto q$, $(p, q) \in \mathcal{U}_N^{lu}$, is proper, hence it is a covering of $\mathbb{R}^n$ and hence a diffeomorphism on $\mathbb{R}^n$. Since $N$ is arbitrary big, we obtain the desired global diffeomorphism of the "less unstable" invariant Lagrangian submanifold of $e^{tH^\alpha}$ on $\mathbb{R}^n$.

We have not yet explained why the problem without discount does not admit optimal trajectories. This follows from the *comparison theorem* (see [3], [1]): positivity of the curvature implies existence of an infinity of conjugate points for any extremal.

# References

1. Agrachev AA (2008) Geometry of Optimal Control Problems and Hamiltonian Systems. C.I.M.E. Lecture Notes in Mathematics, 1932. Springer-Verlag, Berlin
2. Agrachev AA, Gamkrelidze RV (1998) Symplectic methods in optimization and control. In: Jakubczyk B, Respondek W (Eds) Geometry of Feedback and Optimal Control, 19–77. Marcel Dekker
3. Agrachev AA, Gamkrelidze RV (1997) Feedback–invariant optimal control theory and differential geometry, I. Regular extremals. J. Dynamical and Control Systems, 3:343–389.
4. Agrachev AA, Sachkov YuL (2004) Control Theory from the Geometric Viewpoint. Springer-Verlag, Berlin
5. Agrachev AA, Chittaro FC (forthcoming) Smooth optimal synthesis for infinite horizon variational problems. J. ESAIM: Control, Optimisation and Calculus of Variations
6. Katok A, Hasselblatt B (1995) Introduction to Modern Theory of Dynamical Systems. Cambridge University Press, Cambridge
7. Pesin YaB (2004) Lectures on partial hyperbolicity and stable ergodicity, EMS

8. Wojtkovski MP (2000) Magnetic flows and Gaussian thermostats on manifolds of negative curvature. Fundamenta Mathematicæ, 163:177–191.

# Laplace Transforms and the American Call Option

Ghada Alobaidi[1] and Roland Mallier[2]

[1] Department of Mathematics, American University of Sharjah,
Sharjah, United Arab Emirates. `galobaidi@aus.edu`
[2] Department of Applied Mathematics, University of Western Ontario,
London ON Canada. `rolandmallier@hotmail.com`

**Summary.** A partial Laplace transform is used to study the valuation of American call options with constant dividend yield, and to derive an integral equation for the location of the optimal exercise boundary, which is the main result of this paper.

The integral equation differs depending on whether the dividend yield is less than or exceeds the risk-free rate.

## 1 Introduction

One of the defining events in mathematical finance was the publication in 1973 of the Black-Scholes-Merton model for the pricing of equity options, for which the 1997 Nobel Prize in Economics was awarded. Using this model, in which the volatility $\sigma$, interest rate $r$, and dividend yield $D$ were assumed constant, it was possible to obtain closed form expressions for European call and put options, which have pay-offs at expiry of $\max(S - E, 0)$ and $\max(E - S, 0)$ respectively, where $S$ is the price of the stock upon which the option is written and $E$ is the strike price.

Although the Black-Scholes-Merton model made it possible to obtain closed form prices for many European options, which can be exercised only at expiry, American options, which can be exercised at or before expiry at the discretion of the holder, are rather more difficult to price. American call and put options have the same pay-offs when exercised early as when held to expiry, that is $\max(S - E, 0)$ and $\max(E - S, 0)$ respectively. The right to exercise early leads to the issue of when and if an option should be exercised, which leads to an interesting free boundary problem similar to the Stefan problem which arises in melting and solidification, and it is precisely this free boundary problem which makes American options both difficult to price and mathematically interesting. To date, a closed form pricing formula for American options has remained elusive, except in a couple of special cases. One such special case is the American call with either no dividends, when exercise is never optimal so that the value of the option is the same as that of a European call, or discrete dividends [9, 10, 23, 27]. For those cases where exact solutions are not known, practitioners are of course able to price American options numerically, or use one of the approximate or series solutions which appear in the literature.

In the present study, we will consider the free boundary, or *optimal exercise boundary*, for an American call option. This boundary separates the region where it is optimal for an investor to retain an option from that where exercise is optimal, and closed form expressions for the location of the boundary have remained as elusive as a closed form pricing formula for American options, although, as with the price of the option, numerical and approximate and series solutions can be used. The location of the free boundary is key to pricing an American option. In our analysis, we will use a partial Laplace transform to arrive at an integral equation giving the location of the free boundary, which we regard as a curve in $(S,t)$ space, denoted by $S = S_f(t)$ or the inverse relation $t = T_f(S)$.

We are of course not the first to apply integral equation methods to American options: on the contrary, it has been a very popular approach, including such studies as the early work of [18, 26], which drew on the pioneering work of [15] on Stefan problems, the studies by [4, 11, 13, 21], all of which looked at the difference between European and American prices, and the recent work of [1, 8, 14, 16, 25]. We will touch upon the differences between some of those studies and our own in the final section.

The partial Laplace transform approach used here was developed by [7] for diffusion problems, specifically the recrystallization of an infinite metal slab, and an overview of the technique can be found in [6]. The solidification problem considered in [7] was governed by the diffusion equation, and the Black-Scholes-Merton partial differential equation used in our analysis can of course be recast as that equation. [7] were able to use a partial Laplace transform, which we shall define shortly, to give an integral equation formulation of their problem, and were then able to find a series solution of that integral equation. For the problem considered here, the boundary conditions at the free boundary cause the kernel in our integral equation to be much more complicated than that in [7].

## 2 Analysis

Under the Black-Scholes-Merton model, in which the volatility $\sigma$, interest rate $r$, and dividend yield $D$ are assumed constant, the value $V(S,t)$ of an option on an equity obeys the Black-Scholes-Merton partial differential equation or PDE [2, 19],

$$\frac{\partial V}{\partial t} + \frac{\sigma^2 S^2}{2} \frac{\partial^2 V}{\partial S^2} + (r - D) S \frac{\partial V}{\partial S} - rV = 0, \tag{1}$$

where $S$ is the price of the underlying and $t < T$ is the time, with $T$ being the expiry when the holder will receive a pay-off of $\max(S - E, 0)$ for a call with a strike of $E$. To simplify the analysis, we will work in terms of the remaining life of the option, $\tau = T - t$, so that (1) is replaced by

$$\frac{\partial V}{\partial \tau} = \frac{\sigma^2 S^2}{2} \frac{\partial^2 V}{\partial S^2} + (r - D) S \frac{\partial V}{\partial S} - rV. \tag{2}$$

For European options, (2) is valid for $\tau \geq 0$. For options where early exercise is permitted, (2) is only valid when it is optimal to hold the option, and must be solved together with the appropriate conditions at the optimal exercise boundary, whose location is unknown and must be solved for. We will label the position of the free boundary as $S = S_f(\tau)$, which we can invert to give $\tau_f(S)$ as the time at which early exercise should occur.

A review of the properties of the free boundary and the option price can be found in [20]. In our analysis, we will make use of a number of these including:
(i) The price of an American call option is given by a value function. Where it is optimal to hold the option the value function is smooth with $V(S, \tau) > \max(S - E, 0)$ and $0 \leq \partial V/\partial S < 1$ [18, 26].
(ii) There is an optimal exercise policy for American options and an optimal stopping time [12].
(iii) At the free boundary the value of the option is equal to the pay-off from immediate exercise [11, 18], $V_f(S, \tau) = S - E$.
(iv) At the free boundary the value of the option's delta, or derivative of its value with respect to the stock price, is $\partial V_f/\partial S = 1$. This high contact or smooth-pasting condition [24] has been shown to be both a necessary [19] and sufficient [3] condition for the optimality of the boundary.
(v) At expiry [18, 26], the location of the free boundary is

$$S_f(0) = S_0 = \begin{cases} Er/D > E & r > D \\ E & D \geq r, \end{cases}$$

which we can write as $\tau_f(S_0) = 0$.
(vi) As $\tau \to \infty$, from the perpetual American call [18, 19] we know $S_f(\tau) \to S_1 = \frac{E\alpha}{\alpha - 1}$ where $\alpha = \left[ \frac{\sigma^2}{2} - (r - D) + \sqrt{\left(r - D + \frac{\sigma^2}{2}\right)^2 + 2D\sigma^2} \right] / \sigma^2$. We can write this as $\tau_f(S) \to \infty$ as $S \to S_1$ from below. As $D \to 0$, $S_1 \to \infty$ [19] and a perpetual call on a stock with no dividends has the same value as the stock.
(vii) The free boundary is a strictly increasing function of $\tau$ [18, 26], which enables us to define the inverse $\tau_f(S)$ mentioned above. The optimal exercise boundary will move upwards as we move away from the expiration date and will lie between the two limits, $S_0 \leq S_f(\tau) \leq S_1$, with early exercise optimal if $S \geq S_f(\tau)$ and retaining the option optimal if $0 \leq S < S_f(\tau)$.
(viii) The free boundary is a continuous, differentiable function of $\tau$ [18, 26], which enables us to take derivatives of $\tau_f(S)$.

Having formulated the problem, we shall now attempt to solve it using a Laplace transform in time. Because $S_0$ differs depending on whether $r > D$ or $D \geq r$, we will consider these two cases separately. Since (2) only holds where it is optimal to retain the option, we will modify the usual Laplace transform $\mathcal{L}(G)(p) = \int_0^\infty g(\tau)e^{-p\tau}d\tau$ somewhat, and define the *partial* Laplace transform for $S \leq S_1$,

$$\mathcal{V}(S, p) = \int_{\tau_f(S)}^\infty V(S, \tau)e^{-p\tau}d\tau, \tag{3}$$

with the lower limit of $\tau = \tau_f(S)$ rather than $\tau = 0$. As we mentioned in Section 1, the partial Laplace transform is due to [7], and has been successfully used to tackle diffusion problems in the past. This definition of the partial Laplace transform, is of course equivalent to setting $V(S, \tau) = 0$ in the region where it is optimal not to hold. Because of this definition, the price of the option $V(S, \tau)$ will obey (2) everywhere we integrate. We require the real part of $p$ to be positive for the integral in (3) to converge. In addition, we know from the definition that $\mathcal{V}(S, p) \to 0$ as $S \to S_1$. We can also define an inverse transform

$$V(S, \tau) = \frac{1}{2\pi i} \int_{\gamma-i\infty}^{\gamma+i\infty} \mathcal{V}(S, p)e^{p\tau}dp, \tag{4}$$

which of course is only meaningful where it is optimal to retain the option. From our definition, the following transforms can be derived easily,

$$\mathcal{L}\left[\frac{\partial V}{\partial \tau}\right] = p\mathcal{V} - e^{-p\tau_f(S)}V_f(S, \tau_f(S)),$$

$$\mathcal{L}\left[\frac{\partial V}{\partial S}\right] = \frac{d\mathcal{V}}{dS} + e^{-p\tau_f(S)}\tau_f'(S)V_f(S, \tau_f(S)),$$

$$\mathcal{L}\left[\frac{\partial^2 V}{\partial S^2}\right] = \frac{d}{dS}\left(\mathcal{L}\left[\frac{\partial V}{\partial S}\right]\right) + e^{-p\tau_f(S)}\tau_f'(S)\frac{\partial V_f}{\partial S}(S, \tau_f(S)). \tag{5}$$

In the above, we have adopted the convention that $\tau_f(S)$ is the location of the free boundary for $S_0 < S < S_1$, while for $S < S_0$, we set $\tau_f = 0$ since it is optimal to hold the option to expiry. Applying this partial Laplace transform to (2), we arrive at the following (nonhomogeneous Euler) ordinary differential equation for the transform of the option price,

$$\left[\frac{\sigma^2 S^2}{2}\frac{d^2}{dS^2} + (r - D)S\frac{d}{dS} - (p + r)\right]\mathcal{V} + F(S) = 0, \tag{6}$$

where the nonhomogeneous term $F(S)$ takes a different value in various regions, as shown in Table 1:

with $F_0(S) = \left(1 + (r - D)S\tau_f'(S) + \frac{\sigma^2 S^2}{2}\left(\tau_f''(S) + p_0\tau_f'^2(S)\right)\right)(S - E) + \sigma^2 S^2 \tau_f'(S)$ and $F_1(S) = -\frac{\sigma^2 S^2}{2}(S - E)\tau_f'^2(S)$ in region (c) and $p_0 = \frac{(D-r)^2}{2\sigma^2} +$

**Table 1.** The nonhomogeneous term $F(S)$.

| Region | $V(S_f(\tau),\tau)$ | $\frac{\partial V}{\partial S}(S_f(\tau),\tau)$ | $\tau_f(S)$ | $F(S)$ |
|---|---|---|---|---|
| (a) $0 < S < E$ | $0$ | $0$ | $0$ | $0$ |
| (b) $E < S < S_0$ | $S - E$ | $1$ | $0$ | $S - E$ |
| (c) $S_0 < S < S_1$ | $S - E$ | $1$ | $> 0$ | $e^{-p\tau_f(S)}[F_0(S) + (p + p_0) F_1(S)]$ |

$\frac{D+r}{2} + \frac{\sigma^2}{8}$. We would mention that in region (c), $F(S)$ is fairly complicated, being a function of $\tau_f'(S)$ and $\tau_f''(S)$, the derivatives of the inverse of the optimal exercise boundary, as are $F_0(S)$ and $F_1(S)$.

The three regions in Table 1 coexist when $0 < D < r$. When $D \geq r$, the location of the free boundary at expiry is $S_f(0) = S_0 = E$, so that the middle region (b) vanishes and we are left with only two regions, (a) and (c), with the nonhomogeneous terms in these two regions unchanged. For the moment, we will perform our analysis for the case $0 < D < r$, and explain later how the analysis differs when $D \geq r$.

## 2.1 $0 < D < r$

The general solution of (6) is

$$
V = S^{\frac{D-r+\lambda(p)}{\sigma^2} + \frac{1}{2}} \left[ C_1(p) - \int \frac{\tilde{S}^{-\frac{D-r+\lambda(p)}{\sigma^2} - \frac{3}{2}} F(\tilde{S}) d\tilde{S}}{\lambda(p)} \right]
$$
$$
+ S^{\frac{D-r-\lambda(p)}{\sigma^2} + \frac{1}{2}} \left[ C_2(p) + \int \frac{\tilde{S}^{-\frac{D-r-\lambda(p)}{\sigma^2} - \frac{3}{2}} F(\tilde{S}) d\tilde{S}}{\lambda(p)} \right], \tag{7}
$$

where $\lambda(p) = \sqrt{2}\sigma [p + p_0]^{1/2}$, and $C_1$ and $C_2$ are constants of integration, which may depend on the transform variable $p$. Since $r$, $D$ and $\sigma$ are all assumed to be positive, and we assume that $p$ has a positive real part from the definition of the Laplace transform, then the real part of the first exponent $\frac{D-r+\lambda(p)}{\sigma^2} + \frac{1}{2}$ is assumed positive, while the real part of the second exponent, $\frac{D-r-\lambda(p)}{\sigma^2} + \frac{1}{2}$ is assumed negative.

Applying this solution (7) to the three separate regions outlined above, we find that in region (a) we must discard the second solution in order to satisfy the boundary condition on $S = 0$ that $V(0,t) = 0$, and the corollary that $V(0,p) \to 0$ as $p \to \infty$, so in this region we have

$$
V = C_1^{(a)}(p) \left( \frac{S}{E} \right)^{\frac{D-r+\lambda(p)}{\sigma^2} + \frac{1}{2}}. \tag{8}
$$

In region (b), we find

$$\mathcal{V} = \frac{S}{p+D} - \frac{E}{p+r} \tag{9}$$

$$+ C_1^{(b)}(p) \left(\frac{S}{E}\right)^{\frac{D-r+\lambda(p)}{\sigma^2}+\frac{1}{2}} + C_2^{(b)}(p) \left(\frac{S}{E}\right)^{\frac{D-r-\lambda(p)}{\sigma^2}+\frac{1}{2}},$$

and in region (c), we have

$$\mathcal{V} = S^{\frac{D-r+\lambda(p)}{\sigma^2}+\frac{1}{2}} \left[ C_1(p) - \int_{S_0}^{S} \frac{\tilde{S}^{-\frac{D-r+\lambda(p)}{\sigma^2}-\frac{3}{2}} F(\tilde{S}) d\tilde{S}}{\lambda(p)} \right]$$

$$+ S^{\frac{D-r-\lambda(p)}{\sigma^2}+\frac{1}{2}} \left[ C_2(p) + \int_{S_0}^{S} \frac{\tilde{S}^{-\frac{D-r-\lambda(p)}{\sigma^2}-\frac{3}{2}} F(\tilde{S}) d\tilde{S}}{\lambda(p)} \right], \tag{10}$$

with $F$ given in Table 1. Applying the condition that $\mathcal{V}(S,p) \to 0$ as $S \to S_1$, we require

$$C_1^{(c)}(p) = \frac{1}{\lambda(p)} \int_{S_0}^{S_1} \tilde{S}^{-\frac{D-r+\lambda(p)}{\sigma^2}-\frac{3}{2}} F(\tilde{S}) d\tilde{S},$$

$$C_2^{(c)}(p) = -\frac{1}{\lambda(p)} \int_{S_0}^{S_1} \tilde{S}^{-\frac{D-r-\lambda(p)}{\sigma^2}-\frac{3}{2}} F(\tilde{S}) d\tilde{S}, \tag{11}$$

so that the solution in region (c) becomes

$$\mathcal{V} = \int_{S}^{S_1} \left(\frac{\tilde{S}}{S}\right)^{-\frac{D-r}{\sigma^2}-\frac{3}{2}} \left[ \left(\frac{\tilde{S}}{S}\right)^{-\frac{\lambda(p)}{\sigma^2}} - \left(\frac{\tilde{S}}{S}\right)^{\frac{\lambda(p)}{\sigma^2}} \right] \frac{F(\tilde{S})d\tilde{S}}{\lambda(p)S}. \tag{12}$$

We must now match the solutions in these three regions together. We will require that $\mathcal{V}$ and $d\mathcal{V}/dS$ are continuous across $S = E$ and $S = S_0$. Matching regions (b) and (c) together at $S = S_0$, we find we can write $C_1^{(b)}(p)$ and $C_2^{(b)}(p)$ in terms of $C_1^{(c)}(p)$ and $C_2^{(c)}(p)$, which were given in (11) above,

$$C_1^{(b)} = C_1^{(c)} E^{\frac{D-r+\lambda(p)}{\sigma^2}+\frac{1}{2}}$$

$$+ \frac{1}{2\lambda(p)} \left(\frac{E}{S_0}\right)^{\frac{D-r+\lambda(p)}{\sigma^2}+\frac{1}{2}}$$

$$\times \left[ \frac{\left(D-r-\frac{\sigma^2}{2}-\lambda(p)\right) S_0}{p+D} - \frac{\left(D-r+\frac{\sigma^2}{2}-\lambda(p)\right) E}{p+r} \right],$$

$$C_2^{(b)} = C_2^{(c)} E^{\frac{D-r-\lambda(p)}{\sigma^2}+\frac{1}{2}} \tag{13}$$

$$- \frac{1}{2\lambda(p)} \left(\frac{E}{S_0}\right)^{\frac{D-r-\lambda(p)}{\sigma^2}+\frac{1}{2}}$$

$$\times \left[ \frac{\left(D-r-\frac{\sigma^2}{2}+\lambda(p)\right) S_0}{p+D} - \frac{\left(D-r+\frac{\sigma^2}{2}+\lambda(p)\right) E}{p+r} \right].$$

Similarly, matching regions (a) and (b) together at $S = E$, we find we can write $C_1^{(a)}(p)$ in terms of $C_1^{(b)}(p)$ and $C_2^{(b)}(p)$, which we just found,

$$C_1^{(a)}(p) = E\left(\frac{1}{p+D} - \frac{1}{p+r}\right) + C_1^{(b)}(p) + C_2^{(b)}(p), \qquad (14)$$

but we also arrive at another expression for $C_2^{(b)}(p)$,

$$C_2^{(b)}(p) = \frac{E}{2\lambda(p)}\left[\left(\frac{D - r + \frac{\sigma^2}{2} + \lambda(p)}{p+r}\right) - \left(\frac{D - r - \frac{\sigma^2}{2} + \lambda(p)}{p+D}\right)\right], (15)$$

and comparing this to the earlier expression we found for $C_2^{(b)}(p)$, we arrive at the following equation

$$C_2^{(c)} = \frac{1}{2\lambda(p)} S_0^{-\frac{D-r-\lambda(p)}{\sigma^2} - \frac{1}{2}}$$

$$\times \left[\frac{S_0\left(D - r - \frac{\sigma^2}{2} + \lambda(p)\right)}{p+D} - \frac{E\left(D - r + \frac{\sigma^2}{2} + \lambda(p)\right)}{p+r}\right]$$

$$+ \frac{1}{2\lambda(p)} E^{-\frac{D-r-\lambda(p)}{\sigma^2} + \frac{1}{2}}$$

$$\times \left[\frac{D - r + \frac{\sigma^2}{2} + \lambda(p)}{p+r} - \frac{D - r - \frac{\sigma^2}{2} + \lambda(p)}{p+D}\right], \qquad (16)$$

or using (11),

$$\int_{S_0}^{S_1} \tilde{S}^{-\frac{D-r\lambda(p)}{\sigma^2} - \frac{3}{2}} F(\tilde{S}) d\tilde{S}$$

$$= \frac{1}{2} S_0^{-\frac{D-r-\lambda(p)}{\sigma^2} - \frac{1}{2}}$$

$$\times \left[\frac{E\left(D - r + \frac{\sigma^2}{2} + \lambda(p)\right)}{p+r} - \frac{S_0\left(D - r - \frac{\sigma^2}{2} + \lambda(p)\right)}{p+D}\right]$$

$$+ \frac{1}{2} E^{-\frac{D-r-\lambda(p)}{\sigma^2} + \frac{1}{2}}$$

$$\times \left[\frac{D - r - \frac{\sigma^2}{2} + \lambda(p)}{p+D} - \frac{D - r + \frac{\sigma^2}{2} + \lambda(p)}{p+r}\right], \qquad (17)$$

where again $F(S)$ is given in Table 1.

We should comment on why (13), which came from matching regions (b) and (c), appears to be more complicated than (14,15) which came from matching (a) and (b). Because we wrote $S/E$ in the homogeneous terms in (8,9), the matching at $S = E$ was greatly simplified. If we had instead written $S/S_0$ in those terms, the matching at $S = S_0$ would have become much simpler, while that at $S = E$ would have become correspondingly more complex. Another reason for the complexity in (13) is of course that the nonhomogeneous terms in region (c) are fairly lengthy, while those in region (a) vanish.

## 2.2 $D \geq r$

The analysis when $D \geq r$ is very similar to that for $D < r$, so we will merely highlight the differences and give the main results. As mentioned earlier, when $D \geq r$, the location of the free boundary at expiry is $S_f(0) = S_0 = E$. Because of this, instead of the three regions (a)-(c) described above, the middle region (b) vanishes and we are left with only two regions, (a) and (c), with the nonhomogeneous terms in these two regions unchanged, and given in Table 1. The general solutions in these two regions are also the same, namely (8,12). However, the matching process will lead to constants that differ from those found earlier. When $D \geq r$, we have a single boundary to match across, and we require that $\mathcal{V}$ and $\frac{d\mathcal{V}}{dS}$ are continuous across $S = E$. This tells us that

$$C_1^{(a)}(p) = C_1^{(c)}(p)E^{\frac{D-r+\lambda(p)}{\sigma^2}+\frac{1}{2}},$$
$$C_2^{(c)}(p) = 0, \tag{18}$$

the latter of which gives

$$\int_E^{S_1} \tilde{S}^{-\frac{D-r-\lambda(p)}{\sigma^2}-\frac{3}{2}} F(\tilde{S}) d\tilde{S} = 0, \tag{19}$$

with $F(S)$ again given in Table 1.

## 3 The integral equations

The equations, (17) for $0 < D < r$ and (19) for $D \geq r$, are integral equations in transform space for the location of the free boundary $\tau_f(S)$, which appears in equations via the nonhomogeneous term $F(S)$. To be more specific, (17,19) are nonlinear Fredholm integral equations, or to be even more specific, Urysohn equations. When $r = D$, the integral equations for the two cases are the same.

  Each of (17,19) is of course the Laplace transform of an integro-differential equation in physical space, and we can obtain these latter equations by applying the inverse Laplace transform (4) to (17,19). The inversion process is conceptually straightforward, but the algebra is somewhat complicated. To invert (17), we first divide by $(p + p_0)^{3/2} S_0^{-\frac{D-r}{\sigma^2}-\frac{3}{2}} (S_f(\tau))^{\lambda(p)/\sigma^2}$, and rewrite (17) as

$$\int_{S_0}^{S_1} \exp\left[-\frac{\sqrt{2(p+p_0)}}{\sigma} \ln \frac{S_f(\tau)}{\tilde{S}}\right]$$

$$\times \left(\frac{\tilde{S}}{S_0}\right)^{-\frac{D-r}{\sigma^2}-\frac{3}{2}} \frac{e^{-p\tau_f(\tilde{S})}}{\sqrt{p+p_0}} \left[\frac{F_0(\tilde{S})}{p+p_0} + F_1(\tilde{S})\right] d\tilde{S} =$$

$$= \frac{S_0}{2(p+p_0)} \exp\left[-\frac{\sqrt{2(p+p_0)}}{\sigma} \ln \frac{S_f(\tau)}{S_0}\right]$$

$$\times \left(\frac{E}{p+r} \left[\frac{D-r+\frac{\sigma^2}{2}}{(p+p_0)^{1/2}} + \sigma\sqrt{2}\right] - \frac{S_0}{p+D} \left[\frac{D-r-\frac{\sigma^2}{2}}{(p+p_0)^{1/2}} + \sigma\sqrt{2}\right]\right)$$

$$+ \frac{S_0^2}{2(p+p_0)} \left(\frac{E}{S_0}\right)^{-\frac{D-r}{\sigma^2}+\frac{1}{2}} \exp\left[-\frac{\sqrt{2(p+p_0)}}{\sigma} \ln \frac{S_f(\tau)}{E}\right] \qquad (20)$$

$$\times \left(\frac{1}{p+D} \left[\frac{D-r-\frac{\sigma^2}{2}}{(p+p_0)^{1/2}} + \sigma\sqrt{2}\right] - \frac{1}{2p+r} \left[\frac{D-r+\frac{\sigma^2}{2}}{(p+p_0)^{1/2}} + \sigma\sqrt{2}\right]\right),$$

and then use the following standard inverse transforms [22],

$$\mathcal{L}^{-1}\left[e^{-ap}G(p)\right] = H(\tau-a)\,g(\tau-a),$$

$$\mathcal{L}^{-1}\left[G(p+p_0)\right] = e^{-p_0\tau}g(\tau),$$

$$\mathcal{L}^{-1}\left[G_1(p)G_2(p)\right] = \int_0^\tau g_1(\tau-z)g_2(z)dz,$$

$$\mathcal{L}^{-1}\left[p^{-1/2}\exp\left(-ap^{1/2}\right)\right] = \frac{1}{\sqrt{\pi}\tau^{1/2}}\exp\left[-\frac{a^2}{4\tau}\right],$$

$$\mathcal{L}^{-1}\left[p^{-1}\exp\left(-ap^{1/2}\right)\right] = \mathrm{erfc}\left[\frac{a}{2\sqrt{\tau}}\right],$$

$$\mathcal{L}^{-1}\left[p^{-3/2}\exp\left(-ap^{1/2}\right)\right] = \frac{2\tau^{1/2}}{\sqrt{\pi}}\exp\left[-\frac{a^2}{4\tau}\right] - a\,\mathrm{erfc}\left[\frac{a}{2\sqrt{\tau}}\right], \qquad (21)$$

where $H(t)$ is the Heaviside step function, to obtain

$$\int_{S_0}^{S_f(\tau)} \sqrt{\tau - \tau_f(\tilde{S})} \left(\frac{\tilde{S}}{S_0}\right)^{-\frac{D-r}{\sigma^2}-\frac{3}{2}} e^{-p_0(\tau-\tau_f(\tilde{S}))}$$

$$\times \left[\frac{1}{\sqrt{\pi}}\left(2F_0(\tilde{S}) + \frac{F_1(\tilde{S})}{\tau-\tau_f(\tilde{S})}\right) \exp\left(-\frac{\left(\ln(S_f(\tau)/\tilde{S})\right)^2}{2\sigma^2(\tau-\tau_f(\tilde{S}))}\right)\right.$$

$$\left. - \frac{F_0(\tilde{S})\sqrt{2}\ln(S_f(\tau)/\tilde{S})}{\sigma(\tau-\tau_f(\tilde{S}))}\mathrm{erfc}\left(\frac{\ln(S_f(\tau)/\tilde{S})}{\sigma\sqrt{2(\tau-\tau_f(\tilde{S}))}}\right)\right] d\tilde{S} =$$

$$= S_0 \int_0^\tau \left[ E \left( D - r + \frac{\sigma^2}{2} \right) e^{-r(\tau-z)} - S_0 \left( D - r - \frac{\sigma^2}{2} \right) e^{-D(\tau-z)} \right] e^{-p_0 z}$$

$$\times \left( \frac{z^{1/2}}{\sqrt{\pi}} \exp\left[ -\frac{[\ln(S_f(z)/S_0)]^2}{2\sigma^2 z} \right] - \frac{\ln(S_f(z)/S_0)}{\sigma\sqrt{2}} \mathrm{erfc}\left[ \frac{\ln(S_f(z)/S_0)}{\sigma\sqrt{2z}} \right] \right) dz$$

$$+ S_0^2 \left( \frac{E}{S_0} \right)^{-\frac{D-r}{\sigma^2}+\frac{1}{2}}$$

$$\times \int_0^\tau \left[ \left( D - r - \frac{\sigma^2}{2} \right) e^{-D(\tau-z)} - \left( D - r + \frac{\sigma^2}{2} \right) e^{-r(\tau-z)} \right] e^{-p_0 z}$$

$$\times \left( \frac{z^{1/2}}{\sqrt{\pi}} \exp\left[ -\frac{[\ln(S_f(z)/E)]^2}{2\sigma^2 z} \right] - \frac{\ln(S_f(z)/E)}{\sigma\sqrt{2}} \mathrm{erfc}\left[ \frac{\ln(S_f(z)/E)}{\sigma\sqrt{2z}} \right] \right) dz$$

$$+ \frac{S_0 \sigma}{\sqrt{2}} \int_0^\tau \left( E e^{-r(\tau-z)} - S_0 e^{-D(\tau-z)} \right) e^{-p_0 z} \mathrm{erfc}\left[ \frac{\ln(S_f(z)/S_0)}{\sigma\sqrt{2z}} \right] dz$$

$$+ \frac{S_0^2 \sigma}{\sqrt{2}} \left( \frac{E}{S_0} \right)^{-\frac{D-r}{\sigma^2}+\frac{1}{2}}$$

$$\times \int_0^\tau \left( e^{-D(\tau-z)} - e^{-r(\tau-z)} \right) e^{-p_0 z} \mathrm{erfc}\left[ \frac{\ln(S_f(z)/E)}{\sigma\sqrt{2z}} \right] dz, \tag{22}$$

which is an integro-differential equation in physical space for the location of the free boundary for the call with $0 < D < r$, and is the inverse transform of the equation in transform space (17). The equation for $D \geq r$ can be obtained by setting the right-hand side of (22) to zero,

$$\int_{S_0}^{S_f(\tau)} \sqrt{\tau - \tau_f(\tilde{S})} \left( \frac{\tilde{S}}{S_0} \right)^{-\frac{D-r}{\sigma^2}-\frac{3}{2}} e^{-p_0(\tau-\tau_f(\tilde{S}))}$$

$$\times \left[ \frac{1}{\sqrt{\pi}} \left( 2F_0(\tilde{S}) + \frac{F_1(\tilde{S})}{\tau - \tau_f(\tilde{S})} \right) \exp\left( -\frac{\left( \ln(S_f(\tau)/\tilde{S}) \right)^2}{2\sigma^2(\tau - \tau_f(\tilde{S}))} \right) \right.$$

$$\left. - \frac{F_0(\tilde{S})\sqrt{2}\ln(S_f(\tau)/\tilde{S})}{\sigma(\tau - \tau_f(\tilde{S}))} \mathrm{erfc}\left( \frac{\ln(S_f(\tau)/\tilde{S})}{\sigma\sqrt{2(\tau - \tau_f(\tilde{S}))}} \right) \right] d\tilde{S} = 0, \tag{23}$$

which is an integro-differential equation in physical space for the location of the free boundary for the call with $D \geq r$, and is the inverse transform of the equation in transform space (19).

# 4 Discussion

The purpose of this study was to apply one of the tools of classical applied mathematics, the Laplace transform, to the pricing of American options, using the partial Laplace transform method developed by [7] for diffusion problems. The resulting integral equations for the location of the free boundary, (17,19) in transform space and their inverses (22,23) in physical space, form the main result of this paper. These equations were obtained by applying a partial Laplace transform [7] to the Black-Scholes-Merton PDE and solving the resultant ordinary differential equation in transform space. The equations when $D \geq r$ are somewhat simpler than those when $0 < D < r$. It should be recalled that the nonhomogeneous term $F(S)$ in these equations is a function of $\tau_f'(S)$ and $\tau_f''(S)$, the derivatives of the inverse of the optimal exercise boundary, so that (17,19,22,23) involve the first and second derivatives of the unknown boundary and because of this, the integral equations are more complicated than those in [4, 11, 13, 21] which involve the boundary but not the derivatives.

As we mentioned briefly in section 1, integral equation methods have been used to analyze American options before, including the studies of [4, 8, 11, 13, 14, 16, 18, 21, 25, 26]. However, those studies tackled the problem in very different ways to that used here, and ended up with equations of a somewhat different form to those found by us. For example, in their recent studies, [8, 16] used Green's functions to solve the Black-Scholes PDE for American options, and their results involved an integral equation for $S_f(\tau)$, whereas we have an integral equation for the inverse of that function, $\tau_f(S)$. As with our equation, those authors were unable to obtain exact solutions of their integral equations. [25] used a Fourier transform method, while [14] essentially took a Mellin transform with respect to the stock price, and each obtained a (different) integral equation for $S_f(\tau)$. Obviously some connection exists between our results and those other studies, since the integral equations from each study describe the same boundary. It is interesting to note that the Laplace transform, Mellin transform, and Green's function approaches all yield the same expression for the value of a European option but each yields a different integral equation for the free boundary of an American option.

Moving on to the issue of the value of the option, in (8,9,12), we have a series of expressions for $\mathcal{V}(p, S)$, the transform of the option price $V(S, t)$. The constants which appear in these expressions were also given in the previous sections. In theory, given these expressions, we could apply the inverse transform (4), and then we would arrive at the option price itself. Unfortunately, these expressions involve $\tau_f(S)$, the location of the free boundary, which we know only abstractly as the solution of the applicable integral equation; however, if $\tau_f(S)$ were known explicitly, taking the inverse Laplace transform would give the value of the option.

Although the results presented in this study were for the call, it is straightforward to apply them to the put using the well-known put-call 'symmetry' condition of [5, 17], under which the prices of the American call and put are related by

$$C\,[S, E, D, r] = P\,[E, S, r, D]\,, \tag{24}$$

and the positions of the optimal exercise boundary for the call and put are related by

$$S_f^c[t, E, r, D] = E^2 / S_f^p[t, E, D, r].$$
(25)

Of course, (24,25) can also be applied to other studies of American options, such as [4, 11, 13, 21] where the price of the option is given as a closed form expression involving the location of the free boundary.

At this point it behooves us to mention that although we have derived the integro-differential equations (22,23) for the location of the free boundary, we have not addressed either the existence or the regularity or the uniqueness of any solutions to these equations, and these issues remain open, although obviously we would expect the physical free boundary to be a solution. We would suggest that a study addressing these important issues would be a worthwhile endeavor. Indeed, the existence, regularity and uniqueness of solutions remain unresolved for many of the other integral equation formulations of the American option pricing problem mentioned in Section 1, the noticeable exception of course being [4, 11, 13, 21] for which these issues have been successfully resolved. For the integral equations in [4, 11, 13, 21], which involve the boundary but not the derivatives, the existence of solutions follows from the fixed point theorem, while uniqueness has very recently finally been resolved [21].

Finally, we would address the usefulness of the equations (22,23), which describe the location of the free boundary. Although we would not pretend to be proficient in the numerical solution of integral equations, it is reasonable to assume that (22,23) could be used to compute the optimal exercise boundary numerically, as has been done for many of the other integral equation formulations of the American option pricing problem, and we would suggest this, along with a local solution close to expiry, as possible directions for future research. Of course, with such a numerical solution, great care must be taken to verify that any solution of the integral equations corresponds to a solution of the underlying optimal stopping problem, and it would also be of interest to compare the boundary computed using (22,23) with those computed using the other integral equation formulations.

# References

1. Alobaidi G, Mallier R (2006) The American straddle close to expiry. Boundary Value Problems 2006 article ID 32835.
2. Black F, Scholes M (1973) The pricing of options and corporate liabilities. Journal of Political Economy 81:637–659.
3. Brekke KA, Oksendal B (1991) The high contact principle as a sufficiency condition for optimal stopping. In: Lund D, Oksendal B (eds) Stochastic Models and Option Values: Applications to Resources, Environment and Investment Problems. North-Holland, Amsterdam
4. Carr P, Jarrow R, Myneni R (1992) Alternative characterizations of the American put. Mathematical Finance 2:87–106.

5. Chesney M, Gibson R (1993) State space symmetry and two-factor option pricing models. Advances in Futures and Options Research 8:85–112.

6. Crank J (1984) Free and moving boundary problems. Clarendon, Oxford

7. Evans GW, Isaacson E, MacDonald JKL (1950) Stefan-like problems. Quarterly of Applied Mathematics 8:312–319.

8. Evans JD, Kuske RE, Keller JB (2002) American options with dividends near expiry. Mathematical Finance 12:219–237.

9. Geske R (1979) A note on an analytical valuation formula for unprotected American call options on stocks with known dividends. Journal of Financial Economics 7:375–380.

10. Geske R (1981) Comments on Whaley's note. Journal of Financial Economics 9:213-215.

11. Jacka SD (1991) Optimal stopping and the American put. Mathematical Finance 1:1–14.

12. Karatzas I (1988) On the pricing of American options. Applied Mathematics and Optimization 17:37–60.

13. Kim IJ (1990) The analytic valuation of American options. Review of Financial Studies 3:547–572.

14. Knessl C (2001) A note on a moving boundary problem arising in the American put option. Studies in Applied Mathematics 107:157–183.

15. Kolodner II (1956) Free Boundary Problem for the Heat Equation with Applications to Problems of Change of Phase. I. General Method of Solution. Communications in Pure and Applied Mathematics 9:1-31.

16. Kuske RE, Keller JB (1998) Optimal exercise boundary for an American put option. Applied Mathematical Finance 5:107–116.

17. McDonald R, Scroder M (1998) A parity result for American options. Journal of Computational Finance 1:5–13.

18. McKean HP Jr. (1965) Appendix A: A free boundary problem for the heat equation arising from a problem in mathematical economics. Industrial Management Review 6:32–39.

19. Merton RC (1973) The theory of rational option pricing. Bell Journal of Economics and Management Science 4:141–183.

20. Myneni R (1992) The Pricing of the American Option. Annals of Applied Probability 2:1-23.

21. Peskir G (2005) On the American option problem. Mathematical Finance 15:169-181.

22. Polyanin AD, Manzhirov V (1998) Handbook of integral equations. CRC Press, New York

23. Roll R (1977) An analytical formula for unprotected American call options on stocks with known dividends. Journal of Financial Economics 5:251–258.

24. Samuelson PA (1965) Rational theory of warrant pricing. Industrial Management Review 6:13–31.

25. Stamicar R, Sevcovic D, Chadam J (1999) The early exercise boundary for the American put near expiry: numerical approximation. Canadian Applied Mathematics Quarterly 7:427–444.

26. Van Moerbeke J (1976) On optimal stopping and free boundary problems. Archive for Rational Mechanics Analysis 60:101–148.

27. Whaley RE (1981) On the valuation of American call options on stocks with known dividends. Journal of Financial Economics 9:207–211.

# Time Change, Volatility, and Turbulence

Ole E. Barndorff-Nielsen and Jürgen Schmiegel

The T.N. Thiele Centre for Applied Mathematics in Natural Science,
Department of Mathematical Sciences, University of Aarhus,
Ny Munkegade, DK-8000 Aarhus C, Denmark.
oebn@imf.au.dk, schmiegl@imf.au.dk

**Summary.** A concept of volatility modulated Volterra processes is introduced. Apart from some brief discussion of generalities, the paper focusses on the special case of backward moving average processes driven by Brownian motion. In this framework, a review is given of some recent modelling of turbulent velocities and associated questions of time change and universality. A discussion of similarities and differences to the dynamics of financial price processes is included.

## 1 Introduction

Change of time is an important concept in stochastic analysis and some of its applications, especially in mathematical finance and financial econometrics, with quadratic variation and its interpretation as integrated squared volatility playing a key role. (A rather comprehensive treatment of this will be available in [27].) On the other hand there are well known similarities, as well as important differences, between the dynamics of financial markets and of turbulence. From these prospects, the present paper discusses some recent modelling in turbulence and associated questions of time change.

To set the discussion in perspective, a general concept of volatility modulated Volterra processes is introduced. This would seem to be of some rather wide interest in mathematical modelling. Here we focus on its relevance for stochastic modelling of turbulence. For a masterly overview of the main approaches to modelling of turbulence see [51], cf. also [50].

A summary comparison of main stylized features in finance and turbulence is given in the next Section. Of central importance is the fact that volatility is a key concept in turbulence as well as in finance, though in turbulence the phenomenon is referred to as intermittency.

The notion of change of time in mathematical finance and financial econometrics refers to an increasing stochastic process as the time change while in turbulence we have in mind a deterministic time change. We provide empirical and theoretical evidence for the existence of an affine deterministic time change in turbulence in terms of which the main component of the velocity vector in a turbulent flow behaves in a universal way over a wide range of scales. We also discuss the limitations of this type of universality and briefly outline the extension to a non-affine deterministic change of time and its relevance for universality of velocity increments.

Section 2 provides some background on turbulence and the similarities and differences between turbulence and finance, and recalls some features of the Normal inverse Gaussian distribution. Section 3 discusses volatility modulated Volterra processes and their behaviour under change of time. A more general discussion of change of time for stationary processes is presented in Section 4. Section 5 provides empirical and theoretical evidence for the relevance of change of time in turbulence. The potential of Volterra processes for modelling velocity fields in turbulence is outlined in Sections 6 and 7. Section 8 relates the concept of change of time to the particular setting of the proposed modelling frameworks in finance and turbulence. This leads to a primitive and a refined universality statement for turbulence in Section 9. Section 10 concludes.

## 2 Background

The statistics of turbulent flows and financial markets share a number of stylized features ([36], [44], [27] and [7]). The counterpart of the velocity in turbulence is the log price in finance, and velocity increments correspond to log returns. The equivalent of the intermittency of the energy dissipation in turbulence is the strong variability of the volatility in financial markets. Subsection 2.1 briefly summarizes some basic information on turbulence, and subsection 2.2 lists the most important similarities and differences between turbulence and finance. The normal inverse Gaussian laws constitute a useful tool in both fields and some of the properties of these laws are recalled in the Appendix.

### 2.1 Turbulence

There is no generally accepted definition of what should be called a turbulent flow. Turbulent flows are characterized by low momentum diffusion, high momentum convection, and rapid variation of pressure and velocity in space and time. Flow that is not turbulent is called laminar flow. The non-dimensional Reynolds number $R$ characterizes whether flow conditions lead to laminar or turbulent flow. Increasing the Reynolds number increases the turbulent character and the limit of infinite Reynolds number is called the fully developed turbulent state.

The most prominent observable in a turbulent flow is the main component of the velocity field $V_t$ as a function of time $t$ and at a fixed position in space. A derived quantity is the temporal surrogate energy dissipation process

describing the loss of kinetic energy due to friction forces characterized by the
viscosity $\nu$

$$\varepsilon_t \equiv \frac{15\nu}{\overline{V}^2} \left(\frac{dV_t}{dt}\right)^2, \tag{1}$$

where $\overline{V}$ denotes the mean velocity.

The temporal surrogate energy dissipation process takes into account the
experimental condition where only a time series of the main component of the
velocity vector is accessible. The temporal surrogate energy dissipation is a
substitute for the true energy dissipation process (involving the spatial derivatives of all velocity components) for flows which are stationary, homogeneous
and isotropic [34]. In the sequel we call such flows *free turbulent flows*. We
refer to the temporal surrogate energy dissipation as the energy dissipation
unless otherwise stated.

Since the pioneering work of Kolmogorov [39] and Obukhov [43], intermittency of the turbulent velocity field is of major interest in turbulence research.
From a probabilistic point of view, intermittency refers, in particular, to the
increase in the non-Gaussian behaviour of the probability density function
(pdf) of velocity increments

$$\Delta u_s = V_{t+s} - V_t$$

with decreasing time scale $s$. Here we adopt the notation $\Delta u$ for velocity
increments which is traditional in the turbulence literature. A typical scenario
is characterized by an approximate Gaussian shape for the large scales, turning
to exponential tails for the intermediate scales and stretched exponential tails
for dissipation scales ([29] and [52], see also Figure 1).

## 2.2 Stylized features of finance and turbulence

The most important similarities between financial markets and turbulent flows
are semiheavy tails for the distributions of log returns/velocity increments,
the evolution of the densities of log returns/velocity increments across time
scales with the heaviness of the tails decreasing as the time lag increases, and
long range dependence of log returns/velocity increments. It is important to
note that in spite of the long range dependence the autocorrelation of the
log price process is essentially zero whereas the velocity field shows algebraic
decay of the autocorrelation function. Other important differences are the
skewness of the densities of velocity increments in contrast to the symmetry
of the distribution of log returns in FX markets[1] and the different behaviour
of bipower variation [25, 21]. Table 1 gives an overview of the differences and
similarities between turbulence and finance.

---

[1] For stocks, skewness of the distribution of log returns is observed. There, leverage
is believed to be a key mechanism.

## 2.3 Normal inverse Gaussian distributions

Intermittency/volatility is related to the heaviness of the tails and the non-Gaussianity of the distribution of velocity increments and log returns. In this respect, Normal inverse Gaussian (NIG) distributions are a suitable class of probability distributions which fit the empirical densities in both systems to high accuracy ([5, 6], [47], [35], [10], [17]).

Figure 1 shows, as an example, the log densities of velocity increments $\Delta u_s$ measured in the atmospheric boundary layer for various time scales $s$. The solid lines denote the approximation of these densities within the class of NIG distributions. NIG distributions fit the empirical densities equally well for all time scales $s$.

A subsequent analysis of the observed parameters of the NIG distributions from many, widely different data sets with Reynolds numbers ranging from $R_\lambda = 80$ up to $R_\lambda = 17000$ (where $R_\lambda$ is the Taylor based Reynolds number) led to the formulation of a key universality law ([10] and [19]): The temporal development of a turbulent velocity field has an intrinsic clock which depends on the experimental conditions but in terms of which the one-dimensional marginal distributions of the normalized velocity differences become independent of the experimental conditions. Figure 2 provides an empirical validation of this.

# 3 Volatility modulated Volterra processes

This Section is divided into three subsections. The first briefly discusses Volterra type processes, the second introduces the concept of volatility modulated Volterra processes, and the third considers the behaviour of such processes under a change of time.

## 3.1 Volterra type processes

In this paper we shall be referring to processes of the form

$$Y_t = \int_{-\infty}^{\infty} K(t, s) \, dB_s + \chi \int_{-\infty}^{\infty} Q(t, s) \, ds, \qquad (2)$$

as *Brownian Volterra processes* (BVP). Here $K$ and $Q$ are deterministic functions, sufficiently regular to give suitable meaning to the integrals. Furthermore, $\chi$ is a constant and $B$ denotes standard Brownian motion.

*Example 1.* **Fractional Brownian motion**   As is well known (cf, for instance [49]) fractional Brownian motion can be written as

$$B_t^H = \int_{-\infty}^{\infty} \left[ (t - s)_+^{H-1/2} - (-s)_+^{H-1/2} \right] dB_s.$$

Of particular interest are the backward Volterra processes, i.e. where $K(t,s)$ and $Q(t,s)$ are 0 for $s > t$. In this case formula (2) takes the form

$$Y_t = \int_{-\infty}^{t} K(t,s)\, \mathrm{d}B_s + \chi \int_{-\infty}^{t} Q(t,s)\, \mathrm{d}s. \tag{3}$$

*Example 2.* **Fractional Brownian motion**   For $B$ a Brownian motion, the fractional Brownian motion with index $H \in (0,1)$ may alternatively, see [42], be represented as

$$B_t^H = \int_0^t K(t,s)\, \mathrm{d}B_s \tag{4}$$

where

$$K(t,s) = c_H \left\{ \left(\frac{t}{s}\right)(t-s)^{H-1/2} - \left(H - \frac{1}{2}\right) s^{1/2-H} \int_s^t u^{H-3/2}(u-s)^{H-1/2}\, \mathrm{d}u \right\}$$

and

$$c_H = \left( \frac{2\Gamma\left(\frac{3}{2} - H\right)}{\Gamma\left(H + \frac{1}{2}\right)\Gamma(2 - 2H)} \right)^{1/2}.$$

A more general type of Volterra processes are the *Lévy Volterra processes* (LVP), which are of the form

$$Y_t = \int_{-\infty}^{\infty} K(t,s)\, \mathrm{d}L_s + \chi \int_{-\infty}^{\infty} Q(t,s)\, \mathrm{d}s$$

where $L$ denotes a Lévy process on $\mathbb{R}$ and $K$ and $Q$ are deterministic kernels, satisfying certain regularity conditions.

*Example 3.* **Fractional Lévy motion**   [40] introduces Fractional Lévy motion $L^H$ for $H \in \left(\frac{1}{2}, 1\right)$ by the formula

$$L_t^H = \int_{-\infty}^{\infty} \left[ (t-s)_+^{H-1/2} - (-s)_+^{H-1/2} \right] \mathrm{d}L_s$$

where $L$ is a Lévy process.

Stochastic integration in these general settings is discussed for BVP in [37], [31], [32], cf. also [42], and for LVP in [28].

For Brownian Volterra processes we shall refer to the following three conditions: For all $s, t \in \mathbb{R}$

**C1**  $K(t, \cdot) \in L^2(\mathbb{R})$ and $Q(t, \cdot) \in L^2(\mathbb{R})$

**C2**  $K(s,s) = K_0 > 0$ and $K(t,s) = 0$ for $s > t$

$\quad\;\; Q(s,s) = Q_0 > 0$ and $Q(t,s) = 0$ for $s > t$

**C3** $K$ and $Q$ are differentiable with respect to their first arguments and, denoting the derivatives by $\dot{K}$ and $\dot{Q}$, we have

$$\dot{K}(t, \cdot) \in L^2(\mathbb{R}) \text{ and } \dot{Q}(t, \cdot) \in L^2(\mathbb{R}).$$

Under these conditions the covariance function of (3) exists and is, for $s \le t$, given by

$$R(s, t) = \text{Cov}\{Y_s, Y_t\} = \int_{-\infty}^{s} K(t, u) K(s, u) \, du$$

and the autocorrelation function may be written as

$$r(s, t) = \int_{-\infty}^{s} \bar{K}(t, u) \bar{K}(s, u) \, du$$

where

$$\bar{K}(t, u) = K(t, u) / \| K(t, \cdot) \|$$

and $\|\cdot\|$ denotes the $L^2$ norm.

### 3.2 Volatility modulated Volterra processes

For modelling purposes it is of interest to consider *Volatility modulated Volterra processes* (VMVP) which we define (backward case) by

$$Y_t = \int_{-\infty}^{t} K(t, s) \sigma_s dB_s + \chi \int_{-\infty}^{t} Q(t, s) \sigma_s^2 ds \tag{5}$$

where $\sigma > 0$ is a stationary cadlag process on $\mathbb{R}$, embodying the volatility/intermittency.

On the further assumptions that the deterministic kernels $K$ and $Q$ satisfy conditions **C1-C3**, we have that $Y$ is a semimartingale, satisfying the stochastic differential equation

$$dY_t = K_0 \sigma_t dB_t + \chi Q_0 \sigma_t^2 dt + \int_{-\infty}^{t} \dot{K}(t, s) \sigma_s dB_s + \chi \int_{-\infty}^{t} \dot{Q}(t, s) \sigma_s^2 ds.$$

The quadratic variation of $Y$ is then, for $t \ge 0$,

$$[Y]_t = K_0^2 \tau_t \tag{6}$$

where

$$\tau_t = \int_0^t \sigma_s^2 ds \tag{7}$$

is the integrated squared volatility process. For $t < 0$ we define $[Y]_t$ and $\tau_t$ by the same formulae (6) and (7). Then $[Y]$ is a continuous increasing stochastic process with $[Y]_0 = 0$.

Finally, we introduce the inverse process $\theta$ of $\tau$ by

$$\theta_t = \inf \{s : \tau_s \ge t\}. \tag{8}$$

## 3.3 Time change and VMVP

We say that a process $T$ on $\mathbb{R}$ is a time change provided $T$ is increasing with $T_0 = 0$. The time changes on $\mathbb{R}$ we shall be considering are in fact continuous and strictly increasing, with $T \to \pm\infty$ as $t \to \pm\infty$. (This is the case in particular for the processes $\tau$ and $\theta$ defined above.)

For $T$ a time change process on $\mathbb{R}$ and given a Volterra kernel $K$ we define a new Volterra kernel $K \circ T$ by

$$K \circ T (t, s) = K (T_t, T_s). \tag{9}$$

Taking $T = \theta$ as given by (8) we may now rewrite $Y$ of (5) as

$$
\begin{aligned}
Y_t &= \int_{-\infty}^{t} K(t, s)\, \mathrm{d}B_{\tau_s} + \chi \int_{-\infty}^{t} Q(t, s)\, \mathrm{d}\tau_s \\
&= \int_{-\infty}^{\tau_t} K(t, \theta_u)\, \mathrm{d}B_u + \chi \int_{-\infty}^{\tau_t} Q(t, \theta_u)\, \mathrm{d}u
\end{aligned}
$$

implying

$$Y_{\theta_t} = \int_{-\infty}^{t} K \circ \theta\,(t, s)\, \mathrm{d}B_s + \chi \int_{-\infty}^{t} Q \circ \theta\,(t, s)\, \mathrm{d}s. \tag{10}$$

In particular, if the volatility process $\sigma$ is independent of $B$ then, conditional on $[Y]$, $Y_\theta$ is a Volterra process with kernels $(K \circ \theta, Q \circ \theta)$ and the same driving Brownian motion as the VMVP $Y$.

Later in this paper we shall, in the context of turbulence, be concerned with affine time changes.

*Remark 1.* **Affine time change**    Suppose $T_t = ct + c_0$ for some $c > 0$ and a constant $c_0$. Applying this to (5) gives

$$
\begin{aligned}
Y_{ct+c_0} &= \int_{-\infty}^{ct+c_0} K(ct + c_0, s)\, \sigma_s \mathrm{d}B_s + \chi \int_{-\infty}^{ct+c_0} Q(ct + c_0, s)\, \sigma_s^2 \mathrm{d}s \\
&= \int_{-\infty}^{t} K(ct + c_0, cu + c_0)\, \sigma_{cu+c_0} \mathrm{d}B_{cu+c_0} \\
&\qquad\qquad + c\chi \int_{-\infty}^{t} Q(ct + c_0, cu + c_0)\, \sigma_{cu+c_0}^2 \mathrm{d}u
\end{aligned}
$$

i.e.

$$Y_{ct+c_0} = \int_{-\infty}^{t} K_{c,c_0}(t, s)\, \sigma_{cs+c_0} \mathrm{d}\tilde{B}_s + \chi \int_{-\infty}^{t} Q_{c,c_0}(t, s)\, \sigma_{cs+c_0}^2 \mathrm{d}s \tag{11}$$

with $K_{c,c_0}(t, s) = \sqrt{c} K(ct + c_0, cs + c_0)$, $Q_{c,c_0}(t, s) = cQ(ct + c_0, cs + c_0)$ and where $\tilde{B}_s = c^{-1/2} B_{cs}$ is a Brownian motion. Thus the transformed process follows again a VMVP but now with volatility process $\sigma_{c\cdot+c_0}$ (and kernels $K_{c,c_0}$ and $Q_{c,c_0}$).

# 4 Time change in stationary processes

Let $Y$ and $Y^*$ be stationary stochastic processes on $\mathbb{R}$ and let $X$ and $X^*$ be the corresponding increment processes given by $X_t = Y_t - Y_0$ and $X_t^* = Y_t^* - Y_0^*$. The present Section discusses distributional relations between $X$ and $X^*$ under various assumptions. Note first, however, that only affine time changes preserve stationarity in a stationary process.

Assuming $\mathrm{Var}\{Y_t\} = \mathrm{Var}\{Y_t^*\} = \omega^2$, say, and denoting the autocorrelation functions of $Y$ and $Y^*$ respectively by $r$ and $r^*$ we have

$$\mathrm{Var}\{X_t\} = 2\omega^2 \bar{r}(t) \quad \text{and} \quad \mathrm{Var}\{X_t^*\} = 2\omega^2 \bar{r}^*(t) \tag{12}$$

where $\bar{r}(t) = 1 - r(t)$ and $\bar{r}^*(t) = 1 - r^*(t)$.

Let $\psi(t)$ be a time change, and suppose that

$$X_t \overset{law}{=} X_{\psi(t)}^* \quad \text{for every } t \in \mathbb{R}. \tag{13}$$

As discussed in Section 2.3 this type of behaviour has been found in free turbulence.

Assumption (13) and the relation (12) imply

$$r(t) = r^*(\psi(t)),$$

and provided both $r$ and $r^*$ are strictly decreasing and continuous functions we have that the time change $\psi$ is expressible as

$$\psi(t) = \rho^*(r(t)), \tag{14}$$

$\rho^*$ denoting the inverse function of $r^*$

In case the statistical analysis of observations from $X$ and $X^*$ has shown good agreement with the Ansatz (13), it is then natural to ask whether (13) is simply a reflection of the more sweeping Ansatz

$$X_{\cdot} \overset{law}{=} X_{\psi(\cdot)}^*, \tag{15}$$

saying that $X$ and $X^*$ are equal in law as processes and not just pointwise as in (13). This Ansatz implies, in particular, that $\psi$ must be affine since, as mentioned earlier, only affine time changes preserve stationarity.

In fact, the weaker assumption of second order agreement of $X$ and $X^*$ already implies that $\psi$ is affine, as we show now.

Suppose that for all $0 \leq s \leq t$

$$\mathrm{E}\left\{X_{\psi(t)}^{*2}\right\} = \mathrm{E}\left\{X_t^2\right\} \tag{16}$$

$$\mathrm{E}\left\{X_{\psi(s)}^* X_{\psi(t)}^*\right\} = \mathrm{E}\left\{X_s X_t\right\}. \tag{17}$$

(Note that since we have assumed that $Y$ and $Y^*$ are stationary, necessarily $\mathrm{E}\{X_t\} = 0 = \mathrm{E}\left\{X^*_{\psi(t)}\right\}$.) We will then have

$$\mathrm{E}\left\{\left(X^*_{\psi(t)} - X^*_{\psi(s)}\right)^2\right\} = \mathrm{E}\left\{(X_t - X_s)^2\right\}.$$

Further, by (12), on the one hand

$$\mathrm{E}\left\{\left(X^*_{\psi(t)} - X^*_{\psi(s)}\right)^2\right\} = \mathrm{E}\left\{\left(Y^*_{\psi(t)} - Y^*_{\psi(s)}\right)^2\right\} = 2\omega^2\bar{r}^*\left(\psi(t) - \psi(s)\right)$$

while on the other

$$\mathrm{E}\left\{(X_t - X_s)^2\right\} = \mathrm{E}\left\{(Y_t - Y_s)^2\right\} = 2\omega^2\bar{r}(t - s).$$

This implies

$$\bar{r}^*\left(\psi(t) - \psi(s)\right) = \bar{r}(t - s)$$

or, equivalently, by (14),

$$\psi(t) - \psi(s) = \psi(t - s)$$

which can only hold for $\psi$ affine,

$$\psi(t) = ct + c_0$$

for some $c > 0$ and a constant $c_0$.


## 5 Universality in turbulence

In this Section we discuss the empirical support for the existence of affine, intrinsic (one for each experiment), time changes such that for a wide range of time scales the densities of turbulent velocity increments obtained from different experiments collapse. This leads to the formulation of a primitive universality model for turbulence (see subsection 9.1). At very small or very large time scales, deviations from affinity are observed. This then leads to the formulation of a refined universality model (see subsection 9.2). The dynamical aspects of this refined universality model are briefly discussed in the concluding Section 10.

In comparing the equivalence under time change to existing theory of turbulence and to empirical evidence it is illuminating to relate the discussion to the well established fact (cf. Section 2.3) that in free turbulence the distributions of velocity differences over fixed time spans are closely describable by the NIG law.

## 5.1 Empirics

The statistical analysis of a large number of different turbulent data sets revealed the existence of a type of universality which states that the densities of velocity increments are well described within the class of NIG distributions and, moreover, the densities of the increments of the normalized velocity field obtained from different experiments collapse as long as the intrinsic time scales are measured in terms of the scale parameter of the approximate NIG distributions. Here, the velocity field is normalized by its standard deviation. We denote the normalized velocity component by $\tilde{V} = V/\sqrt{\text{Var}(V)}$. Then we have the empirical result, denoting the corresponding velocity increments by $\Delta \tilde{u}$,

$$\Delta \tilde{u}_t \overset{law}{=} \Delta \tilde{u}^*_{\psi(t)}$$

where

$$\psi(t) = \overset{\leftarrow}{\delta}{}^* (\delta(t))$$

and where $\delta(t)$ and $\delta^*(t)$ are the scale parameters of the approximate NIG distributions of $\Delta \tilde{u}_t$ and $\Delta \tilde{u}^*_t$, respectively. Here $\Delta \tilde{u}_t$ refers to the velocity increments for a given turbulent experiment and the superscript $*$ refers to a different independent turbulent experiment, different in Reynolds number and/or experimental set-up. The superscript $\leftarrow$ denotes the inverse function.

Figure 3 shows the estimated time change $\psi$ for a number of independent turbulent experiments. For a wide range of time lags, $\psi$ is essentially affine in a first approximation. The degree of non-affinity increases with increasing difference of the Reynolds numbers.

*Remark 2.* The collapse of the densities of velocity increments implies that the variances $\text{Var}\{\Delta \tilde{u}_t\}$ and $\text{Var}\{\Delta \tilde{u}^*_t\}$ at the corresponding time scales are the same. Denoting the variances by $c_2(t) = \text{Var}\{\Delta \tilde{u}_t\}$ and $c^*_2(t) = \text{Var}\{\Delta \tilde{u}^*_t\}$, the time change $\psi$ can, alternatively be expressed as (c.f. (14))

$$\psi(t) = \overset{\leftarrow}{c}{}^*_2 (c_2(t)).$$

*Remark 3.* It is important to note that the quality of the collapse of the densities of velocity increments does not depend on the degree of non-affinity of the time change $\psi$. Velocity increments of widely different experiments collapse for all amplitudes at time lags at which they have the same variance [17, 20].

## 5.2 Theoretical considerations

The empirically observed approximate affinity of the time change $\psi$ for a range of intermediate time scales can be motivated theoretically for turbulent experiments where a clear Kolmogorov scaling is observed. Such a Kolmogorov scaling is expected for large Reynolds numbers and for a certain range of scales, called the inertial range [38]. In the limit of very large Reynolds numbers, the variance of velocity increments is expected to show a scaling behaviour of the form

$$c_2(t) = at^{2/3} \tag{18}$$

where $a$ is a flow dependent constant and the time scale $t$ is within the inertial range. (In practice, one defines the inertial range as the range of time scales for which (18) holds.) For such flows the expected time change is affine (within the inertial range).

For small and moderate Reynolds numbers, the inertial range is absent or very small. For instance, the examples shown in Figure 3 do not show a clear Kolmogorov scaling for an extended range of time scales. However, the empirically estimated time change appears to be affine for a wide range of time scales. This gives the possibility to define a non-scaling counterpart of Kolmogorov scaling and an associated generalized inertial range where the variances are universal in the sense that

$$\psi(t) = \overleftarrow{c}_2^{*}\, (c_2(t)) = ct + \psi_0,$$

where $\psi_0$ is a constant. A particular example are variances of the form

$$c_2(t) = a\,(t + t_0)^{2/3},$$

where $t_0$ is a constant. In view of Kolmogorov scaling, we then expect $t_0 \to 0$ as the Reynolds number gets very large.

# 6 Modelling frameworks in finance and turbulence

Volatility modulated Volterra processes of the form (5) have found applications in finance as well as for the modelling of the turbulent velocity field. In the turbulence context, these processes capture the main idea of the Reynolds decomposition of the velocity field into a slowly varying component (the second term in Equation (5)) and a rapidly varying component (the first term in Equation (5)) [16, 18].

In the following subsections, we discuss the application of volatility modulated Volterra processes in finance and turbulence with emphasis on the empirical findings concerning time change and universality.

## 6.1 Finance

The basic framework for stochastic volatility modelling in finance is that of Brownian semimartingales

$$Y_t = \int_0^t \sigma_s \mathrm{d}B_s + \int_0^t a_s \mathrm{d}s \tag{19}$$

where $\sigma$ and $a$ are caglad processes and $B$ is Brownian motion, with $\sigma$ expressing the volatility. In general, $Y$, $\sigma$, $B$ and $a$ will be multidimensional but in the present paper we shall only consider one-dimensional processes. Importantly, whatever the process $a$, the quadratic variation of $Y$ satisfies $[Y] = \tau$ with $\tau$ given by (7).

## 6.2 Turbulence

Whereas Brownian semimartingales are 'cumulative' in nature, for free turbulence it is physically natural to model timewise velocity dynamics by stationary processes. In analogy to (19), the following framework for the latter type of dynamics has recently ([16]) been proposed.

At time $t$ and at a fixed position in the turbulent field, the velocity of the main component of the velocity vector (i.e. the component in the mean wind direction) is specified as $V_t = \mu + Y_t$ with

$$Y_t = \int_{-\infty}^{t} g(t-s)\sigma_s \mathrm{d}B_s + \chi \int_{-\infty}^{t} q(t-s)\sigma_s^2 \mathrm{d}s. \tag{20}$$

Here $B$ and $\sigma$ are as above, $\mu$ and $\chi$ are constants and $g$ and $q$ are nonnegative real functions on $(0, \infty)$ satisfying $g(0+) > 0$, $q(0+) > 0$,

$$||g||^2 = \int_0^\infty g^2(t)\,\mathrm{d}t = 1$$

and

$$\int_0^\infty q(t)\,\mathrm{d}t = 1.$$

Furthermore, $g$ and $q$ are assumed to be sufficiently regular to make the integrals in (20) exist, and we require that the derivative $\dot{g}$ of $g$ is square integrable.

Under these conditions, the stationary process $Y$ is a semimartingale and we have

$$[Y] = g^2(0+)\tau$$

where $\tau$ is given by (7).

*Remark 4. Ambit processes*    The model type (20) is a one-dimensional limit of a spatio-temporal modelling framework introduced under the name of *Ambit processes* in [18]. In that more general context, the role of the Brownian motion is taken over by a Gaussian white noise field (or Brownian sheet) and the volatility is expressed as a random field, which may, for instance, be generated from a Lévy basis as in [15]. The paper [18] gives a first discussion of the theoretical properties of such processes and describes some applications to turbulence and cancer growth.

# 7 Increment processes

Both type of processes (19) and (20) have stationary increments. In the latter case, letting $X_t = Y_t - Y_0$ we have

$$
\begin{aligned}
X_t = & \int_{-\infty}^t \left\{ g\left(t-s\right) - 1_{(-\infty,0)}\left(s\right) g\left(-s\right) \right\} \sigma_s dB_s \\
& + \chi \int_{-\infty}^t \left\{ q\left(t-s\right) - 1_{(-\infty,0)}\left(s\right) q\left(-s\right) \right\} \sigma_s^2 ds
\end{aligned}
\tag{21}
$$

which we also write, on VMVP form, as

$$
X_t = \int_{-\infty}^t j\left(t,s\right) \sigma_s dB_s + \chi \int_{-\infty}^t k\left(t,s\right) \sigma_s^2 ds
\tag{22}
$$

where

$$
j\left(t,s\right) = g\left(t-s\right) - 1_{(-\infty,0)}\left(s\right) g\left(-s\right)
$$
$$
k\left(t,s\right) = q\left(t-s\right) - 1_{(-\infty,0)}\left(s\right) q\left(-s\right).
$$

Suppose now that $B$ and $\sigma$ are independent. Clearly, $X|\sigma$ is then a Gaussian process with

$$
\mathrm{E}\{X_t|\sigma\} = \chi \int_{-\infty}^t k\left(t,s\right) \sigma_s^2 ds
$$

$$
\mathrm{Var}\{X_t|\sigma\} = \int_{-\infty}^t j^2\left(t,s\right) \sigma_s^2 ds
$$

and, for $0 \le s \le t$,

$$
\mathrm{Cov}\{X_s X_t|\sigma\} = \int_{-\infty}^s j\left(s,u\right) j\left(t,u\right) \sigma_u^2 du.
$$

We proceed to discuss the conditional law of $X$ given $\sigma$ and its limit behaviour for $t \to 0$ and $t \to \infty$.

Considering first the conditional mean we note that

$$
\begin{aligned}
\int_{-\infty}^t k\left(t,s\right) \sigma_s^2 ds &= \int_0^t q\left(t-s\right) \sigma_s^2 ds + \int_{-\infty}^0 \left\{ q\left(t-s\right) - q\left(-s\right) \right\} \sigma_s^2 ds \\
&= \int_0^t q\left(s\right) \sigma_{t-s}^2 ds + \int_{-\infty}^0 \left\{ q\left(t-s\right) - q\left(-s\right) \right\} \sigma_s^2 ds.
\end{aligned}
$$

From this we find that

$$
\int_{-\infty}^t k\left(t,s\right) \sigma_s^2 ds \sim q\left(0+\right) \sigma_0^2 t + t \int_0^\infty q'(s)\sigma_{-s}^2 ds \quad \text{as} \quad t \downarrow 0
$$

and, under a mild mixing condition on $\sigma$, that

$$\int_{-\infty}^{t} k\left(t,s\right)\sigma_s^2 \mathrm{d}s \sim K - K' \quad \text{as} \quad t \to \infty$$

where $K$ and $K'$ are independent and identically distributed with

$$K' = \int_0^\infty q\left(s\right)\sigma_{-s}^2 \mathrm{d}s.$$

Similarly, for the conditional variance we have

$$\int_{-\infty}^{t} j^2\left(t,s\right)\sigma_s^2 \mathrm{d}s = \int_0^t g^2\left(t-s\right)\sigma_s^2 \mathrm{d}s + \int_{-\infty}^0 \left\{g\left(t-s\right)-g\left(-s\right)\right\}^2 \sigma_s^2 \mathrm{d}s$$

$$= \int_0^t g^2\left(s\right)\sigma_{t-s}^2 \mathrm{d}s + \int_{-\infty}^0 \left\{g\left(t-s\right)-g\left(-s\right)\right\}^2 \sigma_s^2 \mathrm{d}s.$$

Hence

$$\int_{-\infty}^{t} j^2\left(t,s\right)\sigma_s^2 \mathrm{d}s \sim g^2\left(0+\right)\sigma_0^2 t \quad \text{as} \quad t \downarrow 0$$

while

$$\int_{-\infty}^{t} j^2\left(t,s\right)\sigma_s^2 \mathrm{d}s \sim G + G' \quad \text{as} \quad t \to \infty$$

with

$$G' = \int_0^\infty g^2\left(s\right)\sigma_s^2 \mathrm{d}s$$

and $G$ and $G'$ independent and identical in law.

All in all we therefore have, conditionally on the volatility process $\sigma$,

$$\frac{X_t}{g\left(0+\right)\sqrt{t}} \sim N\left(0,\sigma_0^2\right) \quad \text{as} \quad t \downarrow 0$$

while

$$X_t \sim N\left(\chi\left(K-K'\right),G+G'\right) \quad \text{as} \quad t \to \infty.$$

In particular, the law of the increment process $X_t$ will not be normal in the large time scale limit unless the volatility $\sigma$ is constant. In the case of $\sigma_t^2$ following an inverse Gaussian law we get for the small time scale limit of $X_t$ a Normal inverse Gaussian distribution. It has been shown in [16] that for an inverse Gaussian volatility process, the increment process is well fitted by a Normal inverse Gaussian law for all time scales. Moreover, the resulting increment process also reproduces the experimentally observed statistics of the Kolmogorov variable [39].

## 8 Time change in finance and turbulence

The specification, in Section 6, of the modelling frameworks in finance and turbulence as specific types of VMVP now allows to discuss the idea of a time change in more detail.

## 8.1 Finance

Of particular interest are cases where the process $a$ in (19) is of the form $a = \beta\sigma^2$ for some constant $\beta$, i.e.

$$Y_t = \int_0^t \sigma_s \mathrm{d}B_s + \beta\tau_t. \tag{23}$$

For suitable choice of $\sigma$ this type of process is generally capable of modelling the basic dynamics of stock prices and foreign exchange rates while, at the same time, being analytically tractable. More specifically, this is the case when $\sigma^2$ is of supOU type with $\sigma_t^2$ following the inverse Gaussian law; see for instance [23], [26].

Under the specification (23) $Y$ may, by the Dambis-Dubins-Schwartz Theorem, be rewritten as

$$Y_t = B'_{\tau_t} + \chi\tau_t$$

where $B'$ is a Brownian motion, which is a functional of $Y$ itself. Equivalently,

$$Y_{\theta_t} = B'_t + \chi t. \tag{24}$$

In the finance context, $\theta_t$ is thought of as 'operational' or 'business' time. This time is, in principle, known from the quadratic variation process $[Y]$, and that in turn can be estimated by the realised quadratic variation

$$[Y_\delta]_t = \sum_{j=1}^{\lfloor t/\delta \rfloor} \left(Y_{j\delta} - Y_{(j-1)\delta}\right)^2$$

which satisfies

$$[Y_\delta] \xrightarrow{p} [Y]$$

for $t \to \infty$. Equation (24) is interpreted as saying that under (23), log price returns are Gaussian when recorded in operational time. At least as a first approximation this is close to reality, see for instance [1]. Recent, more refined, empirical analysis takes the possibilities of jumps and of microstructure noise, which are not covered by (23), into account; see [2].

## 8.2 Turbulence

In this case, i.e. (20), formula (10) takes the form

$$Y_{\theta_t} = \int_{-\infty}^t g(\theta_t - \theta_s)\mathrm{d}B_s + \chi \int_{-\infty}^t q(\theta_t - \theta_s)\mathrm{d}s.$$

Furthermore, (11) specializes to

$$Y_{ct+c_0} = \int_{-\infty}^t g_c\left(t - u\right)\sigma_{cu+c_0}\mathrm{d}B'_u + \chi \int_{-\infty}^t q_c\left(t - u\right)\sigma_{cu+c_0}^2 \mathrm{d}u$$

where

$$g_c\left(t\right) = \sqrt{c}g\left(ct\right) \quad \text{and} \quad q_c\left(t\right) = cq\left(ct\right).$$

# 9 Universality: Modelling

Let $V = \{V_t\}_{t\in\mathbb{R}}$ denote the time-wise behaviour of the mean component of the velocity vector at a fixed position in an arbitrary free turbulent field and let $U = \{U_t\}_{t\in\mathbb{R}}$ denote the time-wise behaviour of the increment process of $V$.

## 9.1 Primitive universality model

We propose to consider the following as a theoretical model for the empirically observed approximate affinity of the intrinsic time change $\psi$ that results in the collapse of the densities of turbulent velocity increments.

**Primitive universality model** (PUM) Except for a change of location, scale and affine time change $T_t = ct + c_0$ (with $c > 0$ and $c_0$ is a constant), $V$ is equivalent in law to a process $Y$ of the form

$$Y_t = \int_{-\infty}^t g(t-s)\,\sigma_s \mathrm{d}B_s + \chi \int_{-\infty}^t q(t-s)\,\sigma_s^2 \mathrm{d}s$$

with $g, q, \sigma$ and $\chi$ as specified in connection to formula (20) and with these four quantities being universal, i.e. the same for all processes of the type $V$.

*Remark 5.* In this framework, once (20) is specified, an arbitrary process $V$ is characterized by $\eta = \mathrm{E}\{V_0\}$, $\omega^2 = \mathrm{Var}\{Y_0\}$ and the time change constants $c$ and $c_0$.

## 9.2 Refined universality model

The empirically observed deviations from affinity of the time change $\psi$ at small and large time lags are inconsistent with the stationarity of $V$ in free turbulence. To account for this non-affine behaviour, we propose a refined universality model.

**Refined universality model** (RUM) Except for a change of scale and deterministic time change $T$, $U$ is equivalent in law to a process $X$ of the form

$$X_t = \int_{-\infty}^t j(t,s)\,\sigma_s \mathrm{d}B_s + \chi \int_{-\infty}^t k(t,s)\,\sigma_s^2 \mathrm{d}s \tag{25}$$

with $j, k, \sigma$ and $\chi$ as specified in connection to formulae (20) and (22) and with these four quantities being universal, i.e. the same for all increment processes $U$.

*Remark 6.* In this framework, once (25) is specified, an arbitrary increment process $U$ is characterized by $\omega^2 = \dfrac{1}{2} \lim\limits_{t \to \infty} \text{Var}\{Y_t\}$ and the time change $T$.

*Remark 7.* The empirical findings reported here only concern the collapse of the marginal distributions of velocity increments after applying a deterministic time change. The refined universality model goes beyond this equivalence in distribution as it states an equivalence in law of the processes.

# 10 Concluding remarks

In this paper, we presented a review of some recent modelling of turbulent velocities and financial price processes and associated questions of time change. As a preliminary hypothesis, we proposed the existence of an affine time change in terms of which the velocity process is universal in law, except for change of location and scale.

The subsequent empirical findings about the non-affinity of the deterministic time change at very small and very large time scales led to us to propose the existence of a refined universality model for turbulent velocity increments and related to that an intrinsic deterministic time change, capturing the individual characteristics of each turbulent experiment. It is important to note that the empirical verification of the collapse of the densities of the time changed velocity increments is in fact independent of any model specification. Without model specification, the refined universality model can be stated as the equivalence of the law of $U$, except for change of location and scale and the time change.

A natural extension of the empirical results reported here concerns the dynamical aspect of the refined universality model, i.e. whether the empirically observed equivalence in one-dimensional marginal distribution can indeed be extended to an equivalence of the processes.

A first empirical result to clarify this point shows that the conditional distributions $p(\Delta \tilde{u}_t - \Delta \tilde{u}_s | \Delta \tilde{u}_s)$ and $p(\Delta \tilde{u}^*_{\psi(t)} - \Delta \tilde{u}^*_{\psi(s)} | \Delta \tilde{u}^*_{\psi(s)})$ collapse after an appropriate change of scale, for $c_2(\Delta \tilde{u}_s) = c_2(\Delta \tilde{u}^*_{\psi(s)})$ and for a range of time lags at which the time change $\psi$ is essentially affine. For time lags at which the time change $\psi$ is essentially non-affine, i.e. at very small and very large time scales, the conditional densities do not collapse; however, they are only shifted by the conditional means $\text{E}\{\Delta \tilde{u}_t - \Delta \tilde{u}_s | \Delta \tilde{u}_s\}$ and $\text{E}\{\Delta \tilde{u}^*_{\psi(t)} - \Delta \tilde{u}^*_{\psi(s)} | \Delta \tilde{u}^*_{\psi(s)}\}$. A further clarification of this point is outside the scope of the present paper, but will be discussed in an upcoming publication.

The definition of volatility modulated Volterra processes as given by (5) is readily generalized to the multivariate setting, with $B$ being $d$-dimensional Brownian motion and $\sigma$ being a matrix process. It is further of interest to consider cases where processes expressing possible jumps or noise in the dynamics are added.

A central issue in these settings is how to draw inference on the volatility process $\sigma$. In cases where the processes are semimartingales, the theory of multipower variations (see [11], [12] and references given there) provides effective tools for this.

However, VMVP processes are generally not of semimartingale type and the question of how to proceed then is largely unsolved and poses mathematically challenging problems. Some of these problems are presently under study in joint work with José-Manuel Corcuera, Mark Podolski and Neil Shephard.

# A  The normal inverse Gaussian law

The class of NIG distributions equals the family of possible distributions at time $t = 1$ of the NIG Lévy process, which is defined as Brownian motion with drift subordinated by the inverse Gaussian Lévy process, i.e. the Lévy process of first passage times to constant levels of (another, independent) Brownian motion.

The normal inverse Gaussian law, with parameters $\alpha, \beta, \mu$ and $\delta$, is the distribution on the real axis $\mathbf{R}$ having probability density function

$$p(x; \alpha, \beta, \mu, \delta) = a(\alpha, \beta, \mu, \delta)q\left(\frac{x - \mu}{\delta}\right)^{-1}$$

$$\times K_1\left\{\delta\alpha q\left(\frac{x - \mu}{\delta}\right)\right\} e^{\beta x} \qquad (26)$$

where $q(x) = \sqrt{1 + x^2}$ and

$$a(\alpha, \beta, \mu, \delta) = \pi^{-1}\alpha \exp\left\{\delta\sqrt{\alpha^2 - \beta^2} - \beta\mu\right\} \qquad (27)$$

and where $K_1$ is the modified Bessel function of the third kind and index 1. The domain of variation of the parameters is given by $\mu \in \mathbf{R}$, $\delta \in \mathbf{R}_+$, and $0 \leq |\beta| < \alpha$. The distribution is denoted by $\mathrm{NIG}(\alpha, \beta, \mu, \delta)$.

If $X$ is a random variable with distribution $\mathrm{NIG}(\alpha, \beta, \mu, \delta)$ then the cumulant generating function of $X$, i.e. $\mathrm{K}(\theta; \alpha, \beta, \mu, \delta) = \log \mathrm{E}\{e^{\theta X}\}$, has the form

$$\mathrm{K}(\theta; \alpha, \beta, \mu, \delta) = \delta\{\sqrt{\alpha^2 - \beta^2} - \sqrt{\alpha^2 - (\beta + \theta)^2}\} + \mu\theta. \qquad (28)$$

It follows immediately from this that if $x_1, ..., x_m$ are independent normal inverse Gaussian random variables with common parameters $\alpha$ and $\beta$ but individual location-scale parameters $\mu_i$ and $\delta_i$ $(i = 1, ..., m)$ then $x_+ = x_1 + ... + x_m$ is again distributed according to a normal inverse Gaussian law, with parameters $(\alpha, \beta, \mu_+, \delta_+)$.

Furthermore, the first four cumulants of $\mathrm{NIG}(\alpha, \beta, \mu, \delta)$, obtained by differentiation of (28), are found to be

$$\kappa_1 = \mu + \frac{\delta\rho}{\sqrt{1-\rho^2}}, \quad \kappa_2 = \frac{\delta}{\alpha(1-\rho^2)^{3/2}} \tag{29}$$

and

$$\kappa_3 = \frac{3\delta\rho}{\alpha^2(1-\rho^2)^{5/2}}, \quad \kappa_4 = \frac{3\delta(1+4\rho^2)}{\alpha^3(1-\rho^2)^{7/2}}, \tag{30}$$

where $\rho = \beta/\alpha$. Hence, the standardized third and fourth cumulants are

$$\bar{\kappa}_3 = \frac{\kappa_3}{\kappa_2^{3/2}} = 3\frac{\rho}{\{\delta\alpha(1-\rho^2)^{1/2}\}^{1/2}}$$

$$\bar{\kappa}_4 = \frac{\kappa_4}{\kappa_2^2} = 3\frac{1+4\rho^2}{\delta\alpha(1-\rho^2)^{1/2}}. \tag{31}$$

We note that the NIG distribution (26) has semiheavy tails; specifically,

$$p(x;\alpha,\beta,\mu,\delta) \sim \text{const.} \, |x|^{-3/2} \exp\left(-\alpha\,|x| + \beta x\right), \quad x \to \pm\infty \tag{32}$$

as follows from the asymptotic relation

$$K_\nu(x) \sim \sqrt{2/\pi}x^{-1/2}e^{-x} \quad \text{as} \quad x \to \infty. \tag{33}$$

It is often of interest to consider alternative parameterizations of the normal inverse Gaussian laws. In particular, letting $\bar{\alpha} = \delta\alpha$ and $\bar{\beta} = \delta\beta$, we have that $\bar{\alpha}$ and $\bar{\beta}$ are invariant under location—scale changes. Note that $\rho = \bar{\beta}/\bar{\alpha}$.

**NIG shape triangle**     For some purposes it is useful, instead of the classical skewness and kurtosis quantities (31), to work with the alternative asymmetry and steepness parameters $\chi$ and $\xi$ defined by

$$\chi = \rho\xi \tag{34}$$

and

$$\xi = (1+\bar{\gamma})^{-1/2} \tag{35}$$

where $\rho = \beta/\alpha = \bar{\beta}/\bar{\alpha}$ and $\bar{\gamma} = \delta\gamma = \delta\sqrt{\alpha^2 - \beta^2}$. Like $\bar{\kappa}_3$ and $\bar{\kappa}_4$, these parameters are invariant under location-scale changes and the domain of variation for $(\chi,\xi)$ is the *normal inverse Gaussian shape triangle*

$$\{(\chi,\xi) : -1 < \chi < 1, 0 < \xi < 1\}.$$

The distributions with $\chi = 0$ are symmetric, and the normal and Cauchy laws occur as limiting cases for $(\chi,\xi)$ near to $(0,0)$ and $(0,1)$, respectively. Figure 4 gives an impression of the shape of the NIG distributions for various values of $(\chi,\xi)$.

Note in this connection that $\bar{\kappa}_3$ and $\bar{\kappa}_4$ may be reexpressed as

$$\bar{\kappa}_3 = 3\bar{\gamma}^{-1}\frac{\rho}{\{(1+\rho^2)(1-\rho^2)^{1/2}\}^{1/2}}$$

and

$$\bar{\kappa}_4 = 3\bar{\gamma}^{-1} \frac{1 + 4\rho^2}{(1 - \rho^4)^{1/2}}$$

from which it follows that for small $\rho$ we have approximately $\xi \doteq (1+3/\bar{\kappa}_4)^{-1/2}$ and $\bar{\kappa}_3 \doteq \rho\bar{\kappa}_4$ (compare to (34)); Thus the roles of $\chi$ and $\xi$ are rather similar to those of the classical quantities $\bar{\kappa}_3$ and $\bar{\kappa}_4$.

A systematic study of the class of normal inverse Gaussian distributions, and of associated stochastic processes, was begun in [5, 6, 7, 8, 9]. Further theoretical developments and applications are discussed in [47, 48, 45, 33, 46, 22, 23, 24, 14, 13, 3, 30, 35, 41]. As discussed in the papers cited and in references given there, the class of NIG distributions and processes have been found to provide accurate modelling of a great variety of empirical findings in the physical sciences and in financial econometrics. (The wider class of generalized hyperbolic distributions, introduced in [4], provides additional possibilities for realistic modelling of dynamical processes, see references in the papers cited above.)

# References

1. Andersen TG, Bollerslev T, Diebold FX, Ebens H (2001) The distribution of realized stock return volatility. J. Fin. Econometrics, 61:43–76.
2. Andersen TG, Bollerslev T, Frederiksen PH, Nielsen MØ (2006) Continuous-time models, realized volatilities, and testable distributional implications for daily stock returns. Unpublished paper.
3. Asmussen S, Rosinski J (2001) Approximation of small jumps of Lévy processes with a view towards simulation. J. Appl. Probab., 38:482–493.
4. Barndorff-Nielsen OE (1977) Exponentially decreasing distributions for the logarithm of particle size. Proc. R. Soc. London A 353:401–419.
5. Barndorff-Nielsen OE (1995) Normal inverse Gaussian processes and the modelling of stock returns. Research Report 300, Dept. Theor. Statistics, Aarhus University.
6. Barndorff-Nielsen OE (1997) Normal inverse Gaussian distributions and stochastic volatility modelling. Scand. J. Statist., 24:1–14.
7. Barndorff-Nielsen OE (1998) Probability and statistics: self-decomposability, finance and turbulence. In: Acccardi L, Heyde CC (Eds) Probability Towards 2000, 47–57. Proceedings of a Symposium held 2-5 October 1995 at Columbia University. Springer-Verlag, New York
8. Barndorff-Nielsen OE (1998) Processe of normal inverse Gaussian type. Finance and Stochastics, 2:41–68.
9. Barndorff-Nielsen OE (1998) Superposition of Ornstein-Uhlenbeck type processes. Theory Prob. Its Appl., 45:175–194.
10. Barndorff-Nielsen OE, Blæsild P, Schmiegel J (2004) A parsimonious and universal description of turbulent velocity increments. Eur. Phys. J., B 41:345–363.

11. Barndorff-Nielsen OE, Graversen SE, Jacod J, Podolskij M, Shephard N (2006) A central limit theorem for realised power and bipower variations of continuous semimartingales. In: Kabanov Yu, Liptser R, Stoyanov J (Eds) From Stochastic Calculus to Mathematical Finance 33–68. Festschrift in Honour of A.N. Shiryaev. Springer, Heidelberg

12. Barndorff-Nielsen OE, Graversen SE, Jacod J, Shephard N (2006) Limit theorems for bipower variation in financial econometrics. Econometric Theory, 22:677–719.

13. Barndorff-Nielsen OE, Levendorskiĭ SZ (2001) Feller processes of normal inverse Gaussian type. Quantitative Finance, 1:318–331.

14. Barndorff-Nielsen OE, Prause K (2001) Apparent scaling. Finance and Stochastics, 5:103–113.

15. Barndorff-Nielsen OE, Schmiegel J (2004) Lévy-based tempo-spatial modelling; with applications to turbulence. Uspekhi Mat. Nauk., 59:65–91.

16. Barndorff-Nielsen OE, Schmiegel J (2006) A stochastic differential equation framework for the timewise dynamics of turbulent velocities. Theory of Probability and its Applications. (To appear)

17. Barndorff-Nielsen OE, Schmiegel J (2006) Time change and universality in turbulence. (Submitted)

18. Barndorff-Nielsen OE, Schmiegel J. (2007) Ambit processes; with applications to turbulence and cancer growth. Proceedings of the 2005 Abel Symposium on Stochastic Analysis and Applications. Springer, Heidelberg (To appear)

19. Barndorff-Nielsen OE, Schmiegel J (2007) Change of time and universal laws in Turbulence. (Submitted)

20. Barndorff-Nielsen OE, Schmiegel J, Shephard N (2006) Time change and universality in turbulence and finance. (Submitted)

21. Barndorff-Nielsen OE, Schmiegel J, Shephard N (2007) QV, BV and VR under stationary Gaussian processes. (In preparation)

22. Barndorff-Nielsen OE, Shephard N (2001) Modelling by Lévy processes for financial econometrics. In: Barndorff-Nielsen OE, Mikosch T, Resnick S (Eds) Lévy Processes - Theory and Applications, 283–318. Birkhäuser, Boston

23. Barndorff-Nielsen OE, Shephard N (2001) Non-Gaussian Ornstein-Uhlenbeck-based models and some of their uses in financial economics (with Discussion). J. R. Statist. Soc., B 63:167–241.

24. Barndorff-Nielsen OE, Shephard N (2002) Integrated OU processes and non-Gaussian OU-based stochastic volatility. Scand. J. Statist. , 30:277–295.

25. Barndorff-Nielsen OE, Shephard N (2004) Power and bipower variation with stochastic volatility and jumps (with Discussion). J. Fin. Econometrics, 2:1–48.

26. Barndorff-Nielsen OE, Shephard N (2008) Financial Volatility in Continous Time. Cambridge University Press. (To appear)

27. Barndorff-Nielsen OE, Shiryaev AN (2008) Change of Time and Change of Measure. World Scientific, Singapore (To appear)

28. Bender C, Marquardt T (2007) Stochastic calculus for convoluted Lévy processes. (Unpublished manuscript)

29. Castaing B, Gagne Y, Hopfinger EJ (1990) Velocity probability density functions of high Reynolds number turbulence. Physica, D 46:177–200.

30. Cont R, Tankov P (2004) Financial Modelling With Jump Processes. Chapman & Hall/CRC, London

31. Decreusefond L (2005) Stochastic integration with respect to Volterra processes. Ann. I. H. Poincaré, PR41:123–149.

32. Decreusefond L, Savy N (2006) Anticipative calculus with respect to filtered Poisson processes. Ann. I. H. Poincaré, PR42:343–372.
33. Eberlein E (2000) Application of generalized hyperbolic Lévy motion to finance. In: Barndorff-Nielsen OE, Mikosch T, Resnick S (Eds) Lévy Processes - Theory and Applications, 319–336. Birkhäuser, Boston
34. Elsner JW, Elsner W (1996) On the measurement of turbulence energy dissipation. Meas. Sci. Technol., 7:1334–1348.
35. Forsberg L (2002) On the Normal Inverse Gaussian distribution in Modelling Volatility in the Financial Markets. Acta Universitatis Upsaliemsis, Studia Statistica Upsaliensia 5, Uppsala.
36. Ghashgaie S, Breymann W, Peinke J, Talkner P, Dodge Y (1996) Turbulent cascades in foreign exchange markets. Nature 381:767–770.
37. Hult H (2003) Approximating some Volterra type stochastic intergrals with applications to parameter estimation. Stoch. Proc. Appl., 105:1–32.
38. Kolmogorov AN (1941) Dissipation of energy in locally isotropic turbulence. Dokl. Akad. Nauk. SSSR, 32:16–18.
39. Kolmogorov AN (1962) A refinement of previous hypotheses concerning the local structure of turbulence in a viscous incompressible fluid at high Reynolds number. J. Fluid Mech, 13:82–85.
40. Marquardt T (2006) Fractional Lévy processes with an application to long memory moving average processes. Bernoulli, 12:1099–1126.
41. McNeil AJ, Frey R, Embrechts P (2005) Quantitative Risk Management. Princton University Press, Princeton
42. Norros I, Valkeila E, Virtamo J (1999) An elementary approach to a Girsanov formula and other analytic results on fractional Brownian motion. Bernoulli, 5:571–587.
43. Obukhov AM (1962) Some specific features of atmospheric turbulence. J. Fluid Mech., 13:77–81.
44. Peinke J, Bottcher F, Barth S (2004) Anomalous statistics in turbulence, financial markets and other complex systems. Ann. Phys. (Leipzig), 13:450–460.
45. Prause K (1999) The Generalized Hyperbolic Model: Estimation, Financial Derivatives and Risk Measures. Dissertation, Albert-Ludwigs-Universität, Freiburg i. Br.
46. Raible S (2000) Lévy Processes in Finance: Theory, Numerics, Empirical Facts. Dissertation, Albert-Ludwigs-Universität, Freiburg i. Br.
47. Rydberg TH (1997) The normal inverse Gaussian Lévy process: simulation and approximation. Comm. Statist.: Stochastic Models, 13:887–910.
48. Rydberg TH (1999) Generalized hyperbolic diffusions with applications towards finance. Math. Finance, 9:183–201.
49. Samorodnitsky G, Taqqu MS (1994) Stable Non-Gaussian Random Processes. Chapman and Hall, New York
50. Shiryaev AN (2006) Kolmogorov and the turbulence. Research Report 2006-4. Thiele Centre for Applied Mathematics in Natural Science.
51. Shiryaev AN (2007) On the classical, statistical and stochastic approaches to hydrodynamic turbulence. Research Report 2007-2. Thiele Centre for Applied Mathematics in Natural Science.
52. Vincent A, Meneguzzi M (1991) The spatial structure and statistical properties of homogeneous turbulence. J. Fluid Mech., 225:1–25.

|  | Finance | Turbulence |
|---|---|---|
| varying activity | volatility | intermittency |
| semiheavy tails | + | + |
| asymmetry | + | + |
| aggregational Gaussianity | + | + |
| 0 autocorrelation | + | − |
| quasi long range dependence | + | + |
| scaling/selfsimilarity | [+] | [+] |
| leverage | + | − |
| operational time | + | + |
| trend | cumulative | stationary |
| jumps | + | − |

**Table 1.** Stylized features of turbulence and finance.

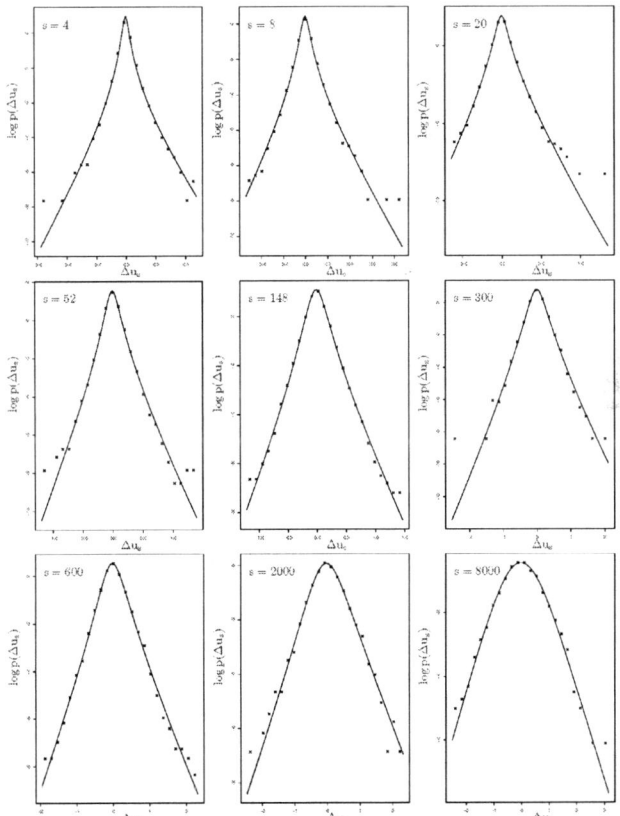

**Fig. 1.** Approximation of the pdf of velocity increments within the class of NIG distributions (solid lines, fitting by maximum likelihood) for data from the atmospheric boundary layer (kindly provided by K.R. Sreenivasan) with $R_\lambda = 17000$ and time scales $s = 4, 8, 20, 52, 148, 300, 600, 2000, 8000$ (in units of the finest resolution).

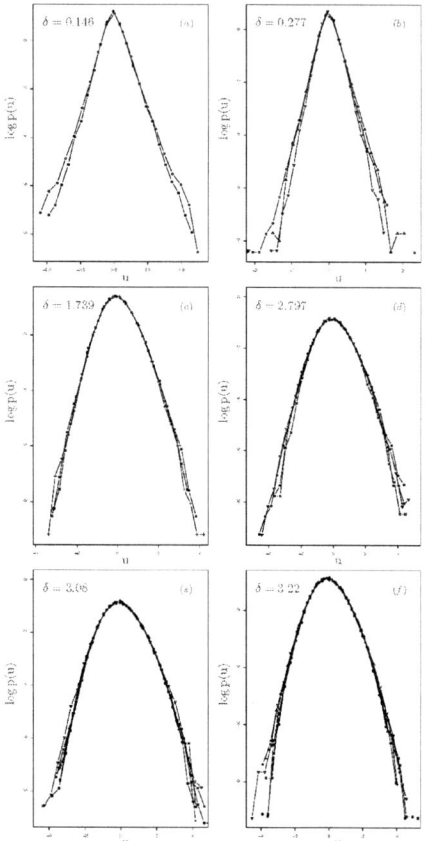

**Fig. 2.** Collapse of the densities of velocity increments at time scale $s$ for various fixed values of the scale parameter $\delta(s)$ of the approximating NIG-distributions. The data are from the atmospheric boundary layer (data set (at) with $R_\lambda = 17000$, kindly provided by K.R. Sreenivasan), from a free jet experiment (data set (j) with $R_\lambda = 190$, kindly provided by J. Peinke), from a wind tunnel experiment (data set (w) with $R_\lambda = 80$, kindly provided by B.R. Pearson) and from a gaseous helium jet flow (data sets (h85), (h124), (h208), (h283), (h352), (h703), (h885), (h929), (h985) and (h1181) with $R_\lambda = 85, 124, 208, 283, 352, 703, 885, 929, 985, 1181$, respectively, kindly provided by B. Chabaud). The corresponding values of the time scales $s$ (in units of the finest resolution of the corresponding data set) and the codes for the data sets are $(a)$ $(s = 116, (\text{at}))$ ($\circ$), $(s = 4, (\text{h}352))$ ($\boxplus$), $(b)$ $(s = 440, (\text{at}))$ ($\circ$), $(s = 8, (\text{j}))$ ($\triangle$), $(s = 8, (\text{h}929))$ ($\triangledown$), $(c)$ $(s = 192, (\text{h}885))$ ($\blacksquare$), $(s = 88, (\text{h}352))$ ($\boxplus$), $(s = 10, (\text{w}))$ ($+$), $(d)$ $(s = 380, (\text{h}885))$ ($\blacksquare$), $(s = 410, (\text{h}929))$ ($\triangledown$), $(s = 350, (\text{h}703))$ ($\times$), $(s = 340, (\text{h}985))$ ($\bullet$), $(e)$ $(s = 420, (\text{h}703))$ ($\times$), $(s = 440, (\text{h}929))$ ($\triangledown$), $(s = 180, (\text{h}352))$ ($\boxplus$), $(s = 270, (\text{h}283))$ ($\bullet$), $(s = 108, (\text{h}124))$ ($*$), $(s = 56, (\text{h}85))$ ($\boxtimes$), $(f)$ $(s = 470, (\text{h}929))$ ($\triangledown$), $(s = 116, (\text{h}124))$ ($*$), $(s = 60, (\text{h}85))$ ($\boxtimes$), $(s = 188, (\text{h}352))$ ($\boxplus$), $(s = 470, (\text{h}1181))$ ($\blacktriangle$), $(s = 140, (\text{h}208))$ ($\blacklozenge$).

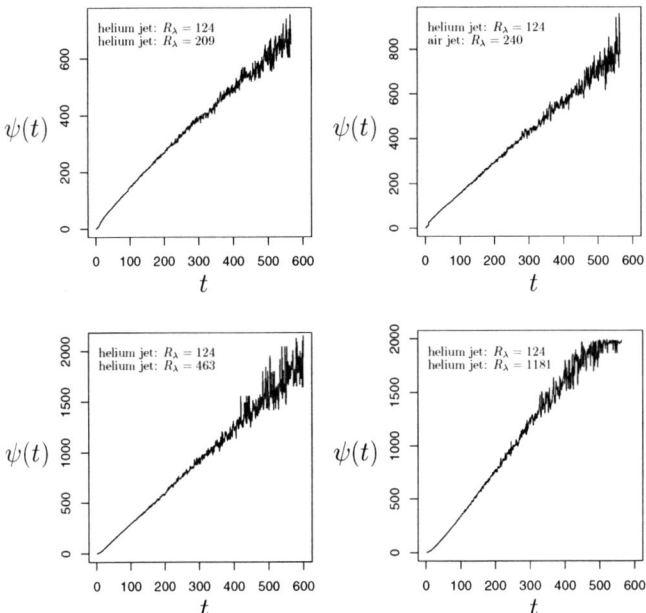

**Fig. 3.** Estimated time change $\psi$ (in units of the finest resolution of the respective data sets) resulting in a collapse of the densities of velocity increments (see Figure 2).

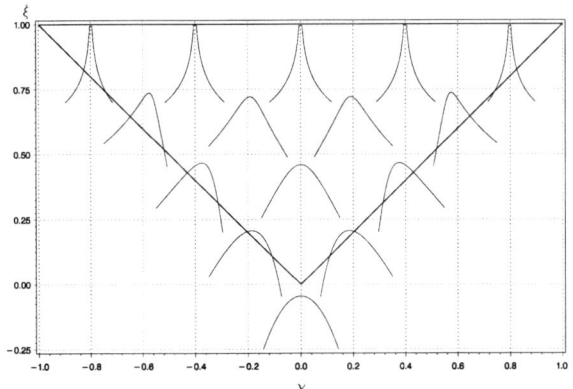

**Fig. 4.** The shape triangle of the NIG distributions with the log density functions of the standardized distributions, i.e. with mean 0 and variance 1, corresponding to the values $(\chi, \xi) = (\pm 0.8, 0.999)$, $(\pm 0.4, 0.999)$, $(0.0, 0.999)$, $(\pm 0.6, 0.75)$, $(\pm 0.2, 0.75)$, $(\pm 0.4, 0.5)$, $(0.0, 0.5)$, $(\pm 0.2, 0.25)$ and $(0.0, 0.0)$. The coordinate system of the log densities is placed at the corresponding value of $(\chi, \xi)$.

# External Dynamical Equivalence of Analytic Control Systems

Zbigniew Bartosiewicz*, Ewa Pawłuszewicz[†]

Białystok Technical University, Faculty of Computer Sciences,
Wiejska 45A, Białystok, Poland. bartos@pb.edu.pl, epaw@pb.edu.pl

**Summary.** Theory of systems on homogeneous time scales unifies theories of continuous-time and discrete-time systems. The characterizations of external dynamical equivalence known for continuous-time and discrete-time systems are extended here to systems on time scales. Under assumption of uniform observability, it is shown that two analytic control systems with output are externally dynamically equivalent if and only if their delta universes are isomorphic. The delta operator associated to the system on a time scale is a generalization of the differential operator associated to a continuous-time system and of the difference operator associated to a discrete-time system.

## 1 Introduction

In 1988, in his Ph.D. thesis [13], Stefan Hilger developed calculus on time scales, which unified the standard differential calculus and the calculus of finite differences. This allowed for a unified treatment of dynamical systems with continuous and discrete time. The book by M. Bohner and A. Peterson [7] contains the most important achievements in this area. But the theory is much richer than just the unification. One can study systems for which time is partly continuous and partly discrete. Many systems appearing in engineering, biology and economy exhibit such features.

In the classical control theory there always have been two parallel areas of research: the continuous-time and the discrete-time systems. Most of the results are similar for both classes of systems, but there are also significant differences. For example, asymptotic stability of linear time-invariant systems is characterized by the condition that the eigenvalues of the system lie in the specific region of the complex plane. However this region depends on the class of systems. Another difference concerns solutions of the differential equations (continuous time) and the difference equations (discrete time). Under some

---

* Supported by Białystok Technical University grant No W/WI/1/07
[†] Supported by Białystok Technical University grant No W/WI/18/07

reasonable conditions ordinary differential equations can be solved forward and backward, while difference equations usually can be solved only forward and some extra assumptions are required to solve them backward. Fortunately, in control theory we are mostly interested in forward solutions, so this difference is less important.

Calculus on time scales entered control theory just a few years ago. First results concerned basic properties of linear systems, like controllability, observability and realizations (see [4, 9, 6]). In [5] we studied dynamical feedback equivalence of nonlinear systems on time scales. The main result of that paper will be used here to show a characterization of external dynamical equivalence for systems on time scales. Another attempt to unify continuous-time and discrete-time systems, without use of calculus on time scales, was made in [11, 18].

Dynamic equivalence for nonlinear continuous-time systems was first studied by B. Jakubczyk [14, 15]. He used a dynamical state feedback to transfer trajectories of one system onto trajectories of the other. His concept of dynamical feedback linearizability was close to the property of flatness introduced earlier by M. Fliess (see e.g. [10, 17] and [22]). The main result of [14] says that two systems are dynamically (state) feedback equivalent if and only if their differential algebras are isomorphic. In [2] dynamical state feedback equivalence of discrete-time systems was studied. In the characterization of this property, difference algebras were used instead of differential algebras. This was one of the examples where the results for continuous time and discrete time are close to each other and a clever change of language is enough to switch between two classes of systems.

In [20, 3] external dynamical equivalence and linearization for discrete-time systems were studied. The systems were equipped with output parts and dynamical output feedback was used instead of dynamical state feedback. Necessary and sufficient criteria of external dynamical equivalence were expressed with the aid of the output difference universe. The concept of (function) universe, introduced by J. Johnson, is a generalization of the notion of (function) algebra [1, 16]. Besides standard algebraic operations possible in algebras, the structure of universe allows for substituting elements of the universe (which are partially defined functions) into real-analytic functions of several variables and amalgamation of partially defined functions from the universe. The results of [20, 3] were then transferred (back) to continuous-time systems, with the output differential universe as the key tool [21].

In [5] we studied dynamical state feedback equivalence for nonlinear systems on homogeneous time scales. The results unified those obtained for continuous-time and discrete-time systems. Instead of differential and difference operators used in earlier works, we introduced so called delta operator, which for a particular time scale would become either a differential operator or a difference one.

In this paper we complete the picture studying external dynamical equivalence for control systems with outputs, defined on homogeneous time scales.

The main result says that two systems are externally dynamically equivalent if and only if their delta universes are isomorphic. This theorem may be seen as an extension of the result of [5] to systems with output or as a unification of our earlier results from [20] and [21]. We assume that the systems are uniformly observable and use the main theorem of [5] to prove the present version. In [3] and [21] we showed that for continuous-time and discrete-time cases one can drop the observability assumptions. Only some regularity of the space obtained by gluing up indistinguishable states of the system is assumed. This suggests that the same could be done for systems on time scales.

The paper is organized as follows. In Section 2 we provide the reader with the necessary background on the calculus on time scales. Section 3 contains setting of the problem and the precise definition of external dynamical equivalence. In Section 4 we recall basic concepts of the theory of universes. The main result of the paper is stated and proved in Section 5.

## 2 Calculus on time scales

We recall here basic concepts and facts of the calculus on time scales. For more information the reader is referred to [7].

A *time scale* $\mathbb{T}$ is an arbitrary nonempty closed subset of the set of real numbers $\mathbb{R}$. The standard examples of time scales are $\mathbb{R}$, $h\mathbb{Z}$, $h > 0$, $\mathbb{N}$, $\mathbb{N}_0$, $2^{\mathbb{N}_0}$ or $\mathbb{P}_{a,b} = \bigcup_{k=0}^{\infty} [k(a+b), k(a+b)+a]$. The time scales $\mathbb{T}$ is a topological space with the relative topology induced from $\mathbb{R}$.

The following operators on $\mathbb{T}$ are often used:

- the *forward jump operator* $\sigma : \mathbb{T} \to \mathbb{T}$, defined by $\sigma(t) := \inf\{s \in \mathbb{T} : s > t\}$ and $\sigma(\sup \mathbb{T}) = \sup \mathbb{T}$, if $\sup \mathbb{T} \in \mathbb{T}$,
- the *backward jump operator* $\rho : \mathbb{T} \to \mathbb{T}$, defined by $\rho(t) := \sup\{s \in \mathbb{T} : s < t\}$ and $\rho(\inf \mathbb{T}) = \inf \mathbb{T}$, if $\inf \mathbb{T} \in \mathbb{T}$,
- the *graininess functions* $\mu, \nu : \mathbb{T} \to [0, \infty)$ defined by $\mu(t) := \sigma(t) - t$ and respectively by $\nu(t) := t - \rho(t)$.

Points from the time scale can be classified as follows: a point $t \in \mathbb{T}$ is called

- *right-scattered* if $\sigma(t) > t$ and *right-dense* if $\sigma(t) = t$,
- *left-scattered* if $\rho(t) < t$ and *left-dense* if $\rho(t) = t$,
- *isolated* if it is both left-scattered and right-scattered,
- *dense* if it is both left-dense and right-dense.

We define also the sets

$$\mathbb{T}^{\kappa} := \begin{cases} \mathbb{T} \setminus \{M\}, & \text{if } M \text{ is the left scattered maximum of } \mathbb{T} \\ \mathbb{T}, & \text{if } \sup \mathbb{T} = \infty. \end{cases}$$

$$\mathbb{T}_\kappa := \begin{cases} \mathbb{T} \setminus \{m\}, & \text{if } m \text{ is the right scattered minimum of } \mathbb{T} \\ \mathbb{T}, & \text{otherwise.} \end{cases}$$

*Example 1.* If $\mathbb{T} = \mathbb{R}$ then $\rho(t) = t = \sigma(t)$ and $\mu(t) = \nu(t) = 0$, for all $t \in \mathbb{R}$.

*Example 2.* If $\mathbb{T} = h\mathbb{Z}$, $h > 0$, then $\rho(t) = t - h$, $\sigma(t) = t + h$, and $\mu(t) = \nu(t) = h$, for all $t \in h\mathbb{Z}$.

*Example 3.* If $\mathbb{T} = \overline{q^{\mathbb{Z}}} := \{q^k : k \in \mathbb{Z}\} \cup \{0\}$, where $q > 1$, then $\rho(t) = \frac{t}{q}$, $\sigma(t) = qt$, $\nu(t) = \frac{t(q-1)}{q}$ and $\mu(t) = (q-1)t$, for all $t \in \mathbb{T}$.

**Definition 1.** *A time scale $\mathbb{T}$ is called* homogeneous *if $\mu$ and $\nu$ are constant on respectively $\mathbb{T}^\kappa$ and $\mathbb{T}_\kappa$.*

The time scales $\mathbb{R}$, $h\mathbb{Z}$, $[0, 1]$ are homogeneous, whereas $\overline{q^{\mathbb{Z}}}$ is not. In this paper we will be interested in homogeneous time scales. Thus we rather want to unify the continuous-time and discrete-time cases, and not to develop a general theory.

**Definition 2.** *Let $f : \mathbb{T} \to \mathbb{R}$ and $t \in \mathbb{T}^k$. Then the number $f^\Delta(t)$ (when it exists), with the property that, for any $\varepsilon > 0$, there exists a neighborhood $U$ of $t$ such that*

$$|[f(\sigma(t)) - f(s)] - f^\Delta(t)[\sigma(t) - s]| \leq \varepsilon|\sigma(t) - s|, \ \forall s \in U,$$

*is called the* delta derivative *of $f$ at $t$.*
*The function $f^\Delta : \mathbb{T}^k \to \mathbb{R}$ is called the* delta derivative *of $f$ on $\mathbb{T}^\kappa$.*
*We say that $f$ is* delta differentiable *on $\mathbb{T}^\kappa$, if $f^\Delta(t)$ exists for all $t \in \mathbb{T}^\kappa$.*

**Definition 3.** *Let $f : \mathbb{T} \to \mathbb{R}$ and $t \in \mathbb{T}_k$. Then the number $f^\nabla(t)$ (when it exists), with the property that, for any $\varepsilon > 0$, there exists a neighborhood $U$ of $t$ such that*

$$|[f(\rho(t)) - f(s)] - f^\nabla(t)[\rho(t) - s]| \leq \varepsilon|\rho(t) - s|, \ \forall s \in U,$$

*is called the* nabla derivative *of $f$ at $t$.*
*The function $f^\nabla : \mathbb{T}_k \to \mathbb{R}$ is called the* nabla derivative *of $f$ on $\mathbb{T}_\kappa$.*
*We say that $f$ is* nabla differentiable *on $\mathbb{T}_\kappa$, if $f^\nabla(t)$ exists for all $t \in \mathbb{T}_\kappa$.*

*Remark 1.* If $\mathbb{T} = \mathbb{R}$, then $f : \mathbb{R} \to \mathbb{R}$ is both delta differentiable and nabla differentiable at $t \in \mathbb{R}$ iff

$$f^\Delta(t) = f^\nabla(t) = \lim_{s \to t} \frac{f(t) - f(s)}{t - s} = f'(t),$$

i.e. $f$ is differentiable in the ordinary sense at $t$.
If $\mathbb{T} = \mathbb{Z}$, then $f : \mathbb{Z} \to \mathbb{R}$ is always delta differentiable and nabla differentiable on $\mathbb{Z}$ and

$$f^{\Delta}(t) = \frac{f(\sigma(t)) - f(t)}{\mu(t)} = f(t+1) - f(t),$$

$$f^{\nabla}(t) = \frac{f(t) - f(\rho(t))}{\nu(t)} = f(t) - f(t-1),$$

for all $t \in \mathbb{Z}$.

**Proposition 1.** *[12]*

 i) *Assume that $f : \mathbb{T} \to \mathbb{R}$ is delta differentiable on $\mathbb{T}^k$. Then $f$ is nabla differentiable at $t$ and*

$$f^{\nabla}(t) = f^{\Delta}(\rho(t)) \tag{1}$$

 *for $t \in \mathbb{T}_k$ such that $\sigma(\rho(t)) = t$. If, in addition, $f^{\Delta}$ is continuous on $\mathbb{T}^k$, then $f$ is nabla differentiable at $t$ and (1) holds for any $t \in \mathbb{T}_k$.*

 ii) *Assume that $f : \mathbb{T} \to \mathbb{R}$ is nabla differentiable on $\mathbb{T}_k$. Then $f$ is delta differentiable at $t$ and*

$$f^{\Delta}(t) = f^{\nabla}(\sigma(t)) \tag{2}$$

 *for $t \in \mathbb{T}^k$ such that $\rho(\sigma(t)) = t$. If, in addition, $f^{\nabla}$ is continuous on $\mathbb{T}_k$, then $f$ is delta differentiable at $t$ and (2) holds for any $t \in \mathbb{T}^k$.*

*Remark 2.* If $t \in \mathbb{T}^{\kappa}$ satisfies $\rho(t) = t < \sigma(t)$, then the forward jump operator $\sigma$ is not delta differentiable at $t$.

*Remark 3.* A function $f : \mathbb{T} \to \mathbb{R}$ is called *regulated* if its right-side limits exist (finite) at all right-dense points in $\mathbb{T}$ and its left-side limits exist (finite) at all left-dense points in $\mathbb{T}$.

**Definition 4.** *A function $f : \mathbb{T} \to \mathbb{R}$ is called* rd-continuous *if it is continuous at the right-dense points in $\mathbb{T}$ and its left-sided limits exist at all left-dense points in $\mathbb{T}$.*

 The set of all rd-continuous functions is denoted by $\mathcal{C}_{rd}$. It may be shown [7] (Theorem 1.60) that

- $f$ is continuous $\Rightarrow$ $f$ is rd-continuous $\Rightarrow$ $f$ is regulated
- $\sigma$ is rd-continuous.

## 3 External dynamical equivalence

Let $\mathbb{T}$ be a homogeneous time scale. Consider a system $\Sigma$ of the following form:

$$x^{\Delta}(t) = f(x(t), u(t))$$
$$y(t) = h(x(t)) \tag{3}$$

where $t \in \mathbb{T}$, $x(t) \in \mathbb{R}^n$, $y(t) \in \mathbb{R}^p$ and $u(t) \in \mathbb{R}^m$. We shall assume that $f$ and $h$ are analytic, and that control $u$ may be infinitely many times differentiated as a function on the time scale $\mathbb{T}$.

By a *trajectory* of the system $\Sigma$ we will mean any triple of functions $(y(\cdot), x(\cdot), u(\cdot))$ defined on some interval $[a, b)$ that satisfies the equations of $\Sigma$. We assume that $a \in \mathbb{T}$, $b \in \mathbb{T}$ or $b = \infty$ and $[a, b)$ contains infinitely many points. The pair $(x(\cdot), u(\cdot))$ is an *inner trajectory* and the pair $(y(\cdot), u(\cdot))$ is an *external trajectory* of $\Sigma$. The set of all inner (external) trajectories of the system $\Sigma$ forms the *inner (external) behavior* of the system. It will be denoted by $B^i(\Sigma)$ (respectively by $B^e(\Sigma)$).

By $J(r)$ we will denote the space of all infinite sequences $S = (s^{(i)})_{i \geq 0}$, $s^{(i)} \in \mathbb{R}^r$. If $z : \mathbb{T} \to \mathbb{R}^r$ has infinitely many delta-derivatives, the map $\bar{Z} : \mathbb{T} \to J(r)$ defined by $Z(t) = (z(t), z^\Delta(t), \dots)$ is called the (infinite) jet of $z$. We shall consider real maps defined on $J(r)$. We assume that each such map $\varphi$ depends only on a finite number of elements of the sequence $S$ (but the number of these elements depends on the given map $\varphi$). In this case we say that the function $\varphi$ is *finitely presented*. A *map* $\phi : J(r) \to \mathbb{R}^{\tilde{r}}$ is *finitely presented* if all its components have this property.

Let us consider two systems on the time scale $\mathbb{T}$:

$$\Sigma : \begin{array}{l} x^\Delta(t) = f(x(t), u(t)) \\ y(t) = h(x(t)) \end{array} \quad \text{and} \quad \tilde{\Sigma} : \begin{array}{l} \tilde{x}^\Delta(t) = \tilde{f}(\tilde{x}(t), \tilde{u}(t) \\ \tilde{y}(t) = \tilde{h}(\tilde{x}(t)) \end{array}$$

where $x(t) \in \mathbb{R}^n$, $\tilde{x}(t) \in \mathbb{R}^{\tilde{n}}$, $y(t) \in \mathbb{R}^p$, $\tilde{y}(t) \in \mathbb{R}^{\tilde{p}}$, $u(t), \tilde{u}(t) \in \mathbb{R}^m$, $t \in \mathbb{T}$.

Consider the following dynamical transformations:

$$y = \phi^e(\tilde{Y}, \tilde{U}), \quad u = \psi^e(\tilde{Y}, \tilde{U}) \tag{4}$$

$$\tilde{y} = \tilde{\phi}^e(Y, U), \quad \tilde{u} = \tilde{\psi}^e(Y, U) \tag{5}$$

where $Y \in J(p)$, $\tilde{Y} \in J(\tilde{p})$, $U, \tilde{U} \in J(m)$, and $\phi^e, \tilde{\phi}^e, \psi^e, \tilde{\psi}^e$ are finitely presented maps of the class $C^\omega$.

We say that $\Sigma$ and $\tilde{\Sigma}$ are *externally dynamically equivalent*, if there exist transformations (4) and (5) mutually inverse on systems' behaviors that induce the following relations between the external trajectories of both systems:

$$y(t) = \phi^e(\tilde{y}(t), \tilde{y}^\Delta(t), \dots, \tilde{y}^{\Delta^k}(t), \tilde{u}(t), \tilde{u}^\Delta(t), \dots, \tilde{u}^{\Delta^k}(t))$$
$$u(t) = \psi^e(\tilde{y}(t), \tilde{y}^\Delta(t), \dots, \tilde{y}^{\Delta^k}(t), \tilde{u}(t), \tilde{u}^\Delta(t), \dots, \tilde{u}^{\Delta^k}(t)) \tag{6}$$

for some integer $k \geq 0$, and similarly for $(\tilde{y}(t), \tilde{u}(t))$.

*Example 4.* Let $\Sigma$ be given by the equations:

$$x_1^\Delta = \int_0^1 \cosh(\sinh^{-1}(x_1) + \mu h x_2) dh \cdot x_2$$
$$x_2^\Delta = u$$
$$y = x_1.$$

Then $\Sigma$ is externally dynamically equivalent to the linear system $\tilde{\Sigma}$

$$\tilde{x}^{\Delta} = \tilde{u}, \quad \tilde{y} = \tilde{x}.$$

The equivalence transformations are given by

$$\tilde{y} = \sinh^{-1}(y)$$

$$\tilde{u} = \frac{y^{\Delta}}{\cosh(\sinh^{-1}(y))}$$

and

$$y = \sinh(\tilde{y})$$

$$u = \tilde{u}^{\Delta}.$$

For discrete-time systems the dynamical transformations (4) and (5) depend on forward differences, which in fact means dependence on future values of the output and the control. This is inconvenient from the practical point of view. However the same effect may by achieved by transformations that depend on past values of the output and the control (see [20]). This may be extended to homogeneous time scales with the following definition.

Assume that the trajectories of $\Sigma$ and $\tilde{\Sigma}$ are related by

$$y(t) = \gamma^e(\tilde{y}(t), \tilde{y}^{\nabla}(t), \ldots, \tilde{y}^{\nabla^l}(t), \tilde{u}(t), \tilde{u}^{\nabla}(t), \ldots, \tilde{u}^{\nabla^l}(t))$$
$$u(t) = \eta^e(\tilde{y}(t), \tilde{y}^{\nabla}(t), \ldots, \tilde{y}^{\nabla^l}(t), \tilde{u}(t), \tilde{u}^{\nabla}(t), \ldots, \tilde{u}^{\nabla^l}(t)) \qquad (7)$$

for some integer $l \geq 0$ and

$$\tilde{y}(t) = \tilde{\gamma}^e(y(t), y^{\nabla}(t), \ldots, y^{\nabla^{\tilde{l}}}(t), u(t), u^{\nabla}(t), \ldots, u^{\nabla^{\tilde{l}}}(t))$$
$$\tilde{u}(t) = \tilde{\eta}^e(y(t), y^{\nabla}(t), \ldots, y^{\nabla^{\tilde{l}}}(t), u(t), u^{\nabla}(t), \ldots, u^{\nabla^{\tilde{l}}}(t)) \qquad (8)$$

for some $\tilde{l}$, i.e. substituting an external trajectory of one system on the right-hand side, we obtain an external trajectory of the other system on the left-hand side. We say that $\Sigma$ and $\tilde{\Sigma}$ are *externally dynamically delay equivalent* if applying transformations (7) to $(\tilde{y}, \tilde{u})$ and then transformations (8) to the resulting pair $(y, u)$, we finally obtain $(\tilde{y}, \tilde{u}) \circ \rho^{l+\tilde{l}}$ and the same holds the pair $(y, u)$ with the transformations applied in the reverse order. Thus the transformations (7) and (8) are mutually inverse on the external trajectories modulo the backward time shift $\rho^{l+\tilde{l}}$.

It can be shown that $\Sigma$ and $\tilde{\Sigma}$ are externally dynamically equivalent if and only if they are externally dynamically delay equivalent. This follows from the fact that on a homogeneous time scale we have $z^{\Delta^r}(t) = z^{\nabla^r}(\sigma^r(t))$ and $z^{\nabla^r}(t) = z^{\Delta^r}(\rho^r(t))$. Thus one can replace delta derivatives with nabla derivatives and vice versa, but some forward or backward shifts are involved in this operation. Moreover, the backward shift may be expressed with the aid of the nabla derivative. As the external dynamical equivalence is simpler from the mathematical point of view than the external delay dynamical equivalence, we shall concentrate on the former property in the rest of the paper.

External dynamical equivalence is an equivalence relation in the set of all control systems with output. It is a natural generalization of similar concepts for continuous-time and discrete-time systems.

We say that the system $\Sigma$ is *externally dynamically linearizable* if it is externally dynamically equivalent to a linear minimal one (i.e. controllable and observable).

To state a characterization of the external dynamical equivalence, we need the concept of function universe.

## 4 Function universes

Let $X, Y$ be sets. A *partially defined function* on $X$ with values in $Y$ is any map $\varphi : A \to Y$, where $A \subseteq X$ is called domain of $\varphi$ and denoted by $\mathrm{dom}\varphi$. If $\mathrm{dom}\varphi = X$ then $\varphi$ is *global*. Let $Y_X$ be the set of all partially defined functions on $X$. One can extend any $\varphi \in Y_X$ to one defined on $X$ by assigning $\varphi(x) = \emptyset_0$ for $x \notin \mathrm{dom}\varphi$. We call $\emptyset_0$ the *phantom*. Now $\mathrm{dom}\varphi = \{x \in X : \varphi(x) \neq \emptyset_0\}$. If $a \in Y$, $x \in X$ then we set $a_X(x) := a$.

Let $A_n$ denote the set of functions of class $C^\omega$, partially defined on open subsets in $\mathbb{R}^n$ with values in $\mathbb{R}$. In particular, $A_0$ can be identified with $\mathbb{R} \cup \emptyset_0$. The topology in $A_0$ can be defined as follows: a subset $B \subset A_0$ is open if $B = A_0$ or $B$ is an open subset in $\mathbb{R}$.

Functions $\varphi, \psi \in Y_X$ are *matching*, if they take on the same values on $\mathrm{dom}\varphi \cap \mathrm{dom}\psi$. Let us consider a set $M \subseteq Y_X$ of functions that are matching and define a function $\underline{M} \in Y_X$: $\underline{M}(x) = \emptyset_0$ if no function in $M$ is defined at $x$ and $\underline{M}(x) = \varphi(x)$ for any function $\varphi \in M$ defined at $x$. The process of constructing $\underline{M}$ is called *amalgamation* of the functions of $M$.

Let $\varphi_1, \ldots, \varphi_k \in \mathbb{R}_X$ and $F \in A_k$. Then $F \circ (\varphi_1, \ldots, \varphi_k)$ is a partially defined function on $X$ given by

$$(F \circ (\varphi_1, \ldots, \varphi_k))(x) = F(\varphi_1(x), \ldots, \varphi_k(x))$$

for $x \in X$. If $\varphi_i(x) = \emptyset_0$ or $(\varphi_1(x), \ldots, \varphi_k(x)) \notin \mathrm{dom}F$ then $F(\varphi_0(x), \ldots, \varphi_n(x)) = \emptyset_0$. The map

$$(\varphi_1, \ldots, \varphi_k) \mapsto F \circ (\varphi_1, \ldots, \varphi_k)$$

is called a *substitution*.

A set $\mathcal{U} \subseteq \mathbb{R}_X$ containing $0_X$ and closed under substitutions and amalgamation is called a *function universe on the set $X$* [16]. A *function subuniverse* of the universe $\mathcal{U}$ is a subset $\hat{U} \subset \mathcal{U}$ that is a function universe on $X$. If $\mathcal{H} \subset \mathcal{U}$, then function *subuniverse generated by* $\mathcal{H}$ is the smallest subuniverse of $\mathcal{U}$ containing $\mathcal{H}$ [1].

In a natural way a function universe $\mathcal{U}$ on $X$ induces a topology on $X$: the open sets have the form $\mathrm{dom}\varphi : \varphi \in \mathcal{U}$.

Let $\mathcal{U}_1, \mathcal{U}_2$ be function universes on $X_1$ and $X_2$ respectively. A map $\tau : \mathcal{U}_1 \to \mathcal{U}_2$ is a *homomorphism* of function universes $\mathcal{U}_1$ and $\mathcal{U}_2$ if

1. $\tau(F \circ (\varphi_1, \dots, \varphi_k)) = F(\tau\varphi_1, \dots, \tau\varphi_k)$ for $\varphi_1, \dots, \varphi_k \in \mathcal{U}_1$, $F \in A_k$
2. $\tau(\underline{M}) = \underline{\tau(M)}$ for any matching set $M \subset \mathcal{U}_1$
3. $\tau(0_{X_1}) = 0_{X_2}$

If a homomorphism $\tau$ is a bijective map then it is an *isomorphism* of function universes.

## 5 Conditions of equivalence

We shall assume the following conditions on the dynamics of the system (3):

**Condition D1.** For every $x, y \in \mathbb{R}^n$ there is at most one $u$ that satisfies the equation

$$y = f(x, u). \tag{9}$$

**Condition D2.** For any $x$ and $u$ the rank of the matrix

$$\frac{\partial f}{\partial u}(x, u)$$

is full (i.e. equal $m$).

**Condition D3.** The map

$$\mathbb{R}^n \times \mathbb{R}^m \to \mathbb{R}^n \times \mathbb{R}^n : (x, u) \mapsto (x, f(x, u))$$

is proper, i.e. the inverse image of a compact set in $\mathbb{R}^n \times \mathbb{R}^n$ is a compact set in $\mathbb{R}^n \times \mathbb{R}^m$.

Conditions D1-D3 were first introduced in [14, 15] to prove criteria of dynamical feedback equivalence for nonlinear continuous-time systems. Later there were used in [2] to show a similar result for nonlinear discrete-time systems, and recently, under the same assumptions, we have unified these characterizations stating the dynamical feedback equivalence criteria for nonlinear systems on homogeneous time scales [5].

The dynamical feedback equivalence concerns only the dynamical part of the system $\Sigma$, i.e. the equation

$$x^{\Delta}(t) = f(x(t), u(t)).$$

Two systems are said to be *dynamically feedback equivalent* if they are externally dynamically equivalent after replacing their output functions by the identity maps (i.e. setting $y(t) = x(t)$). Then the dynamical transformations maps $\phi^e$ and $\psi^e$, denoted now by $\phi$ and $\psi$, depend rather on delta-derivatives of $x$ and $u$ and not of $y$ and $u$.

Let $T$ be the map $\mathbb{R}^n \times J(m) \to J(n)$ defined by

$$T(x_0, U) := X = (x^{(0)}, x^{(1)}, \ldots),$$

where $(X, U)$ is the infinite jet at $t = 0$ of the inner trajectory $(x, u)$ of $\Sigma$ that satisfies the initial condition $x(0) = x_0$. One can show that such $X$ is unique which means that $T$ is well defined. Moreover, Conditions D1 and D2 imply that $T$ is left invertible. Thus, if $X$ is the jet of the solution $x$ corresponding to a control $u$ (and the initial condition $x(0) = x_0 = x^{(0)}(0)$), then the jet of $u$ can be expressed as $U = S(X)$, where components of $S$ are finitely presented and analytic. This means that the dynamical feedback equivalence is given by the equations

$$x(t) = \phi(\tilde{x}(t), \tilde{x}^{\Delta}(t), \ldots, \tilde{x}^{\Delta^k}(t))$$
$$u(t) = \psi(\tilde{x}(t), \tilde{x}^{\Delta}(t), \ldots, \tilde{x}^{\Delta^k}(t), \tilde{u}(t), \tilde{u}^{\Delta}(t), \ldots, \tilde{u}^{\Delta^k}(t)) \qquad (10)$$

and similarly for $(\tilde{x}(t), \tilde{u}(t))$.

Let $A(n, m)$ denote the algebra of all finitely presented analytic functions $\varphi : \mathbb{R}^n \times J(m) \to \mathbb{R}$ and let $\Sigma$ be given by (3). Let us define the operator $\delta_\Sigma : A(n, m) \to A(n, m)$, associated with $\Sigma$, by

$$(\delta_\Sigma \varphi)(x, U) :=$$
$$\int_0^1 \frac{\partial \varphi}{\partial x}(x + h\mu f(x, u^{(0)}), U) dh \cdot f(x, u^{(0)}) +$$
$$\sum_{i=0}^{\infty} \int_0^1 \frac{\partial \varphi}{\partial u^{(i)}}(x, U + h\mu U_1)) dh \cdot u^{(i+1)} \qquad (11)$$

where $U_1 = (u^{(1)}, u^{(2)}, \ldots)$. It is clear that this operator depends only on the dynamical part of $\Sigma$. It will be called the delta operator of the system $\Sigma$.

*Remark 4.* The delta operator has the following interpretation. Let $U(\cdot) = (u(\cdot), u^{\Delta}(\cdot), \ldots)$ be the infinite jet of control $u$ and let $x(\cdot)$ be the solution of (3) corresponding to $u$ and the initial condition $x(0) = x_0$. Then the delta derivative at $t = 0$ of $t \to \varphi(x(t), U(t))$ is equal to $(\delta_\Sigma \varphi)(x_0, U(0))$. If $\mathbb{T} = \mathbb{R}$, then $\delta_\Sigma$ is a derivation of the algebra $A(n, m)$, i.e. it is linear and satisfies the Leibniz rule. This is not the case for other time scales. But because of the above interpretation we shall often call $\delta_\Sigma \varphi$ the $\delta_\Sigma$ derivative of $\varphi$.

The algebra $A(n, m)$ together with the operator $\delta_\Sigma$ is called the *delta algebra of the system* $\Sigma$ and denoted by $A_\Sigma$. A homomorphism of delta algebras $A_\Sigma$ and $A_{\tilde{\Sigma}}$ is a homomorphism $\tau : A(n, m) \to A(\tilde{n}, m)$ of algebras that satisfies the condition $\delta_{\tilde{\Sigma}} \circ \tau = \tau \circ \delta_\Sigma$. An *isomorphism* of the delta algebras $A_\Sigma$ and $A_{\tilde{\Sigma}}$ is a homomorphism that is a bijective map.

The main result of [5] says the following

**Theorem 1.** *Systems $\Sigma$ and $\tilde{\Sigma}$ are dynamically feedback equivalent iff their delta algebras $A_\Sigma$ and $A_{\tilde{\Sigma}}$ are isomorphic.*

Now we are going to show a similar characterization of external dynamical equivalence for systems on homogeneous time scales.

Let $\mathcal{A}(n,m)$ denote the function universe of all partially defined and finitely presented analytic functions on $\mathbb{R}^n \times J(m)$. It is generated by the algebra $A(n,m)$. Observe that the operator $\delta_\Sigma$ can naturally be extended to $\mathcal{A}(n,m)$. For a system $\Sigma$ given by (3) we define the *observation universe* $\mathcal{U}_\Sigma$ to be the smallest subuniverse of $\mathcal{A}(n,m)$ containing the components $h_i$, $i = 1, \ldots, r$, of the map $h$, the coordinate functions $u_j$, $j = 1, \ldots, m$, and invariant under the action of $\delta_\Sigma$.

The observation universe $\mathcal{U}_\Sigma$ together with the operator $\delta_\Sigma$ is called the *delta universe of the system $\Sigma$* (and will be denoted by the same symbol $\mathcal{U}_\Sigma$). A morphism of delta universes $\mathcal{U}_\Sigma$ and $\mathcal{U}_{\tilde{\Sigma}}$ is a morphism $\tau : \mathcal{U}_\Sigma \to \mathcal{U}_{\tilde{\Sigma}}$ of function universes that satisfies the condition

$$\delta_{\tilde{\Sigma}} \circ \tau = \tau \circ \delta_\Sigma.$$

An *isomorphism* of the delta universes $\mathcal{U}_\Sigma$ and $\mathcal{U}_{\tilde{\Sigma}}$ is a morphism that is a bijective map.

The system $\Sigma$ is called *uniformly observable* if every coordinate function $x_i$, $i = 1, \ldots, n$, belongs to $\mathcal{U}_\Sigma$.

*Remark 5.* Uniform observability means that locally one can express coordinate functions as analytic functions of the output function $h$, the control $u$ and their $\delta_\Sigma$ 'derivatives'. This means that for every control function $u$, any two distinct initial states can be distinguished by observing the output. Moreover, we can recover the state from the output, the control and their derivatives. As in [20] we could assume specific conditions on $h$ and $f$ that guarantee uniform observability of the system. However we want to concentrate here on the problem of dynamical equivalence, so we put aside all the details concerning observability.

**Proposition 2.** *The system $\Sigma$ is uniformly observable iff $\mathcal{U}_\Sigma = \mathcal{A}(n,m)$.*

*Proof.* If $\Sigma$ is uniformly observable, then for every $i = 1, \ldots, n$, $x_i$ belongs to $\mathcal{U}_\Sigma$. By definition, $\mathcal{U}_\Sigma$ contains also $u_j$ for $j = 1, \ldots, m$ and all $\delta_\Sigma^k u_j = u_j^{(k)}$. Substituting $x_i$, $i = 1, \ldots, n$, and $u_j^{(k)}$, $j = 1, \ldots, m$, $k \geq 0$, into analytic partially defined functions we get all the elements of $\mathcal{A}(n,m)$. On the other hand, if $\mathcal{U}_\Sigma = \mathcal{A}(n,m)$, then every $x_i$, $i = 1, \ldots, n$, belongs to $\mathcal{U}_\Sigma$.

**Corollary 1.** *Assume that two systems $\Sigma$ and $\tilde{\Sigma}$ are uniformly observable. Then, $A_\Sigma$ and $A_{\tilde{\Sigma}}$ are isomorphic iff $\mathcal{U}_\Sigma$ and $\mathcal{U}_{\tilde{\Sigma}}$ are isomorphic.*

*Proof.* From uniform observability we get $\mathcal{U}_\Sigma = \mathcal{A}(n,m)$ and $\mathcal{U}_{\tilde{\Sigma}} = \mathcal{A}(\tilde{n},m)$. As $A(n,m)$ and $A(\tilde{n},m)$ are the algebras of global functions of the universes

$\mathcal{A}(n,m)$ and $\mathcal{A}(\tilde{n},m)$, respectively, then the restriction of an isomorphism of $\mathcal{U}_\Sigma$ and $\mathcal{U}_{\tilde\Sigma}$ gives an isomorphism of $A_\Sigma$ and $A_{\tilde\Sigma}$. The other implication follows from the specific form of any isomorphism $\tau$ of $A_\Sigma$ and $A_{\tilde\Sigma}$ (see [5]). It is a pullback of a certain map, so it commutes with substitutions and amalgamations. Thus the isomorphism $\tau$ of $A_\Sigma$ and $A_{\tilde\Sigma}$ is also an isomorphism of the function universes $\mathcal{U}_\Sigma$ and $\mathcal{U}_{\tilde\Sigma}$. This also allows to show that the condition $\delta_{\tilde\Sigma} \circ \tau = \tau \circ \delta_\Sigma$ can be extended to the universes $\mathcal{U}_\Sigma$ and $\mathcal{U}_{\tilde\Sigma}$.

If $\theta$ is a real function on $\mathbb{R}^n \times J(m)$, then by $\Theta$ we denote its infinite *delta-jet*: $(\theta, \delta_\Sigma(\theta), \delta_\Sigma^2(\theta), \ldots)$. This, in particular, concerns the coordinate functions $x_i$ and $u_j$ and their aggregations $x$ and $u$, whose delta-jets are denoted by $X$ and $U$.

Uniform observability of $\Sigma$ means that $x = \varrho(Y, U)$ for some finitely presented analytic map $\varrho$. Thus $X = R(Y, U)$, where $R$ is the infinite delta-jet of $\varrho$. This equation may be rewritten on the level of jets of functions of time: $X(t) = R(Y(t), U(t))$. On the other hand, from the relations $y = h(x)$, after applying the operator $\delta_\Sigma$, we get $Y = \Gamma(X, U)$, for some map $\Gamma$ with finitely presented components.

**Proposition 3.** *Assume that two systems $\Sigma$ and $\tilde\Sigma$ are uniformly observable. Then, $\Sigma$ and $\tilde\Sigma$ are dynamically feedback equivalent iff they are externally dynamically equivalent.*

*Proof.* Assume first that $\Sigma$ and $\tilde\Sigma$ are dynamically feedback equivalent. Thus there exist transformations $\phi$, $\psi$, $\tilde\phi$ and $\tilde\psi$ such that for an inner trajectory $(\tilde{x}, \tilde{u})$ of $\tilde\Sigma$ the equations

$$x(t) = \phi(\tilde{X}(t))$$
$$u(t) = \psi(\tilde{X}(t), \tilde{U}(t))$$

define an inner trajectory of $\Sigma$, and similarly for $(\tilde{x}(t), \tilde{u}(t))$. Thus

$$y(t) = h(\phi(\tilde{X}(t)))$$
$$u(t) = \psi(\tilde{X}(t), \tilde{U}(t))$$

define an external trajectory of $\Sigma$. From uniform observability, $\tilde{X} = R(\tilde{Y}, \tilde{U})$. Thus

$$y(t) = h(\phi(R(\tilde{Y}(t), \tilde{U}(t))))$$
$$u(t) = \psi(R(\tilde{Y}(t), \tilde{U}(t)), \tilde{U}(t))$$

define the transformations $\phi^e$ and $\psi^e$. Similarly we get the transformations $\tilde\phi^e$ and $\tilde\psi^e$, which gives external dynamical equivalence of $\Sigma$ and $\tilde\Sigma$.

Now assume that $\Sigma$ and $\tilde\Sigma$ are externally dynamically equivalent. Thus external trajectories of both systems are related by equations

$$y(t) = \phi^e(\tilde{Y}(t), \tilde{U}(t))$$
$$u(t) = \psi^e(\tilde{Y}(t), \tilde{U}(t)).$$

They can be extended to jets

$$Y(t) = \Phi^e(\tilde{Y}(t), \tilde{U}(t))$$
$$U(t) = \Psi^e(\tilde{Y}(t), \tilde{U}(t)).$$

Again from uniform observability, $x = \varrho(Y, U)$, so

$$x(t) = \varrho(\Phi^e(\tilde{Y}(t), \tilde{U}(t)), \Psi^e(\tilde{Y}(t), \tilde{U}(t))).$$

However $Y = \Gamma(X, U)$ and $U = S(X)$, so we finally obtain

$$x(t) = \phi(\tilde{X}(t))$$
$$u(t) = \psi(\tilde{X}(t), \tilde{U}(t)).$$

for some finitely presented maps $\phi$ and $\psi$. Similarly for $\tilde{x}$ and $\tilde{u}$.

We can state now the main result of this paper.

**Theorem 2.** *Two uniformly observable systems $\Sigma$ and $\tilde{\Sigma}$ are externally dynamically equivalent if and only if their delta universes $\mathcal{U}_\Sigma$ and $\mathcal{U}_{\tilde{\Sigma}}$ are isomorphic.*

*Proof.* The theorem follows from Corollary 1 and Proposition 3.

*Example 5.* The concept of external dynamical equivalence of systems on time scales may be used to give a unified setting for external dynamical linearization. In [21] and [19] it was shown that a continuous-time (discrete-time) system is externally dynamically linearizable if and only if its differential output universe (difference output universe, respectively) is free. To have this result on an arbitrary homogeneous time scale, it is enough to replace differential and difference output universes by the delta universe of the system on the time scale.

## 6 Conclusions and future works

### 6.1 Conclusions

The theory of control systems on time scales allows to use a common language for continuous-time and discrete-time systems. The concept of external dynamical equivalence of control systems with output unifies the earlier notions introduced separately for continuous-time and discrete-time systems. It was shown that two uniformly observable systems on a time scale are externally dynamically equivalent if and only if their delta universes are isomorphic. This extends corresponding results for continuous-time and discrete-time systems, where differential and difference output universes were used instead of the delta universe. The language introduced here allows to study external dynamical linearizations for systems on time scales.

## 6.2 Future works

The main assumption on the systems in this paper is uniform observability. It was shown in [21] and [3] that for continuous-time and discrete-time systems the criterion of external dynamical equivalence holds without this assumption. Thus one can hope that also for systems on a time scale observability can be dropped. This, however, requires a thorough study of observability of systems on time scales. A more serious challenge is to extend the result to nonhomogeneous time scales. In this case the delta operator may depend on time and this requires a new language. Another interesting extension should lead to modification of the concept of dynamical equivalence by allowing a change of time or, more generally, time scale.

# References

1. Bartosiewicz Z, Johnson J (1994) Systems on universe spaces. Acta Applicandae Mathematicae 34.
2. Bartosiewicz Z, Jakubczyk B, Pawłuszewicz E (1994) Dynamic feedback equivalence of nonlinear discrete-time systems. In: Proceedings of First International Symposium on Mathematical Models in Automation and Robotics, Sept. 1-3, 1994, Międzyzdroje, Poland, Tech. Univ. of Szczecin Press, Szczecin
3. Bartosiewicz Z, Pawłuszewicz E (1998) External equivalence of unobservable discrete-time systems. In: Proceedings of NOLCOS'98, Enschede, Netherlands, July 1998.
4. Bartosiewicz Z, Pawłuszewicz E (2004) Unification of continuous-time and discrete-time systems: the linear case. In: Proceedings of Sixteenth International Symposium on Mathematical Theory of Networks and Systems (MTNS2004) Katholieke Universiteit Leuven, Belgium July 5-9, 2004, Leuven
5. Bartosiewicz Z, Pawłuszewicz E (2005) Dynamic feedback equivalence of nonlinear systems on time scales. In: Proceedings of 16th IFAC Congress, Prague (CD-ROM)
6. Bartosiewicz Z, Pawłuszewicz E (2006) Realizations of linear control systems on time scales. Control & Cybernetics, 35(4).
7. Bohner M, Peterson A (2001) Dynamic Equation on Time Scales. Birkhäuser, Boston Basel Berlin
8. Charlet B, Levine J, Marino R (1991) Sufficient conditions for dynamic state feedback linearization. SIAM J. Control and Optimization 29:38–57.
9. Fausett LV, Murty KN (2004) Controllability, observability and realizability criteria on time scale dynamical systems. Nonlinear Stud. 11:627–638.
10. Fliess M, Lévine J, Martin Ph, Rouchon P (1992) Sur les systèmes non linéaires différentiellement plats. C. R. Acad. Sci. Paris Sér. I Math., 315(5):619–624.
11. Goodwin GC, Graebe SF, Salgado ME (2001) Control System Design. Prentice Hall International
12. Gürses M, Guseinov GSh, Silindir B (2005) Integrable equations on time scales. Journal of Mathematical Physics 46:113–510.
13. Hilger S (1988) Ein Maßkettenkalkülmit Anwendung auf Zentrumsmannigfaltigkeiten. Ph.D. thesis, Universität Würzburg, Germany

14. Jakubczyk B (1992) Remarks on equivalence and linearization of nonlinear systems. In: Proceedings of the 2nd IFAC NOLCOS Symposium, 1992, Bordeaux, France, 393–397.
15. Jakubczyk B (1992) Dynamic feedback equivalence of nonlinear control systems. Preprint
16. Johnson J (1986) A generalized global differential calculus I. Cahiers Top. et Geom. Diff. XXVII
17. Martin Ph, Murray RM, Rouchon P (1997) In: Bastin G, Gevers M (Eds) Flat systems. Plenary Lectures and Mini-Courses, European Control Conference ECC'97, Brussels
18. Middleton RH, Goodwin GC (1990) Digital Control and Estimation: A Unified Approach. Englewood Cliffs, NJ: Prentice Hall
19. Pawłuszewicz E (1996) Dynamic linearization of input-output discrete-time systems. In: Proceedings of the International Conference UKACC Control'96, Exeter, UK
20. Pawłuszewicz E, Bartosiewicz Z (1999) External Dynamic Feedback Equivalence of Observable Discrete-time Control Systems. Proc. of Symposia in Pure Mathematics, vol.64, American Mathematical Society, Providence, Rhode Island
21. Pawłuszewicz E, Bartosiewicz Z (2003) Differential universes in external dynamic linearization. Proceedings of European Control Conference ECC'03, Cambridge, UK
22. Pomet J-B (1995) A differential geometric setting for dynamic equivalence and dynamic linearization, in: Geometry in Nonlinear Control and Differential Inclusions, Banach Center Publications, vol.32, Institute of Mathematics, Polish Academy of Sciences, Warsaw, Poland

# On Option-Valuation in Illiquid Markets: Invariant Solutions to a Nonlinear Model

Ljudmila A. Bordag

Halmstad University, Box 823, 301 18 Halmstad, Sweden. `Ljudmila.Bordag@hh.se`

**Summary.** The present model describes a perfect hedging strategy for a large trader. In this case the hedging strategy affects the price of the underlying security. The feedback-effect leads to a nonlinear version of the Black-Scholes partial differential equation. Using Lie group theory we reduce in special cases the partial differential equation to some ordinary differential equations. The Lie group found for the model equation gives rise to invariant solutions. Families of exact invariant solutions for special values of parameters are described.

## 1 Introduction

One of the basic assumptions of the Black–Scholes option theory is that all participants act on the market as price takers. Recently a series of papers [6], [7], [15], [17] appeared in which this assumption has been relaxed. In these models the hedging strategy affects the price of the underlying security. For a large trader a hedge-cost of the claim differs from the price of the option for price takers.

In [6], [7] Frey developed a model of market illiquidity. He describes the asset price dynamics which results if a large trader chooses a given stock-trading strategy $(\alpha_t)_t$. The admissible class of stock trading strategies includes the left-continuous stock-holdings $(\alpha_t)_t$ for which the right-continuous process with $\alpha_t^+ = lim_{s \to t, s > t} \alpha_s$ is a semimartingale. If the large trader uses a particular trading strategy $\alpha^+$ then, as expected, $d\alpha_t^+ > 0$ leads to $dS_t > 0$ and correspondingly $d\alpha_t^+ < 0$ leads to $dS_t < 0$, where $S_t$ denotes the asset price process. Frey's model has the form of the stochastic differential equation

$$dS_t = \sigma S_{t-} dW_t + \rho S_{t-} d\alpha_t^+, \tag{1}$$

where $\sigma > 0$ is a constant volatility, $W_t$ is a standard Brownian motion, $S_{t-}$ denotes the left limit $lim_{s \to t, s < t} S_t$, and $\rho$ is the market illiquidity parameter with $0 < \rho < 1$. If we denote the filtration associated to the Brownian motion by $\mathcal{F}(t)$ than a hedge cost of the claim $u(S,t)$ is given by a conditional expectation $u(S,t) = \mathrm{E}[h(S)|\mathcal{F}(t)]$ for any Borel–measurable function $h(S)$. The function $h(S)$ is the payoff at time $T$ of a derivative security $u(S,t)$ whose underlying asset price process is (1). After the Feynman–Kac theorem [16] together with a fixed point argument [7] $u(S,t)$ satisfies the corresponding partial differential equation

$$u_t + \frac{\sigma^2 S^2}{2} \frac{u_{SS}}{(1 - \rho S u_{SS})^2} = 0, \tag{2}$$

and the terminal condition $u(S,T) = h(S)$ for all $S$ where $S \geq 0$, $t \in [0,T]$, $T > 0$.

The value $1/(\rho S_t)$ is called the depth of the market at time $t$. If $\rho \to 0$ then simultaneously $\mathrm{d}\alpha_t^+ \to 0$ and the equation (2) reduces to the Black–Scholes case.

In model (1) the parameter $\rho$ is a characteristics of the market and does not depend on the payoff of hedged derivatives. The value of $\rho$ is fixed during the trading process and can be estimated in different ways, in the papers [6], [7] the value $\rho$ is equal to 0.1, 0.2, 0.3, 0.4.

In the paper by Frey and Stremme, [9], a heterogeneous portfolio was introduced, and a model with variable volatility was studied.

Later on, in [8] Frey and Patie followed another approach and examined the feedback effect of the option replication strategy of the large trader on the asset price process. They obtain a new model by introduction of a liquidity coefficient which depends on the current stock price.

It means the stochastic equation now takes the form

$$\mathrm{d}S_t = \sigma S_{t-}\mathrm{d}W_t + \rho\lambda(S_{t-})S_{t-}\mathrm{d}\alpha_t^+, \tag{3}$$

where $\lambda : R^+ \to R^+$ is a continuous bounded function for all $S \geq 0$. In this model the depth of the market at time $t$ is defined as $1/(\rho\lambda(S_t)S_t)$; $\lambda(S)$ is called *level-depended liquidity profile*.

The Feynman–Kac theorem together with a fixed point argument [8] leads in case of (3) to a partial differential equation which is similar to the equation (2) where the constant $\rho$ is replaced by $\rho\lambda(S)$. The values of $\rho$ and $\lambda(S)$ may be estimated from the observed option prices and depend on the payoff $h(S)$.

The feedback-effect described in (3) leads to a nonlinear version of the Black-Scholes partial differential equation,

$$u_t + \frac{\sigma^2 S^2}{2} \frac{u_{SS}}{(1 - \rho S \lambda(S) u_{SS})^2} = 0, \tag{4}$$

with $S \in [0,\infty)$, $t \in [0,T]$. As usual, $S$ denotes the price of the underlying asset and $u(S,t)$ denotes the hedge-cost of the claim with a payoff $h(S)$ which will be defined later. The hedge-cost is different from the price of the derivatives product in illiquid markets. In the sequel $t$ is the time variable, $\sigma$ defines the volatility of the underlying asset.

In [8] some examples for estimates of $\rho$ and $\lambda(S)$ were matched for the call spread and for the European call. The values $\rho$ are equal to 0.01, 0.02, 0.05, 0.1, 0.2, 0.4 and the level-depended liquidity profile $\lambda(S)$ is approximated by a quadratic function of $S$. We shall concentrate our investigations on the non-trivial case $\rho \neq 0$.

Depending on the assumptions concerning the market, different variations of the Black-Scholes formula can arise like in the well known models [15], [17]. Usually the volatility term in the Black-Scholes formula will be replaced in order to fit the behavior of the price on the market. The modeling process is not concluded now and new models can appear. For those of the type (2), (4), or in [17] the appearance of the small parameter multiplying the highest derivative is crucial. The nonlinear parabolic equations with a singular perturbation describe usually a richer class of phenomena than equations with a regular perturbation. An analytical study of these equations may be useful for an easier classification of models.

Frey and co-authors studied equation (4) under some restrictions and did some numerical simulations. Our goal is to investigate this equation using analytical methods.

We study the model equation (4) by means of Lie group theory. We used the method of Lie group analysis to investigate other financial models given by nonlinear partial differential equations in [3], [1], [5] [2], [4]. This method has a long tradition beginning with the work of S. Lie [12]. Its applications are based on local symmetries. A modern description of the method as well as a large number of applications can be found in [14], [13], [18], [10], [11].

In Section 2 we find the Lie algebra and global equations for the symmetry group of equation (4). For a special form of the function $\lambda(S)$ it is possible to find two functionally independent invariants of the symmetry group. Using the symmetry group and its invariants we reduce the partial differential equation (4) in some special cases to ordinary differential equations in Section 3. We shell study singular points of the reduced equations in Section 4 and describe the behavior of the invariant solutions. For a fixed set of parameters the complete set of exact invariant solutions is given. It is important to obtain exact solutions for the model (4) to use invariant solutions for instance as a test case for various numerical methods.

Further, in Section 5 we study different properties of the invariant solutions and their sensitivity with respect to the illiquidity parameter. In particular, if any terminal conditions are fixed, $u(S,T) = h(S)$, then the value $u(S,t)$ will increase if the value of the parameter $\rho$ increases, i.e. hedge costs of the large trader on the market depends in expected way on the position of the large trader. We obtain a typical terminal payoff function for these solutions if we just fix $t = T$. By changing the value of the integrating constants and by adding a linear function of $S$ we are able to modify terminal payoff functions for the solutions. Hence we can approximate typical payoff profiles of some financial derivatives quite well.

## 2 Lie group symmetries

Let us introduce a two-dimensional space $X$ of independent variables $(S, t) \in X$ and a one-dimensional space of dependent variables $u \in U$. We consider the space $U_{(1)}$ of the first derivatives of the variable $u$ on $S$ and $t$, i.e., $(u_S, u_t) \in U_{(1)}$ and analogously we introduce the space $U_{(2)}$ of the second order derivatives $(u_{SS}, u_{St}, u_{tt}) \in U_{(2)}$. We denote by $M = X \times U$ a base space which is a Cartesian product of pairs $(x, u)$ with $x = (S, t) \in X$, $u \in U$. The differential equation (4) is of the second order and to represent this equation as an algebraic equation we introduce a second order jet bundle $M^{(2)}$ of the base space $M$, i.e.,

$$M^{(2)} = X \times U \times U_{(1)} \times U_{(2)} \qquad (5)$$

with a natural contact structure. We label the coordinates in the jet bundle $M^{(2)}$ by
$w = (S, t, u, u_S, u_t, u_{SS}, u_{St}, u_{tt}) \in M^{(2)}$.

In the space $M^{(2)}$ equation (4) is equivalent to the relation

$$\Delta(w) = 0, \quad w \in M^{(2)}, \qquad (6)$$

where we denote by $\Delta$ the following function

$$\Delta(S, t, u, u_S, u_t, u_{SS}, u_{St}, u_{tt}) = u_t + \frac{\sigma^2 S^2}{2} \frac{u_{SS}}{(1 - \rho S \lambda(S) u_{SS})^2}. \qquad (7)$$

We identify the algebraic equation (6) with its solution manifold $L_\Delta$ defined by

$$L_\Delta = \{w \in M^{(2)} | \Delta(w) = 0\} \subset M^{(2)}. \qquad (8)$$

Let us consider an action of a Lie-point group on our differential equation and its solutions. We define a symmetry group $G_\Delta$ of equation (6) by

$$G_\Delta = \{g \in \text{Diff}(M^{(2)}) | \ g : \ L_\Delta \to L_\Delta\}, \qquad (9)$$

consequently we are interested in a subgroup of $\text{Diff}(M^{(2)})$ which is compatible with the structure of $L_\Delta$.

As usual we first find the corresponding symmetry Lie algebra $\mathcal{D}iff_\Delta(M^{(2)}) \subset \mathcal{D}iff(M^{(2)})$ and then use the main Lie theorem to obtain $G_\Delta$ and its invariants.

We denote an element of a Lie-point vector field on $M$ by

$$V = \xi(S, t, u) \frac{\partial}{\partial S} + \tau(S, t, u) \frac{\partial}{\partial t} + \phi(S, t, u) \frac{\partial}{\partial u}, \qquad (10)$$

where $\xi(S, t, u), \tau(S, t, u)$ and $\phi(S, t, u)$ are smooth functions of their arguments, $V \in \mathcal{D}iff(M)$. The operators (10) are called as well infinitesimal generators.

If the infinitesimal generators of $g \in G_\Delta(M)$ exist then they have the structure of type (10) and form an Lie algebra $\mathcal{D}iff_\Delta(M)$.

A Lie group of transformations $G_\Delta(M)$ acting on the base space $M$ induce as well the transformations on $M^{(2)}$ which we denoted by $G_\Delta$.

The corresponding Lie algebra $\mathcal{D}iff_\Delta(M^{(2)})$ will be composed of vector fields

$$
\begin{aligned}
pr^{(2)}V = {} & \xi(S,t,u)\frac{\partial}{\partial S} + \tau(S,t,u)\frac{\partial}{\partial t} + \phi(S,t,u)\frac{\partial}{\partial u} \\
& + \phi^S(S,t,u)\frac{\partial}{\partial u_S} + \phi^t(S,t,u)\frac{\partial}{\partial u_t} \\
& + \phi^{SS}(S,t,u)\frac{\partial}{\partial u_{SS}} + \phi^{St}(S,t,u)\frac{\partial}{\partial u_{St}} + \phi^{tt}(S,t,u)\frac{\partial}{\partial u_{tt}},
\end{aligned}
\tag{11}
$$

which are the second prolongation of vector fields $V$. Here the smooth functions
$\phi^S(S,t,u)$, $\phi^t(S,t,u)$, $\phi^{SS}(S,t,u)$, $\phi^{St}(S,t,u)$ and $\phi^{tt}(S,t,u)$ are uniquely defined by the functions $\xi(S,t,u)$, $\tau(S,t,u)$ and $\phi(S,t,u)$ using the prolongation procedure (see [14], [13], [18], [10], [11]).

**Theorem 1.** *The differential equation (4) with an arbitrary function $\lambda(S)$ possesses a trivial three dimensional Lie algebra $Diff_\Delta(M)$ spanned by infinitesimal generators*

$$
V_1 = \frac{\partial}{\partial t}, \quad V_2 = S\frac{\partial}{\partial u}, \quad V_3 = \frac{\partial}{\partial u}.
$$

*Only for the special form of the function $\lambda(S) \equiv \omega S^k$, where $\omega, k \in R$ equation (4) admits a non-trivial four dimensional Lie algebra spanned by generators*

$$
V_1 = \frac{\partial}{\partial t}, \quad V_2 = S\frac{\partial}{\partial u}, \quad V_3 = \frac{\partial}{\partial u}, V_4 = S\frac{\partial}{\partial S} + (1-k)u\frac{\partial}{\partial u}.
$$

*Proof.* The symmetry algebra $\mathcal{D}iff_\Delta(M^{(2)})$ of the second order differential equation (6) can be found as a solution to the determining equations

$$
pr^{(2)}V(\Delta) = 0 \; (mod(\Delta = 0)),
\tag{12}
$$

i.e., the equation (12) should be satisfied on the solution manifold $L_\Delta$.

For our calculations we will use the exact form of the coefficients $\phi^t(S,t,u)$ and $\phi^{SS}(S,t,u)$ only. The coefficient $\phi^t(S,t,u)$ can be defined by the formula

$$
\phi^t(S,t,u) = \phi_t + u_t\phi_u - u_S\xi_t - u_Su_t\xi_u - u_t\tau_t - (u_t)^2\tau_u,
\tag{13}
$$

and the coefficient $\phi^{SS}(S,t,u)$ by the expression

$$
\begin{aligned}
\phi^{SS}(S,t,u) = {} & \phi_{SS} + 2u_S\phi_{Su} + u_{SS}\phi_u \\
& + (u_S)^2\phi_{uu} - 2u_{SS}\xi_S - u_S\xi_{SS} - 2(u_S)^2\xi_{Su} \\
& - 3u_Su_{SS}\xi_u - (u_S)^3\xi_{uu} - 2u_{St}\tau_S - u_t\tau_{SS} \\
& - 2u_Su_t\tau_{Su} - (u_tu_{SS} + 2u_Su_{St})\tau_u - (u_S)^2u_t\tau_{uu},
\end{aligned}
\tag{14}
$$

where the subscripts by $\xi, \tau, \phi$ denote corresponding partial derivatives.

The first equations of the set (12) imply that if $V \in \mathcal{D}iff_{\Delta}(M)$ then

$$\xi(S, t, u) = a_1 S, \quad \tau(S, t, u) = a_2, \quad \phi(S, t, u) = a_3 S + a_4 + a_5 u, \qquad (15)$$

where $a_1, a_2, a_3, a_4, a_5$ are arbitrary constants and $\xi, \tau, \phi$ are coefficients in the expression (10).

The remaining equation has a form

$$a_1 S \lambda_S(S) - (a_1 - a_5) \lambda(S) = 0. \qquad (16)$$

Because this equation should be satisfied for all $S$ identically we obtain for an arbitrary function $\lambda(S)$

$$a_1 = a_5 = 0, \quad \rightarrow \xi(S, t, u) = 0, \quad \tau(S, t, u) = a_2, \quad \phi(S, t, u) = a_3 S + a_4. \quad (17)$$

Finally, $\mathcal{D}iff_{\Delta}(M)$ admits the following infinitesimal generators

$$V_1 = \frac{\partial}{\partial t}, \quad V_2 = S\frac{\partial}{\partial u}, \quad V_3 = \frac{\partial}{\partial u}, \qquad (18)$$

with commutator relations

$$[V_1, V_2] = [V_1, V_3] = [V_2, V_3] = 0. \qquad (19)$$

If the function $\lambda(S)$ has a special form

$$\lambda(S) \equiv \omega S^k, \quad \omega, k \in R \qquad (20)$$

then the equation (16) on the coefficients of (10) is less restrictive and we obtain

$$\xi(S, t, u) = a_1 S, \quad \tau(S, t, u) = a_2, \quad \phi(S, t, u) = (1 - k)a_1 u + a_3 S + a_4. \quad (21)$$

Now the symmetry algebra $\mathcal{D}iff_{\Delta}(M)$ admits four generators

$$V_1 = \frac{\partial}{\partial t}, \quad V_2 = S\frac{\partial}{\partial u}, \quad V_3 = \frac{\partial}{\partial u}, V_4 = S\frac{\partial}{\partial S} + (1 - k)u\frac{\partial}{\partial u}, \qquad (22)$$

with commutator relations

$$[V_1, V_2] = [V_1, V_3] = [V_1, V_4] = [V_2, V_3] = 0,$$
$$[V_2, V_4] = -kV_2, \quad [V_3, V_4] = (1 - k)V_3. \qquad (23)$$

*Remark 1.* In the general case the algebra (22) possesses a three dimensional Abelian sub-algebra and one dimensional center. For the cases $k = 0, 1$ the center is two dimensional (23) and we see later that the corresponding equations (44) became autonomous.

The symmetry algebra $Diff_\Delta(M)$ defines by the Lie equations [12] the corresponding symmetry group $G_\Delta$ of the equation (6). To find the global form of transformations for the solutions to equation (4) corresponding to this symmetry group we just integrate the system of ordinary differential equations

$$\frac{d\tilde{S}}{d\epsilon} = \xi(\tilde{S}, \tilde{t}, u), \tag{24}$$

$$\frac{d\tilde{t}}{d\epsilon} = \tau(\tilde{S}, \tilde{t}, \tilde{u}), \tag{25}$$

$$\frac{d\tilde{u}}{d\epsilon} = \phi(\tilde{S}, \tilde{t}, \tilde{u}), \tag{26}$$

with initial conditions

$$\tilde{S}|_{\epsilon=0} = S, \ \tilde{t}|_{\epsilon=0} = t, \ \tilde{u}|_{\epsilon=0} = u. \tag{27}$$

Here the variables $\tilde{S}, \tilde{t}$ and $\tilde{u}$ denote values $S, t, u$ after a symmetry transformation. The parameter $\epsilon$ describes a motion along an orbit of the group.

**Theorem 2.** *The action of the symmetry group $G_\Delta(M)$ of (4) with an arbitrary function $\lambda(S)$ is given by (28)–(30). If the function $\lambda(S)$ has the special form (20) then the action of the symmetry group $G_\Delta(M)$ is represented by (31)-(33).*

*Proof.* The solutions to the system of ordinary differential equations (24) with functions $\xi, \tau, \phi$ defined by (17) and initial conditions (27) have the form

$$\tilde{S} = S, \tag{28}$$

$$\tilde{t} = t + a_2\epsilon, \tag{29}$$

$$\tilde{u} = u + a_3 S\epsilon + a_4\epsilon, \ \epsilon \in (-\infty, \infty). \tag{30}$$

The equations (28)–(30) are the global representation of the symmetry group $G_\Delta$ which corresponds to the symmetry algebra defined by (18) in case of an arbitrary function $\lambda(S)$.

If the function $\lambda(S)$ has a special form given by (20) we obtain a richer symmetry group. The solution to the system of equations (24) with the functions $\xi, \tau, \phi$ defined by (21) and initial conditions (27) have the form

$$\tilde{S} = Se^{a_1\epsilon}, \ \epsilon \in (-\infty, \infty), \tag{31}$$

$$\tilde{t} = t + a_2\epsilon,$$

$$\tilde{u} = ue^{a_1(1-k)\epsilon} + \frac{a_3}{a_1 k} S\epsilon e^{a_1\epsilon}(1 - e^{-a_1 k\epsilon})$$

$$+ \frac{a_4}{a_1(1-k)}(e^{a_1(1-k)\epsilon} - 1), \ k \neq 0, \ k \neq 1 \tag{32}$$

$$\tilde{u} = ue^{a_1\epsilon} + a_3 S\epsilon e^{a_1\epsilon} + \frac{a_4}{a_1}(e^{a_1\epsilon} - 1), \ k = 0,$$

$$\tilde{u} = u + \frac{a_3}{a_1} S(e^{a_1\epsilon} - 1) + a_4\epsilon, \ k = 1 \tag{33}$$

where we assume that $a_1 \neq 0$ because the case $a_1 = 0$ coincides with the former case (28)-(30).

We will use the symmetry group $G_\Delta$ to construct invariant solutions to equation (4). To obtain the invariants of the symmetry group $G_\Delta$ we exclude $\epsilon$ from the equations (28)–(30) or in the special case from equations (31)–(33).

In the first case the symmetry group $G_\Delta$ is very poor and we can obtain just the following invariants

$$inv_1 = S, \tag{34}$$
$$inv_2 = u - (a_3 S + a_4)/a_2, \quad a_2 \neq 0.$$

These invariants are useless because they do not lead to any reduction of (4).

In the special case (20) the symmetry group admits two functionally independent invariants of the form

$$inv_1 = \log S + at, \quad a = a_1/a_2, \ a_2 \neq 0 \tag{35}$$
$$inv_2 = u \ S^{(k-1)}. \tag{36}$$

In general the form of invariants is not unique because each function of invariants is an invariant. But it is possible to obtain just two non-trivial functionally independent invariants which we take in the form (35), (36). The invariants can be used as new independent and dependent variables in order to reduce the partial differential equation (4) with the special function $\lambda(S)$ defined by (20) to an ordinary differential equation.

## 3 The special case $\lambda(S) = \omega S^k$

Let us study a special case of equation (4) with $\lambda(S) = \omega S^k, \omega, k \in R$. The equation under investigation is now

$$u_t + \frac{\sigma^2 S^2}{2} \frac{u_{SS}}{(1 - bS^{k+1}u_{SS})^2} = 0 \tag{37}$$

with the constant $b = \rho\omega$. As usual we assume that $\rho \in (0, 1)$. The value of the constant $\omega$ depends on the corresponding option type and in our investigation it can be assumed that $\omega$ is an arbitrary constant, $\omega \neq 0$. The variables $S, t$ are in the intervals

$$S > 0, \quad t \in [0, T], \quad T > 0. \tag{38}$$

*Remark 2.* The case $b = 0$, i.e. $\rho = 0$ or $\omega = 0$ leads to the well known linear Black-Scholes model and we will exclude this case from our investigations.

We will suppose that the denominator in equation (4) (correspondingly (37)) is non equal to zero identically.

Let us study the denominator in the second term of the equation (37). It will be equal to zero if the function $u(S, t)$ satisfies the equation

$$1 - bS^{k+1}u_{SS} = 0. \tag{39}$$

The solution to this equation is a function $u_0(S, t)$

$$
\begin{aligned}
u_0(S, t) &= \frac{1}{bk(k-1)}S^{1-k} + Sc_1(t) + c_2(t), \quad b \neq 0, k \neq 0, 1, \\
u_0(S, t) &= -\frac{1}{b}\log S + Sc_1(t) + c_2(t), \quad b \neq 0, k = 1, \\
u_0(S, t) &= \frac{1}{b}S\log S + Sc_1(t) + c_2(t), \quad b \neq 0, k = 0,
\end{aligned}
\tag{40}
$$

where the functions $c_1(t)$ and $c_2(t)$ are arbitrary functions of the variable $t$.

In the sequel we will assume that the denominator in the second term of the equation (37) is not identically zero, i.e., a solution $u(S, t)$ is not equal to the function $u_0(S, t)$ (40) except in a discrete set of points.

Let us now introduce new invariant variables

$$
\begin{aligned}
z &= \log S + at, \quad a \in R, a \neq 0, \\
v &= u\, S^{(k-1)}.
\end{aligned}
\tag{41}
$$

After this substitution equation (37) will be reduced to an ordinary differential equation

$$av_z + \frac{\sigma^2}{2}\frac{v_{zz} + (1 - 2k)v_z - k(1 - k)v}{(1 - b(v_{zz} + (1 - 2k)v_z - k(1 - k)v))^2} = 0, \quad a, b \neq 0. \tag{42}$$

Elementary solutions to this equation we obtain if we assume that $v = $ const. or $v_z = $ const.. It is easy to prove that there exists the trivial solution $v = 0$ for any $k$ and $v_z = 0$ if $k \neq 0, 1$, and the solutions $v = $ const. $\neq 0$, $v_z = $ const. $\neq 0$ if $k = 0, 1$ only. The condition that the denominator in (42) is non equal to zero, i.e.,

$$(1 - b(v_{zz} + (1 - 2k)v_z - k(1 - k)v))^2 \neq 0 \tag{43}$$

corresponds to equation (39) in new variables $z, v$.

If the function $v(z)$ satisfies the inequality (43) then we can multiply both terms of equation (42) with the denominator of the second term. In equation (42) all coefficients are constants hence we can reduce the order of the equation. We assume that $v, v_z \neq $ const. We now choose as a new independent variable $v$ and introduce as a new dependent variable $x(v) = v_z(z)$. This variable substitution reduces equation (42) to a first order differential equation which is second order polynomial corresponding to the function $x(v)_v$. Under assumption (43) the set of solutions to equation (42) is equivalent to a union of solution sets of the following equations

$$x = 0 , k = 0, 1 \tag{44}$$

$$x_v = -1 + 2\,k - \frac{\sigma^2}{4\,a\,b^2\,x^2} + \frac{1}{b\,x} + \frac{k(1-k)v}{x} - \frac{\sqrt{\sigma^2\,(\sigma^2 - 8\,a\,b\,x)}}{4\,a\,b^2\,x^2},$$

$$x_v = -1 + 2\,k - \frac{\sigma^2}{4\,a\,b^2\,x^2} + \frac{1}{b\,x} + \frac{k(1-k)v}{x} + \frac{\sqrt{\sigma^2\,(\sigma^2 - 8\,a\,b\,x)}}{4\,a\,b^2\,x^2}.$$

Equations (44) are of an autonomous type if the parameter $k$ is equal to $k = 0, 1$ only. We see that these are exactly the cases in which the corresponding Lie-algebra (23) has a three dimensional Abelian sub-algebra with two-dimensional centrum. The case $k = 0$ was studied earlier in [3], [5] and in [4]. In the next section we will study the case $k = 1$.

## 4 The special case $\lambda = \omega S$

If we put $k = 1$ in (20) then equation (42) takes the form

$$v_z + q\frac{v_{zz} - v_z}{(1 - b(v_{zz} - v_z))^2} = 0, \tag{45}$$

where $q = \frac{\sigma^2}{2a}$, $a, b \neq 0$. It is an autonomous equation which possesses a simple structure. We will use this structure and introduce a more simple substitution as described at the end of the previous section to reduce the order of equation.

One family of solutions to this equation is very easy to find. We just suppose that the value $v_z(z)$ is equal to a constant. The equation (45) admits as a solution the value $v_z = (-1 \pm \sqrt{q})/b$ consequently the corresponding solution $u(s, t)$ to (37) with $\lambda = \omega S$ can be represented by the formula

$$u(S, t) = \frac{1}{\rho\omega}\left(-1 \pm \sqrt{q}\right)(\log S + at) + c, \ a > 0, \tag{46}$$

where $c$ is an arbitrary constant.

To find other families of solutions we introduce a new dependent variable

$$y(z) = v_z(z) \tag{47}$$

and assume that the denominator of the equation (45) is not equal to zero, i.e.

$$v(z) \neq -\frac{z}{b} + c_1\,e^z + c_2, \ \text{i.e.} \ y(z) \neq -\frac{1}{b} + c_1\,e^z, \tag{48}$$

where $c_1, c_2$ are arbitrary constants.

We multiply both terms of equation (45) by the denominator of the second term and obtain

$$yy_z^2 - 2\left(y^2 + \frac{1}{b}y - \frac{q}{2b^2}\right)y_z + \left(y^2 + \frac{2}{b}y + \left(\frac{1-q}{b^2}\right)\right)y = 0, \ b \neq 0. \tag{49}$$

We denote the left hand side of this equation by $F(y, y_z)$. The equation (49) can possess exceptional solutions which are the solutions to a system

$$\frac{\partial F(y, y_z)}{\partial y_z} = 0, \quad F(y, y_z) = 0. \tag{50}$$

The first equation in this system defines a discriminant curve which has the form

$$y(z) = \frac{q}{4b}. \tag{51}$$

If this curve is also a solution to the original equation (49) then we obtain an exceptional solution. We obtain an exceptional solution if $q = 4$, i.e. $a = \sigma^2/8$. It has the form

$$y(z) = \frac{1}{b}. \tag{52}$$

This solution belongs to the family of solutions (54) by the specified value of the parameter $q$. In all other cases the equation (49) does not possess any exceptional solutions.

Hence the set of solutions to equation (49) is a union of solution sets of following equations

$$y = 0, \tag{53}$$

$$y = (-1 \pm \sqrt{q})/b, \tag{54}$$

$$y_z = \left(y^2 + \frac{1}{b}y - \frac{q}{2b^2} - \sqrt{\frac{\sigma^2}{2ab^3}\left(\frac{q}{4b} - y\right)}\right)\frac{1}{y}, \quad y \neq 0, \tag{55}$$

$$y_z = \left(y^2 + \frac{1}{b}y - \frac{q}{2b^2} + \sqrt{\frac{q}{b^3}\left(\frac{q}{4b} - y\right)}\right)\frac{1}{y}, \quad y \neq 0, \tag{56}$$

where one of the solutions (54) is an exceptional solution (52) by $q = 4$. We denote the right hand side of equations (55), (56) by $f(y)$. The Lipschitz condition for equations of type $y_z = f(y)$ is satisfied in all points where the derivative $\frac{\partial f}{\partial y}$ exists and is bounded. It is easy to see that this condition will not be satisfied by

$$y = 0, \quad y = \frac{q}{4b}, \quad y = \infty. \tag{57}$$

It means that on the lines (57) the uniqueness of solutions to equations (55), (56) can be lost. We will study in detail the behavior of solutions in the neighborhood of lines (57). For this purpose we look at the equation (49) from another point of view. If we assume now that $z, y, y_z$ are complex variables and denote

$$y(z) = \zeta, \quad y_z(z) = w, \quad \zeta, w \in C, \tag{58}$$

then the equation (49) takes the form

$$F(\zeta, w) = \zeta w^2 - 2\left(\zeta^2 + \frac{1}{b}\zeta - \frac{q}{2b^2}\right)w + \left(\zeta^2 + \frac{2}{b}\zeta + \frac{1-q}{b^2}\right)\zeta = 0, \tag{59}$$

where $b \neq 0$. The equation (59) is an algebraic relation in $C^2$ and defines a plane curve in this space. The polynomial $F(\zeta, w)$ is an irreducible polynomial if at all roots $w_r(z)$ of $F(\zeta, w_r)$ either the partial derivative $F_\zeta(\zeta, w_r)$ or $F_w(\zeta, w_r)$ are non equal to zero. It is easy to prove that the polynomial (59) is irreducible.

We can treat equation (59) as an algebraic relation which defines a Riemann surface $\Gamma : F(\zeta, w) = 0$ of $w = w(\zeta)$ as a compact manifold over the $\zeta$-sphere. The function $w(\zeta)$ is uniquely analytically extended over the Riemann surface $\Gamma$ of two sheets over the $\zeta-$sphere. We find all singular or branch points of $w(\zeta)$ if we study the roots of the first coefficient of the polynomial $F(\zeta, w)$, the common roots of equations

$$F(\zeta, w) = 0, \quad F_w(\zeta, w) = 0, \quad \zeta, w \in C \cup \infty. \tag{60}$$

and the point $\zeta = \infty$. The set of singular or branch points consists of the points

$$\zeta_1 = 0, \quad \zeta_2 = \frac{q}{4b}, \quad \zeta_3 = \infty. \tag{61}$$

As expected we got the same set of points as in real case (57) by the study of the Lipschitz condition but now the behavior of solutions at the points is more visible.

The points $\zeta_2, \zeta_3$ are the branch points at which two sheets of $\Gamma$ are glued on. We remark that

$$w(\zeta_2) = \frac{1}{b}(q - 4) + t\frac{1}{4\sqrt{-bq}} + \cdots, \quad t^2 = \zeta - \frac{q}{4b}, \tag{62}$$

where $t$ is a local parameter in the neighborhood of $\zeta_2$. For the special value of $q = 4$ the value $w(\zeta_2)$ is equal to zero.

At the point $\zeta_3 = \infty$ we have

$$w(\zeta) = \frac{1}{t^2} + \frac{1}{b} + t\sqrt{\frac{-q}{4b^3}}, \quad t^2 = \frac{1}{\zeta}, \quad \zeta \to \infty,$$

where $t$ is a local parameter in the neighborhood of $\zeta_3$. At the point $\zeta_1 = 0$ the function $w(\zeta)$ has the following behavior

$$w(\zeta) \sim -\frac{q}{b^2}\frac{1}{\zeta}, \quad \zeta \to \zeta_1 = 0, \quad \text{on the principal sheet,} \tag{63}$$

$$w(\zeta) \sim (1 - q)\zeta, \quad \zeta \to \zeta_1 = 0, \quad q \neq 1, \quad \text{on the second sheet,} \tag{64}$$

$$w(\zeta) \sim -2b^2\zeta^2, \quad \zeta \to \zeta_1 = 0, \quad q = 1, \quad \text{on the second sheet.} \tag{65}$$

Any solution $w(\zeta)$ to an irreducible algebraic equation (59) is meromorphic on this compact Riemann surface $\Gamma$ of the genus 0 and has a pole of the order one correspondingly (63) over the point $\zeta_1 = 0$ and the pole of the second order over $\zeta_3 = \infty$. It means also that the meromorphic function $w(\zeta)$ cannot be defined on a manifold of less than 2 sheets over the $\zeta$ sphere.

To solve differential equations (55) and (56) from this point of view it is equivalent to integrate on $\Gamma$ a differential of type $\frac{d\zeta}{w(\zeta)}$ and then to solve an Abel's inverse problem of degenerated type

$$\int \frac{d\zeta}{w(\zeta)} = z + \text{const.} \tag{66}$$

The integration can be done very easily because we can introduce a uniformizing parameter on the Riemann surface $\Gamma$ and represent the integral (66) in terms of rational functions merged possibly with logarithmic terms.

To realize this program we introduce a new variable (our uniformizing parameter $p$) in the way

$$\zeta = \frac{q(1 - p^2)}{4b}, \tag{67}$$

$$w = \frac{(1 - p)(q(1 + p)^2 - 4)}{4b(p + 1)}. \tag{68}$$

Then the equations (55) and (56) will take the form

$$2q \int \frac{p(p + 1)dp}{(p - 1)(q(p + 1)^2 - 4)} = z + \text{const}, \tag{69}$$

$$2q \int \frac{p(p - 1)dp}{(p + 1)(q(p - 1)^2 - 4)} = z + \text{const.} \tag{70}$$

The integration procedure of equation (69) gives rise to the following relations

$$2q \log (p - 1) + (q - \sqrt{q} - 2) \log ((p + 1)\sqrt{q} - 2) \tag{71}$$
$$+ (q + \sqrt{q} - 2) \log ((p + 1)\sqrt{q} + 2) = 2(q - 1)z + c, \quad q \neq 1, q > 0,$$

$$\frac{1}{1 - p} + \frac{1}{4} \log \frac{(p + 3)^3}{(p - 1)^5} = z + c, \quad q = 1. \tag{72}$$

$$2\sqrt{(-q)} \arctan ((p + 1)\sqrt{(-q)}/2) - 2q \log (p - 1)$$
$$+ (2 - q) \log (4 - q(p + 1)^2) = 2(1 - q)z + c, \quad q < 0, \tag{73}$$

where $c$ is an arbitrary constant. The equation (70) leads to

$$2q \log (p + 1) + (q + \sqrt{q} - 2) \log ((p - 1)\sqrt{q} - 2) \tag{74}$$
$$+ (q - \sqrt{q} - 2) \log ((p - 1)\sqrt{q} + 2) = 2(q - 1)z + c, \quad q \neq 1, q > 0,$$

$$\frac{1}{p + 1} + \frac{1}{4} \log \frac{(p - 3)^3}{(p + 1)^5} = z + c, \quad q = 1. \tag{75}$$

$$-2\sqrt{(-q)} \arctan ((p - 1)\sqrt{(-q)}/2) - 2q \log (1 + p)$$
$$+ (2 - q) \log (4 - q(p - 1)^2) = 2(1 - q)z + c, \quad q < 0. \tag{76}$$

where $c$ is an arbitrary constant.

The relations (71)-(76) are first order ordinary differential equations because of the substitutions (58) and (47) we have

$$p = \sqrt{1 - \frac{4b}{q}v_z}. \tag{77}$$

All these results can be collected to the following theorem.

**Theorem 3.** *The equation (45) for arbitrary values of the parameters $q, b \neq 0$ can be reduced to the set of first order differential equations which consists of the equations*

$$v_z = 0, \quad v_z = (-1 \pm \sqrt{q})/b \tag{78}$$

*and equations (71)-(76). The complete set of solutions to the equation (45) coincides with the union of solutions to these equations.*

To solve equations (71)-(76) exactly we should first invert these formulas in order to obtain an exact representation $p$ as a function of $z$. If an exact formula for the function $p = p(z)$ is found we can use the substitution (77) to obtain an explicit ordinary differential equation of the type $v_z(z) = f(z)$ or another suitable type and if it possible then to integrate this equation.

But even in the first step we would not be able to do this for an arbitrary value of the parameter $q$. It means we have just implicit representations for the solutions to the equation (45) as solutions to the implicit first order differential equations (71)-(76).

## 4.1 Exact invariant solutions in case of a fixed relation between variables $S$ and $t$

For a special value of the parameter $q$ we can invert the equations (71) and (74). Let us take $q = 4$, i.e., the relation between variables $S, t$ is fixed in the form

$$z = \log S + \frac{\sigma^2}{8}t. \tag{79}$$

In this case the equation (71) takes the form

$$(p - 1)^2(p + 2) = 2\,c\,\exp(3z/2) \tag{80}$$

and correspondingly the equation (74) the form

$$(p + 1)^2(p - 2) = 2\,c\,\exp(3z/2), \tag{81}$$

where $c$ is an arbitrary constant. It is easy to see that the equations (80) and (81) are connected by a transformation

$$p \to -p, \quad c \to -c. \tag{82}$$

This symmetry arises from the symmetry of the underlining Riemann surface $\Gamma$ (59) and corresponds to a change of the sheets on $\Gamma$.

**Theorem 4.** *The second order differential equation*

$$v_z + 4 \frac{v_{zz} - v_z}{(1 - b(v_{zz} - v_z))^2} = 0, \qquad (83)$$

*is exactly integrable for an arbitrary value of the parameter $b$. The complete set of solutions for $b \neq 0$ is given by the union of solutions (87), (89) -(92) and solutions*

$$v(z) = d, \quad v(z) = -\frac{3}{b}z + d, \quad v(z) = \frac{1}{b}z + d, \qquad (84)$$

*where $d$ is an arbitrary constant. The last solution in (84) corresponds to the exceptional solution to equation (49).*
*For $b = 0$ equation (83) is linear and its solutions are given by $v(z) = d_1 + d_2 \exp(3z/4)$, where $d_1, d_2$ are arbitrary constants.*

*Proof.* Because of the symmetry (82) it is sufficient to study either the equations (80) or (81) for $c \in R$ or both these equations for $c > 0$. The value $c = 0$ can be excluded because it complies with the constant value of $p(z)$ and correspondingly constant value of $v_z(z)$, but all such cases are studied before and the solutions are given by (84).

We will study equation (81) in case $c \in R \setminus \{0\}$ and obtain on this way the complete class of exact solutions for equations (80)-(81).

Equation (81) for $c > 0$ has a one real root only. It leads to an ordinary differential equation of the form

$$v_z(z) = -\frac{1}{b}\left(1 + \left(1 + ce^{\frac{3z}{2}} + \sqrt{2ce^{\frac{3z}{2}} + c^2 e^{3z}}\right)^{-\frac{2}{3}}\right.$$

$$\left. + \left(1 + ce^{\frac{3z}{2}} + \sqrt{2ce^{\frac{3z}{2}} + c^2 e^{3z}}\right)^{\frac{2}{3}}\right), \quad c > 0. \qquad (85)$$

Equation (85) can be exactly integrated if we use an Euler substitution and introduce a new independent variable

$$\tau = 2\left(1 + ce^{\frac{3z}{2}} + \sqrt{2ce^{\frac{3z}{2}} + c^2 e^{3z}}\right). \qquad (86)$$

The corresponding solution is given by

$$
v_r(z) = -\frac{1}{b}\left(\left(1 + c\,e^{\frac{3z}{2}} + \sqrt{2\,c\,e^{\frac{3z}{2}} + c^2\,e^{3z}}\right)^{-\frac{2}{3}}\right.
$$

$$
+ \left(1 + c\,e^{\frac{3z}{2}} + \sqrt{2\,c\,e^{\frac{3z}{2}} + c^2\,e^{3z}}\right)^{\frac{2}{3}}
$$

$$
+ 2\log\left(\left(1 + c\,e^{\frac{3z}{2}} + \sqrt{2\,c\,e^{\frac{3z}{2}} + c^2\,e^{3z}}\right)^{-\frac{1}{3}}\right.
$$

$$
\left.\left.+ \left(1 + c\,e^{\frac{3z!}{2}} + \sqrt{2\,c\,e^{\frac{3z}{2}} + c^2\,e^{3z}}\right)^{\frac{1}{3}} - 2\right)\right) + d, \tag{87}
$$

where $d \in R$ is an arbitrary constant.

If in the right hand side of equation (81) the parameter $c$ satisfies the inequality $c < 0$ and the variable $z$ chosen in the region

$$
z \in \left(-\infty, -\frac{4}{3}\ln|c|\right) \tag{88}
$$

then the equation on $p$ possesses maximal three real roots.

These three roots of cubic equation (81) give rise to three differential equations of type $v_z = (1 - p^2(z))/b$. The equations can be exactly solved and we find correspondingly three solutions $v_i(z)$, $i = 1, 2, 3$.

The first solution is given by the expression

$$
v_1(z) = \frac{z}{b} - \frac{2}{b}\cos\left(\frac{2}{3}\arccos\left(1 - |c|\,e^{\frac{3z}{2}}\right)\right) \tag{89}
$$

$$
- \frac{4}{3b}\log\left(1 + 2\cos\left(\frac{1}{3}\arccos\left(1 - |c|\,e^{\frac{3z}{2}}\right)\right)\right)
$$

$$
- \frac{16}{3b}\log\left(\sin\left(\frac{1}{6}\arccos\left(1 - |c|\,e^{\frac{3z}{2}}\right)\right)\right) + d,
$$

where $d \in R$ is an arbitrary constant. The second solution is given by the formula

$$
v_2(z) = \frac{z}{b} - \frac{2}{b}\cos\left(\frac{2}{3}\pi + \frac{2}{3}\arccos\left(-1 + |c|\,e^{\frac{3z}{2}}\right)\right) \tag{90}
$$

$$
- \frac{4}{3b}\log\left(1 + 2\cos\left(\frac{1}{3}\pi + \frac{1}{3}\arccos\left(-1 + |c|\,e^{\frac{3z}{2}}\right)\right)\right)
$$

$$
- \frac{16}{3b}\log\left(\sin\left(\frac{1}{6}\pi + \frac{1}{6}\arccos\left(-1 + |c|\,e^{\frac{3z}{2}}\right)\right)\right) + d,
$$

where $d \in R$ is an arbitrary constant. The first and second solutions are defined up to the point $z = -\frac{4}{3}\ln|c|$ where they coincide (see Fig. 1).

The third solution for $z < -\frac{4}{3}\ln|c|$ is given by the formula

$$v_{3,1}(z) = \frac{z}{b} - \frac{2}{b} \cos\left(\frac{2}{3} \arccos\left(-1 + |c| e^{\frac{3z}{2}}\right)\right) \tag{91}$$

$$- \frac{4}{3b} \log\left(-1 + 2 \cos\left(\frac{1}{3} \arccos\left(-1 + |c| e^{\frac{3z}{2}}\right)\right)\right)$$

$$- \frac{16}{3b} \log\left(\cos\left(\frac{1}{6} \arccos\left(-1 + |c| e^{\frac{3z}{2}}\right)\right)\right) + d,$$

where $d \in R$ is an arbitrary constant. In case $z > -\frac{4}{3} \ln|c|$ the polynomial (81) has a one real root and the corresponding solution can be represented by the formula

$$v_{3,2}(z) = \frac{z}{b} - \frac{2}{b} \cosh\left(\frac{2}{3}\text{arccosh}\left(-1 + |c| e^{\frac{3z}{2}}\right)\right) \tag{92}$$

$$- \frac{16}{3b} \log\left(\cosh\left(\frac{1}{6}\text{arccosh}\left(-1 + |c| e^{\frac{3z}{2}}\right)\right)\right)$$

$$- \frac{4}{3b} \log\left(-1 + 2 \cosh\left(\frac{1}{3}\text{arccosh}\left(-1 + |c| e^{\frac{3z}{2}}\right)\right)\right) + d.$$

The third solution is represented by formulas $v_{3,2}(z)$ and $v_{3,1}(z)$ for different values of the variable $z$.

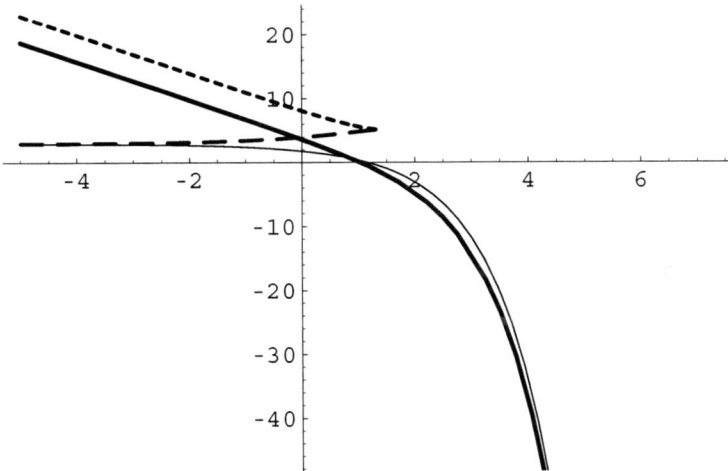

**Fig. 1.** Plot of the solution $v_r(z)$, (87), (thick solid line), $v_1(z)$, (89), (short dashed line), $v_2(z)$, (90), (long dashed line) and the third solution $v_{3,1}(z), v_{3,2}(z)$, (91), (92), which is represented by the thin solid line. The parameters take the values $|c| = 0.5, q = 4, d = 0, b = 1$ and the variable $z \in (-5, 4.5)$.

One of the sets of solutions (87), (89) -(92) for fixed parameters $b, c, d$ is represented in Fig. 1. The solution $v_r(z)$ (87) and the third solution given by both (91) and (92) are defined for any values of $z$. The solutions $v_1(z)$ and $v_2(z)$ cannot be continued after the point $z = -\frac{4}{3}\ln|c|$ where they coincide.

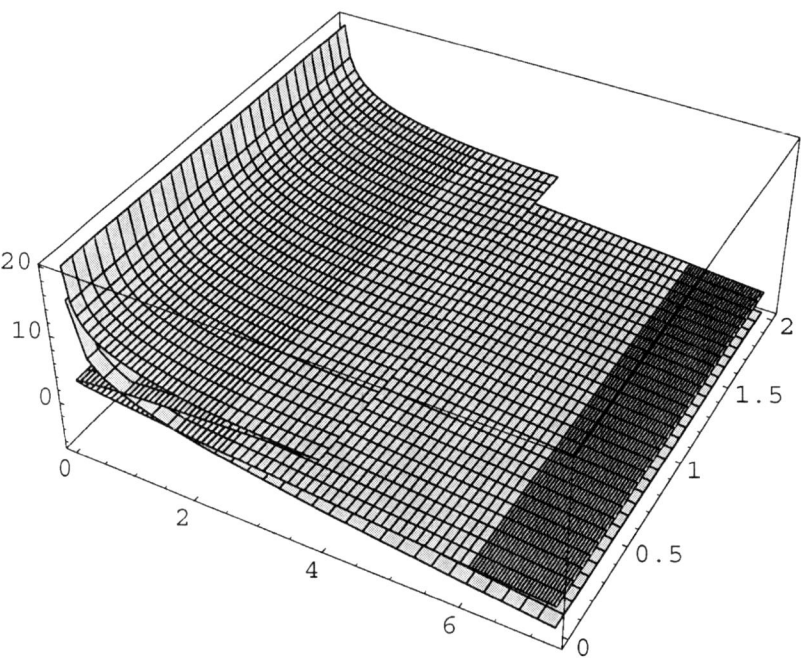

**Fig. 2.** Plot of solutions $u_r(S,t), u_1(S,t), u_2(S,t), u_{3,1}(S,t), u_{3,2}(S,t)$ for the parameters $|c| = 0.25, q = 4, b = 1.0, d = 0$. The variables $S, t$ lie in intervals $S \in (0,9)$ and $t \in [0,2.0]$. All invariant solutions change slowly in $t$-direction.

If we keep in mind that $z = \log S + \frac{\sigma^2}{8}t$ and $v(z) = u(S,t)$ we can represent exact invariant solutions to equation (37). The solution (87) gives rise to an invariant solution $u_r(S,t)$ in the form

$$
u_r(S,t) = -\frac{1}{\omega\rho}\left(1 + cS^{\frac{3}{2}}e^{\frac{3\sigma^2}{16}t} + \sqrt{2cS^{\frac{3}{2}}e^{\frac{3\sigma^2}{16}t} + c^2S^3e^{\frac{3\sigma^2}{8}t}}\right)^{-\frac{2}{3}}
$$
$$
-\frac{1}{\omega\rho}\left(1 + cS^{\frac{3}{2}}e^{\frac{3\sigma^2}{16}t} + \sqrt{2cS^{\frac{3}{2}}e^{\frac{3\sigma^2}{16}t} + c^2S^3e^{\frac{3\sigma^2}{8}t}}\right)^{\frac{2}{3}} \tag{93}
$$
$$
-\frac{2}{\omega\rho}\log\left(\left(1 + cS^{\frac{3}{2}}e^{\frac{3\sigma^2}{16}t} + \sqrt{2cS^{\frac{3}{2}}e^{\frac{3\sigma^2}{16}t} + c^2S^3e^{\frac{3\sigma^2}{8}t}}\right)^{-\frac{1}{3}}\right.
$$
$$
\left. +\left(1 + cS^{\frac{3}{2}}e^{\frac{3\sigma^2}{16}t} + \sqrt{2cS^{\frac{3}{2}}e^{\frac{3\sigma^2}{16}t} + c^2S^3e^{\frac{3\sigma^2}{8}t}}\right)^{\frac{1}{3}} - 2\right) + d
$$

where $d \in R$, $c > 0$.

In case $c < 0$ we can obtain correspondingly three solutions if

$$0 < S \le |c|^{-\frac{4}{3}} \exp\left(-\frac{\sigma^2}{8} t\right). \tag{94}$$

The first solution is represented by

$$
\begin{aligned}
u_1(S,t) &= \frac{1}{\omega\rho}\left(\log S + \frac{\sigma^2}{8} t\right) - \frac{2}{\omega\rho}\cos\left(\frac{2}{3}\arccos\left(1 - |c|\, S^{\frac{3}{2}} e^{\frac{3\sigma^2}{16} t}\right)\right) \\
&\quad - \frac{4}{3\omega\rho}\log\left(1 + 2\cos\left(\frac{1}{3}\arccos\left(1 - |c|\, S^{\frac{3}{2}} e^{\frac{3\sigma^2}{16} t}\right)\right)\right) \tag{95} \\
&\quad - \frac{16}{3\omega\rho}\log\left(\sin\left(\frac{1}{6}\arccos\left(1 - |c|\, S^{\frac{3}{2}} e^{\frac{3\sigma^2}{16} t}\right)\right)\right) + d,
\end{aligned}
$$

where $d \in R$, $c < 0$. The second solution is given by the formula

$$
\begin{aligned}
u_2(S,t) &= \frac{1}{\omega\rho}\left(\log S + \frac{\sigma^2}{8} t\right) - \frac{2}{\omega\rho}\cos\left(\frac{2}{3}\pi + \frac{2}{3}\arccos\left(-1 + |c|\, S^{\frac{3}{2}} e^{\frac{3\sigma^2}{16} t}\right)\right) \\
&\quad - \frac{4}{3\omega\rho}\log\left(1 + 2\cos\left(\frac{1}{3}\pi + \frac{1}{3}\arccos\left(-1 + |c|\, S^{\frac{3}{2}} e^{\frac{3\sigma^2}{16} t}\right)\right)\right) \tag{96} \\
&\quad - \frac{16}{3\omega\rho}\log\left(\sin\left(\frac{1}{6}\pi + \frac{1}{6}\arccos\left(-1 + |c|\, S^{\frac{3}{2}} e^{\frac{3\sigma^2}{16} t}\right)\right)\right) + d,
\end{aligned}
$$

where $d \in R$, $c < 0$. The first and second solutions are defined for the variables under conditions (94). They coincide along the curve

$$S = |c|^{-4/3} \exp\left(-\frac{\sigma^2}{8} t\right)$$

and cannot be continued further.

The third solution is defined by

$$
\begin{aligned}
u_{3,1}(S,t) &= \frac{1}{\omega\rho}\left(\log S + \frac{\sigma^2}{8} t\right) - \frac{2}{\omega\rho}\cos\left(\frac{2}{3}\arccos\left(-1 + |c|\, S^{\frac{3}{2}} e^{\frac{3\sigma^2}{16} t}\right)\right) \\
&\quad - \frac{4}{3\omega\rho}\log\left(-1 + 2\cos\left(\frac{1}{3}\arccos\left(-1 + |c|\, S^{\frac{3}{2}} e^{\frac{3\sigma^2}{16} t}\right)\right)\right) \tag{97} \\
&\quad - \frac{16}{3\omega\rho}\log\left(\cos\left(\frac{1}{6}\arccos\left(-1 + |c|\, S^{\frac{3}{2}} e^{\frac{3\sigma^2}{16} t}\right)\right)\right) + d,
\end{aligned}
$$

where $d \in R$ and $S, t$ satisfied the condition (94).

In case $\log S + \frac{\sigma^2}{8} t > -\frac{4}{3}\ln|c|$ the third solution can be represented by the formula

$$
\begin{aligned}
u_{3,2}(S,t) &= \frac{1}{\omega\rho}\left(\log S + \frac{\sigma^2}{8} t\right) - \frac{2}{\omega\rho}\cosh\left(\frac{2}{3}\operatorname{arccosh}\left(-1 + |c|\, S^{\frac{3}{2}} e^{\frac{3\sigma^2}{16} t}\right)\right) \\
&\quad - \frac{16}{3\,\omega\rho}\log\left(\cosh\left(\frac{1}{6}\operatorname{arccosh}\left(-1 + |c|\, S^{\frac{3}{2}} e^{\frac{3\sigma^2}{16} t}\right)\right)\right) \tag{98} \\
&\quad - \frac{4}{3\,\omega\rho}\log\left(-1 + 2\cosh\left(\frac{1}{3}\operatorname{arccosh}\left(-1 + |c|\, S^{\frac{3}{2}} e^{\frac{3\sigma^2}{16} t}\right)\right)\right) + d.
\end{aligned}
$$

The solution $u_r(S,t)$ (93) and the third solution given by $u_{3,1}$, $u_{3,2}$ (97),(98) are defined for all values of variables $t$ and $S > 0$. They have a common intersection curve of type $S = $ const. $\exp(-\frac{\sigma^2}{8}t)$. The typical behavior of all these invariant solutions is represented in Fig. 2.

It should be noted that because of the symmetry properties (see Theorem 2) any solution remains a solution if we add to each solution a linear function of $S$,

$$u(S,t) \rightarrow u(S,t) + d_1 S + d_2, \tag{99}$$

with arbitrary constants $d_1$, $d_2$.

It means we have two additional constants to model boundary and terminal conditions.

We first study the non-trivial solutions, i.e., $u(S,t) \neq d_1 S + d_2$.

Previous results can be summed up in the following theorem describing the set of non-trivial invariant solutions to equation (4).

**Theorem 5.** *1. The equation (4) possesses non-trivial invariant solutions for the only special form of the function $\lambda(S)$ given by (20).*

*2. In case (20) the invariant solutions to equation (4) are defined by ordinary differential equations (44). In special cases $k = 0, 1$ equations (44) are of an autonomous type.*

*3. If $\lambda(S) = \omega S$, i.e. $k = 1$, then the invariant solutions to equation (4) can be defined by the set of first order ordinary differential equations (71)–(76) and equation (78).*

*If additionally the parameter $q = 4$, or equivalent in the first invariant (35) we chose $a = \sigma^2/8$ then the complete set of invariant solutions (4) can be found exactly. This set of invariant solutions is given by formulas (93)–(98) and by solutions*

$$u(S,t) = d, \quad u(S,t) = -3/b \, (\log S + \sigma^2 t/8), \quad u(S,t) = 1/b \, (\log S + \sigma^2 t/8),$$

*where $d$ is an arbitrary constant. This set of invariant solutions is unique up to the transformations of the symmetry group $G_\Delta$ given by theorem 2*

## 5 Properties of invariant solutions

All solutions (93)-(98) have the form
$u(S,t) = w(S,t)/(\omega\rho)$, where $w(S,t)$ is a smooth function of $S,t$. It means that the function $w(S,t)$ solves the equation (37) with $b = \omega\rho = 1$, i.e.

$$w_t + \frac{\sigma^2 S^2}{2} \frac{w_{SS}}{(1 - S^2 w_{SS})^2} = 0. \tag{100}$$

In other words, if we find any solution to the above equation (100) for any fixed boundary and terminal conditions, we can immediately obtain the corresponding solution to the equation (37) if we just divide the function $w(S,t)$ by $b = \omega\rho$ (the solution $u(S,t)$ satisfies the boundary and terminal conditions which we obtain if just divide the corresponding conditions on the function $w(S,t)$ by $b$). Therefor any $\rho$-dependence of a solution to (37) is trivial. It means as well that if the terminal conditions are fixed $u(S,T) = h(S)$ then the value $u(S,t)$ will increase if the value of the parameter $\rho$ increases. This dependence of hedge costs on the position of the large trader on market is very natural.

If the parameter $\rho \to 0$ then equation (4) and correspondingly equation (37) will be reduced to the linear Black-Scholes equation but solutions (93)-(98) which we obtained here will be completely blown up by $\rho \to 0$ because of the factor $1/b = 1/(\omega\rho)$ in the formulas (93)-(98). It means that the solutions $u(S,t)$ (93), $u_1(S,t)$ (95), $u_2(S,t)$ (96), $u_{3,1}(S,t)$ (97), $u_{3,2}(S,t)$ (98), have no one counterpart in a linear case.

This phenomena is rather typical for nonlinear partial differential equations with singular perturbations and was described as well in [3], [5], [4] for the invariant solutions to equation (37) with $k = 0$.

The families of exact solutions reflect the nonlinearity of this equation in a essential way. If we take a numerical method which was developed for the linear Black-Scholes equation or other types of parabolic equations and will test it for a new type of a nonlinear equation we should take, if possible, the solutions which reflect this nonlinearity in the most complete way. The existence of non-trivial explicit solutions allows to test different numerical methods usually used to calculate hedge-costs of derivatives (see [3]).

We obtain a typical terminal payoff function for the solutions (93)-(98) if we just fix $t = T$. All these payoffs are smooth functions. Other sides a typical payoff of derivatives similar to combination of calls and puts is a continuous piece-wise linear function. The smooth payoffs are more convenient for numerical calculations and usually one replaced the standard payoffs by the corresponding solution to the Black-Scholes model for a very small time interval. If also in that case the numerical method does not work properly, there is no hope that it works better in a worse case and the fact that the payoffs are smooth functions is immaterial for these purposes.

In order to illustrate how we can model some typical payoffs we investigate asymptotic properties of solutions (93)-(98) as $S \to 0$ and as $S \to \infty$.

Using the exact formulas for solutions we retain the first two terms and obtain as $S \to 0$

$$u_1(S,t) \sim$$
$$-\frac{1}{b}\left(2 + \frac{4}{3}\log\left(|c|^2 2^{-2} 3^{-3}\right) + \frac{3}{8}\sigma^2 t + 3\log S + \mathcal{O}(S^{3/2})\right), \quad (101)$$

$$u_2(S,t) \sim$$
$$\frac{1}{b}\left(1 + \frac{1}{3}\log\left(2^8 3^{-6}|c|^{-2}\right) + \frac{2^3\sqrt{2|c|}}{3^{4/3}}e^{\frac{3}{32}\sigma^2 t}S^{3/4} + \mathcal{O}(S^{3/2})\right), \quad (102)$$

$$u_{3,1}(S,t) \sim$$
$$\frac{1}{b}\left(1 + \frac{1}{3}\log\left(2^8 3^{-6}|c|^{-2}\right) - \frac{2^3\sqrt{2|c|}}{3^{4/3}}e^{\frac{3}{32}\sigma^2 t}S^{3/4} + \mathcal{O}(S^{3/2})\right), \quad (103)$$

$$u_r(S,t) \sim -\frac{1}{b}\left(2 + 2\log\left(2\ 3^{-2}\ c\right) + \frac{3}{8}\sigma^2 t + 3\log S + \mathcal{O}(S^{5/4})\right). \quad (104)$$

Plot of solutions (93)-(98) by $\sigma = 0.4$, $b = 1$, $|c| = 0.5$, $d_1 = 0$, $d_2 = 0$.

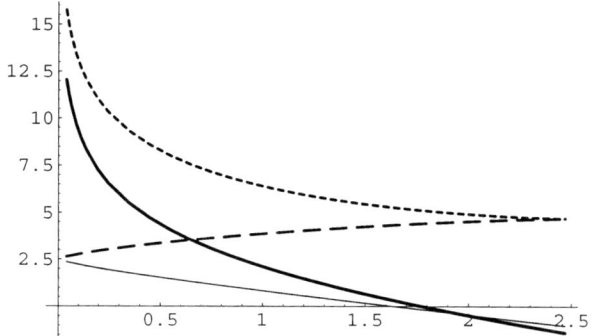

**Fig. 3.**  Behavior of solutions $u_r(z)$, (93), (thick solid line), $u_1(z)$, (95), (short dashed line), $u_2(z)$, (96), (long dashed line) and $u_{3,1}(z)$, (97) (thin solid line) in the neighborhood of $S \sim 0$, for $t = 0$, where $S$ in (94).

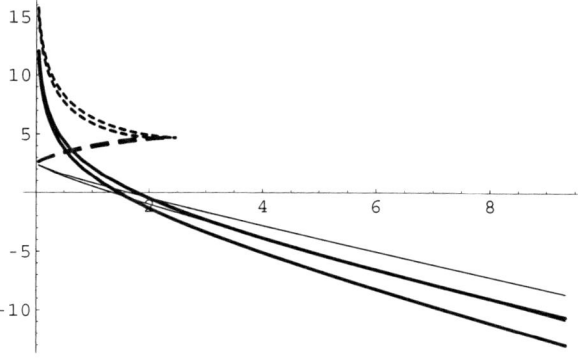

**Fig. 4.**  The same solutions for $t = 0.$ and $t = 10.$ for $S \in (0.04, 9]$, $u_r(z)$, (93), (thick solid line), $u_1(z)$, (95), (short dashed line), $u_2(z)$, (96), (long dashed line) and $u_{3,1}(z), u_{3,2}(z)$, (97), (98) (thin solid line).

In Fig. 5 we present solutions (93)-(98) in a case where both additional constants $d_1 = d_2$ in (99) are equal to zero, i.e. without linear background.

If $S$ is large enough we have just two solutions. The asymptotic behavior both solutions $u_r(S,t)$,(93), and $u_{3,2}$,(98), coincides in the main terms as $S \to \infty$ and is given by formula

$$u_r(S,t),\ u_{3,2}(S,t) \sim -\frac{1}{b}\left((2\,|c|)^{2/3}e^{\frac{3\sigma^2}{8}t}S + \log S + \mathcal{O}(1)\right),\quad S \to \infty. \quad (105)$$

The main term in formulas (101)-(105) depends on the time and on the constant $c$.

The families of exact solutions (93)-(98) have the following parameters $\omega,\ c,\ d_1,\ d_2$ which can be used to match suitable boundary and terminal conditions to a desired accuracy. The formulas (101)-(105) can be useful for these purposes. In Fig. 5 we represent as an example a long strip payoff and the solution $u_r(S,t)$, (93), which partly matches this payoff.

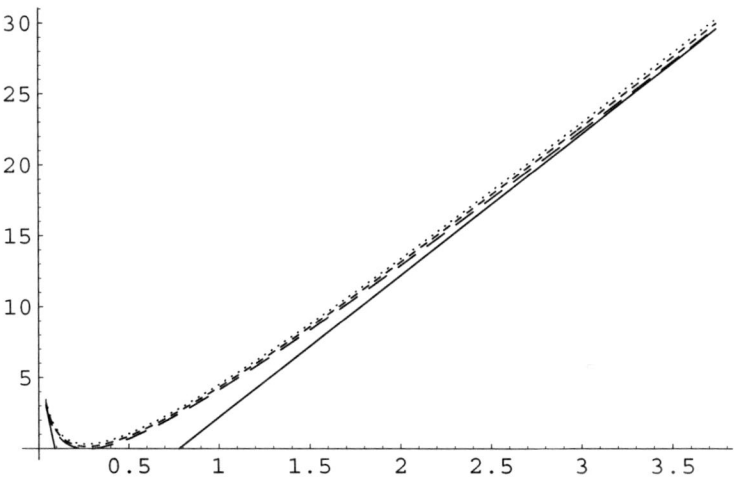

**Fig. 5.** Plot of the solution $u_r(S,t)$ for the parameters $|c| = 0.5$, $b = 1.0$, $\sigma = 0.3$, $d1 = 11.5$, $d2 = -9.0$. The variables $S,t$ lie in intervals $S \in (0.04, 3.7)$ and $t = 0.$, (doted line), $t = 5.$, (short dashed line) and $t = 10.$, (long dashed line). Payoff for a long strip with 60 Puts and 10 Calls with exercise price 0.2 marked by thin solid line.

## Acknowledgments

The author is grateful to Albert N. Shiryaev for the interesting and fruitful discussions.

# References

1. Bordag LA (2005) New types of solutions for nonlinear Black-Scholes model. In: Abstracts of the twelfth General Meeting of EWM, Volgograd, September 18-24, 2005

2. Bordag LA (2006) Symmetry reductions of a nonlinear model of financial derivatives. arXiv:math.AP/0604207, 09.04.2006

3. Bordag LA, Chmakova AY (2007) Explicit solutions for a nonlinear model of financial derivatives. International Journal of Theoretical and Applied Finance 10:1–21

4. Bordag LA, Frey R (2008) Nonlinear option pricing models for illiquid markets: scaling properties and explicit solutions. In: Ehrhardt M (ed) Nonlinear models in mathematical finance: Research trends in option pricing. Nova Science Publishers Inc.

5. Chmakova AY (2005) Symmetriereduktionen und explicite Lösungen für ein nichtlineares Modell eines Preisbildungsprozesses in illiquiden Märkten. PhD thesis, BTU Cottbus.

6. Frey R (1998) Perfect option replication for a large trader. Finance Stochastic 2:115–148

7. Frey R (2000) Market illiquidity as a source of model risk in dynamic hedging. In: Gibson R (ed) Model Risk, Risk Publications, London

8. Frey R, Patie P (2002) Risk management for derivatives in illiquid markets: a simulation study. In: Sandmann K, Schönbucher P (eds) Advances in Finance and Stochastics, Springer, Berlin

9. Frey R, Stremme A (1997) Market volatility and feedback effect from dynamic hedging. Mathematical Finance, 7(4):351–374

10. Gaeta G (1994) Nonlinear Symmetries and Nonlinear Equations, volume 299 of Mathematics and its Applications. Kluwer Academic Publishers, Dordrecht Boston London

11. Ibragimov NH (1999) Elementary Lie Group Analysis and Ordinary Differential Equations. John Wiley&Sons, Chischester New York Weinheim Brisbane Singapore Toronto etc.

12. Lie S (1912) Vorlesungen über Differentialgleichungen mit bekannten infinitesimalen Transformationen. Teubner, Leipzig

13. Olver PJ (1986) Application of Lie groups to differential equations. Springer-Verlag, New York

14. Ovsiannikov LV (1982) Group Analysis of Differential Equations. Academic Press, New York

15. Schonbucher P, Wilmott P (2000) The feedback effect of hedging in illiquid markets. SIAM J. Appl. Math. 61:232–272

16. Shreve SE (2004) Stochastic Calculus for Finance II: Continuous–Time Models. Springer Finance. Springer-Verlag, New York

17. Sircar K, Papanicolaou G (1998) General Black-Scholes models accounting for increased market volatility from hedging strategies. Appl. Math. Finance 5:45–82

18. Stephani H (1994) Differential Gleichungen: Symmetrien und Lösungsmethoden Spektrum Akademischer Verlag GmbH, Heidelberg

# Predicting the Time of the Ultimate Maximum for Brownian Motion with Drift

Jacques du Toit and Goran Peskir

School of Mathematics, The University of Manchester,
Oxford Road, Manchester M13 9PL, United Kingdom.
Jacques.Du-Toit@postgrad.manchester.ac.uk, goran@maths.man.ac.uk

**Summary.** Given a standard Brownian motion $B^\mu = (B_t^\mu)_{0 \le t \le 1}$ with drift $\mu \in \mathbb{R}$, letting $S_t^\mu = \max_{0 \le s \le t} B_s^\mu$ for $t \in [0,1]$, and denoting by $\theta$ the time at which $S_1^\mu$ is attained, we consider the optimal prediction problem

$$V_* = \inf_{0 \le \tau \le 1} \mathsf{E}|\theta - \tau|$$

where the infimum is taken over all stopping times $\tau$ of $B^\mu$. Reducing the optimal prediction problem to a parabolic free-boundary problem and making use of local time-space calculus techniques, we show that the following stopping time is optimal:

$$\tau_* = \inf \{ 0 \le t \le 1 \,|\, S_t^\mu - B_t^\mu \ge b(t) \}$$

where $b : [0,1] \to \mathbb{R}$ is a continuous decreasing function with $b(1) = 0$ that is characterized as the unique solution to a nonlinear Volterra integral equation. This also yields an explicit formula for $V_*$ in terms of $b$. If $\mu = 0$ then there is a closed form expression for $b$. This problem was solved in [14] and [4] using the method of time change. The latter method cannot be extended to the case when $\mu \ne 0$ and the present paper settles the remaining cases using a different approach. It is also shown that the shape of the optimal stopping set remains preserved for all Lévy processes.

**Key words:** Brownian motion, optimal prediction, optimal stopping, ultimate-maximum time, parabolic free-boundary problem, smooth fit, normal reflection, local time-space calculus, curved boundary, nonlinear Volterra integral equation, Markov process, diffusion process, Lévy process.

*Mathematics Subject Classification 2000.* Primary 60G40, 62M20, 35R35. Secondary 60J65, 60G51, 45G10.

## 1 Introduction

Stopping a stochastic process as close as possible to its ultimate maximum is of great practical and theoretical interest. It has numerous applications in the fields of engineering, physics, finance and medicine, for example determining the best time to sell an asset or the optimal time to administer a drug. In recent years the area has attracted considerable interest (and has yielded some counter-intuitive results), and the problems have collectively become known as *optimal prediction problems* (within optimal stopping).

In particular, a number of different variations on the following prediction problem have been studied: let $B = (B_t)_{0 \leq t \leq 1}$ be a standard Brownian motion started at zero, set $S_t = \max_{0 \leq s \leq t} B_s$ for $t \in [0,1]$, and consider the optimal prediction problem

$$\inf_{0 \leq \tau \leq 1} \mathsf{E}(S_1 - B_\tau)^2 \tag{1}$$

where the infimum is taken over all stopping times $\tau$ of $B$. This problem was solved in [4] where the optimal stopping time was found to be

$$\tau_* = \inf \left\{ 0 \leq t \leq 1 \,|\, S_t - B_t \geq z_* \sqrt{1-t} \right\} \tag{2}$$

with $z_* > 0$ being a specified constant. The result was extended in [8] where two different formulations were considered: firstly the problem (1) above for $p > 1$ in place of 2, and secondly a probability formulation maximizing $\mathsf{P}(S_1 - B_\tau \leq \varepsilon)$ for $\varepsilon > 0$. Both these were solved explicitly: in the first case, the optimal stopping time was shown to be identical to (2) except that the value of $z_*$ was now dependent on $p$, and in the second case the optimal stopping time was found to be $\tau_* = \inf \{ t_* \leq t \leq 1 \,|\, S_t - B_t = \varepsilon \}$ where $t_* \in [0,1]$ is a specified constant.

Setting $B_t^\mu = B_t + \mu t$ and $S_t^\mu = \max_{0 \leq s \leq t} B_s^\mu$ for $t \in [0,1]$ and $\mu \in \mathbb{R}$, one can formulate the analogue of the problem (1) for Brownian motion with drift, namely

$$\inf_{0 \leq \tau \leq 1} \mathsf{E}(S_1^\mu - B_\tau^\mu)^2 \tag{3}$$

where the infimum is taken over all stopping times $\tau$ of $B^\mu$. This problem was solved in [2] where it was revealed that (3) is fundamentally more complicated than (1) due to its highly nonlinear time dependence. The optimal stopping time was found to be $\tau_* = \inf \{ 0 \leq t \leq 1 \,|\, b_1(t) \leq S_t^\mu - B_t^\mu \leq b_2(t) \}$ where $b_1$ and $b_2$ are two specified functions of time giving a more complex shape to the optimal stopping set (which moreover appears to be counter-intuitive when $\mu > 0$).

The variations on the problem (1) summarized above may all be termed *space domain* problems, since the measures of error are all based on a distance from $B_\tau$ to $S_1$. In each case they lead us to stop as 'close' as possible to the maximal value of the Brownian motion. However, the question can also be asked in the *time domain*: letting $\theta$ denote the time at which $B$ attains its maximum $S_1$, one can consider

$$\inf_{0 \leq \tau \leq 1} \mathsf{E}|\theta - \tau| \tag{4}$$

where the infimum is taken over all stopping times $\tau$ of $B$. This problem was first considered in [12] and then further examined in [14] where the following identity was derived

$$E(B_\theta - B_\tau)^2 = E|\theta - \tau| + \tfrac{1}{2} \tag{5}$$

for any stopping time $\tau$ of $B$ satisfying $0 \leq \tau \leq 1$. Recalling that $B_\theta = S_1$ it follows that the time domain problem (4) is in fact equivalent to the space domain problem (1). Hence stopping optimally in time is the same as stopping optimally in space (when distance is measured in mean square). This fact, although intuitively appealing, is mathematically quite remarkable.

It is interesting to note that (with the exception of the probability formulation) all the space domain problems above have trivial solutions when distance is measured in mean. Indeed, in (1) (with 1 in place of 2) any stopping time is optimal, while in (3) one either waits until time 1 or stops immediately (depending on whether $\mu > 0$ or $\mu < 0$ respectively). The error has to be distorted to be seen by the expectation operator, and this introduces a parameter dependence into these problems. While the mean square distance may seem a natural setting (due to its close links with the conditional expectation), there is no reason a priori to prefer one penalty measure over any other. The problems are therefore all based on *parameterized measures of error*, and the solutions are similarly parameter dependent.

The situation becomes even more acute when one extends these space domain problems to other stochastic processes, since there are many processes for which the higher order moments simply do not exist. Examples of these include stable Lévy processes of index $\alpha \in (0, 2)$, for which (1) would only make sense for powers $p$ strictly smaller than $\alpha$. This leads to further loss of transparency in the problem formulation. By contrast, the time domain formulation is free from these difficulties as it deals with bounded random variables. One may therefore use any measure of error, including mean itself, and it is interesting to note that even in this case the problem (4) above yields a non-trivial solution.

Motivated by these facts, our aim in this paper will be to study the analogue of the problem (4) for Brownian motion with drift, namely

$$\inf_{0 \leq \tau \leq 1} E|\theta^\mu - \tau| \tag{6}$$

where $\theta^\mu$ is the time at which $B^\mu = (B_t^\mu)_{0 \leq t \leq 1}$ attains its maximal value $S_1^\mu$, and the infimum is taken over all stopping times $\tau$ of $B^\mu$. This problem is interesting not only because it is a *parameter free* measure of optimality, but also because it is unclear what form the solution will take: whether it will be similar to that of (1) or (3), or whether it will be something else entirely. There are also several applications where stopping close to the maximal *value* is less important than detecting the *time* at which this maximum occurred as accurately as possible.

Our main result (Theorem 1) states that the optimal stopping time in (6) is given by

$$\tau_* = \inf \left\{ 0 \leq t \leq 1 \,|\, S_t^\mu - B_t^\mu > b(t) \right\} \tag{7}$$

where $b : [0, 1] \to \mathbb{R}$ is a continuous decreasing function with $b(1) = 0$ that is characterized as the unique solution to a nonlinear Volterra integral equation.

The shape of the optimal stopping set is therefore quite different from that of the problem (3), and more closely resembles the shape of the optimal stopping set in the problem (1) above. This result is somewhat surprising and it is not clear how to explain it through simple intuition.

However, by far the most interesting and unexpected fact to emerge from the proof is that this problem considered for any *Lévy process* will yield a similar solution. That is, for *any* Lévy process $X$, the problem (6) of finding the closest stopping time $\tau$ to the time $\theta$ at which $X$ attains its supremum (approximately), will have a solution

$$\tau_* = \inf\left\{0 \leq t \leq 1 \,|\, S_t - X_t \geq c(t)\right\} \tag{8}$$

where $S_t = \sup_{0 \leq s \leq t} X_s$ and $c : [0,1] \to \overline{\mathbb{R}}$ is a decreasing function with $c(1) = 0$. This result is remarkable indeed considering the breadth and depth of different types of Lévy processes, some of which have extremely irregular behavior and are analytically quite unpleasant. In fact, an analogous result holds for a certain broader class of Markov processes as well, although the state space in this case is three-dimensional (time-space-supremum). Our aim in this paper will not be to derive (8) in all generality, but rather to focus on Brownian motion with drift where the exposition is simple and clear. The facts indicating the general results will be highlighted as we progress (cf. Lemma 1 and Lemma 2).

## 2 Reduction to standard form

As it stands, the optimal prediction problem (6) falls outside the scope of standard optimal stopping theory (see e.g. [11]). This is because the gain process $(|\theta - t|)_{0 \leq t \leq 1}$ is not adapted to the filtration generated by $B^\mu$. The first step in solving (6) aims therefore to reduce it to a standard optimal stopping problem. It turns out that this reduction can be done not only for the process $B^\mu$ but for any Markov process.

1. To see this, let $X = (X_t)_{t \geq 0}$ be a right-continuous Markov process (with left limits) defined on a filtered probability space $(\Omega, \mathcal{F}, (\mathcal{F}_t)_{t \geq 0}, \mathsf{P}_x)$ and taking values in $\mathbb{R}$. Here $\mathsf{P}_x$ denotes the measure under which $X$ starts at $x \in \mathbb{R}$. Define the stochastic process $S = (S_t)_{t \geq 0}$ by $S_t = \sup_{0 \leq s \leq t} X_s$ and the random variable $\theta$ by

$$\theta = \inf\left\{0 \leq t \leq 1 \,|\, S_t = S_1\right\} \tag{9}$$

so that $\theta$ denotes the first time that $X$ "attains" its ultimate supremum over the interval $[0,1]$. We use "attains" rather loosely here, since if $X$ is discontinuous the supremum might not actually be attained, but will be approached arbitrarily closely. Indeed, since $X$ is right-continuous it follows that $S$ is right-continuous, and hence $S_\theta = S_1$ so that $S$ attains its ultimate supremum over the interval $[0,1]$ at $\theta$. This implies that either $X_\theta = S_1$ or $X_{\theta-} = S_1$ depending on whether $X_\theta \geq X_{\theta-}$ or $X_\theta < X_{\theta-}$ respectively.

The reduction to standard form may now be described as follows (stopping times below refer to stopping times with respect to the filtration $(\mathcal{F}_t)_{t \geq 0}$).

**Lemma 1.** *Let $X$, $S$ and $\theta$ be as above. Define the function $F$ by*

$$F(t, x, s) = \mathsf{P}_x(S_t \leq s) \tag{10}$$

*for $t \in [0,1]$ and $x \leq s$ in $\mathbb{R}$. Then the following identity holds:*

$$\mathsf{E}_x|\theta - \tau| = \mathsf{E}_x\left(\int_0^\tau \left(2F(1-t, X_t, S_t) - 1\right) dt\right) + \mathsf{E}_x(\theta) \tag{11}$$

*for any stopping time $\tau$ satisfying $0 \leq \tau \leq 1$.*

*Proof.* Recalling the argument from the proof of Lemma 1 in [14] (cf. [12], [13]) one has

$$|\theta - \tau| = (\tau - \theta)^+ + (\tau - \theta)^- = (\tau - \theta)^+ + \theta - \theta \wedge \tau \tag{12}$$

$$= \int_0^\tau I(\theta \leq t)\, dt + \theta - \int_0^\tau I(\theta > t)\, dt$$

$$= \theta + \int_0^\tau \left(2I(\theta \leq t) - 1\right) dt.$$

Examining the integral on the right hand side, one sees by Fubini's theorem that

$$\mathsf{E}_x\left(\int_0^\tau I(\theta \leq t)\, dt\right) = \mathsf{E}_x\left(\int_0^\infty I(\theta \leq t)\, I(t < \tau)\, dt\right) \tag{13}$$

$$= \int_0^\infty \mathsf{E}_x\left(I(t < \tau)\, \mathsf{E}_x\left(I(\theta \leq t)\,\big|\,\mathcal{F}_t\right)\right) dt$$

$$= \mathsf{E}_x\left(\int_0^\tau \mathsf{P}_x(\theta \leq t\,|\,\mathcal{F}_t)\, dt\right).$$

We can now use the Markov structure of $X$ to evaluate the conditional probability. For this note that if $S_t = \sup_{0 \leq s \leq t} X_s$, then $\sup_{t \leq s \leq 1} X_s = S_{1-t} \circ \theta_t$ where $\theta_t$ is the shift operator at time $t$. We therefore see that

$$\mathsf{P}_x(\theta \leq t\,|\,\mathcal{F}_t) = \mathsf{P}_x\left(\sup_{0 \leq s \leq t} X_s \geq \sup_{t \leq s \leq 1} X_s\,\Big|\,\mathcal{F}_t\right) \tag{14}$$

$$= \mathsf{P}_x\left(S_{1-t} \circ \theta_t \leq s\,|\,\mathcal{F}_t\right)\big|_{s=S_t}$$

$$= \mathsf{P}_{X_t}\left(S_{1-t} \leq s\right)\big|_{s=S_t} = F(1-t, X_t, S_t)$$

where $F$ is the function defined in (10) above. Inserting these identities back into (12) after taking $\mathsf{E}_x$ on both sides, we conclude the proof.   □

Lemma 1 reveals the rich structure of the optimal prediction problem (6). Two key facts are to be noted. Firstly, for any Markov process $X$ the problem is inherently three dimensional and has to be considered in the time-space-supremum domain occupied by the stochastic process $(t, X_t, S_t)_{t \geq 0}$. Secondly, for any two values $x \leq s$ fixed, the map $t \mapsto 2F(1-t, x, s) - 1$ is increasing. This fact is important since we are considering a minimization problem so that the passage of time incurs a hefty penalty and always forces us to stop sooner rather than later. This property will be further explored in Section 4 below.

2. If $X$ is a Lévy process then the problem reduces even further.

**Lemma 2.** *Let $X$, $S$ and $\theta$ be as above, and let us assume that $X$ is a Lévy process. Define the function $G$ by*

$$G(t, z) = \mathsf{P}_0(S_t \leq z) \tag{15}$$

*for $t \in [0, 1]$ and $z \in \mathbb{R}_+$. Then the following identity holds:*

$$\mathsf{E}_x |\theta - \tau| = \mathsf{E}_x \left( \int_0^\tau \left( 2\,G(1-t, S_t - X_t) - 1 \right) dt \right) + \mathsf{E}_x(\theta) \tag{16}$$

*for any stopping time $\tau$ satisfying $0 \leq \tau \leq 1$.*

*Proof.* This result follows directly from Lemma 1 above upon noting that

$$F(1-t, X_t, S_t) = \mathsf{P}_0 \left( \sup_{0 \leq s \leq 1-t} (x + X_s) \leq s \right) \Big|_{x = X_t,\, s = S_t} \tag{17}$$

$$= \mathsf{P}_0(S_{1-t} \leq z) \big|_{z = S_t - X_t}$$

since $X$ under $\mathsf{P}_x$ is realized as $x + X$ under $\mathsf{P}_0$.  □

If $X$ is a Lévy process then the reflected process $(S_t - X_t)_{0 \leq t \leq 1}$ is Markov. This is not true in general and means that for Lévy processes the optimal prediction problem is inherently two-dimensional (rather than three-dimensional as in the general case). It is also important to note that for a Lévy process we have the additional property that the map $z \mapsto 2\,G(1-t, z) - 1$ is increasing for any $t \in [0, 1]$ fixed. Further implications of this will also be explored in Section 4 below.

# 3 The free-boundary problem

Let us now formally introduce the setting for the optimal prediction problem (6). Let $B = (B_t)_{t \geq 0}$ be a standard Brownian motion defined on a probability space $(\Omega, \mathcal{F}, \mathsf{P})$ with $B_0 = 0$ under $\mathsf{P}$. Given $\mu \in \mathbb{R}$ set $B_t^\mu = B_t + \mu t$ and $S_t^\mu = \max_{0 \leq s \leq t} B_s^\mu$ for $t \in [0, 1]$. Let $\theta$ denote the first time at which the process $B^\mu = (B_t^\mu)_{0 \leq t \leq 1}$ attains its maximum $S_1^\mu$.

Consider the optimal prediction problem

$$V_* = \inf_{0 \le \tau \le 1} \mathsf{E}|\theta - \tau| \tag{18}$$

where the infimum is taken over all stopping times $\tau$ of $B^\mu$. By Lemma 2 above this problem is equivalent to the standard optimal stopping problem

$$V = \inf_{0 \le \tau \le 1} \mathsf{E}\left( \int_0^\tau H(t, X_t)\, dt \right) \tag{19}$$

where the process $X = (X_t)_{0 \le t \le 1}$ is given by $X_t = S_t^\mu - B_t^\mu$, the infimum is taken over all stopping times $\tau$ of $X$, and the function $H : [0,1] \times \mathbb{R}_+ \to [-1,1]$ is computed as

$$H(t, x) = 2\,\mathsf{P}(S_{1-t}^\mu \le x) - 1 \tag{20}$$

$$= 2\left[ \Phi\left( \frac{x - \mu(1-t)}{\sqrt{1-t}} \right) - e^{2\mu x}\, \Phi\left( \frac{-x - \mu(1-t)}{\sqrt{1-t}} \right) \right] - 1$$

using the well-known identity for the law of $S_{1-t}^\mu$ (cf. [1, p. 397], [7, p. 526]). Note that $V_* = V + \mathsf{E}(\theta)$ where

$$\mathsf{E}(\theta) = \int_0^1 \mathsf{P}(\theta > t)\, dt \tag{21}$$

$$= \sqrt{\frac{2}{\pi}} \int_0^1 \int_0^\infty \int_{-\infty}^s \frac{2s - b}{t^{3/2}} \left[ 1 - \Phi\left( \frac{s - b - \mu(1-t)}{\sqrt{1-t}} \right) \right.$$

$$\left. + e^{2\mu(s-b)}\, \Phi\left( \frac{b - s - \mu(1-t)}{\sqrt{1-t}} \right) \right] e^{-\frac{(2s-b)^2}{2t} + \mu(b - \frac{\mu t}{2})}\, db\, ds\, dt$$

which is readily derived using (14), (17), (20) and (28) below.

It is known (cf. [3]) that the strong Markov process $X$ is equal in law to $|Y| = (|Y_t|)_{0 \le t \le 1}$ where $Y = (Y_t)_{0 \le t \le 1}$ is the unique strong solution to $dY_t = -\mu\, \mathrm{sign}(Y_t)\, dt + dB_t$ with $Y_0 = 0$. It is also known (cf. [3]) that under $Y_0 = x$ the process $|Y|$ has the same law as a Brownian motion with drift $-\mu$ started at $|x|$ and reflected at 0. Hence the infinitesimal generator $\mathbb{L}_X$ of $X$ acts on functions $f \in C_b^2\big([0, \infty)\big)$ satisfying $f'(0) = 0$ as $\mathbb{L}_X f(x) = -\mu f'(x) + \frac{1}{2} f''(x)$. Since the infimum in (19) is attained at the first entry time of $X$ to a closed set (this follows from general theory of optimal stopping and will be made more precise below) there is no restriction to replace the process $X$ in (19) by the process $|Y|$.

It is especially important for the analysis of optimal stopping to see how $X$ depends on its starting value $x$. Although the equation for $Y$ is difficult to solve explicitly, it is known (cf. [2, Lemma 2.2], [10, Theorem 2.1]) that the Markov process $X^x = (X_t^x)_{0 \le t \le 1}$ defined under $\mathsf{P}$ as $X_t^x = x \vee S_t^\mu - B_t^\mu$ also realizes a Brownian motion with drift $-\mu$ started at $x \ge 0$ and reflected at 0. Following the usual approach to optimal stopping for Markov processes (see e.g. [11]) we may therefore extend the problem (19) to

$$V(t,x) = \inf_{0 \le \tau \le 1-t} \mathsf{E}_{t,x} \left( \int_0^\tau H(t+s, X_{t+s})\, ds \right) \tag{22}$$

where $X_t = x$ under $\mathsf{P}_{t,x}$ for $(t,x) \in [0,1] \times \mathbb{R}_+$ given and fixed, the infimum is taken over all stopping times $\tau$ of $X$, and the process $X$ under $\mathsf{P}_{t,x}$ can be identified with either $|Y|$ (under the same measure) or $X_{t+s}^x = x \vee S_s^\mu - B_s^\mu$ under the measure $\mathsf{P}$ for $s \in [0, 1-t]$. We will freely use either of these representations in the sequel without further mention.

We will show in the proof below that the value function $V$ is continuous on $[0,1] \times \mathbb{R}_+$. Defining $C = \{(t,x) \in [0,1] \times \mathbb{R}_+ | V(t,x) < 0\}$ to be the (open) continuation set and $D = \{(t,x) \in [0,1] \times \mathbb{R}_+ | V(t,x) = 0\}$ to be the (closed) stopping set, standard results from optimal stopping theory (cf. [11, Corollary 2.9]) indicate that the stopping time

$$\tau_D = \inf\{0 \le t \le 1 | (t, X_t) \in D\} \tag{23}$$

is optimal for the problem (19) above. We will also show in the proof below that the value function $V : [0,1] \times \mathbb{R}_+ \to \mathbb{R}$ solves the following free-boundary problem:

$$(V_t - \mu V_x + \tfrac{1}{2} V_{xx} + H)(t,x) = 0 \quad \text{for } (t,x) \in C \tag{24}$$

$$V(t,x) = 0 \quad \text{for } (t,x) \in D \text{ (instantaneous stopping)} \tag{25}$$

$$x \mapsto V_x(t,x) \quad \text{is continuous over } \partial C \text{ for } t \in [0,1) \text{ (smooth fit)} \tag{26}$$

$$V_x(t, 0+) = 0 \quad \text{for } t \in [0,1) \text{ (normal reflection).} \tag{27}$$

Our aim will not be to tackle this free-boundary problem directly, but rather to express $V$ in terms of the boundary $\partial C$, and to derive an analytic expression for the boundary itself. This approach dates back to [6] in a general setting (for more details see [11]).

## 4 The result and proof

The function $H$ from equation (20) above and the set $\{H \ge 0\} := \{(t,x) \in [0,1] \times \mathbb{R}_+ | H(t,x) \ge 0\}$ will play prominent roles in our discussion. A direct examination of $H$ reveals the existence of a continuous decreasing function $h : [0,1] \to \mathbb{R}_+$ satisfying $h(1) = 0$ such that $\{H \ge 0\} = \{(t,x) \in [0,1] \times \mathbb{R}_+ | x \ge h(t)\}$. Recall (see e.g. [5, p. 368]) that the joint density function of $(B_t^\mu, S_t^\mu)$ under $\mathsf{P}$ is given by

$$f(t,b,s) = \sqrt{\frac{2}{\pi}} \frac{(2s-b)}{t^{3/2}} e^{-\frac{(2s-b)^2}{2t} + \mu(b - \frac{\mu t}{2})} \tag{28}$$

for $t > 0$, $s \ge 0$ and $b \le s$. Define the function

$$K(t, x, r, z) = \mathsf{E}\Big(H(t+r, X_r^x)\, I(X_r^x < z)\Big) \qquad (29)$$

$$= \int_0^\infty \int_{-\infty}^s H\big(t + r, x \vee s - b\big)\, I\big(x \vee s - b < z\big)\, f(r, b, s)\, db\, ds$$

for $t \in [0, 1]$, $r \in [0, 1-t]$ and $x, z \in \mathbb{R}_+$. We may now state the main result of this paper.

**Theorem 1.** *Consider the optimal stopping problem* (22). *Then there exists a continuous decreasing function* $b : [0, 1] \to \mathbb{R}_+$ *satisfying* $b(1) = 0$ *such that the optimal stopping set is given by* $D = \{(t, x) \in [0, 1] \times \mathbb{R}_+ \,|\, x \geq b(t)\}$. *Furthermore, the value function* $V$ *defined in* (22) *is given by*

$$V(t, x) = \int_0^{1-t} K\big(t, x, r, b(t+r)\big)\, dr \qquad (30)$$

*for all* $(t, x) \in [0, 1] \times \mathbb{R}_+$, *and the optimal boundary* $b$ *itself is uniquely determined by the nonlinear Volterra integral equation*

$$\int_0^{1-t} K\big(t, b(t), r, b(t+r)\big)\, dr = 0 \qquad (31)$$

*for* $t \in [0, 1]$, *in the sense that it is the unique solution to* (31) *in the class of continuous functions* $t \mapsto b(t)$ *on* $[0, 1]$ *satisfying* $b(t) \geq h(t)$ *for all* $t \in [0, 1]$. *It follows therefore that the optimal stopping time in* (22) *is given by*

$$\tau_D := \tau_D(t, x) = \inf \big\{ 0 \leq r \leq 1 - t \,|\, x \vee S_r^\mu - B_r^\mu \geq b(t+r) \big\} \qquad (32)$$

*for* $(t, x) \in [0, 1] \times \mathbb{R}_+$. *Finally, the value* $V_*$ *defined in* (18) *equals* $V(0, 0) + \mathsf{E}(\theta)$ *where* $\mathsf{E}(\theta)$ *is given in* (21), *and the optimal stopping time in* (18) *is given by* $\tau_D(0, 0)$ *(see Figure 1).*

**Proof.** *Step 1.* We first show that an optimal stopping time for the problem (22) exists, and then determine the shape of the optimal stopping set $D$. Since $H$ is continuous and bounded, and the flow $x \mapsto x \vee S_t^\mu - B_t^\mu$ of the process $X^x$ is continuous, it follows that for any stopping time $\tau$ the map $(t, x) \mapsto \mathsf{E}\big(\int_0^\tau H(t + s, x \vee S_s^\mu - B_s^\mu)\, ds\big)$ is continuous and thus upper semicontinuous (usc) as well. Hence we see that $V$ is usc (recall that the infimum of usc functions is usc) and so by general results of optimal stopping (cf. [11, Corollary 2.9]) it follows that $\tau_D$ from (23) is optimal in (22) with $C$ open and $D$ closed.

Next we consider the shape of $D$. Our optimal prediction problem has a particular internal structure, as was noted in the comments following Lemmas 1 and 2 above, and we now expose this structure more fully. Take any point $(t, x) \in \{H < 0\}$ and let $U \subset \{H < 0\}$ be an open set containing $(t, x)$. Define $\sigma_U$ to be the first exit time of $X$ from $U$ under the measure $\mathsf{P}_{t,x}$ where $\mathsf{P}_{t,x}(X_t = x) = 1$. Then clearly

$$V(t, x) \leq \mathsf{E}_{t,x}\left( \int_0^{\sigma_U} H(t+s, X_{t+s})\, ds \right) < 0 \qquad (33)$$

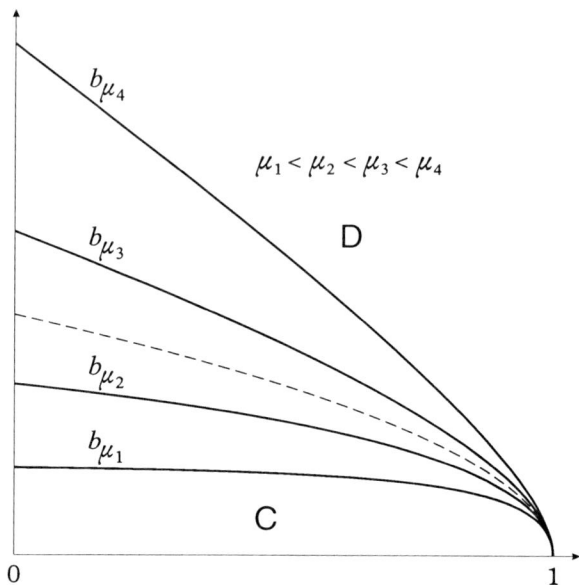

**Fig. 1.** A computer drawing of the optimal stopping boundaries for Brownian motion with drift $\mu_1 = -1.5$, $\mu_2 = -0.5$, $\mu_3 = 0.5$ and $\mu_4 = 1.5$. The dotted line is the optimal stopping boundary for Brownian motion with zero drift.

showing that it is not optimal to stop at $(t, x)$. Hence the entire region below the curve $h$ is contained in $C$.

As was observed following Lemma 1, the map $t \mapsto H(t, x)$ is increasing, so that taking any $s < t$ in $[0, 1]$ and setting $\tau_s = \tau_D(s, x)$ and $\tau_t = \tau_D(t, x)$, it follows that

$$V(t, x) - V(s, x) \tag{34}$$
$$= \mathsf{E}\left(\int_0^{\tau_t} H(t+r, X_r^x)\, dr\right) - \mathsf{E}\left(\int_0^{\tau_s} H(s+r, X_r^x)\, dr\right)$$
$$\geq \mathsf{E}\left(\int_0^{\tau_t} H(t+r, X_r^x) - H(s+r, X_r^x)\, dr\right) \geq 0$$

for any $x \geq 0$. Hence $t \mapsto V(t, x)$ is increasing so that if $(t, x) \in D$ then $(t + s, x) \in D$ for all $s \in [0, 1 - t]$ when $x \geq 0$ is fixed. Similarly, since $x \mapsto H(t, x)$ is increasing we see for any $x < y$ in $\mathbb{R}_+$ that

$$V(t, y) - V(t, x) \tag{35}$$
$$= \mathsf{E}\left(\int_0^{\tau_y} H(t+s, X_s^y)\, ds\right) - \mathsf{E}\left(\int_0^{\tau_x} H(t+s, X_s^x)\, ds\right)$$
$$\geq \mathsf{E}\left(\int_0^{\tau_y} H(t+s, X_s^y) - H(t+s, X_s^x)\, ds\right) \geq 0$$

for all $t \in [0, 1]$ where $\tau_x = \tau_D(t, x)$ and $\tau_y = \tau_D(t, y)$. Hence $x \mapsto V(t, x)$ is increasing so that if $(t, x) \in D$ then $(t, y) \in D$ for all $y \geq x$ when $t \in [0, 1]$ is fixed. We therefore conclude that in our problem (and indeed for any Lévy process) there is a single boundary function $b : [0, 1] \to \overline{\mathbb{R}}$ separating the sets $C$ and $D$ where $b$ is formally defined as

$$b(t) = \inf \{ x \geq 0 \, | \, (t, x) \in D \} \tag{36}$$

for all $t \in [0, 1]$, so that $D = \{ (t, x) \in [0, 1] \times \mathbb{R}_+ \, | \, x \geq b(t) \}$. It is also clear that $b$ is decreasing with $b \geq h$ and $b(1) = 0$.

We now show that $b$ is finite valued. For this, define $t_* = \sup \{ t \in [0, 1] \, | \, b(t) = \infty \}$ and suppose that $t_* \in (0, 1)$ and that $b(t_*) < \infty$ (the cases $t_* = 1$ and $b(t_*) = \infty$ follow by a small modification of the argument below). Let $\tau_x = \tau_D(0, x)$ and set $\sigma_x = \inf \{ t \in [0, 1] \, | \, X_t^x \leq h(0) + 1 \}$. Then from the properties of $X^x$ we have $\tau_x \to t_*$ and $\sigma_x \to 1$ as $x \to \infty$. On the other hand, from the properties of $H$ and $h$ we see that there exists $\varepsilon > 0$ such that $H(t, x) \geq \varepsilon$ for all $x \geq h(0) + 1$ and all $t \in [0, t_*]$. Hence we find that

$$0 \geq \lim_{x \to \infty} V(0, x) = \lim_{x \to \infty} \mathsf{E} \left[ \int_0^{\tau_x} H(t, X_t^x) \, dt \, I(\tau_x \leq \sigma_x) \right] \tag{37}$$

$$+ \lim_{x \to \infty} \mathsf{E} \left[ \int_0^{\tau_x} H(t, X_t^x) \, dt \, I(\tau_x > \sigma_x) \right] \geq \varepsilon t_* > 0$$

which is a contradiction. The case $t_* = 0$ can be disproved similarly by enlarging the horizon from 1 to a strictly greater number. Hence $b$ must be finite valued as claimed.

*Step 2.* We show that the value function $(t, x) \mapsto V(t, x)$ is continuous on $[0, 1] \times \mathbb{R}_+$. For this, take any $x \leq y$ in $\mathbb{R}_+$ and note by the mean value theorem that for every $t \in [0, 1]$ there is $z \in (x, y)$ such that

$$0 \leq H(t, y) - H(t, x) = (y - x) H_x(t, z) \tag{38}$$

$$= 2(y - x) \left[ \frac{2}{\sqrt{1-t}} \, \varphi \left( \frac{z - \mu(1-t)}{\sqrt{1-t}} \right) - 2\mu e^{2\mu z} \, \Phi \left( \frac{-z - \mu(1 - t)}{\sqrt{1 - t}} \right) \right]$$

$$\leq 4(y - x) \left[ \frac{1}{\sqrt{1-t}} + |\mu| \right] .$$

Since $0 \leq y \vee S_t^\mu - x \vee S_t^\mu \leq y - x$ it follows that

$$0 \leq V(t, y) - V(t, x) \leq \mathsf{E} \left( \int_0^{\tau_x} H(t+s, X_s^y) - H(t+s, X_s^x) \, ds \right) \tag{39}$$

$$\leq \mathsf{E} \left( 4 \int_0^{\tau_x} (X_s^y - X_s^x) \left[ \frac{1}{\sqrt{1-t-s}} + |\mu| \right] ds \right)$$

$$\leq \mathsf{E} \left( 8(y - x) \left[ \sqrt{1-t} - \sqrt{1-t-\tau_x} + |\mu| \, \tau_x \right] \right)$$

$$\leq 8(y - x)(1 + |\mu|)$$

where we recall that $\tau_x = \tau_D(t, x)$. Letting $y - x \to 0$ we see that $x \mapsto V(t, x)$ is continuous on $\mathbb{R}_+$ uniformly over all $t \in [0, 1]$.

To conclude the continuity argument it is enough to show that $t \mapsto V(t, x)$ is continuous on $[0, 1]$ for every $x \in \mathbb{R}_+$ given and fixed. To do this, take any $s \le t$ in $[0, 1]$ and let $\tau_s = \tau_D(s, x)$. Define the stopping time $\sigma = \tau_s \wedge (1 - t)$ so that $0 \le \sigma \le 1 - t$. Note that $0 \le \tau_s - \sigma \le t - s$ so that $\tau_s - \sigma \to 0$ as $t - s \to 0$. We then have

$$0 \le V(t, x) - V(s, x) \qquad (40)$$

$$\le \mathsf{E}\left(\int_0^\sigma H(t+r, X_r^x)\, dr\right) - \mathsf{E}\left(\int_0^{\tau_s} H(s+r, X_r^x)\, dr\right)$$

$$= \mathsf{E}\left(\int_0^\sigma H(t+r, X_r^x) - H(s+r, X_r^x)\, dr\right) - \mathsf{E}\left(\int_\sigma^{\tau_s} H(s+r, X_r^x)\, dr\right).$$

Letting $t - s \to 0$ and using the fact that $|H| \le 1$ it follows by the dominated convergence theorem that both expectations on the right-hand side of (40) tend to zero. This shows that the map $t \mapsto V(t, x)$ is continuous on $[0, 1]$ for every $x \in \mathbb{R}_+$, and hence the value function $(t, x) \mapsto V(t, x)$ is continuous on $[0, 1] \times \mathbb{R}_+$ as claimed.

*Step 3.* We show that $V$ satisfies the smooth fit condition (26). Take any $t$ in $[0, 1)$, set $c = b(t)$ and define the stopping time $\tau_\varepsilon = \tau_D(t, c-\varepsilon)$ for $\varepsilon > 0$. Then from the second last inequality in (39) we see that

$$0 \le \frac{V(t, c) - V(t, c-\varepsilon)}{\varepsilon} \le 8\, \mathsf{E}\left(\sqrt{1-t} - \sqrt{1-t-\tau_\varepsilon} + |\mu|\, \tau_\varepsilon\right) \qquad (41)$$

for $\varepsilon > 0$. We now show that $\tau_\varepsilon \to 0$ as $\varepsilon \downarrow 0$. To see this, consider the stopping time $\sigma = \inf\{0 \le s \le 1-t \mid X_{t+s} \ge c\}$ under the measure $\mathsf{P}_{t, c-\varepsilon}$. The process $X$ started at $c - \varepsilon$ at time $t$ will always hit the boundary $b$ before hitting the level $c$ since $b$ is decreasing. Hence $0 \le \tau_\varepsilon \le \sigma$ and thus it is enough to show that $\sigma \to 0$ under $\mathsf{P}_{t, c-\varepsilon}$ as $\varepsilon \downarrow 0$. For this, note by the Itô-Tanaka formula that

$$dX_t = -\mu\, dt + \mathrm{sign}(Y_t)\, dB_t + d\ell_t^0(Y) \qquad (42)$$

where $\ell^0(Y) = (\ell_t^0(Y))_{0 \le t \le 1}$ is the local time of $Y$ at zero. It follows that

$$\sigma = \inf\{0 \le s \le 1-t \mid c - \varepsilon - \mu s + \beta_s + \ell_s^0(Y) \ge c\} \qquad (43)$$

$$\le \inf\{s \ge 0 \mid -\varepsilon + \beta_s \ge \mu s\}$$

where $\beta_s = \int_0^s \mathrm{sign}(Y_r)\, dB_r$ is a standard Brownian motion for $s \ge 0$. Letting $\varepsilon \downarrow 0$ and using the fact that $s \mapsto \mu s$ is a lower function of $\beta = (\beta_s)_{s \ge 0}$ at $0+$, we see that $\sigma \to 0$ under $\mathsf{P}_{t, c-\varepsilon}$ and hence $\tau_\varepsilon \to 0$ as claimed. Passing to the limit in (41) for $\varepsilon \downarrow 0$, and using the dominated convergence theorem, we conclude that $x \mapsto V(t, x)$ is differentiable at $c$ and $V_x(t, c) = 0$. Moreover, a small modification of the preceding argument shows that $x \mapsto V(t, x)$ is continuously differentiable at $c$. Indeed, for $\delta > 0$ define the stopping time $\tau_\delta = \tau_D(t, c-\delta)$. Then as in (41) we have

$$0 \leq \frac{V(t, c-\delta+\varepsilon) - V(t, c-\delta)}{\varepsilon} \leq 8\,\mathsf{E}\left(\sqrt{1-t} - \sqrt{1-t-\tau_\delta} + |\mu|\,\tau_\delta\right) \qquad (44)$$

for $\varepsilon > 0$. Letting first $\varepsilon \downarrow 0$ (upon using that $V_x(t, x - \delta)$ exists) and then $\delta \downarrow 0$ (upon using that $\tau_\delta \to 0$) we see as above that $x \mapsto V_x(t, x)$ is continuous at $c$. This establishes the smooth fit condition (26).

Standard results on optimal stopping for Markov processes (see e.g. [11, Section 7]) show that $V$ is $C^{1,2}$ in $C$ and satisfies $V_t + \mathbb{L}_X V + H = 0$ in $C$. This, together with the result just proved, shows that $V$ satisfies (24)–(26) of the free-boundary problem (24)–(27). We now establish the last of these conditions.

*Step 4.* We show that $V$ satisfies the normal reflection condition (27). For this, note first since $x \mapsto V(t, x)$ is increasing on $\mathbb{R}_+$ that $V_x(t, 0+) \geq 0$ for all $t \in [0, 1]$ where the limit exists since $V$ is $C^{1,2}$ on $C$. Suppose now that there exists $t \in [0, 1)$ such that $V_x(t, 0+) > 0$. Recalling again that $V$ is $C^{1,2}$ on $C$ so that $t \mapsto V_x(t, 0+)$ is continuous on $[0, 1)$, we see that there exist $\delta > 0$ and $\varepsilon > 0$ such that $V_x(t+s, 0+) \geq \varepsilon$ for all $s \in [0, \delta]$ where $t + \delta < 1$. Setting $\tau_\delta = \tau_D(t, 0) \wedge \delta$ we see by Itô's formula using (42) and (24) that

$$V(t+\tau_\delta, X_{t+\tau_\delta}) = V(t, 0) + \int_0^{\tau_\delta} (V_t - \mu V_x + \tfrac{1}{2} V_{xx})(t+r, X_{t+r})\, dr \qquad (45)$$

$$+ \int_0^{\tau_\delta} V_x(t+r, X_{t+r})\, \mathrm{sign}(Y_{t+r})\, dB_{t+r}$$

$$+ \int_0^{\tau_\delta} V_x(t+r, X_{t+r})\, d\ell_{t+r}^0(Y)$$

$$\geq V(t, 0) - \int_0^{\tau_\delta} H(t+r, X_{t+r})\, dr + M_{\tau_\delta} + \varepsilon\,\ell_{t+\tau_\delta}^0(Y)$$

where $M_s = \int_0^s V_x(t+r, X_{t+r})\, \mathrm{sign}(Y_{t+r})\, dB_{t+r}$ is a continuous martingale for $s \in [0, 1-t]$ (note from (44) that $V_x$ is uniformly bounded). By general theory of optimal stopping for Markov processes (see e.g. [11]) we know that $V(t + s \wedge \tau_\delta, X_{t+s\wedge\tau_\delta}) + \int_0^{s\wedge\tau_\delta} H(t + r, X_{t+r})\, dr$ is a martingale starting at $V(t, 0)$ under $\mathsf{P}_{t,0}$ for $s \in [0, 1-t]$. Hence by taking $\mathsf{E}_{t,0}$ on both sides of (45) and using the optional sampling theorem to deduce that $\mathsf{E}_{t,0}(M_{\tau_\delta}) = 0$, we see that $\mathsf{E}_{t,0}(\ell_{t+\tau_\delta}^0(Y)) = 0$. Since the properties of the local time clearly exclude this possibility, we must have $V_x(t, 0+) = 0$ for all $t \in [0, 1]$ as claimed.

*Step 5.* We show that the boundary function $t \mapsto b(t)$ is continuous on $[0, 1]$. Let us first show that $b$ is right-continuous. For this, take any $t \in [0, 1)$ and let $t_n \downarrow t$ as $n \to \infty$. Since $b$ is decreasing we see that $\lim_{n\to\infty} b(t_n) =: b(t+)$ exists, and since each $(t_n, b(t_n))$ belongs to $D$ which is closed, it follows that $(t, b(t+))$ belongs to $D$ as well. From the definition of $b$ in (36) we must therefore have $b(t) \leq b(t+)$. On the other hand, since $b$ is decreasing, we see that $b(t) \geq b(t_n)$ for all $n \geq 1$, and hence $b(t) \geq b(t+)$. Thus $b(t) = b(t+)$ and consequently $b$ is right-continuous as claimed.

We now show that $b$ is left-continuous. For this, let us assume that there is $t \in (0,1]$ such that $b(t-) > b(t)$, and fix any $x \in (b(t), b(t-))$. Since $b \geq h$ we see that $x > h(t)$ so that by continuity of $h$ we have $h(s) < x$ for all $s \in [s_1, t]$ with some $s_1 \in (0,t)$ sufficiently close to $t$. Hence $c := \inf \{ H(s,y) | \, s \in [s_1, t], \, y \in [x, b(s)] \} > 0$ since $H$ is continuous. Moreover, since $V$ is continuous and $V(t, y) = 0$ for all $y \in [x, b(t-)]$, it follows that

$$|\mu V(s, y)| \leq \frac{c}{4} \big( b(t-) - x \big) \tag{46}$$

for all $s \in [s_2, t]$ and all $y \in [x, b(s)]$ with some $s_2 \in (s_1, t)$ sufficiently close to $t$. For any $s \in [s_2, t]$ we then find by (24) and (26) that

$$V(s, x) = \int_x^{b(s)} \int_y^{b(s)} V_{xx}(s, z) \, dz \, dy \tag{47}$$

$$= 2 \int_x^{b(s)} \int_y^{b(s)} \big( -V_t + \mu V_x - H \big)(s, z) \, dz \, dy$$

$$\leq 2 \int_x^{b(s)} \big[ -\mu V(s, y) - c(b(s) - y) \big] \, dy$$

$$\leq \frac{c}{2} \big( b(t-) - x \big) \big( b(s) - x \big) - c \big( b(s) - x \big)^2$$

where in the second last inequality we use that $V_t \geq 0$ and in the last inequality we use (46). Letting $s \uparrow t$ we see that $V(t, x) \leq (-c/2)(b(t-) - x)^2 < 0$ which contradicts the fact that $(t, x)$ belongs to $D$. Thus $b$ is left-continuous and hence continuous on $[0, 1]$. Note that the preceding proof also shows that $b(1) = 0$ since $h(1) = 0$ and $V(1, x) = 0$ for $x \geq 0$.

*Step 6.* We may now derive the formula (30) and the equation (31). From (39) we see that $0 \leq V_x(t, x) \leq 8(1 + |\mu|) =: K$ for all $(t, x) \in [0, 1] \times \mathbb{R}_+$. Hence by (24) we find that $V_{xx} = 2(-V_t + \mu V_x - H) \leq 2(|\mu| K - H)$ in $C$. Thus if we let

$$f(t, x) = 2 \int_0^x \int_0^y \big( 1 + |\mu| K - H(t, z) \big) \, dz \, dy \tag{48}$$

for $(t, x) \in [0, 1] \times \mathbb{R}_+$, then $V_{xx} \leq f_{xx}$ on $C \cup D^o$ since $V \equiv 0$ in $D$ and $|H| \leq 1$. Defining the function $F : [0, 1] \times \mathbb{R}_+ \to \mathbb{R}$ by $F = V - f$ we see by (26) that $x \mapsto F(t, x)$ is concave on $\mathbb{R}_+$ for every $t \in [0, 1]$. Moreover, it is evident that (i) $F$ is $C^{1,2}$ on $C \cup D^o$ and $F_x(t, 0+) = V_x(t, 0+) = f_x(t, 0+) = 0$; (ii) $F_t - \mu F_x + \frac{1}{2} F_{xx}$ is locally bounded on $C \cup D^o$; and (iii) $t \mapsto F_x(t, b(t) \pm) = -f_x(t, b(t) \pm)$ is continuous on $[0, 1]$. Since $b$ is decreasing (and thus of bounded variation) on $[0, 1]$ we can therefore apply the local time-space formula [9] to $F(t+s, X_{t+s})$, and since $f$ is $C^{1,2}$ on $[0, 1] \times \mathbb{R}_+$ we can apply Itô's formula to

$f(t+s, X_{t+s})$. Adding the two formulae, making use of (42) and the fact that $F_x(t, 0+) = f_x(t, 0+) = 0$, we find by (24)–(26) that

$$V(t+s, X_{t+s}) = V(t, x) \tag{49}$$
$$+ \int_0^s \left(V_t - \mu V_x + \frac{1}{2}V_{xx}\right)(t+r, X_{t+r}) \, I\big(X_{t+r} \neq b(t+r)\big) \, dr$$
$$+ \int_0^s V_x(t+r, X_{t+r}) \operatorname{sign}(Y_{t+r}) \, I\big(X_{t+r} \neq b(t+r)\big) \, dB_{t+r}$$
$$+ \frac{1}{2} \int_0^s \Big(V_x(t+r, X_{t+r}+) - V_x(t+r, X_{t+r}-)\Big)$$
$$\times \, I\big(X_{t+r} = b(t+r)\big) \, d\ell^b_{t+r}(X)$$
$$= V(t, x) - \int_0^s H\big(t+r, X_{t+r}\big) I\big(X_{t+r} < b(t+r)\big) \, dr$$
$$+ \int_0^s V_x(t+r, X_{t+r}) \operatorname{sign}(Y_{t+r}) \, dB_{t+r}$$

under $\mathsf{P}_{t,x}$ for $(t, x) \in [0, 1] \times \mathbb{R}_+$ and $s \in [0, 1-t]$, where $\ell^b_{t+r}(X)$ is the local time of $X$ on the curve $b$ for $r \in [0, 1-t]$. Inserting $s = 1-t$ and taking $\mathsf{E}_{t,x}$ on both sides, we see that

$$V(t, x) = \mathsf{E}_{t,x} \left( \int_0^{1-t} H\big(t+s, X_{t+s}\big) I\big(X_{t+s} < b(t+s)\big) \, ds \right) \tag{50}$$

which after exchanging the order of integration is exactly (30). Setting $x = b(t)$ in the resulting identity we obtain

$$\int_0^{1-t} \mathsf{E}_{t,b(t)} \Big( H\big(t+s, X_{t+s}\big) I\big(X_{t+s} < b(t+s)\big) \Big) \, ds = 0 \tag{51}$$

which is exactly (31) as claimed.

*Step 7.* We show that $b$ is the unique solution to the integral equation (31) in the class of continuous functions $t \mapsto b(t)$ on $[0, 1]$ satisfying $b(t) \geq h(t)$ for all $t \in [0, 1]$. This will be done in the four steps below.

Take any continuous function $c$ satisfying $c \geq h$ which solves (51) on $[0, 1]$, and define the continuous function $U^c : [0, 1] \times \mathbb{R}_+ \to \mathbb{R}$ by

$$U^c(t, x) = \mathsf{E}_{t,x} \left( \int_0^{1-t} H\big(t+s, X_{t+s}\big) I\big(X_{t+s} < c(t+s)\big) \, ds \right). \tag{52}$$

Note that $c$ solving (51) means exactly that $U^c\big(t, c(t)\big) = 0$ for all $t \in [0, 1]$. We now define the closed set $D_c := \{ (t, x) \in [0, 1] \times \mathbb{R}_+ \,|\, x \geq c(t) \}$ which will play the role of a "stopping set" for $c$. To avoid confusion we will denote by $D_b$ our original optimal stopping set $D$ defined by the function $b$ in (36).

We show that $U^c = 0$ on $D_c$. The Markovian structure of $X$ implies that the process

$$N_s := U^c(t+s, X_{t+s}) + \int_0^s H(t+s, X_{t+s}) I(X_{t+s} < c(t+s)) \, ds \qquad (53)$$

is a martingale under $\mathsf{P}_{t,x}$ for $s \in [0, 1-t]$ and $(t, x) \in [0, 1] \times \mathbb{R}_+$. Take any point $(t, x) \in D_c$ and consider the stopping time

$$\begin{aligned} \sigma_c &= \inf \{ 0 \le s \le 1-t \,|\, X_{t+s} \notin D_c \} \qquad (54) \\ &= \inf \{ 0 \le s \le 1-t \,|\, X_{t+s} \le c(t+s) \} \end{aligned}$$

under the measure $\mathsf{P}_{t,x}$. Since $U^c(t, c(t)) = 0$ for all $t \in [0, 1]$ and $U^c(1, x) = 0$ for all $x \in \mathbb{R}_+$, we see that $U^c(t+\sigma_c, X_{t+\sigma_c}) = 0$. Inserting $\sigma_c$ in (53) above and using the optional sampling theorem (upon recalling that $H$ is bounded), we get

$$U^c(t, x) = \mathsf{E}_{t,x}(U^c(t+\sigma_c, X_{t+\sigma_c})) = 0 \qquad (55)$$

showing that $U^c = 0$ on $D_c$ as claimed.

*Step 8.* We show that $U^c(t, x) \ge V(t, x)$ for all $(t, x) \in [0, 1] \times \mathbb{R}_+$. To do this, take any $(t, x) \in [0, 1] \times \mathbb{R}_+$ and consider the stopping time

$$\tau_c = \inf \{ 0 \le s \le 1-t \,|\, X_{t+s} \in D_c \} \qquad (56)$$

under $\mathsf{P}_{t,x}$. We then claim that $U^c(t + \tau_c, X_{t+\tau_c}) = 0$. Indeed, if $(t, x) \in D_c$ then $\tau_c = 0$ and we have $U^c(t, x) = 0$ by our preceding argument . Conversely, if $(t, x) \notin D_c$ then the claim follows since $U^c(t, c(t)) = U(1, x) = 0$ for all $t \in [0, 1]$ and all $x \in \mathbb{R}_+$. Therefore inserting $\tau_c$ in (53) and using the optional sampling theorem, we see that

$$\begin{aligned} U^c(t, x) &= \mathsf{E}_{t,x} \int_0^{\tau_c} H(t+s, X_{t+s}) I(X_{t+s} \notin D_c) \, ds \qquad (57) \\ &= \mathsf{E}_{t,x} \int_0^{\tau_c} H(t+s, X_{t+s}) \, ds \ge V(t, x) \end{aligned}$$

where the second identity follows by the definition of $\tau_c$. This shows that $U^c \ge V$ as claimed.

*Step 9.* We show that $c(t) \le b(t)$ for all $t \in [0, 1]$. Suppose that this is not the case and choose a point $(t, x) \in [0, 1) \times \mathbb{R}_+$ so that $b(t) < c(t) < x$. Defining the stopping time

$$\sigma_b = \inf \{ 0 \le s \le 1-t \,|\, X_{t+s} \notin D_b \} \qquad (58)$$

under $\mathsf{P}_{t,x}$ and inserting it into the identities (49) and (53), we can take $\mathsf{E}_{t,x}$ on both sides and use the optional sampling theorem to see that

$$\mathsf{E}_{t,x}\big(V(t+\sigma_b, X_{t+\sigma_b})\big) = V(t,x) \tag{59}$$

$$\mathsf{E}_{t,x}\big(U^c(t+\sigma_b, X_{t+\sigma_b})\big) = U^c(t,x) \tag{60}$$

$$- \mathsf{E}_{t,x}\left(\int_0^{\sigma_b} H(t+s, X_{t+s}) I(X_{t+s} \notin D_c)\, ds\right).$$

The fact that the point $(t,x)$ belongs to both sets $D_c$ and $D_b$ implies that $V(t,x) = U^c(t,x) = 0$, and since $U^c \geq V$ we must have $U^c(t+\sigma_b, X_{t+\sigma_b}) \geq V(t+\sigma_b, X_{t+\sigma_b})$. Hence we find that

$$\mathsf{E}_{t,x}\left(\int_0^{\sigma_b} H(t+s, X_{t+s}) I(X_{t+s} \notin D_c)\, ds\right) \leq 0. \tag{61}$$

The continuity of $b$ and $c$, however, implies that there is a small enough $[t, u] \subseteq [t, 1]$ such that $b(s) < c(s)$ for all $s \in [t, u]$. With strictly positive probability, therefore, the process $X$ will spend non-zero time in the region between $b(s)$ and $c(s)$ for $s \in [t, u]$, and this combined with the fact that both $D_c$ and $D_b$ are contained in $\{H \geq 0\}$, forces the expectation in (61) to be strictly positive and provides a contradiction. Hence we must have $c \leq b$ on $[0, 1]$ as claimed.

*Step 10.* We finally show that $c = b$ on $[0, 1]$. Suppose that this is not the case. Choose a point $(t, x) \in [0, 1) \times \mathbb{R}_+$ such that $c(t) < x < b(t)$ and consider the stopping time

$$\tau_D = \inf\{0 \leq s \leq 1 - t \mid X_{t+s} \in D_b\} \tag{62}$$

under the measure $\mathsf{P}_{t,x}$. Inserting $\tau_D$ in (49) and (53), taking $\mathsf{E}_{t,x}$ on both sides and using the optional sampling theorem, we see that

$$\mathsf{E}_{t,x}\left(\int_0^{\tau_D} H(t+s, X_{t+s})\, ds\right) = V(t,x) \tag{63}$$

$$\mathsf{E}_{t,x}\big(U^c(t+\tau_D, X_{t+\tau_D})\big) = U^c(t,x) \tag{64}$$

$$- \mathsf{E}_{t,x}\left(\int_0^{\tau_D} H(t+s, X_{t+s}) I(X_{t+s} \notin D_c)\, ds\right).$$

Since the set $D_b$ is contained in the set $D_c$ and $U^c = 0$ on $D_c$, we must have $U^c(t+\tau_D, X_{t+\tau_D}) = 0$, and using the fact that $U^c \geq V$ we get

$$\mathsf{E}_{t,x}\left(\int_0^{\tau_D} H(t+s, X_{t+s}) I(X_{t+s} \in D_c)\, ds\right) \leq 0. \tag{65}$$

Then, as before, the continuity of $b$ and $c$ implies that there is a small enough $[t, u] \subseteq [t, 1]$ such that $c(s) < b(s)$ for all $s \in [t, u]$. Since with strictly positive probability the process $X$ will spend non-zero time in the region between $c(s)$ and $b(s)$ for $s \in [t, u]$, the same argument as before forces the expectation in (65) to be strictly positive and provides a contradiction. Hence we conclude that $b(t) = c(t)$ for all $t \in [0, 1]$ completing the proof.  □

# References

1. Doob JL (1948) Heuristic approach to the Kolmogorov-Smirnov theorems. Ann. Math. Statist. 20:393–403.
2. Du Toit J, Peskir G (2007) The trap of complacency in predicting the maximum. Ann. Probab. 35:340–365.
3. Graversen SE, Shiryaev AN (2000) An extension of P. Lévy's distributional properties to the case of a Brownian motion with drift. Bernoulli 6:615–620.
4. Graversen SE, Peskir G, Shiryaev AN (2001) Stopping Brownian motion without anticipation as close as possible to its ulimate maximum. Theory Probab. Appl. 45:125–136.
5. Karatzas I, Shreve SE (1998) Methods of Mathematical Finance. Springer.
6. Kolodner II (1956) Free boundary problem for the heat equation with applications to problems of change of phase I. General method of solution. Comm. Pure Appl. Math. 9:1–31.
7. Malmquist S (1954) On certain confidence contours for distribution functions. Ann. Math. Statist. 25:523–533.
8. Pedersen JL (2003) Optimal prediction of the ultimate maximum of Brownian motion. Stoch. Stoch. Rep. 75:205–219.
9. Peskir G (2005) A change-of-variable formula with local time on curves. J. Theoret. Probab. 18:499–535.
10. Peskir G (2006) On reflecting Brownian motion with drift. Proc. Symp. Stoch. Syst. (Osaka, 2005), ISCIE Kyoto 1–5.
11. Peskir G, Shiryaev AN (2006) Optimal Stopping and Free-Boundary Problems. Lectures in Mathematics, ETH Zürich, Birkhäuser.
12. Shiryaev AN (2002) Quickest detection problems in the technical analysis of the financial data. Proc. Math. Finance Bachelier Congress (Paris, 2000), Springer 487–521.
13. Shiryaev AN (2004) A remark on the quickest detection problems. Statist. Decisions 22:79–82.
14. Urusov MA (2005) On a property of the moment at which Brownian motion attains its maximum and some optimal stopping problems. Theory Probab. Appl. 49:169–176.

# A Stochastic Demand Model for Optimal Pricing of Non-Life Insurance Policies

Paul Emms

Faculty of Actuarial Science and Insurance, Cass Business School, City University, 106 Bunhill Row, London EC1Y 8TZ, United Kingdom. p.emms@city.ac.uk

**Summary.** A model for non-life insurance pricing is developed which is a stochastic version of that given in [4]. Two forms of stochasticity are considered: the uncertainty in the future market average premium and the uncertainty in the change of exposure from a given relative premium level. The optimal premium strategy is determined using dynamic programming, and this is compared with the deterministic model both analytically and numerically. If the market average premium is stochastic then the optimization problem reduces to a set of characteristic strip equations, which are analyzed using the phase diagram. If the change in exposure is stochastic then an analytical expression is found for the optimal premium strategy when the objective is to maximize the expected terminal wealth in an infinite insurance market with an exponential utility function. As the volatility is increased the optimal strategy changes to the breakeven premium strategy for both forms of stochasticity and positive risk aversion. However, the terminal optimal premium is given by the deterministic problem if the market average premium is stochastic.

## 1 Introduction

The daily change in the exposure of a non-life insurer increases as policies are sold and decreases as policies are not renewed or canceled. In a highly competitive price-conscious market the insurer's premium relative to the rest of the insurance market is an important factor in policy sales. The size of the insurer as measured by its current exposure is also important, since larger insurers tend to attract greater volumes of business than small insurers with comparable premium rates [11]. However, there are many other factors which influence demand: the marketing of the policies, the need for insurance, the reputation of the insurer and the capacity of the insurer to underwrite policies. These factors are too numerous to incorporate into a simple non-life insurance model and they are hard to quantify, yet they all contribute to the uncertainty in how much exposure a given pricing strategy will generate.

Taylor [11] develops an insurance model where each insurance policy generates a certain number of units of exposure and prices are measured per unit of this exposure. Since an insurer sells many policies, one can consider the exposure of the insurer as a continuous stochastic process. The change in exposure is then described by a stochastic differential equation, whose drift term reflects the demand for policies. The focus of this paper is not the uncertainty in the frequency and size of claims on policies, but the uncertainty of gaining or losing exposure through the setting of a premium for a non-life insurance policy. In the management science literature these are called stochastic demand models and they are usually formulated with a Poisson process representing the demand for retail products with its intensity taken as a regular demand function [6]. However, in contrast to retail sales, the exposure of an insurer decreases when policies lapse.

We study two forms of uncertainty in the demand for policies: the uncertainty of the future market average premium for comparable policies, and the uncertainty of gaining or losing exposure from a fixed relative premium. For the first form of uncertainty there is some analytical reduction, which enables us to characterize the optimal control in the same manner as the deterministic case. For the second form of uncertainty, the control (here the premium) explicitly scales the volatility of wealth of the insurer as in the optimal asset allocation problem [10]. Consequently the volatility of the exposure is expected to significantly alter the optimal pricing strategy in contrast to the deterministic case.

The simple model of a non-life insurance market given in [3] is the basis for this study. Emms & Haberman [4] extend this model to a finite deterministic insurance market and characterize the optimal control using the phase diagram. Representative results are reproduced in Figure 1 for a terminal wealth objective, a linear demand law and an infinite and a finite insurance market. Graph (a) is a plot of the insurer's exposure $q$ against the adjoint variable $\Lambda$ for a number of trajectories computed using different initial conditions. It is shown in [4] that the relative premium increases as $\Lambda$ decreases, and only those trajectories which intersect the line $\Lambda = 0$ are optimal since this is the transversality condition from Pontryagin's Maximum Principle. For the chosen parameters, phase diagram (a) shows that it is optimal to build up exposure and possibly loss-lead if the time horizon is sufficiently large: the arrows on the plots (indicating the direction of increasing time) are at half-year intervals and trajectories which originate in the phase diagram above $\Lambda \sim 0.5$ are loss-leading. For the finite market case shown in phase plane 1(b), the optimal strategy is strongly dependent on the position of the equilibrium point $(q_+, \Lambda_+)$ and expansion or market withdrawal is possible dependent on the initial exposure $q$.

The phase diagrams in Figure 1 characterize the optimal premium strategy for the deterministic version of the model presented in this paper. However, when the state equations are stochastic such a characterization is not always possible since the optimal state trajectory and the optimal control can be stochastic. In this paper we give a qualitative description of the optimal premium strategy for two different stochastic demand models, and we compare

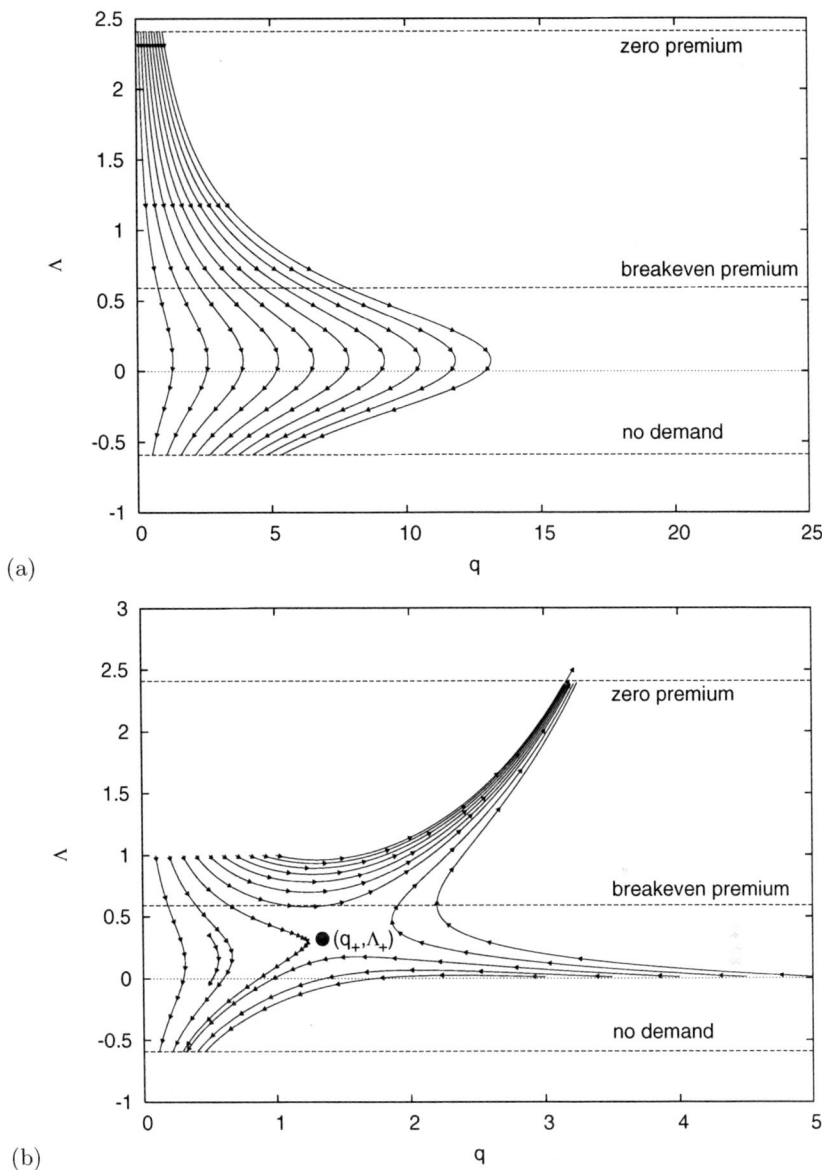

**Fig. 1.** Phase diagrams reproduced from [4] showing the optimal premium strategy for a terminal wealth objective in (a) an infinite market and (b) a finite market. The upper and lower dashed lines bound the domain of the demand law. For numerical values of the parameter set for these diagrams see [4].

how the optimal premium strategy differs in each case from the deterministic problem. In particular, we describe how the phase diagram of the deterministic model changes into a state diagram characterizing the stochastic control.

We introduce the stochastic insurance market model in the next section and suggest suitable parameterizations for this model in Section 2.1. Section 3 formulates the pricing of non-life insurance policies as a stochastic optimal control problem and uses dynamic programming for the solution. In Section 3.1 we relate the deterministic dynamic programming problem to the equivalent formulation in terms of Pontryagin's Maximum Principle. We consider the two stochastic components of the model in Sections 4 & 5, and assess how these forms of uncertainty alter the optimal premium strategy. Finally, conclusions are given in Section 6.

## 2 Stochastic demand model

Suppose the premium per unit exposure of a non-life insurance policy is $p_t$ and the corresponding breakeven premium is $\pi_t$. The premium $p_t$ is a continuously adapted process set by the insurer for a policy of fixed length $\kappa^{-1}$. A policyholder must pay this premium at the start of the policy for the fixed period of cover. The breakeven premium (per unit exposure) is determined by an insurer from historical claims data and represents the cost of selling an insurance policy including expenses. Let the market average premium for an equivalent insurance policy be $\bar{p}_t$ per unit of exposure. Emms & Haberman [4] use these variables in order to calculate the optimal premium strategy using a deterministic state model. Here, we extend that model by incorporating two stochastic factors.

First, we suppose there is some uncertainty as to the actual exposure the insurer gains in $\Delta t$:

$$dq_t = q_t(G - \kappa)\,dt + \sigma_1\,dW_{1,t}, \tag{1}$$

where $G = G(q_t, k_t)$ is the demand function, $\sigma_1 = \sigma_1(q_t) \geq 0$ is the volatility of policy sales, $W_{1,t}$ is a standard Brownian motion and $k_t = p_t/\bar{p}_t$ is the premium relative to the market average premium. The term $q_t G$ in (1) models the increase in exposure from selling insurance at relative premium $k_t$, while the relative rate of loss of exposure $\kappa$ models the non-renewal of policies of length $\kappa^{-1}$.

The second stochastic factor models the uncertainty in the market average premium $\bar{p}$:

$$d\bar{p}_t = \mu\bar{p}_t\,dt + \sigma_2\sqrt{\bar{p}_t}\,dW_{2,t}, \tag{2}$$

where the drift $\mu$ and volatility $\sigma_2$ are constant and the standard Brownian motion $W_{2,t}$ is uncorrelated with $W_{1,t}$. We have chosen this form of stochastic process since it yields a non-negative market average premium and leads to state space reduction in the forthcoming HJB equation. Other parameterizations are possible but our aim is to assess how the uncertainty in the market average premium affects the optimal premium strategy for the insurer.

We suppose that the market prices non-life insurance policies according to the expected value principle. Consequently, we define the breakeven premium relative to the market average premium as

$$\gamma = \frac{\pi_t}{\bar{p}_t} = \frac{1}{1 + \theta},$$

where $\theta > 0$ is the constant market loading and generates a profit for the market. Therefore, the market average premium and the breakeven premium are perfectly correlated. A further stochastic factor could be incorporated to model the uncertainty in future loading as in [3], but for simplicity this is not attempted here.

The wealth of the insurer changes as policies are sold:

$$dw_t = (p_t - \pi_t)(q_t G \, dt + \sigma_1 \, dW_{1,t}). \tag{3}$$

Notice that the wealth and the exposure of the insurer are correlated stochastic processes and the volatility of the wealth is $\sigma_1(p_t - \pi_t)$.

We aim to find the optimal relative premium strategy $k_t^*$ which maximizes the expected utility of terminal wealth:

$$\mathbb{E}\left\{U_2(w_T) | \mathbf{x}_t = x_0\right\}, \tag{4}$$

where $U_2$ is the utility function, $\mathbf{x}_t = (q_t, \bar{p}_t, w_t)^T$ is the state vector, and $T$ is the planning horizon. Emms & Haberman [4] also consider a total wealth objective, which introduces an additional utility function $U_1$. We shall concentrate on the terminal wealth problem since Emms & Haberman [4] find that the qualitative form of the optimal control is similar for comparable utility functions.

## 2.1 Parameterisations

In general, there are no analytical solutions of the optimization problem and the numerical solution is complicated by the number of state variables in the model. In order to progress we must make a judicious choice of model parameterization with the aim of reducing the dimension of the optimization problem. If the state space is of two dimensions or more it is difficult to visualize the optimal control and hard to categorise the qualitative optimal behaviour for the insurer. In this section we collect together those parameterizations which allow us to simplify the problem.

We call the demand function separable if

$$G(q, k) = f(q)g(k), \tag{5}$$

where $f, g$ are called the exposure and price functions respectively [7]. Both of these parameterizations are decreasing functions of their respective arguments. Following the terminology in [4], we call the market infinite if $f = a = \text{const.}$ or finite with saturation exposure $q_m$ if

$$f(q) = a \left( 1 - \frac{q}{q_m} \right)^+ , \tag{6}$$

where $a > 0$ is a constant.

There are many possible parameterizations for the price function [9], but the simplest is a linear relationship:

$$g(k) = b - k, \tag{7}$$

where $k = b > 0$ is the relative premium for which there is no demand for insurance policies. This parameterization has the benefit of leading to analytical optimal premium strategies in some cases because there is an explicit expression for the control in terms of the value function. We require that $b > \gamma$, otherwise from (3) there is no relative premium which generates wealth for the insurer. This is a slightly different parameterization from the deterministic model in Emms & haberman [4], who take $g(k) = (b - k)^+$ since new customers can only generate exposure. However, in a stochastic model if $k \gg b$ then exposure is always generated by the stochastic term in (1), which means the insurer can gain wealth no matter how high its premium. In order to prevent this behaviour we adopt (7), which has the effect of decreasing the exposure rapidly if the relative premium is significantly above $b$.

If the market is finite we require the volatility of exposure $\sigma_1$ to be zero when the insurer has no exposure or when it reaches its saturation exposure $q_m$:

$$\sigma_1(0) = \sigma_1(q_m) = 0. \tag{8}$$

These requirements ensure that the exposure remains in the range $0 \le q_t \le q_m$. A parameterization for a finite market model which satisfies these requirements is

$$\sigma_1(q) = \sigma_0 \sqrt{q \left( 1 - \frac{q}{q_m} \right)^+} . \tag{9}$$

For an infinite market we require $\sigma_1(0) = 0$ and an attractive model is

$$\sigma_1(q) = \sigma_0 \sqrt{q}, \tag{10}$$

which is a similar parameterization to that employed in the Cox-Ingersoll-Ross model of the short interest rate. This leads to model simplification because the state variable $q$ appears linearly in the resulting HJB equation.

The exponential terminal utility function is

$$U_2(w) = \frac{1}{c} \left( 1 - e^{-cw} \right), \tag{11}$$

**Table 1.** Sample data set where $[q]$ represents the units chosen for the exposure.

| | |
|---|---|
| Time horizon $T$ | 3.0 yr |
| Demand parameterization $a$ | 3 p.a. |
| Demand parameterization $b$ | 1.5 |
| Market loading $\theta$ | 0.1 |
| Length of policy $\tau = \kappa^{-1}$ | 1 yr |
| Market average premium growth $\mu$ | 0.1 p.a. |
| Saturation exposure | 5.0[q] |

where $c \geq 0$ is the constant risk aversion. In the limit $c \to 0$ we recover the linear utility function as a special case. The advantage of this utility function is that it is easier to determine the form of the value function, and this leads to problems which are independent of the current wealth $w$.

In the forthcoming numerical work we adopt the parameter values given in Table 1. These values correspond to those given in [4].

## 3 Optimization problem

In Emms & Haberman [4] the optimal pricing strategy was calculated using Pontryagin's Maximum Principle. Here, the tool for calculating the optimal control is dynamic programming.

In order to make further progress we incorporate some of the parameterizations given in Section 2.1. Let us suppose the demand function $G$ is separable (5) and the premium dependence is linear (7). Based on the objective function, an appropriate definition of the value function is

$$V(x,t) = \sup_k \mathbb{E}\left\{U_2(\mathbf{w}_T)|\mathbf{x}_t = x\right\},$$

where $x$ is the current state vector. If $V$ is sufficiently smooth then it satisfies the HJB equation

$$V_t + \tfrac{1}{2}\sigma_1^2 V_{qq} + \mu\bar{p}V_{\bar{p}} + \tfrac{1}{2}\sigma_2^2\bar{p}V_{\bar{p}\bar{p}} + \sup_k \{q(f(b-k)-\kappa)V_q+$$

$$\bar{p}qf(k-\gamma)(b-k)V_w + \tfrac{1}{2}\sigma_1^2\bar{p}^2(k-\gamma)^2 V_{ww} + \bar{p}\sigma_1^2(k-\gamma)V_{qw}\} = 0, \quad (12)$$

with boundary condition $V(t = T) = U_2(w)$.

Let us define the interior relative premium $k^i$ by first-order condition:

$$k^i = \frac{qf(V_q - \bar{p}(b+\gamma)V_w) + \sigma_1^2\bar{p}(\gamma\bar{p}V_{ww} - V_{qw})}{\bar{p}(\sigma_1^2\bar{p}V_{ww} - 2qfV_w)}. \quad (13)$$

The corresponding second-order condition for a maximum is

$$\bar{p}\sigma_1^2 V_{ww} \leq 2qfV_w. \quad (14)$$

If the second-order condition is satisfied and the partial derivatives of $V$ are finite then the interior control $k^i$ yields the supremum in the HJB equation.

## 3.1 Deterministic case

Emms & Haberman [4] take the deterministic form of the model ($\sigma_1 = \sigma_2 = 0$) and calculate the optimal premium strategy using the Maximum Principle. They find that the optimal strategy is independent of the parameterization of the utility function $U_2(w)$ in non-degenerate cases. The Maximum Principle leads to a set of ODEs which are the characteristics of the HJB equation [13] and it is not immediately clear that both formulations lead to the same premium strategy. The deterministic version of the HJB equation (12) is

$$V_t + \mu \bar{p} V_{\bar{p}} + q \left( f(b - k^i) - \kappa \right) V_q + \bar{p} q f \left( k^i - \gamma \right) (b - k^i) V_w = 0, \qquad (15)$$

and the first-order condition reduces to

$$k^i = \frac{1}{2} \left( b + \gamma - \frac{V_q}{\bar{p} V_w} \right). \qquad (16)$$

Now if we partially differentiate (15) with respect to $w$ and use the first-order condition then we find $DV_w/Dt = 0$ where we have used the convective differential operator

$$\frac{D}{Dt} := \frac{\partial}{\partial t} + \frac{dq}{dt} \frac{\partial}{\partial q} + \frac{dw}{dt} \frac{\partial}{\partial w}, \qquad (17)$$

to denote the derivative along the state trajectory [1]. Consequently $V_w = U_2'(w(T))$ is constant along an optimal state trajectory. By analogy with the adjoint variable in [4] we write

$$\Lambda(q, w, t) = \frac{V_q}{\bar{p} V_w}, \qquad (18)$$

so that the interior control is just a function of one adjoint variable. Now if we partially differentiate (15) with respect to $q$ we find that

$$\frac{DV_q}{Dt} + \frac{\partial}{\partial q} \left[ q \left( f (b - k) - \kappa \right) \right] V_q + \bar{p} \frac{\partial}{\partial q} \left[ q f (k - \gamma)(b - k) \right] V_w =$$

$$\frac{DV_q}{Dt} + \left[ \tfrac{1}{2}(qf)'(b - \gamma + \Lambda) - \kappa \right] V_q + \tfrac{1}{4}(qf)' \left( (b - \gamma)^2 - \Lambda^2 \right) V_w = 0.$$

Consequently along an optimal state trajectory

$$\frac{D\Lambda}{Dt} = \frac{1}{\bar{p} V_w} \frac{DV_q}{Dt} - \mu \Lambda$$

$$= -\tfrac{1}{4}(qf)'(b - \gamma)^2 - \left( \tfrac{1}{2}(qf)'(b - \gamma) - \kappa + \mu \right) \Lambda - \tfrac{1}{4}(qf)'\Lambda^2, \qquad (19)$$

and we arrive back at equation (25) in Emms & Haberman [4]. This equation and the corresponding deterministic version of (1) yield the phase diagrams in Figure 1.

A verification theorem such as that in [5] proves that the control $k^i$ is optimal providing solutions to (15) are sufficiently smooth. The phase diagrams demonstrate that a smooth solution to the coupled $(q, \Lambda)$ equations exists (dependent on the model parameters), which also satisfies the transversality condition $\Lambda = 0$. Therefore, the phase diagram describes the qualitative features of the optimal control in the deterministic case providing we accept a robust numerical solution as indicative of existence.

If $\sigma_1 = 0$, $\sigma_2 \neq 0$ and $U_2 = w$ then it is easy to see that the value function is of the form

$$V = w + \bar{p}F(q,t); \quad F(q,T) = 0,$$

so that $V$ is linear in $\bar{p}$ and the stochastic term disappears from the HJB equation (12). Consequently, the optimal premium strategy is the certainty equivalent control given by the phase diagram. We consider the case that the utility function is nonlinear in the next section.

# 4 Stochastic market average premium

Suppose $\sigma_1 = 0$, $\sigma_2 \neq 0$ and the terminal utility function is exponential (11). Emms [3] examines the case that the risk aversion $c$ is small and the loss ratio $\gamma$ is stochastic. Here we relax the assumption that the risk aversion is small, but consider a fixed loss ratio.

The HJB equation (12) contains only linear $\bar{p}$ terms. This observation and the form of the objective suggests we look for a value function of the form

$$V(t, \bar{p}, q, w, t) = \frac{1}{c}\left(1 - e^{-c(w+\bar{p}F(q,t))}\right),$$

with $F(q,T) = 0$. This candidate value function satisfies the second-order condition (14). Substituting $V$ into the HJB equation yields

$$F_t - \kappa q F_q + \mu F - \tfrac{1}{2}c\sigma_2^2 F^2 + \tfrac{1}{4}qf\left(b - \gamma + F_q\right)^2 = 0, \qquad (20)$$

which is a first-order nonlinear PDE. Let us introduce the notation $P = F_t$ and

$$\Lambda = F_q = \frac{V_q}{\bar{p}V_w},$$

which is consistent with the definition (18). Using this notation the PDE becomes

$$\mathcal{F}(t, q, F, P, \Lambda) = P + \tfrac{1}{4}(qf)(b - \gamma + \Lambda)^2 - \kappa q \Lambda + \mu F - \tfrac{1}{2}\sigma_2^2 c F^2 = 0.$$

The characteristic strip equations [2, p.78] for this PDE are then

$$\frac{dt}{du} = \mathcal{F}_P, \quad \frac{dq}{du} = \mathcal{F}_\Lambda, \quad \frac{dF}{du} = P\mathcal{F}_P + \Lambda\mathcal{F}_\Lambda,$$

$$\frac{dP}{du} = -\mathcal{F}_t - P\mathcal{F}_F, \quad \frac{d\Lambda}{du} = -\mathcal{F}_q - \Lambda\mathcal{F}_F,$$

where the initial conditions for these equations are at $t = T$:

$$u = 0, \; q = s, \; F = 0, \; \Lambda = 0, \; P = -\tfrac{1}{4}(qf)'(b - \gamma)^2.$$

We interpret $u$ as the variable along a characteristic trajectory, while $s$ parameterizes the initial curve which in this case is the line $t = T$.

If we substitute in the functional $\mathcal{F}$ then this system can be reduced to three ODEs along the characteristics

$$\frac{dq}{dt} = \tfrac{1}{2}(qf)(b - \gamma + \Lambda) - \kappa q,$$

$$\frac{d\Lambda}{dt} = -\tfrac{1}{4}(qf)'(b - \gamma + \Lambda)^2 + (\kappa - \mu)\Lambda + c\sigma_2^2 F\Lambda, \tag{21}$$

$$\frac{dF}{dt} = \tfrac{1}{2}c\sigma_2^2 F^2 - \mu F + \tfrac{1}{4}(qf)\left(\Lambda^2 - (b - \gamma)^2\right),$$

which we integrate backwards from $t = T$, and then substitute the adjoint solution into the first-order condition

$$k^i = \tfrac{1}{2}(b + \gamma - \Lambda), \tag{22}$$

in order to find the interior control.

We observe that if $c\sigma_2^2 = 0$ then these equations uncouple and we retrieve the deterministic optimal control. The transversality condition ($\Lambda = 0$) means that the terminal interior premium is $\tfrac{1}{2}(b + \gamma)$ and this value is independent of the volatility $\sigma_2$ and the current exposure $q$. Consequently, the terminal interior control is the same as the deterministic problem and its value is known at time $t = 0$.

In general, (21) is a coupled system of nonlinear ODEs and they must be solved numerically. Their solution leads to an open loop interior deterministic control $k^i(t)$, which depends on the volatility of the market average premium $\sigma_2$. If $q^i(t), k^i(t) \in C^1[0, T]$ then we can apply the verification theorem in [5] since the control does not affect the SDE (2). Under these conditions, the interior control $k^i$ is the optimal control $k^*$.

The system (21) is non-autonomous so we can study the phase space $(q, \Lambda, F)$.

## 4.1 Infinite market

If the market is infinite ($f = a$) then the equilibrium points of (21) are given by

$$q = 0, \qquad F\left(\tfrac{1}{2}c\sigma_2^2 F - \mu\right) = 0,$$

$$(\kappa - \mu)\Lambda + c\sigma_2^2\Lambda F - \tfrac{1}{4}a(b - \gamma + \Lambda)^2 = 0.$$

If $F = 0$ then there are at most two equilibrium points $(0, \Lambda_{\pm}^0 (\kappa, \mu, a, b))$ given by the real roots of

$$(\kappa - \mu)\Lambda - \tfrac{1}{4}a (b - \gamma + \Lambda)^2 = 0.$$

These are the equilibrium points reported in Emms & Haberman [4] for the deterministic problem. In addition, if $F = 2\mu/c\sigma_2^2$ then there are up to two more equilibrium points $(0, \Lambda_{\pm}^0 (\kappa, -\mu, a, b))$. Thus, the behaviour of the optimal control may differ from the deterministic case.

In fact, the properties of the optimal control are similar for both the deterministic and stochastic model. In the case that the deterministic optimal strategy is to withdraw then there are two equilibrium points in the phase plane (corresponding to Figure 4(a) of [4]). For the stochastic case there are two more equilibrium points, but they do not interact significantly with the optimal trajectories because they lie in the plane $F = 2\mu/c\sigma_2^2$. One finds that withdrawal is still optimal but the change in the optimal premium is diminished as the volatility of the market average premium increases. These results are available on request, but are merely summarized here for brevity.

The case that there are no equilibrium points in the phase plane for the deterministic problem is shown in Figure 1(a) using the parameters in Table 1. The optimal strategy is to build up exposure and loss-lead if the time horizon $T$ is sufficiently large irrespective of the initial exposure. For the stochastic case the corresponding plot is shown in Figure 2. One can see that building up exposure is still optimal but that the optimal trajectories are squeezed towards the axis $\Lambda = 0$. This means the change in the optimal relative premium over the time horizon decreases, and from (22) the relative premium is higher than in the deterministic case.

## 4.2 Finite market

If the market is finite and we adopt the exposure function (6) then the equilibrium points of the infinite market case are also equilibrium points of the finite market case and lie on the plane $q = 0$ provided that they exist. If we suppose $q \neq 0$ then any remaining equilibrium points are given by the coupled equations

$$\Lambda^2 + (2\phi - 4\zeta)\Lambda + \phi(\phi - 4\psi) + 4\beta F\Lambda = 0,$$

$$\tfrac{1}{2}\beta F^2 - \zeta F + \tfrac{1}{2}q_m\psi \frac{(\Lambda + \phi - 2\psi)(\Lambda - \phi)}{\Lambda + \phi} = 0, \qquad (23)$$

where we have used the same non-dimensional parameters given in Emms & Haberman [4]:

$$\phi = b - \gamma, \ \psi = \frac{\kappa}{a}, \ \zeta = \frac{\mu}{a},$$

In addition, we have introduced the parameter

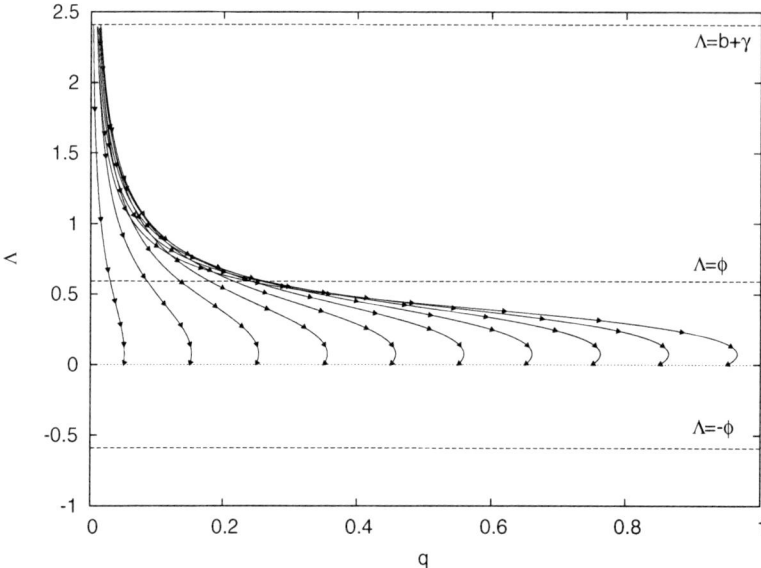

**Fig. 2.** Projected phase plane if the market average premium is stochastic ($c\sigma_2^2 = 2$) and the market is infinite. This figure should be compared with the phase plane for the deterministic model shown in Figure 1(a). Note that we have substantially reduced the $q$ scale on this plot.

$$\beta = \frac{c\sigma_2^2}{a}, \tag{24}$$

which has units per-unit exposure and determines the effect of stochasticity on the equilibrium points. The term $\frac{1}{2}\beta F^2$ suggests we write $\hat{F} = \beta F$ and $\hat{q}_m = \beta q_m$, which are both non-dimensional, so that the non-dimensional equilibrium equations are

$$\Lambda^2 + (2\phi - 4\zeta)\Lambda + \phi(\phi - 4\psi) + 4\hat{F}\Lambda = 0,$$
$$\frac{1}{2}\hat{F}^2 - \zeta\hat{F} + \frac{1}{2}\hat{q}_m\psi\frac{(\Lambda + \phi - 2\psi)(\Lambda - \phi)}{\Lambda + \phi} = 0. \tag{25}$$

The existence of equilibrium points depends on the values of the four parameters $\phi$, $\psi$, $\zeta$, $\hat{q}_m$ and a simple asymptotic analysis reveals their structure. If $\hat{q}_m \ll 1$ corresponding to an almost deterministic system and all other parameters are $O(1)$ then $\hat{F} \sim O(\hat{q}_m)$ or $\hat{F} \sim 2\zeta$. In the first case the adjoint equation becomes at leading order

$$\Lambda^2 + (2\phi - 4\zeta)\Lambda + \phi(\phi - 4\psi) = 0.$$

Let us denote these equilibrium points by $(q_\pm(\phi, \zeta, \psi), \Lambda_\pm(\phi, \zeta, \psi), O(\hat{q}_m))$ where

$$\Lambda_\pm(\phi, \zeta, \psi) = 2\zeta - \phi \pm 2\left(\zeta^2 - \phi\zeta + \phi\psi\right)^{1/2},$$

and $q_\pm(\phi, \zeta, \psi)$ denotes the corresponding exposure. These are the points identified by Emms & Haberman [4] in the deterministic finite market case if they are projected onto the $(q, \Lambda)$ plane. Emms & Haberman [4] show that only one equilibrium point $(q_+, \Lambda_+)$ exists in the relevant part of the phase plane for the deterministic problem and an appropriate parameter set, and that this point determines the qualitative features of the optimal control.

If $\hat{F} \sim 2\zeta$ (that is $F = 2\mu/c\sigma_2^2$) then the adjoint equation becomes

$$\Lambda^2 + (2\phi + 4\zeta)\,\Lambda + \phi\,(\phi - 4\psi) = 0,$$

at leading order. Accordingly, we denote the corresponding equilibrium points by $(q_\pm(\phi, -\zeta, \psi), \Lambda_\pm(\phi, -\zeta, \psi), 2\zeta)$. Thus, in $(q, \Lambda, F)$ space there are up to four additional equilibrium points if $\hat{q}_m \ll 1$. We expect that only two of these points $(q_+(\phi, \zeta, \psi), \Lambda_+(\phi, \zeta, \psi), O(\hat{q}_m))$ and $(q_+(\phi, -\zeta, \psi), \Lambda_+(\phi, -\zeta, \psi), 2\zeta)$ are located in the part of phase space for the stochastic case which affects the optimal premium strategies.

When $\hat{q}_m \sim O(1)$ we must use numerical methods to find the equilibrium points. The position of the two equilibrium points $(q_+^{1,2}, \Lambda_+^{1,2}, F^{1,2})$, which correspond to $(q_+, \Lambda_+)$, are shown in Figure 3 as the length of the policies is varied using the parameter set in Table 1. In general, the $(q_+^1, \Lambda_+^1, F^1)$ equilibrium point lies close to the $F = 0$ plane as $\beta \to 0$, whilst the $(q_+^2, \Lambda_+^2, F^2)$ point tends towards the plane $F = \infty$: these are the two cases described by the asymptotics above. Thus, the roots are not coincident if $\beta = 0$, and there is not a bifurcation in the usual sense as $\beta$ is increased from zero. Numerical experiments reveal that both these equilibrium points are unstable for the range of parameters considered here. These results are not shown here, but this behaviour is consistent with the forthcoming phase diagrams.

Now we describe Figure 3 in greater detail. Figure 3(a) shows the position of both equilibrium points in the three-dimensional phase space $(q, \Lambda, F)$ as $\kappa$ varies from 2.0 to 0.3 with $\beta = 1/6$. We do this to compare the position of the equilibrium points with the phase diagrams in the deterministic case shown in Figure 5 of Emms & Haberman [4]. In order to interpret Figure 3(a), one should note that the equilibrium points initially increase in $q$ as $\kappa$ is decreased, which corresponds to increasing the length of policies. The projection of these two curves onto the $(q, \Lambda)$ plane is shown in graph (b) as the two outer labeled curves. The inner (almost linear) curves show the projected position of the equilibrium points $(q_+^{1,2}, \Lambda_+^{1,2}, F^{1,2})$ when $\beta = 0$. We can only show the projection because $F^2 = \infty$ when $\beta = 0$. Thus we see how in the stochastic model the equilibrium points appear pushed apart in the projection of the phase diagram. Optimal trajectories in phase space must intersect the line $F = \Lambda = 0$ and it is not clear how the interaction of the two equilibrium points affect these trajectories. Thus we plot the optimal trajectories for fixed policy lengths next.

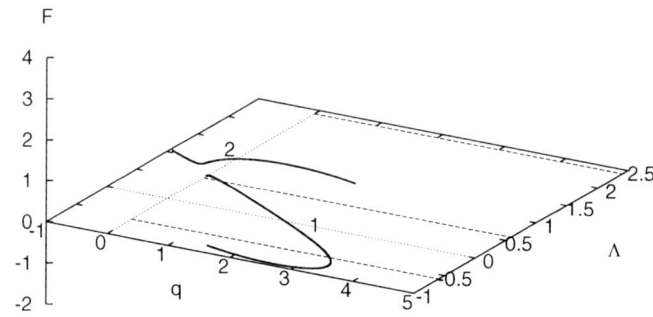

(a)

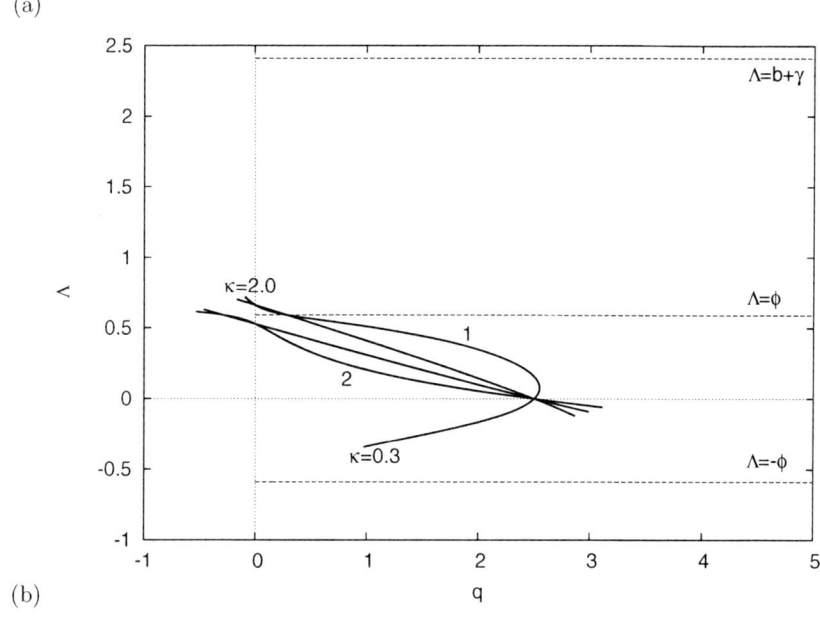

(b)

**Fig. 3.** Position of two equilibrium points $(q_+^{1,2}, \Lambda_+^{1,2}, F^{1,2})$ in phase space shown as solid lines as $\kappa$ is decreased from $\kappa = 2.0 \ldots 0.3$ for $\beta = 0$ and $\beta = \frac{1}{6}$. The upper graph (a) shows the full three-dimensional phase space $(q, \Lambda, F)$ with $\beta = \frac{1}{6}$, while (b) shows the *projected* phase plane $(q, \Lambda)$. The $\beta = 0$ curves are the inner apparently straight solid lines on graph (b).

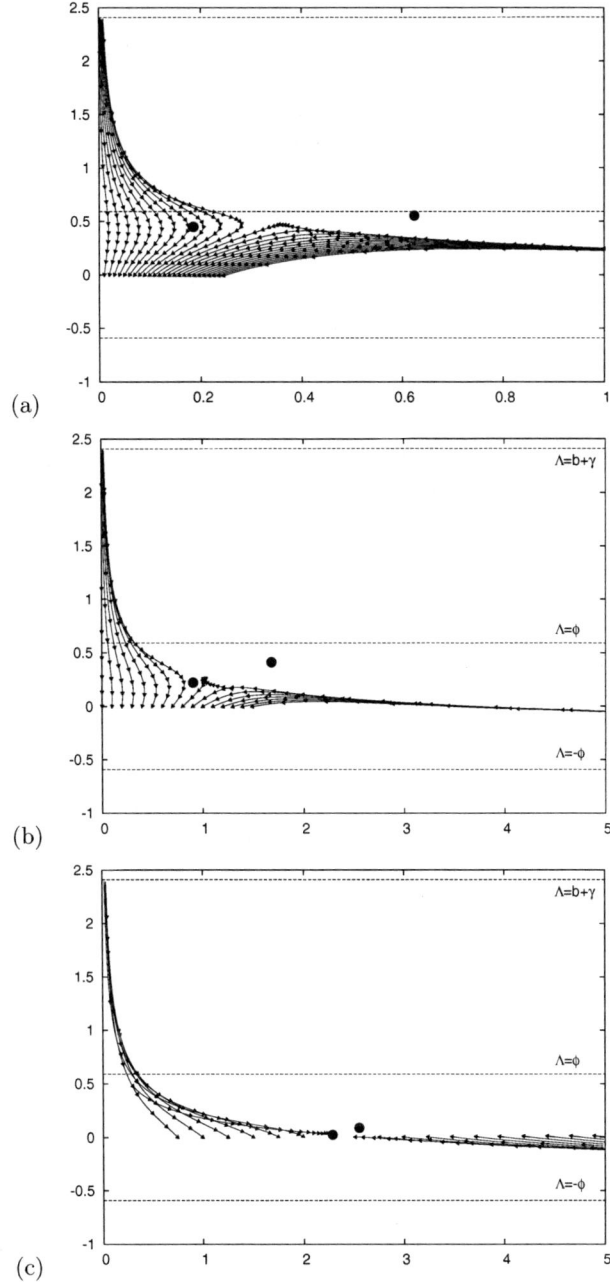

**Fig. 4.** Optimal trajectories projected onto the $(q, \Lambda)$ plane if the market average premium is stochastic and the market is finite. Equilibrium points are denoted by solid dots and the numerical parameters are taken as $\beta = \frac{1}{6}$ where (a) $\kappa = 1.5$, (b) $\kappa = 1.0$, and $\kappa = 0.5$. Notice that on plane (a) the $q$ scale has been reduced.

Figure 4 shows three projected phase diagrams with superimposed optimal trajectories and equilibrium points. From the first phase plane (a) we can see that the broad structure of the phase diagram is similar to the deterministic case: the optimal strategy is to either expand or withdraw from the insurance market. Phase plane (b) is the stochastic version of the deterministic phase plane shown in Figure 1(b). The interaction of the trajectories with the equilibrium points alters the conditions for when it is optimal to build-up exposure or leave the market. Notice in plane (a) trajectories can cross each other in the projection and that the position of the equilibrium point may be unimportant if the optimal trajectory traverses through a high $F$ value. In plane (b) we see an optimal trajectory appear to circle near the $(q_+^2, \Lambda_+^2)$ equilibrium point. The circling behaviour does not necessarily occur near the equilibrium points but is a feature of the interaction between them and it implies that the optimal premium is not easy to categorise in a withdrawal strategy. The plane (c) seems to indicate that it is the position of the $(q_+^2, \Lambda_+^2)$ equilibrium point which determines the type of optimal premium strategy at least in this part of parameter space.

## 5 Stochastic exposure

Next, we examine the problem when there is uncertainty in the exposure obtained for a given premium so that $\sigma_1, \sigma_2 \neq 0$. Again, consider the exponential utility function (11).

Let us look for a solution of the HJB equation (12) of the form

$$V = \frac{1}{c}\left(1 - e^{-c(w+F(\bar{p},q,t))}\right),$$

which requires that $F$ satisfy

$$F_t + \tfrac{1}{2}\sigma_1^2\left(F_{qq} - cF_q^2\right) + \mu\bar{p}F_{\bar{p}} + \tfrac{1}{2}\sigma_2^2\bar{p}\left(F_{\bar{p}\bar{p}} - cF_{\bar{p}}^2\right) + q\left(f(b - k^i) - \kappa\right)F_q +$$
$$\bar{p}qf\left(k^i - \gamma\right)\left(b - k^i\right) - \tfrac{1}{2}\sigma_1^2\bar{p}^2c\left(k^i - \gamma\right)^2 - \sigma_1^2\bar{p}c\left(k^i - \gamma\right)F_q = 0, \quad (26)$$

and $F(\bar{p}, q, T) = 0$. Here, the interior control is

$$k^i = \frac{qf\left(\bar{p}(b + \gamma) - F_q\right) + \sigma_1^2\bar{p}c\left(\gamma\bar{p} - F_q\right)}{\bar{p}\left(\sigma_1^2\bar{p}c + 2qf\right)},$$

and $V$ satisfies the second-order condition (14).

In general, there is no analytical solution of this problem and the numerical solution is difficult to visualize because there are two independent state variables $\bar{p}$ and $q$. Consequently, we focus on two specific cases which highlight the features of the optimal control. We separate the analysis into sections on linear and exponential utility functions because in the linear case one can incorporate stochastic $\bar{p}$.

## 5.1 Linear utility function

The case that $U_2(w)$ is linear in wealth corresponds to $c \to 0^+$. The HJB equation (26) now contains terms linear in $\bar{p}$, which suggests we look for a solution of the form

$$F = \bar{p}H(q,t),$$

so that the problem simplifies to

$$H_t + \mu H - \kappa q H_q + \tfrac{1}{2}\sigma_1^2 H_{qq} + qf(b - k^i)(k^i - \gamma + H_q) = 0, \qquad (27)$$
$$k^i = \tfrac{1}{2}(b + \gamma - H_q).$$

If we substitute the interior control back into the reduced HJB equation we find the nonlinear second-order PDE

$$H_t + \mu H + \tfrac{1}{2}\sigma_1^2 H_{qq} - \kappa q H_q + \tfrac{1}{4}qf(b - \gamma + H_q)^2 = 0, \qquad (28)$$

which has boundary condition $H(q,T) = 0$, from the terminal wealth condition. In general, it is clear that the volatility of the exposure $\sigma_1$ affects the optimal control whereas the volatility of the market average premium $\sigma_2$ does not.

In an infinite market $(f = a)$ $H$ is linear in $q$, and so we retrieve the deterministic, open-loop, interior control. If the market is finite then we must solve the HJB equation (28) numerically. Notice that in this case the HJB equation is second-order in contrast to (20). It is easier to solve this equation numerically using the uncoupled form (27) because then one can account for the direction of the characteristics as $\sigma_2 \to 0$.

We adopt a backward first-order time step for the time derivative, a centered second-order difference for the diffusion term, and a one-sided derivative for the advective term. The interior control gives the sign of the coefficient of the advective term $H_q$ in (27) and determines which one-sided difference we use to compute this spatial derivative. Thus we adopt the numerical parameters given in Table 1, and assume $f(q)$ is given by (6), while the volatility $\sigma_2(q)$ is parameterized by (9).

There is a complication with the numerical scheme at $q = q_m$ since we do not know the derivative $H_q$. Consequently, we extend the domain of the exposure to $q = q_{max}$, suppose $\sigma_1(q) = 0$ for $q_m \leq q \leq q_{max}$ and take $H_q(q_{max}, t) = 0$. The reasoning for this boundary condition is as follows. For $q(T) > q_m$ the volatility $\sigma_2(q(T)) = 0$ which means the HJB equation is deterministic at termination:

$$H_t + \mu H - \kappa q H_q = 0.$$

Thus, for sufficiently large $q$ we obtain from the termination condition that $H \equiv 0$ so that $k^i \equiv \tfrac{1}{2}(b + \gamma)$ and the exposure decays exponentially. The periphery of this region is determined by the characteristic which originates from the point $q = q_m$, $t = T$. Provided we take $q_{max}$ sufficiently large that it

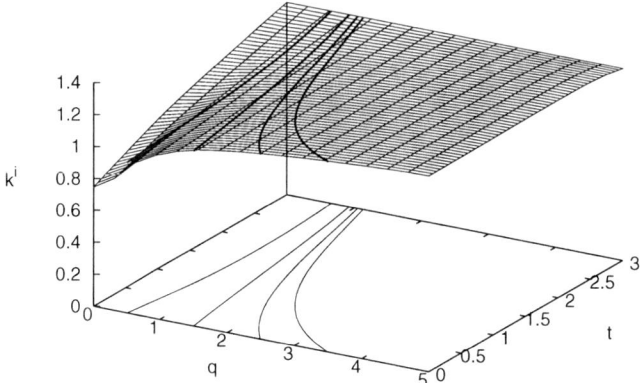

**Fig. 5.** State diagram of the interior premium strategy for the deterministic model in a finite insurance market calculated numerically using a finite difference scheme. Optimal trajectories computed using shooting are superimposed on the plot in order to illustrate the change in exposure for given initial exposure. These trajectories are also projected on to the $(q, t)$ plane.

lies in the region where $H \equiv 0$ then we need not explicitly calculate the line on which $H = 0$. Moreover, rather than take $H = 0$ as the boundary condition we use $H_q = 0$ because then numerical oscillations do not interact with the fictitious boundary. Wang *et al.* [12] employ a similar extended domain when solving deterministic optimal control problems.

The integration proceeds backwards from $t = T$ using the terminal boundary condition, and spatial conditions $H(0, t) = H_q(q_{max}, t) = 0$. First we show the results with no volatility in order to compare against the deterministic results. Figure 5 shows the results from the finite difference scheme superimposed with optimal trajectories computed from the maximum principle using shooting. We call this figure the state diagram: for state $q$ at time $t$ we can read off the interior control $k^i(q, t)$. It is reassuring to note that both numerical schemes lead to the same interior control $k^i$. It can be seen that low initial exposure leads to exposure generating strategies, while high initial exposure leads to strategies which represent market withdrawal. Notice that the state trajectories are functions of time whereas in the phase diagram of Figure 1(b), which is for an identical parameter set, time *parameterizes* the curves.

When the exposure is stochastic $\sigma_1 \neq 0$ we use the finite difference scheme to calculate the interior control. However, the results are qualitatively similar to the deterministic case in the sense that the computed surface for the interior control is only slightly perturbed for reasonable values of the volatility. The interpretation of the optimal control differs because now the state equation is stochastic so one must read off the current value of the state variable from the plot in order to determine the current interior control. This means the optimal control $k^* = k^*(q,t)$ is in feedback form rather than open-loop. Since $0 < q_t^* < q_m$ we can safely say the optimal state trajectory is well-defined.

Figure 6 shows numerical results from the finite difference scheme for the finite market case. The first pair of graphs (a) show the interior premium $k^i$ and the adjoint variable $\Lambda$ when the volatility $\sigma_0$ is small. The shape of the surface for the interior control is similar to the deterministic case (Figure 5) implying that for low values of the current exposure the insurer should set a low premium in order to build-up exposure. The instantaneous drift in the exposure is proportional to $G^i - \kappa = \frac{1}{2}a\left(\phi + \Lambda^i\right) - \kappa$, so that the drift is zero when

$$\Lambda^i = \Lambda_0 := 2\psi - \phi.$$

On the plot for $\Lambda$ we have superimposed the contour $\Lambda = \Lambda_0$, which partitions the $(q,t)$ plane into two regions. We can use this partition to describe the strategy based on the current value of the insurer's exposure. If $\Lambda_0 > 0$ then we say the current optimal strategy is expansion, while if $\Lambda < \Lambda_0$ then we say the current optimal strategy is withdrawal. Depending on the path of the Brownian motion, the optimal strategy can change from expansion to withdrawal an infinite number of times along the optimal state trajectory.

Figure 6(b) shows the corresponding result if we adopt a very large volatility $\sigma_0 = 5.0$. Now the interior relative premium is larger and above the market average premium $\left(k^i > 1\right)$. Indeed, since $k^i > \gamma = 0.91$ then no value of the current exposure leads to loss-leading. If the insurer loss-leads but does not actually achieve the expected increase in exposure then significant losses will occur. Consequently, when the volatility is high it is optimal to set a much higher premium, in this case above the market average premium, so that the change in exposure is much smaller. The contour $\Lambda = \Lambda_0$ moves to smaller values of exposure, so that the withdrawal strategy is enhanced as the volatility of the exposure is increased. We also observe that the profile of $k^i(q)$ for fixed small $t$ is almost linear in $q$. This is consistent with the diffusive term dominating the HJB equation when the volatility is very large.

## 5.2 Exponential utility function

Now we focus on just the stochastic exposure equation and take $\sigma_2 = 0$. Thus, if we set $F = \bar{p}H(q,t)$ then the reduced HJB equation becomes

$$H_t + \mu H + \frac{1}{2}\sigma_1^2\left(H_{qq} - \bar{p}cH_q^2\right) + q\left(f(b-k^i) - \kappa\right)H_q + qf(k^i - \gamma)(b - k^i)$$
$$- \frac{1}{2}c\sigma_1^2\bar{p}\left(k^i - \gamma\right)^2 - c\sigma_1^2\bar{p}\left(k^i - \gamma\right)H_q = 0, \quad (29)$$

with boundary condition $H(q,T) = 0$. The interior control is

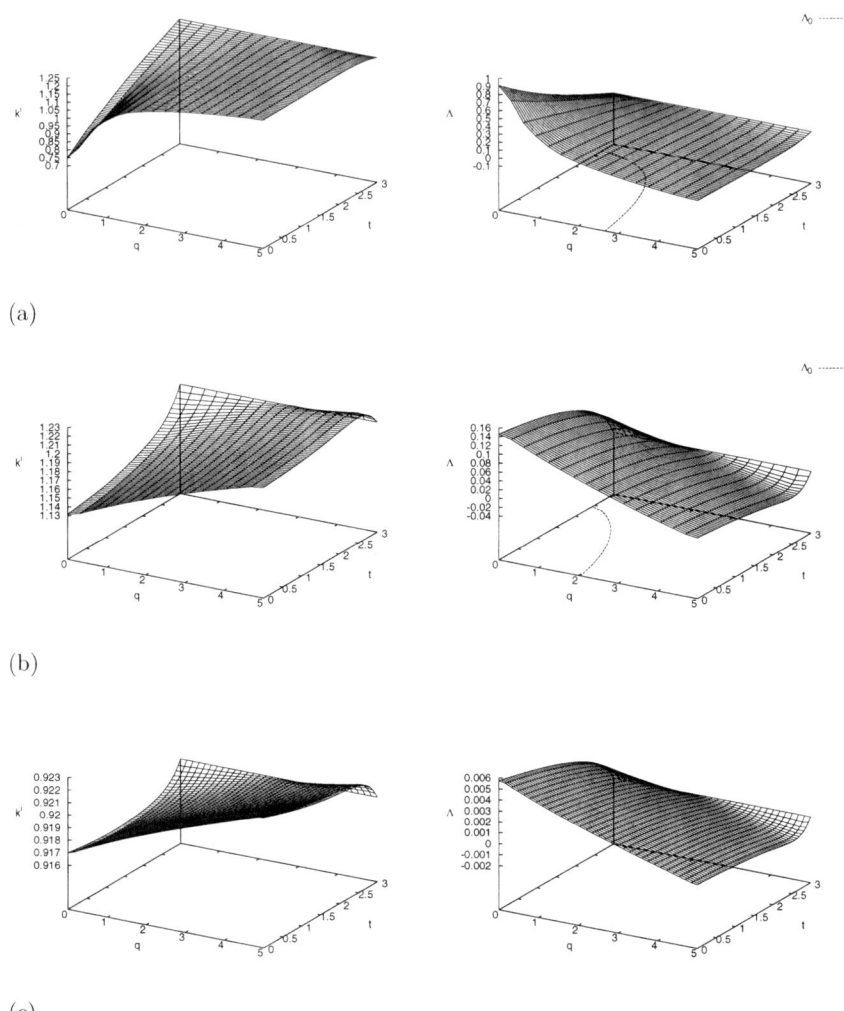

(a)

(b)

(c)

**Fig. 6.** Numerical solution of the HJB equation for a terminal wealth objective and a finite market. The plots show the interior control $k^i$ and the adjoint variable $\Lambda = H_q$ for (a) $\sigma_0 = 0.1$, $c = 0$, (b) $\sigma_0 = 5.0$, $c = 0$ (c) $\sigma_0 = 5.0$, $c = 5.0$.

$$k^i = \frac{qf(b + \gamma - H_q) + c\sigma_1^2 \bar{p}(\gamma - H_q)}{2qf + c\sigma_1^2 \bar{p}}.$$

At termination the interior control is

$$k^i = \frac{qf(b + \gamma) + c\sigma_1^2 \bar{p}\gamma}{2qf + c\sigma_1^2 \bar{p}}.$$

If $c\sigma_1^2 = 0$ then the interior premium is $\frac{1}{2}(b + \gamma)$, that is the average of the relative premium which leads to no demand, $b$, and the ratio of the breakeven premium $\pi$ to the market average premium $\bar{p}$. If $c\sigma_1^2 \gg 1$ then the interior premium is $\gamma$ so that the interior premium is the breakeven premium. This behaviour of the interior control at termination is independent of the exposure function $f$.

## Infinite market

In an infinite market ($f = a$) with volatility parameterized by (10) we look for a solution of the form

$$H(q, t) = qI(t),$$

which reduces the HJB equation to a Riccati equation:

$$\frac{dI}{dt} + a_0(t) + a_1(t)I + a_2(t)I^2 = 0; \qquad I(T) = 0,$$

with coefficients

$$a_0(t) = \frac{\frac{1}{2}a(b - \gamma)^2}{2 + \beta_0 \bar{p}}, \quad a_1(t) = \frac{a(1 + \beta_0 \bar{p})(b - \gamma)}{2 + \beta_0 \bar{p}} + \mu - \kappa - a\beta_0 \bar{p}, \quad a_2(t) = \frac{\frac{1}{2}a}{2 + \beta_0 \bar{p}},$$

and $\beta_0 = c\sigma_0^2/a$. The interior control becomes

$$k^i(t) = \frac{(b + \gamma) + \beta_0 \bar{p}\gamma - I(t)(1 + \beta_0)}{2 + \beta_0 \bar{p}},$$

which is an open-loop control. Consequently, the optimal state equation is

$$dq_t = q_t\left(\tfrac{1}{2}a(b - k^i(t)) - \kappa\right)dt + \sigma_0 \sqrt{q\left(1 - \frac{q}{q_m}\right)^+}\, dW_{1,t}.$$

Provided that $k^i(t) \in C^1[0, T]$ then this equation has a strong solution in $[0, T]$ [8]. This means we can apply the verification theorem in [5] and $k^i(t)$ is the optimal control.

An analytical expression is available for the optimal control only if $\bar{p} = $ const., that is $\mu = 0$, because then the coefficients $a_2, a_1, a_0$ are constant. Notice that $a_0, a_2 \to 0$ as $\beta_0 \to \infty$ so that $I \equiv 0$ satisfies the Riccati equation, and so $k^i \equiv \gamma$. Consequently, as gaining insurance becomes more uncertain it is optimal for the insurer to adopt the breakeven premium over the *entire time horizon* in order not to make significant losses.

**Finite market**

In a finite market we adopt the parameterizations (6) and (9) so that at termination the interior relative premium is

$$k^i = \frac{b + \gamma + \beta \overline{p} \gamma}{2 + \beta \overline{p}},$$

where $\beta$ is given by (24). In contrast to the case when the market average premium is stochastic (Section 4), the volatility changes the terminal interior control. There is no analytical solution of (29) and therefore we calculate the interior control numerically.

If the risk aversion $c$ is small then the results are similar to Figure 6(a). If the volatility is small then the problem almost deterministic and so the optimal control is independent of the utility function and also similar to that given in Figure 6(a). Graphs for these cases are not shown.

Numerical results for the parameters given in Table 1 are shown in Figure 6(c) for large values of the risk aversion $c$ and volatility of exposure $\sigma_0$. We can see the contour $\Lambda = \Lambda_0 = 0.076$ no longer appears in the plots indicating that the interior premium always leads to an expected instantaneous loss of exposure. In addition, there is little variation of the interior premium from $k = \gamma = 0.91$ over the entire time horizon. Consequently, when the demand for policies is uncertain and the insurance company is risk averse it is optimal to set premium rates at breakeven and this leads to expected market withdrawal irrespective of the current exposure of the insurer.

# 6 Conclusions

We have determined the relationship between the deterministic optimal premium strategies given by the two-dimensional phase diagram shown in Figure 1, and the optimal premium strategy of the generalized stochastic model. We have focused on the terminal wealth objective, a linear price function and an exponential utility function. These simplifications allow us to describe the optimal control either through a three-dimensional phase diagram or a state diagram of one state dimension plus time.

In the deterministic problem the optimal control is independent of the utility function if we ignore degenerate cases. This is not true for the stochastic model: for example, Figure 4 shows how the optimal pricing strategy changes if the risk aversion of the insurer is non-zero.

In a finite market there is a limit to the number of policies that the insurer can sell, which also alters the qualitative form of the optimal strategy. If the uncertainty in the model is derived solely from the market average premium then the optimal control can be described by three-dimensional phase space and a pair of equilibrium points. The optimal control is either expansion (with possible loss-leading) or withdrawal and the criterion for which strategy is optimal is determined by the position of the two equilibrium points and the initial conditions. The optimal relative premium is deterministic and in open loop form because there is no explicit analytical relationship between the optimal control and the current state.

If instead, the change in demand is explicitly stochastic then the type of the HJB equation is parabolic so the phase space cannot be used to study the optimal control. We cannot reduce the HJB equation to a finite system of ODEs. Instead we examine the state diagram, which describes the evolution of the feedback control and for which we must explicitly state the time horizon. In this diagram the expansion strategy is delineated from the withdrawal strategy by the premium which yields no instantaneous expected change in the exposure. On a typical optimal state trajectory the insurer can jump from one region to the other due to the volatility of the exposure. One can only say at the current time whether it is optimal to expand and set a relatively low premium or withdraw with a higher premium. The optimal relative premium is a stochastic process.

The two forms of stochasticity lead to two similar types of optimal premium strategy when the market is very volatile. As the volatility of the market average premium, $\sigma_1$, is increased then a breakeven premium strategy is optimal up until near termination of the planning horizon where the optimal premium is independent of this volatility. In contrast, as the volatility of the demand for policies, $\sigma_2$, is increased then the breakeven premium strategy is optimal over the entire planning horizon if there is positive risk aversion.

# References

1. Batchelor GK (1967) An introduction to fluid dynamics. Cambridge University Press.
2. Courant R, Hilbert D (1966) Methods of Mathematical Physics, Vol. II. Wiley.
3. Emms P (2007) Dynamic pricing of general insurance in a competitive market. ASTIN Bulletin, 37(1):1–34.
4. Emms P, Haberman S (2006) Optimal management of an insurer's exposure in a competitive general insurance market. Cass Business School, London. Available at SSRN: http://ssrn.com/abstract=988901.
5. Fleming W, Rishel R (1975) Deterministic and Stochastic Optimal Control. Springer Verlag, New York
6. Gallego G, van Ryzin G (1994) Optimal Dynamic Pricing of Inventories with Stochastic Demand over Finite Horizons. Management Science, 40(8):999–1020.
7. Kalish S (1983) Monopolistic pricing with dynamic demand and production cost. Marketing Science, 2:135–160.
8. Kloeden PE, Platen E (1999) Numerical Solution of Stochastic Differential Equations. Springer-Verlag, Berlin
9. Lilien GL, Kotler P (1983) Marketing Decision Making. Harper & Row
10. Merton RC (1990) Continuous-time Finance. Blackwell
11. Taylor GC (1986) Underwriting strategy in a competitive insurance environment. Insurance: Mathematics and Economics, 5(1):59–77.

12. Wang S, Gao F, Teo KL (2000) Solving Hamiltonian-Jacobi-Bellman equations by an upwind finite difference scheme. In: Mees AI, Fisher ME, Jennings LS (eds) Progress in Optimisation II, 255–268. Kluwer Academic Publishers
13. Yong J, Zhou XY (1999) Stochastic controls: Hamiltonian systems and HJB equations. Springer-Verlag.

# Optimality of Deterministic Policies for Certain Stochastic Control Problems with Multiple Criteria and Constraints

Eugene A. Feinberg

State University of New York at Stony Brook,
Stony Brook, NY 11794-3600, USA. Eugene.Feinberg@sunysb.edu

**Summary.** For single-criterion stochastic control and sequential decision problems, optimal policies, if they exist, are typically nonrandomized. For problems with multiple criteria and constraints, optimal nonrandomized policies may not exist and, if optimal policies exist, they are typically randomized. In this paper we discuss certain conditions that lead to optimality of nonrandomized policies. In the most interesting situations, these conditions do not impose convexity assumptions on the action sets and reward functions.

## 1 Introduction

In many applications, the system performance is measured by multiple criteria. For example, in finance, such criteria measure returns and risks, in manufacturing such criteria may be production volumes, quality of outputs, and costs, in service operations performance criteria include service levels and operating costs.

For problems with multiple criteria, the natural approach is to optimize one of the criteria subject to the inequality constraints on the other criteria. In other words, for a problem with $K + 1$ criteria $W_0(\pi), W_1(\pi), \ldots, W_K(\pi)$, where $\pi$ is a policy, the natural approach is to find a policy $\pi$ that is a solution to the following problem

$$\text{maximize } W_0(\pi) \tag{1}$$

subject to

$$W_k(\pi) \geq C_k, \qquad k = 1, \ldots, K, \tag{2}$$

where $C_1, \ldots, C_K$ are given numbers. For example, since it is possible to consider $W_{k+1}(\pi) = -W_k(\pi)$, this approach can be used to find policies satisfying interval constraints $a \leq W_k(\pi) \leq b$.

Optimal solutions of problem (1, 2), if they exist, are typically randomized with the number of randomization procedures limited by the number of constraints $K$; see [1, 16]. If there are no constraints, i.e. $K = 0$, optimal policies are nonrandomized. The following simple example illustrates that it is possible that any optimal policy for a constrained problem is randomized.

Consider a one-step problem when a decision-maker chooses among two decisions $a$ and $b$. There are two reward functions $r_0$ and $r_1$ defined as $r_0(a) = r_1(b) = 0$ and $r_1(a) = r_0(b) = 1$. The decision-maker selects action $a$ with probability $\pi(a)$ and action $b$ with probability $\pi(b)$, where $\pi(a) + \pi(b) = 1$. The criteria are $W_k(\pi) = \pi(a)r_k(a) + \pi(b)r_k(b)$, $k = 0, 1$. Then the problem

$$\text{maximize } W_0(\pi) \tag{3}$$

subject to

$$W_1(\pi) \geq 1/2 \tag{4}$$

is equivalent to the following linear program (LP)

$$\text{maximize } \pi(b)$$

subject to

$$\pi(a) \geq 1/2,$$

$$\pi(a) + \pi(b) = 1,$$

$$\pi(a) \geq 0, \pi(b) \geq 0.$$

This LP has the unique optimal solution $\pi(a) = \pi(b) = 1/2$. Therefore, the optimal policy is randomized.

In many applications, implementation of randomized policies is not natural. In many cases, it is more natural to apply nonrandomized policies when they are optimal. In addition, it appears that the use of randomization procedures increases the variance of the performance criteria. Also, from the computational point of view, finding the best randomized policy in many cases is easy, because this can be done by using linear programming. Finding the best nonrandomized policy may be a computationally intractable problem. For example, finding the best nonrandomized stationary policy for constrained dynamic programming is an NP-hard problem [5]. Typically, when nonrandomized policies are optimal, an optimal nonrandomized policy can be computed by a simple transformation of an optimal randomized policy. Thus, computing optimal nonrandomized policies becomes a computationally tractable problem when nonrandomized policies are optimal.

In this article, we discuss the situations when nonradomized policies are optimal for problems with multiple criteria and constraints. Of course, nonrandomized policies are at least as good as randomized policies when actions sets are convex subsets of a linear space and reward functions are concave. This situation is trivial and we concentrate on the models when neither the convexity of the action sets nor the concavity of reward functions is assumed. In particular, we consider the following three cases: (i) unichain Markov Decision Processes (MDPs) with average rewards per unit time, (ii) continuous-time

MDPs, and (iii) nonatomic dynamic programming. In case (i), it is possible to achieve optimal state-action frequencies by selecting different actions at different visits to the same state, in case (ii) it is possible to change actions between jumps instead of choosing actions randomly, and in case (iii) the non-atomicity of initial and transition probabilities implies that the for any randomized policy there exists a nonrandomized policy with the same performance vector.

## 2 Discrete-time MDPs with average rewards per unit time

Consider a discrete-time MDP with finite state and action sets and with average rewards per unit time [1, 16, 17]. This model is defined by the objects $\{X, A, A(\cdot), p, K, r.\}$, where

(i) $X$ is a finite *state space*;
(ii) $A$ is a finite *action set*;
(iii) $A(x)$ are *the sets of available actions* at state $x \in X$, where $A(x) \subseteq A$;
(iv) $p(y|x, a)$ is the one-step *transition probability*, i.e. $p(\cdot|x, a) \geq 0$ is a probability distribution on $X$ for each $x \in X$ and $a \in A(x)$;
(v) a finite nonnegative integer $K$ is *the number of constraints*;
(vi) $r_k(x, a)$, $k = 0, \ldots, K$, is the one-step *reward* according to the $k^{\text{th}}$ criterion if an action $a \in A(x)$ is selected at a state $x \in X$.

Let $H_n = X \times (A \times X)^n$ be the set of *trajectories* up to time $n = 0, 1, \ldots$. A policy $\pi$ is a sequence $(\pi_0, \pi_1, \ldots)$ of transition probabilities from $H_n$ to $A$ such that $\pi_n(A(x_n)|x_0, a_0, \ldots, x_n) = 1$. A policy $\pi$ is called *nonrandomized* if $\pi_n(a|h_n) \in \{0, 1\}$ for all $n = 0, 1, \ldots$, $h_n \in H_n$, and $a \in A$. Equivalently, nonrandomized policy $\sigma$ is defined by a sequence of measurable mappings $\sigma_n : H_n \to A$ such that $\sigma_n(x_0, a_0, \ldots, x_n) \in A(x_n)$, $n = 0, 1, \ldots$. A policy $\pi$ is called *randomized stationary* if $\pi_n(a_n|x_0, a_0, \ldots x_n)$ depends only on the value of $x_n$, i.e. there exists a transition probability $\pi$ from $X$ to $A$ such that $\pi_n(a_n|x_0, a_0, \ldots x_n) = \pi(a_n|x_n)$. A nonrandomized stationary policy is called *stationary*.

Any initial distribution $\mu$ and any policy $\pi$ define a probability measure $P_\mu^\pi$ on the set of infinite trajectories $H_\infty$ endowed with its Borel $\sigma$-field. We denote by $E_\mu^\pi$ expectations with respect to this measure. We shall write $P_x^\pi$ and $E_x^\pi$ instead of $P_\mu^\pi$ and $E_\mu^\pi$ if $x$ is the initial state, i.e. $\mu(x) = 1$.

For an initial distribution $\mu$, and policy $\pi$, *the average rewards per unit time* are defined as

$$W_k(\mu, \pi) = \liminf_{N \to \infty} \frac{1}{N} E_\mu^\pi \sum_{n=0}^{N-1} r_k(x_n, a_n), \qquad k = 0, 1, \ldots, K.$$

For a fixed initial distribution $\mu$, consider problem (1, 2) when $W_k(\pi) = W_k(\mu, \pi)$, $k = 0, \ldots, K$. So, the goal is to maximize the expected total rewards

$W_0(\mu,\pi)$ subject to the constraints on the expected total rewards $W_k(\mu,\pi)$, $k = 1,\ldots,K$. Though an optimal policy exists for this problem when the state and action spaces are finite, it may have a complicated form. In particular, it may be nonstationary [12, 14]. However, optimal randomized stationary policies exist under the following condition.

**Unichain Condition.** *Any nonrandomized stationary policy defines a Markov chain on $X$ with one recurrent class (and possible transient states).*

According to [12, 18], if the Unichain Condition holds then there exists a randomized stationary policy that is optimal for any initial distribution. To find such a policy, we need to solve the following LP with variables $u_{x,a}$, where $x \in X$, $a \in A(x)$ :

$$\text{maximize} \sum_{x\in X}\sum_{a\in A(i)} r_0(x,a)u_{x,a}$$

subject to

$$\sum_{x\in X}\sum_{a\in A(x)} r_k(x,a)u_{x,a} \geq C_k, \qquad k = 1,\ldots,K,$$

$$\sum_{a\in A(y)} u_{y,a} - \sum_{x\in X}\sum_{a\in A(x)} p(y,a,x)u_{x,a} = 0, \qquad y \in X,$$

$$\sum_{x\in X}\sum_{a\in A(x)} u_{x,a} = 1,$$

$$u_{x,a} \geq 0, \qquad x \in X, \ a \in A(x).$$

Let $u$ be an optimal solution of the above LP. Let

$$u_x = \sum_{a\in A(x)} u_{x,a}, \qquad x \in X.$$

Then, if the Unichain Condition holds, the following formula defines an optimal randomized stationary policy $\pi$

$$\pi(a|x) = \begin{cases} \frac{u_{x,a}}{u_x} & \text{if } u_x > 0, \\ \text{arbitrary} & \text{if } u_x = 0. \end{cases}$$

The policy $\pi$ may not be nonrandomized. If $u_x = 0$, it is natural to select $\pi$ being nonrandomized at $x$. If $u$ is a basic optimal solution of the above LP, there are at most $|X| + K$ nonzero state-action pairs $(x,a)$ such that $\pi(a|x) > 0$ [18], where $|E|$ is the number of elements of a finite set $E$. For a nonrandomized stationary policy, the number of such pairs is $|X|$.

Ross [18] and Altman and Shwartz [2, 3] investigated approaches to construct a nonrandomized stationary policy $\sigma$ such that $W_k(\mu,\sigma) = W_k(\mu,\pi)$ for any reward function $r_k$. If such a policy is constructed, then it is optimal. Ross [18] studied the case $K = 1$ when the Unichain Condition holds. Altman and Shwartz [2, 3] also assumed the Unichain Condition and considered the case of an arbitrary finite $K$. In addition, they studied MDPs with a countable state space $X$ and finite state MDPs with finite action sets. Altman and Shwartz [2, 3] introduced the concept of time-sharing policies. These policies combine several nonrandomized policies.

For an initial state distribution $\mu$ and a policy $\sigma$, a state-action frequency is defined as

$$f_\mu^\sigma(x, a) = \lim_{N \to \infty} \frac{1}{N} \sum_{n=0}^{N-1} P_\mu^\sigma \{x_n = x, a_n = a\}, \qquad x \in X, a \in A(x), \quad (5)$$

if this limit exists for all $x \in X$ and for all $a \in A(x)$. In particular, state-action frequencies exist for randomized stationary policies for countable-state MDPs with countable action sets. If for two policies $\sigma_1$ and $\sigma_2$ the limits in the definition of the vectors of state-action frequencies (5) exist for the initial distribution $\mu$ and these vectors are equal for $\sigma_1$ and $\sigma_2$, then $W(\mu, \sigma_1) = W(\mu, \sigma_2)$ for any bounded or nonnegative reward function $r$. In [18, 2, 3], the policy $\sigma$ is constructed in the way that the state-action frequencies for $\sigma$ exist and equal to state-action frequencies for an optimal randomized stationary policy $\pi$, and these frequencies do not depend on the initial distribution $\mu$.

For a finite history $h_n = x_0, a_0, \ldots, x_n$, define the empirical frequencies

$$N(h_n; x, a) = \sum_{t=0}^{n-1} I\{x_t = x, a_t = a\}, \qquad x \in X, a \in A(x),$$

$$N(h_n, x) = \sum_{t=0}^{n-1} I\{x_t = x\}, \qquad x \in X.$$

For the optimal solution $u$ of the LP, define the nonrandomized policy $\sigma$ by

$$\sigma_n(h_n) = \operatorname{argmax}_{a \in A(i)} \{u_{i,a} - \frac{N(h_n; x, a) + 1}{N(h_n; x) + 1}\},$$

where ties are broken arbitrarily. Then for finite $X$ and $A$, if the Unichain Condition holds, then $\sigma$ has the same state-action frequencies as $\pi$ and therefore $\sigma$ is an optimal nonrandomized policy. Feinberg and Curry [8] used the above form of the optimal nonrandomized policy $\sigma$ in a heuristic algorithm for a so-called Generalized Pinwheel problem, which is an NP-hard scheduling problem.

## 3 Continuous-time MDPs

Consider a continuous-time MDP with a finite or countable state space [6, 7, 13]. This model is defined by the objects $\{X, A, A(\cdot), p, K, r., R.\}$, where

(i) $X$ is a finite or countable *state space*;

(ii) $A$ is a Borel *action set* (a measurable subset of a complete separable metric space);

(iii) $A(x)$ are *the sets of available actions* at state $x \in X$, where $A(x)$ are measurable subsets of $A$;

(iv) $q(y|x,a)$ are *jump intensities*, i.e. $q(y|x,a) \geq 0$, $q(x|x,a) = 0$, and $q(x,a) := \sum_{y \in X} q(y|x,a) \leq C < \infty$ for all $x \in X$ and $a \in A(x)$, and, in addition, the functions $q(y|x,a)$ are measurable on $A(x)$ for all $x, y \in X$;

(v) a finite nonnegative integer $K$ is *the number of constraints*;

(vi) $r_k(x,a)$, $k = 0, \ldots, K$, is the *reward rate* according to the $k^{\text{th}}$ criterion if an action $a \in A(x)$ is selected at a state $x \in X$;

(vii) $R_k(x,a)$, $k = 0, \ldots, K$, is the *reward* according to the $k^{\text{th}}$ criterion if a jump from state $x \in X$ to state $y \in Y$ occurs while an action $a \in A(x)$ is selected at a state $x \in X$.

Let $\mathbb{R}_+ = [0, \infty)$ and $H_n = X \times (\mathbb{R}_+ \times X)^n$, $n = 0, 1, \ldots$ . Let $H = \cup_{0 \leq n < \infty} H_n$. The sets $H_n$ are endowed with the $\sigma$-fields $\mathcal{H}_n$ which are the products of the $\sigma$-fields on $X$ and on $\mathbb{R}_+$, where the $\sigma$-field on $X$ is the set of the subsets of $X$ and the $\sigma$-field on $\mathbb{R}_+$ is the Borel $\sigma$-field on $\mathbb{R}_+$. Let $\mathcal{H}$ be the minimal $\sigma$-field on $H$ that contains $\mathcal{H}_n$, $n = 0, 1, \ldots$ . We interpret $x_0, \xi_0, \ldots x_n \in H_n$ as the sequence of the first $n + 1$ states $x_0, \ldots, x_n$ and the first $n$ sojourn times $\xi_0, \ldots, \xi_{n-1}$ of a multivariate point process with the state space $X$. The jump epochs are $t_m = \sum_{i=0}^{m-1} \xi_i$, $i = 1, \ldots, n$, with $t_0 = 0$.

A *randomized strategy* is a regular transition probability from $H \times \mathbb{R}_+$ to $A$ such that $\pi(A(x_n)|x_0, \xi_0, \ldots, x_n, s) = 1$ for any $x_0, \xi_0, \ldots, x_n \in H$, $n = 0, 1, \ldots$ and for any $s \geq 0$. In other words, $\pi(\cdot|x_0, \xi_0, \ldots, x_n, s)$ is a probability measure on $A(x_n)$ for each $x_0, \xi_0, \ldots, x_n \in H$ and $0 \leq s < \infty$, and $\pi(E|\cdot)$ is a Borel function on $H \times \mathbb{R}_+$ for any measurable subset $B$ of $A$. We interpret $s$ as the time elapsed since the last jump. Thus, the decision at any time depends on the previous states and jump epochs, and on the time elapsed since the last jump.

Let $H_\infty = X \times \mathbb{R}_+ \times X \times \mathbb{R}_+ \ldots$ be the infinite product of $X$ and $\mathbb{R}_+$. Let $\mathcal{H}_n$ be the countable product of the $\sigma$-fields on $X$ and $\mathbb{R}_+$. We define $\Omega = \cup_{0 \leq n \leq \infty} H_n$ and $\mathcal{F}$ as the minimal $\sigma$-field containing $\{\mathcal{H}_n, 0 \leq n \leq \infty\}$.

We interpret $\Omega$ as the sample space of the multivariate point process without accumulation points. In particular, $H_n$ represent the sets of trajectories with $n$ jumps, $n = 0, 1, \ldots$, or $n = \infty$. We set $t_m = \infty$ when $m > n$ and $\omega \in H_n$ with $n < \infty$. In particular, the value $X(t)$ of the process at time $t > 0$ is defined as $X(t) = \sum_{0 \leq n < \infty} I\{t_n < t \leq t_{n+1}\} x_n$.

A randomized strategy $\pi$ and an initial state distribution $\mu$ define a unique probability measure $P_\mu^\pi$ on $(\Omega, \mathcal{F})$ such that $P_\mu^\pi(x_0 = x) = \mu(x)$, $x \in X$, and the compensator of the random measure corresponding to $P_\mu^\pi$ is

$$\nu^\pi(\omega; dt, j) = \sum_{n \geq 0} I\{t_n < t \leq t_{n+1}\} \int_A q(j|x_n, a) \pi(da|x_0, \xi_0, \ldots, x_n, t - t_n) dt,$$

where $t_n = \sum_{i=0}^{n-1} t_i$; see [11, 13, 15]. The assumption $q(x, a) \leq C < \infty$ implies that the process defined by $P_\mu^\pi$ does not have accumulation points.

A *nonrandomized strategy* $\phi$ is defined by a measurable mapping from $H \times \mathbb{R}_+$ to $A$ such that $\phi(x_0, \xi_0, \ldots, x_n, s) \in A(x_n)$. For a nonrandomized policy, the measure $P_\mu^\phi$ is defined by $P_\mu^\phi(x_0 = x) = \mu(x)$, $x \in X$, and by the compensator of the random measure corresponding to $P_\mu^\pi$ equal

$$\nu^\phi(\omega; dt, j) = \sum_{n \geq 0} I\{t_n < t \leq t_{n+1}\} q(j|x_n, \phi(x_0, \xi_0, \ldots, x_n, t - t_n)) dt.$$

A nonrandomized strategy is called *switching stationary* if $\phi(x_0, \xi_0, \ldots, x_n, s) = \phi(x_n, s)$ and, in addition, the function $\phi : X \to \mathbb{R}_+$ is piecewise-constant in $s$ and has a finite number of discontinuity points, where the discrete topology is considered on $X$. In other words, there is a finite subset $Y$ of $X$ such that the function $\phi(x, s)$ is constant in $s$ for each $x \in X \setminus Y$ and $\phi(x, s)$ is piecewise-constant in $s$ and has a finite number of jumps for each $x \in Y$. So, the switching stationary policy may change actions between jumps only at a finite number of states and for each such state $x$, where it changes actions, there is a finite number of times $0 < S_1(x) < S_2(x) < \ldots < S_j(x)$ such that the policy changes actions when the time $S_i(x)$ elapsed since the last jump.

Let $E_\mu^\pi$ be the expectation with respect to the measure $P_\mu^\pi$. Let $N(t)$ be the number of jumps up to time $t$, $N(t) = \max\{n \geq 0 | t_n \leq t\}$. Then the assumption $q(x, a) \leq C < \infty$ implies that $N(t) < \infty$ for all $0 \leq t < \infty$.

**Discounted total rewards.** Let $\alpha > 0$ be the discount rate. For a randomized strategy $\pi$, the expected total rewards are

$$W_k(\mu, \pi) = E_\mu^\pi \sum_{n=0}^\infty [e^{-\alpha t_n} \int_{A(x_n)} R_k(x_n, a, x_{n+1}) \pi(da|\omega, t_{n+1})$$

$$+ \int_{t_n}^{t_{n+1}} e^{-\alpha t} \int_{A(x_n)} r_k(x_n, a) \pi(da|\omega, t)], \qquad k = 0, \ldots, K,$$

where $\pi(da|\omega, t) = \sum_{n \geq 0} I\{t_n < t \leq t_{n+1}\} \pi(da|x_0, \xi_0, \ldots, x_n, t - t_n) dt$.

Consider problem $(\bar{1}, 2)$ with $W_k(\pi) = W_k(\mu, \pi)$ when all the action sets $A(x)$ are compact, the functions $R_k$ and $r_k$, $k = 0, \ldots, K$, are bounded above and continuous in $a$, and the functions $q(y|x, a)$ and $q(x, a)$ are continuous in $a$. This problem was studied in [7]. Three natural forms of optimal policies were described in [7] when this problem is feasible. In particular, for a feasible problem, there exists an optimal strategy that randomly selects actions at jump epoch and keeps them unchanged between jumps. The strategies of this type are called randomized stationary policies. A randomized stationary policy is defined by the distributions $\sigma(da|x)$ concentrated on $A(x)$, where the action $a$ is always selected with the probability $\pi(da|x_n)$ when the process jumps to a state $x_n$ and this action is used to control the process until the next jump epoch. Formula (8.6) in [7] shows how a randomized stationary policy can be presented as a randomized strategy $\pi$ described above. A *nonrandomized stationary policy* $\phi$, where $\phi$ is a function from $X$ to $A$ satisfying $\phi(x) \in A(x)$ for all $x \in X$, always uses action $\phi(x)$ at state $x \in X$.

As was shown in [7], any $m$-randomized stationary policy $\sigma$ can be transformed into a (nonrandomized) switching stationary strategy $\phi$ such that $W(\mu, \phi) = W(\mu, \sigma)$ for any bounded above or bounded below reward functions $r$ and $R$. For $x \in X$ consider the actions $a^1, \ldots, a^j$ from $A(x)$ such that $\sigma(a^i | x) > 0$, $i = 1, \ldots, j$, and $\sum_{i=1}^j \sigma(a^j | x) = 1$. Then $\phi(x, s) = a^1$ if $j = 1$ and for the case $j > 1$

$$\phi(x, s) = a_i \qquad \text{for } S_{i-1}(x) < s \le S_i(x), \quad i = 1, \ldots, j, \qquad (6)$$

where $S_0(x) = 0$ and $S_i(x) = \sum_{\ell=1}^i s_i(x)$, $i = 1, \ldots, j$, with

$$s_i(x) = -\frac{1}{\alpha + q(x, a^i)} \ln \left( 1 - \frac{\sigma(a^i | x)}{\sum_{\ell=i}^j \sigma(a^j | x)} \right). \qquad (7)$$

The equality $W(\mu, \phi) = W(\mu, \sigma)$ is based on the fact that the occupation measures for continuous-time MDPs introduced in [7] are equal for $\sigma$ and $\phi$ under any initial state distribution. The optimality of nonrandomized strategies can be summarized in the following theorem based on the results from [7].

**Theorem 1.** *Consider a discounted continuous-time MDP. Let $\sigma$ be an $m$-randomized stationary policy, $m = 0, 1, \ldots$, and $\phi$ be an $m$-switching stationary strategy defined in (6, 7). Then $W_k(\mu, \phi) = W_k(\mu, \sigma)$ for all $k = 0, \ldots, K$ and for all initial state distributions $\mu$.*

*If problem (1, 2) is feasible, all the action sets $A(x)$, $x \in X$, are compact, the reward functions $R_k$ and $r_k$, $k = 0, \ldots, K$ are bounded above and upper semi-continuous in $a$, and the functions $q(y|x, a)$ and $q(x, a)$ are continuous in $a$, then there exist optimal $k$-randomized stationary policies and optimal $k$-switching stationary strategies.*

Similar to the discrete time, LPs can be used to compute optimal $k$-randomized stationary policies and optimal $k$-switching stationary strategies; see [7] for details.

**Average rewards per unit time.** Let the sets $X$ and $A$ be finite. Consider average rewards per unit time

$$W_k(\mu, \pi) = \liminf_{T \to \infty} \frac{1}{T} E_\mu^\pi \left[ \sum_{n=0}^{N(T)} \int_{A(x_n)} R_k(x_n, a, x_{n+1}) \pi(da|\omega, t_{n+1}) \right.$$

$$\left. + \int_0^T \int_{A(x_n)} r_k(X(t), a) \pi(da|\omega, t) \right], \qquad k = 0, \ldots, K.$$

Let the Unichain Condition hold. Then for any randomized stationary policy $\pi$ and for any switching stationary strategy $\pi$, $W_k(\mu, \pi)$ does not depend on $\mu$. So, for such policies, we set $W_k(\pi) = W_k(\mu, \pi)$, $k = 0, \ldots, K$.

For an $m$-randomized stationary policy $\sigma$ we can consider the $m$-switching policy $\phi$ defined by (6, 7) with $\alpha = 0$. Then, according to [6], $W(\phi) = W(\sigma)$ for any rewards $r$ and $R$. The following theorem summarizes the results on average rewards per unit time from [6].

**Theorem 2.** *Let $X$ and $A$ be finite sets and let the Unichain Condition hold. Consider average rewards per unit time.*

*(i) For an $m$-randomized stationary policy $\sigma$ consider an $m$-switching stationary strategy $\phi$ defined in (6, 7) with $\alpha = 0$. Then $W_k(\phi) = W_k(\sigma)$ for all $k = 0, \ldots, K$.*

*(ii) If problem (1, 2) is feasible then there exists an optimal $K$-switching stationary strategy.*

*(iii) If problem (1, 2) is feasible and $q(x, a) > 0$ for all $x \in X$ and for all $a \in A(x)$ then there exists an optimal $K$-randomized stationary policy.*

If $q(x, a) = 0$ for some $x$ and $a$ then optimal randomized stationary policies may not exist even if the problem is feasible [6, Example 3.1]. Optimal $K$-randomized stationary policies and optimal $K$-switching stationary strategies can be found by using LPs described in [6].

# 4 Nonatomic MDPs

Consider an MDP $\{X, A, A(\cdot), p, K, r.\}$, introduced in Section 2 for finite $X$ and $A$. In this section we consider the situation when $X$ and $A$ are Borel spaces (measurable subsets of complete separable metric spaces) and the standard measurability conditions are satisfied. These conditions are:

(a) the graph of the set-valued mapping $x \rightarrow A(x)$ of $X$ is a measurable subset of $X \times A$ allowing a measurable selection, i.e. there exists a measurable function $\varphi : X \rightarrow A$ such that $\varphi(x) \in A(x)$ for all $x \in X$;
(b) $p$ is a regular transition probability from $X \times A$ to $X$;
(c) $r_k(x, a)$ are measurable functions on $X \times A$, $k = 0, \ldots, K$.

Let $C^+ = \max\{C, 0\}$, $C^- = \min\{C, 0\}$ and let $\beta$ be a discount factor, $0 \leq \beta \leq 1$. For a policy $\pi$ and for an initial probability measure $\mu$ on X, we define for $k = 0, \ldots, K$ the expected total rewards

$$W_{k,+}(\mu, \pi) = E_\mu^\pi \left[ \sum_{n=0}^\infty \beta^n (r_k(x_n, a_n))^+ \right],$$

$$W_{k,-}(\mu, \pi) = E_\mu^\pi \left[ \sum_{n=0}^\infty \beta^n (r_k(x_n, a_n))^- \right],$$

$$W_k(\mu, \pi) = W_{k,+}(\mu, \pi) + W_{k,-}(\mu, \pi), \tag{8}$$

where $+\infty - \infty = -\infty$.

We fix the initial probability distribution $\mu$ on $X$. We recall that a probability measure $P$ on a Borel space $Y$ is called *nonatomic* if $P(\{y\}) = 0$ for any $y \in Y$.

**Nonatomicity Assumption.** *The initial probability measure $\mu$ and all the transition probabilities $p(\cdot|x,a)$, $x \in X$ and $a \in A(x)$ are nonatomic.*

A nonrandomized policy $\phi$ is called *Markov* if $\phi_n(x_0, a_0, \ldots, x_n) = \phi_n(x_n)$ for all $n = 0, 1, \ldots$ .

The following statement is a particular case of [9, Theorem 2.1], where a nonhomogeneous MDP (the transition probabilities and rewards may depend on time) was considered.

**Theorem 3.** *If the Nonatomicity Assumption holds then for any policy $\pi$ there exists a Markov policy $\phi$ such that $W_k(\mu, \phi) = W_k(\mu, \pi)$ for all $k = 0, \ldots, K$.*

Unlike the results in Sections 2 and 3, the Nonatomicity Assumption does not imply that the occupation measures are equal and it is essential that $K$ is finite. The proof of Theorem 3 uses Lyapunov's theorem; see [9] for details. Combined with sufficient conditions for the existence of optimal solutions to problem (1, 2) for $W_k(\pi) = W_k(\mu, \pi)$, Theorem 3 implies the existence of optimal (nonrandomized) Markov policies; see [9, Section 5]

**Application to Statistical Decision Theory.** Let $X$ and $A$ be Borel sets. Similar to MDPs, $X$ is the state spaces and $A$ is the decision space. For each $x \in X$, a Borel subset $A(x)$ of $A$ (the set of decisions available at $x$) is given. The sets $A(x)$ satisfy the conditions described above in this section.

The initial distribution $\mu$ is not known, but it is known that it is equal to one of the measures $\mu_1, \ldots, \mu_N$, where $N = 1, 2, \ldots$ . For each measure $\mu_n$, $n = 1, \ldots, N$, there is a gain function $\rho(\mu_n, x, a) = (\rho_1(\mu_n, x, a), \ldots, \rho_M(\mu_n, x, a)$, $M = 1, 2, \ldots$, which is a Borel mapping of $X \times A$ to $\mathbb{R}^M$.

A *decision rule* $\pi$ is a regular transition probability from $X$ to $A$ such that $\pi(A(x)|x) = 1$ for all $x \in X$. A decision rule is called *nonrandomized* if for each $x \in X$ the measure $\pi(\cdot|x)$ is concentrated at one point. A nonrandomized decision rule $\pi$ is defined by a measurable mapping $\varphi : X \to A$ such that $\varphi(x) \in A(x)$ and $\pi(\varphi(x)|x) = 1$, $x \in X$. We call such a mapping *a decision function* and denote it by $\varphi$.

Using the same agreements as in the definitions of integrals as in (8), for any decision rule $\pi$ define

$$\mathcal{R}(\mu_n, \pi) = \int_X \int_A \rho(\mu_n, x, a)\pi(da|x)\mu_n(dx), \quad n = 1, \ldots, N.$$

**Theorem 4.** *[10, Theorem 1] If all the measures $\mu_1, \ldots, \mu_N$ are nonatomic then for each decision rule $\pi$ there exists a decision function $\varphi$ such that $\mathcal{R}(\mu_n, \varphi) = \mathcal{R}(\mu_n, \pi)$ for all $n = 1, \ldots, N$.*

Dworetzky, Wald, and Wolfowitz [4] proved Theorem 4 when $A$ is a finite set and weaker results were obtained in [4] for an infinite set $A$.

**Financial Application.** The following example from [9] is based on the version of Theorem 3 when transition probabilities and rewards may depend on the time parameter.

An investor has an option to sell a portfolio at epoch $t = 1, \ldots, T$, where $T = 1, 2, \ldots$ . The value of the portfolio at epoch $t = 0, 1, 2, \ldots$ is $z_t \in \mathbb{R}_+$. Suppose that $z_0 \geq 0$ is given and the value of $z_{t+1}$ is defined by transition probabilities $q_t(dz_{t+1}|z_t)$, $t = 0, 1, \ldots$ . We assume that $q_t(\cdot|z_t)$ are nonatomic and weakly continuous.

At each epoch $t = 1, 2, \ldots, T$, the investor has two options: to sell the whole portfolio or to keep it. We construct a Markov decision process for this problem. Let $X = \{0, 1\} \times \mathbb{R}$ and $A = \{0, 1\}$. Action 0 (1) means to hold (to sell) the portfolio. The state of the system is $x_t = (0, z_{t+1})$ ($x_t = (1, z_{t+1})$) if the portfolio has not been sold (has been sold). In particular, $x_0 = (0, z_1)$ has a nonatomic distribution $\mu$ defined by $\mu(0 \times B) = q_0(B|x_0)$ for any measurable subset $B$ of $\mathbb{R}_+$. For $t = 0, 1, \ldots$ we set $A_t((0, z)) = \{0, 1\}$; $A_t((1, z)) = \{0\}$. If at epoch $t = 0, 1, \ldots$ the system is in state $x_t = (0, z)$ and action $a_t = 0$ is selected then the next state is $x_{t+1} = (0, y)$, where $y$ has the distribution $q_{t+1}(dy|z)$. It does not matter what going on with the system after the portfolio is sold. To satisfy the Nonatomicity Assumption, we set that the system moves from state $x_t$ to state $(1, y)$, where $y$ has the distribution $q_{t+1}(dy|z)$, when $x_t = (1, z)$ or $x_t = (0, z)$ and $a_t = 1$.

Suppose that $N$-dimensional vectors $r_t(x, a)$ of measurable additive rewards (or losses) with values in $[-\infty, \infty]$ at steps $t = 0, 1, \ldots$ are given. Theorem 4 implies that (nonrandomized) Markov policies for this multicriterion problem are as good as general policies.

We consider the problem when the investor's goal is to maximize the expected discounted value of the sold portfolio under the constraint that with at least probability $P > 0$ the discounted value of the sold portfolio is greater (or equal) than the given level $C$. For $t = 1, \ldots, T-1$, we define $r_t^1((i, z), a) = \beta^t \cdot (1-i)az$; $r_t^2((i, z), a) = (1-i)a \cdot I\{z \geq C/\beta^t\}$, $i = 0, 1$; $\beta \in (0, 1]$ is the given discount factor. We set $r_t^n(x, a) = 0$ when $t \geq T$. Then the problem can be formulated as

$$\text{maximize}_\pi \ W_1(\mu, \pi) \qquad \text{s.t.} \qquad W_2(\mu, \pi) \geq P. \tag{9}$$

Suppose that $q_t(\cdot|z)$ are concentrated on a finite interval $[0, D]$. Then we can set $X = \{0, 1\} \times [0, D]$ and [9, Condition 5.3] holds. Therefore, [9, Corollary 5.2] implies that if this problem is feasible then there exists an optimal (nonrandomized) Markov policy.

**Acknowledgement**

This research was partially supported by NSF grant DMI-0600538.

# References

1. Altman E (1999) Constrained Markov Decision Processes. Chapman & Hall, Boca Raton, London, New York, Washington, D.C.
2. Altman E, Shwartz A (1991) Markov decision problems and state-action frequencies. SIAM J. Control and Optimization, 29:786–809.
3. Altman E, Shwartz A (1993) Time-sharing policies for controlled Markov chains. Operations Research, 41:1116–1124.
4. Dvoretzky A, Wald A, Wolfowitz J (1951) Elimination of randomization in certain statistical decision procedures and zero-sum two-person games. Ann. Math. Stat., 22:1–21.
5. Feinberg EA (2000) Constrained discounted Markov decision processes and Hamiltonian cycles. Mathematics of Operations Research, 25:130–140.
6. Feinberg EA (2002) Constrained finite continuous-time Markov decision processes with average rewards. In: Proceedings of IEEE 2002 Conference on Decisions and Control (Las Vegas, December 10-13, 2002), 3805–3810.
7. Feinberg EA (2004) Continuous time discounted jump Markov decision processes: a discrete-event approach. Mathematics of Operations Research, 29:492–524.
8. Feinberg EA, Curry MT (2005) Generalized pinwheel problem. Mathematical Methods of Operations Research, 62:99–122.
9. Feinberg EA, Piunovskiy AB (2002) Nonatomic total reward Markov decision processes with multiple criteria. Journal of Mathematical Analysis and Applications, 273:93–111.
10. Feinberg EA, Piunovskiy AB (2006) On Dvoretzky-Wald-Wolfowitz theorem on nonrandomized statistical decisions. Theory Probability and its Applications, 50:463–466.
11. Jacod J (1975) Multivariate point processes: predictable projection, Radon-Nikodym derivatives, representatioin of martingales. Z. Wahrscheinlichkeittheorie verw. Gebite, 31:235–253.
12. Kallenberg LCM (1983) Linear Programming and Finite Markovian Control Problems. Mathematical Center Tracts 148, Amsterdam, The Netherlands
13. Kitaev MYu, Rykov VV (1995) Controlled Queueing Systems. CRC Press, Boca Raton, New York, London, Tokyo
14. Krass D, Vrieze OJ (2002) Achieving target state-action frequencies in multichain average-reward Markov decision processes. Mathematics of Operations Research, 27:545–566.
15. Liptser RS, Shiryaev AN (1989) Theory of Maringales. Kluwer Academic Publishers, Dordrecht, Boston, London
16. Piunovskiy AB (1997) Optimal Control of Random Sequences in Problems with Constraints. Kluwer Academic Publishers, Dordrecht, Boston, London
17. Puterman ML (1994) Markov Decision Processes. John Wiley & Sons, New York, Chichester, Brisbane, Toronto, Singapure
18. Ross KW (1989) Randomized and past-dependent policies for Markov decision processes with multiple constraints, Operations Research, 37:474–477.

# Higher-Order Calculus of Variations on Time Scales

Rui A. C. Ferreira and Delfim F. M. Torres

Department of Mathematics, University of Aveiro,
3810-193 Aveiro, Portugal. `ruiacferreira@yahoo.com`, `delfim@ua.pt`

**Summary.** We prove a version of the Euler-Lagrange equations for certain problems of the calculus of variations on time scales with higher-order delta derivatives.

**Keywords:** time scales, calculus of variations, delta-derivatives of higher-order, Euler-Lagrange equations.

**2000 Mathematics Subject Classification:** 39A12, 49K05.

## 1 Introduction

Calculus of variations on time scales (we refer the reader to Section 2 for a brief introduction to time scales) has been introduced in 2004 in the papers by Bohner [2] and Hilscher and Zeidan [4], and seems to have many opportunities for application in economics [1]. In both works of Bohner and Hilscher&Zeidan, the Euler-Lagrange equation for the fundamental problem of the calculus of variations on time scales,

$$\mathcal{L}[y(\cdot)] = \int_a^b L(t, y^\sigma(t), y^\Delta(t))\Delta t \longrightarrow \min, \quad y(a) = y_a, \, y(b) = y_b, \quad (1)$$

is obtained (in [4] for a bigger class of admissible functions and for problems with more general endpoint conditions). Here we generalize the previously obtained Euler-Lagrange equation for variational problems involving delta derivatives of more than the first order, i.e. for *higher-order problems*.

We consider the following extension to problem (1):

$$\mathcal{L}[y(\cdot)] = \int_a^{\rho^{r-1}(b)} L(t, y^{\sigma^r}(t), y^{\sigma^{r-1}\Delta}(t), \dots, y^{\sigma\Delta^{r-1}}(t), y^{\Delta^r}(t))\Delta t \longrightarrow \min,$$

$$y(a) = y_a, \quad y\left(\rho^{r-1}(b)\right) = y_b,$$

$$\vdots \qquad\qquad\qquad\qquad (2)$$

$$y^{\Delta^{r-1}}(a) = y_a^{r-1}, \quad y^{\Delta^{r-1}}\left(\rho^{r-1}(b)\right) = y_b^{r-1},$$

where $y^{\sigma^i \Delta^{r-i}}(t) \in \mathbb{R}^n$, $i \in \{0, \ldots, r\}$, $n$, $r \in \mathbb{N}$, and $t$ belongs to a time scale $\mathbb{T}$. Assumptions on the time scale $\mathbb{T}$ are stated in Section 2; the conditions imposed on the Lagrangian $L$ and on the admissible functions $y$ are specified in Section 3. For $r = 1$ problem (2) is reduced to (1); for $\mathbb{T} = \mathbb{R}$ we get the classical problem of the calculus of variations with higher-order derivatives.

While in the classical context of the calculus of variations, i.e. when $\mathbb{T} = \mathbb{R}$, it is trivial to obtain the Euler-Lagrange necessary optimality condition for problem (2) as soon as we know how to do it for (1), this is not the case on the time scale setting. The Euler-Lagrange equation obtained in [2, 4] for (1) follow the classical proof, substituting the usual integration by parts formula by integration by parts for the delta integral (Lemma 2). Here we generalize the proof of [2, 4] to the higher-order case by successively applying the delta-integration by parts and thus obtaining a more general delta-differential Euler-Lagrange equation. It is worth to mention that such a generalization poses serious technical difficulties and that the obtained necessary optimality condition is not true on a general time scale, being necessary some restrictions on $\mathbb{T}$. Proving an Euler-Lagrange necessary optimality condition for a completely arbitrary time scale $\mathbb{T}$ is a deep and difficult open question.

The paper is organized as follows: in Section 2 a brief introduction to the calculus of time scales is given and some assumptions and basic results provided. Then, under the assumed hypotheses on the time scale $\mathbb{T}$, we obtain in Section 3 the intended higher-order delta-differential Euler-Lagrange equation.

## 2 Basic definitions and results on time scales

A nonempty closed subset of $\mathbb{R}$ is called a *time scale* and it is denoted by $\mathbb{T}$.

The *forward jump operator* $\sigma : \mathbb{T} \to \mathbb{T}$ is defined by

$$\sigma(t) = \inf \{s \in \mathbb{T} : s > t\}, \text{ for all } t \in \mathbb{T},$$

while the *backward jump operator* $\rho : \mathbb{T} \to \mathbb{T}$ is defined by

$$\rho(t) = \sup \{s \in \mathbb{T} : s < t\}, \text{ for all } t \in \mathbb{T},$$

with $\inf \emptyset = \sup \mathbb{T}$ (i.e. $\sigma(M) = M$ if $\mathbb{T}$ has a maximum $M$) and $\sup \emptyset = \inf \mathbb{T}$ (i.e. $\rho(m) = m$ if $\mathbb{T}$ has a minimum $m$).

A point $t \in \mathbb{T}$ is called *right-dense, right-scattered, left-dense* and *left-scattered* if $\sigma(t) = t$, $\sigma(t) > t$, $\rho(t) = t$ and $\rho(t) < t$, respectively.

Throughout the paper we let $\mathbb{T} = [a, b] \cap \mathbb{T}_0$ with $a < b$ and $\mathbb{T}_0$ a time scale containing $a$ and $b$.

*Remark 1.* The time scales $\mathbb{T}$ considered in this work have a maximum $b$ and, by definition, $\sigma(b) = b$. For example, let $[a, b] = [1, 5]$ and $\mathbb{T}_0 = \mathbb{N}$: in this case $\mathbb{T} = [1, 5] \cap \mathbb{N} = \{1, 2, 3, 4, 5\}$ and one has $\sigma(t) = t + 1$, $t \in \mathbb{T} \backslash \{5\}$, $\sigma(5) = 5$.

Following [3, pp. 2 and 11], we define $\mathbb{T}^k = \mathbb{T}\backslash(\rho(b), b]$, $\mathbb{T}^{k^2} = (\mathbb{T}^k)^k$ and, more generally, $\mathbb{T}^{k^n} = (\mathbb{T}^{k^{n-1}})^k$, for $n \in \mathbb{N}$. The following standard notation is used for $\sigma$ (and $\rho$): $\sigma^0(t) = t$, $\sigma^n(t) = (\sigma \circ \sigma^{n-1})(t)$, $n \in \mathbb{N}$.

The *graininess function* $\mu : \mathbb{T} \to [0, \infty)$ is defined by

$$\mu(t) = \sigma(t) - t, \text{ for all } t \in \mathbb{T}.$$

We say that a function $f : \mathbb{T} \to \mathbb{R}$ is *delta differentiable* at $t \in \mathbb{T}^k$ if there is a number $f^\Delta(t)$ such that for all $\varepsilon > 0$ there exists a neighborhood $U$ of $t$ (i.e. $U = (t - \delta, t + \delta) \cap \mathbb{T}$ for some $\delta > 0$) such that

$$|f(\sigma(t)) - f(s) - f^\Delta(t)(\sigma(t) - s)| \le \varepsilon |\sigma(t) - s|, \text{ for all } s \in U.$$

We call $f^\Delta(t)$ the *delta derivative* of $f$ at $t$.

If $f$ is continuous at $t$ and $t$ is right-scattered, then (see Theorem 1.16 (ii) of [3])

$$f^\Delta(t) = \frac{f(\sigma(t)) - f(t)}{\mu(t)}. \tag{3}$$

Now, we define the $r^{th}-$*delta derivative* $(r \in \mathbb{N})$ of $f$ to be the function $f^{\Delta^r} : \mathbb{T}^{k^r} \to \mathbb{R}$, provided $f^{\Delta^{r-1}}$ is delta differentiable on $\mathbb{T}^{k^r}$.

For delta differentiable functions $f$ and $g$, the next formulas hold:

$$f^\sigma(t) = f(t) + \mu(t)f^\Delta(t),$$
$$(fg)^\Delta(t) = f^\Delta(t)g^\sigma(t) + f(t)g^\Delta(t) \tag{4}$$
$$= f^\Delta(t)g(t) + f^\sigma(t)g^\Delta(t),$$

where we abbreviate here and throughout $f \circ \sigma$ by $f^\sigma$. We will also write $f^{\Delta^\sigma}$ as $f^{\Delta\sigma}$ and all the possible combinations of exponents of $\sigma$ and $\Delta$ will be clear from the context.

The following lemma will be useful for our purposes.

**Lemma 1.** *Let $t \in \mathbb{T}^k$ ($t \ne \min \mathbb{T}$) satisfy the property $\rho(t) = t < \sigma(t)$. Then, the jump operator $\sigma$ is not delta differentiable at $t$.*

*Proof.* We begin to prove that $\lim_{s \to t^-} \sigma(s) = t$. Let $\varepsilon > 0$ and take $\delta = \varepsilon$. Then for all $s \in (t - \delta, t)$ we have $|\sigma(s) - t| \le |s - t| < \delta = \varepsilon$. Since $\sigma(t) > t$, this implies that $\sigma$ is not continuous at $t$, hence not delta-differentiable by Theorem 1.16 (i) of [3]. $\qquad\square$

A function $f : \mathbb{T} \to \mathbb{R}$ is called *rd-continuous* if it is continuous in right-dense points and if its left-sided limit exists in left-dense points. We denote the set of all rd-continuous functions by $C_{rd}$ and the set of all differentiable functions with rd-continuous derivative by $C_{rd}^1$.

It is known that rd-continuous functions possess an *antiderivative*, i.e. there exists a function $F$ with $F^\Delta = f$, and in this case an *integral* is defined by $\int_a^b f(t)\Delta t = F(b) - F(a)$. It satisfies

$$\int_t^{\sigma(t)} f(\tau)\Delta\tau = \mu(t)f(t). \tag{5}$$

We now present the integration by parts formulas for the delta integral:

**Lemma 2.** *(Theorem 1.77 (v) and (vi) of [3]) If $a, b \in \mathbb{T}$ and $f, g \in C_{rd}^1$, then*

1. $\int_a^b f(\sigma(t))g^\Delta(t)\Delta t = [(fg)(t)]_{t=a}^{t=b} - \int_a^b f^\Delta(t)g(t)\Delta t$;
2. $\int_a^b f(t)g^\Delta(t)\Delta t = [(fg)(t)]_{t=a}^{t=b} - \int_a^b f^\Delta(t)g(\sigma(t))\Delta t$.

The main result of the calculus of variations on time scales for problem (1) is given by the following necessary optimality condition.

**Theorem 1.** *([2]) If $y_* \in C_{rd}^1$ is a weak local minimum of the problem*

$$\mathcal{L}[y(\cdot)] = \int_a^b L(t, y^\sigma(t), y^\Delta(t))\Delta t \longrightarrow \min, \ y(a) = y_a, \ y(b) = y_b,$$

*then the Euler-Lagrange equation*

$$L_{y^\Delta}^\Delta(t, y_*^\sigma(t), y_*^\Delta(t)) = L_{y^\sigma}(t, y_*^\sigma(t), y_*^\Delta(t)),$$

$t \in \mathbb{T}^{k^2}$, *holds.*

*Remark 2.* In Theorem 1, and in what follows, the notation conforms to that used in [2]. Expression $L_{y^\Delta}^\Delta$ denotes the $\Delta$ derivative of a composition.

We will assume from now on that the time scale $\mathbb{T}$ has sufficiently many points in order for all the calculations to make sense (with respect to this, we remark that Theorem 1 makes only sense if we are assuming a time scale $\mathbb{T}$ with at least three points). Further, we consider time scales such that:

(H) $\sigma(t) = a_1 t + a_0$ for some $a_1 \in \mathbb{R}^+$ and $a_0 \in \mathbb{R}$, $t \in [a, b)$.

Under hypothesis (H) we have, among others, the differential calculus ($\mathbb{T}_0 = \mathbb{R}$, $a_1 = 1$, $a_0 = 0$), the difference calculus ($\mathbb{T}_0 = \mathbb{Z}$, $a_1 = a_0 = 1$) and the quantum calculus ($\mathbb{T}_0 = \{q^k : k \in \mathbb{N}_0\}$, with $q > 1$, $a_1 = q$, $a_0 = 0$).

*Remark 3.* From assumption (H) it follows by Lemma 1 that it is not possible to have points which are simultaneously left-dense and right-scattered. Also points that are simultaneously left-scattered and right-dense do not occur, since $\sigma$ is strictly increasing.

**Lemma 3.** *Under hypothesis (H), if $f$ is a two times delta differentiable function, then the next formula holds:*

$$f^{\sigma\Delta}(t) = a_1 f^{\Delta\sigma}(t), \ t \in \mathbb{T}^{k^2}. \tag{6}$$

*Proof.* We have $f^{\sigma\Delta}(t) = \left[f(t) + \mu(t)f^\Delta(t)\right]^\Delta$ by formula (4). By the hypothesis on $\sigma$, $\mu$ is delta differentiable, hence $\left[f(t) + \mu(t)f^\Delta(t)\right]^\Delta = f^\Delta(t) + \mu^\Delta(t)f^{\Delta\sigma}(t) + \mu(t)f^{\Delta^2}(t)$ and applying again formula (4) we obtain $f^\Delta(t) + \mu^\Delta(t)f^{\Delta\sigma}(t) + \mu(t)f^{\Delta^2}(t) = f^{\Delta\sigma}(t) + \mu^\Delta(t)f^{\Delta\sigma}(t) = (1 + \mu^\Delta(t))f^{\Delta\sigma}(t)$. Now we only need to observe that $\mu^\Delta(t) = \sigma^\Delta(t) - 1$ and the result follows.    □

## 3 Main results

Assume that the Lagrangian $L(t, u_0, u_1, \ldots, u_r)$ of problem (2) has (standard) partial derivatives with respect to $u_0, \ldots, u_r$, $r \geq 1$, and partial delta derivative with respect to $t$ of order $r + 1$. Let $y \in C^{2r}$, where

$$C^{2r} = \left\{ y : \mathbb{T} \to \mathbb{R} : y^{\Delta^{2r}} \text{ is continuous on } \mathbb{T}^{k^{2r}} \right\}.$$

We say that $y_* \in C^{2r}$ is a *weak local minimum* for (2) provided there exists $\delta > 0$ such that $\mathcal{L}(y_*) \leq \mathcal{L}(y)$ for all $y \in C^{2r}$ satisfying the constraints in (2) and $\|y - y_*\|_{r,\infty} < \delta$, where

$$\|y\|_{r,\infty} := \sum_{i=0}^{r} \|y^{(i)}\|_{\infty},$$

with $y^{(i)} = y^{\sigma^i \Delta^{r-i}}$ and $\|y\|_{\infty} := \sup_{t \in \mathbb{T}^{k^r}} |y(t)|$.

**Definition 1.** *We say that $\eta \in C^{2r}$ is an admissible variation for problem (2) if*

$$\eta(a) = 0, \quad \eta\left(\rho^{r-1}(b)\right) = 0$$

$$\vdots$$

$$\eta^{\Delta^{r-1}}(a) = 0, \quad \eta^{\Delta^{r-1}}\left(\rho^{r-1}(b)\right) = 0.$$

For simplicity of presentation, from now on we fix $r = 3$.

**Lemma 4.** *Suppose that $f$ is defined on $[a, \rho^6(b)]$ and is continuous. Then, under hypothesis (H), $\int_a^{\rho^5(b)} f(t)\eta^{\sigma^3}(t)\Delta t = 0$ for every admissible variation $\eta$ if and only if $f(t) = 0$ for all $t \in [a, \rho^6(b)]$.*

*Proof.* If $f(t) = 0$, then the result is obvious.

Now, suppose without loss of generality that there exists $t_0 \in [a, \rho^6(b)]$ such that $f(t_0) > 0$. First we consider the case in which $t_0$ is right-dense, hence left-dense or $t_0 = a$ (see Remark 3). If $t_0 = a$, then by the continuity of $f$ at $t_0$ there exists a $\delta > 0$ such that for all $t \in [t_0, t_0 + \delta)$ we have $f(t) > 0$. Let us define $\eta$ by

$$\eta(t) = \begin{cases} (t - t_0)^8 (t - t_0 - \delta)^8 & \text{if } t \in [t_0, t_0 + \delta); \\ 0 & \text{otherwise.} \end{cases}$$

Clearly $\eta$ is a $C^6$ function and satisfy the requirements of an admissible variation. But with this definition for $\eta$ we get the contradiction

$$\int_a^{\rho^5(b)} f(t)\eta^{\sigma^3}(t)\Delta t = \int_{t_0}^{t_0+\delta} f(t)\eta^{\sigma^3}(t)\Delta t > 0.$$

Now, consider the case where $t_0 \neq a$. Again, the continuity of $f$ ensures the existence of a $\delta > 0$ such that for all $t \in (t_0 - \delta, t_0 + \delta)$ we have $f(t) > 0$. Defining $\eta$ by

$$\eta(t) = \begin{cases} (t - t_0 + \delta)^8 (t - t_0 - \delta)^8 & \text{if } t \in (t_0 - \delta, t_0 + \delta); \\ 0 & \text{otherwise,} \end{cases}$$

and noting that it satisfy the properties of an admissible variation, we obtain

$$\int_a^{\rho^5(b)} f(t)\eta^{\sigma^3}(t)\Delta t = \int_{t_0-\delta}^{t_0+\delta} f(t)\eta^{\sigma^3}(t)\Delta t > 0,$$

which is again a contradiction.

Assume now that $t_0$ is right-scattered. In view of Remark 3, all the points $t$ such that $t \geq t_0$ must be isolated. So, define $\eta$ such that $\eta^{\sigma^3}(t_0) = 1$ and is zero elsewhere. It is easy to see that $\eta$ satisfies all the requirements of an admissible variation. Further, using formula (5)

$$\int_a^{\rho^5(b)} f(t)\eta^{\sigma^3}(t)\Delta t = \int_{t_0}^{\sigma(t_0)} f(t)\eta^{\sigma^3}(t)\Delta t = \mu(t_0)f(t_0)\eta^{\sigma^3}(t_0) > 0,$$

which is a contradiction. $\qquad\square$

**Theorem 2.** *Let the Lagrangian $L(t, u_0, u_1, u_2, u_3)$ satisfy the conditions in the beginning of the section. On a time scale $\mathbb{T}$ satisfying (H), if $y_*$ is a weak local minimum for the problem of minimizing*

$$\int_a^{\rho^2(b)} L\left(t, y^{\sigma^3}(t), y^{\sigma^2\Delta}(t), y^{\sigma\Delta^2}(t), y^{\Delta^3}(t)\right) \Delta t$$

*subject to*

$$y(a) = y_a, \; y\left(\rho^2(b)\right) = y_b,$$
$$y^{\Delta}(a) = y_a^1, \; y^{\Delta}\left(\rho^2(b)\right) = y_b^1,$$
$$y^{\Delta^2}(a) = y_a^2, \; y^{\Delta^2}\left(\rho^2(b)\right) = y_b^2,$$

*then $y_*$ satisfies the Euler-Lagrange equation*

$$L_{u_0}(\cdot) - L_{u_1}^{\Delta}(\cdot) + \frac{1}{a_1}L_{u_2}^{\Delta^2}(\cdot) - \frac{1}{a_1^3}L_{u_3}^{\Delta^3}(\cdot) = 0, \quad t \in [a, \rho^6(b)],$$

*where $(\cdot) = (t, y_*^{\sigma^3}(t), y_*^{\sigma^2\Delta}(t), y_*^{\sigma\Delta^2}(t), y_*^{\Delta^3}(t))$.*

*Proof.* Suppose that $y_*$ is a weak local minimum of $\mathcal{L}$. Let $\eta \in C^6$ be an admissible variation, i.e. $\eta$ is an arbitrary function such that $\eta$, $\eta^{\Delta}$ and $\eta^{\Delta^2}$ vanish at $t = a$ and $t = \rho^2(b)$. Define function $\Phi : \mathbb{R} \to \mathbb{R}$ by $\Phi(\varepsilon) = \mathcal{L}(y_* + \varepsilon\eta)$. This function has a minimum at $\varepsilon = 0$, so we must have (see [2, Theorem 3.2])

$$\Phi'(0) = 0. \tag{7}$$

Differentiating $\Phi$ under the integral sign (we can do this in virtue of the conditions we imposed on $L$) with respect to $\varepsilon$ and setting $\varepsilon = 0$, we obtain from (7) that

$$0 = \int_a^{\rho^2(b)} \left\{ L_{u_0}(\cdot)\eta^{\sigma^3}(t) + L_{u_1}(\cdot)\eta^{\sigma^2\Delta}(t) \right.$$
$$\left. + L_{u_2}(\cdot)\eta^{\sigma\Delta^2}(t) + L_{u_3}(\cdot)\eta^{\Delta^3}(t) \right\} \Delta t. \tag{8}$$

Since we will delta differentiate $L_{u_i}$, $i = 1, 2, 3$, we rewrite (8) in the following form:

$$0 = \int_a^{\rho^3(b)} \left\{ L_{u_0}(\cdot)\eta^{\sigma^3}(t) + L_{u_1}(\cdot)\eta^{\sigma^2\Delta}(t) \right.$$
$$\left. + L_{u_2}(\cdot)\eta^{\sigma\Delta^2}(t) + L_{u_3}(\cdot)\eta^{\Delta^3}(t) \right\} \Delta t$$
$$+ \mu(\rho^3(b)) \left\{ L_{u_0}\eta^{\sigma^3} + L_{u_1}\eta^{\sigma^2\Delta} + L_{u_2}\eta^{\sigma\Delta^2} + L_{u_3}\eta^{\Delta^3} \right\} (\rho^3(b)). \tag{9}$$

Integrating (9) by parts gives

$$0 = \int_a^{\rho^3(b)} \left\{ L_{u_0}(\cdot)\eta^{\sigma^3}(t) - L_{u_1}^{\Delta}(\cdot)\eta^{\sigma^3}(t) \right.$$
$$\left. - L_{u_2}^{\Delta}(\cdot)\eta^{\sigma\Delta\sigma}(t) - L_{u_3}^{\Delta}(\cdot)\eta^{\Delta^2\sigma}(t) \right\} \Delta t$$
$$+ \left[ L_{u_1}(\cdot)\eta^{\sigma^2}(t) \right]_{t=a}^{t=\rho^3(b)} + \left[ L_{u_2}(\cdot)\eta^{\sigma\Delta}(t) \right]_{t=a}^{t=\rho^3(b)} + \left[ L_{u_3}(\cdot)\eta^{\Delta^2}(t) \right]_{t=a}^{t=\rho^3(b)}$$
$$+ \mu(\rho^3(b)) \left\{ L_{u_0}\eta^{\sigma^3} + L_{u_1}\eta^{\sigma^2\Delta} + L_{u_2}\eta^{\sigma\Delta^2} + L_{u_3}\eta^{\Delta^3} \right\} (\rho^3(b)). \tag{10}$$

Now we show how to simplify (10). We start by evaluating $\eta^{\sigma^2}(a)$:

$$\eta^{\sigma^2}(a) = \eta^{\sigma}(a) + \mu(a)\eta^{\sigma\Delta}(a)$$
$$= \eta(a) + \mu(a)\eta^{\Delta}(a) + \mu(a)a_1\eta^{\Delta\sigma}(a) \tag{11}$$
$$= \mu(a)a_1 \left( \eta^{\Delta}(a) + \mu(a)\eta^{\Delta^2}(a) \right)$$
$$= 0,$$

where the last term of (11) follows from (6). Now, we calculate $\eta^{\sigma\Delta}(a)$. By (6) we have $\eta^{\sigma\Delta}(a) = a_1\eta^{\Delta\sigma}(a)$ and applying (4) we obtain

$$a_1\eta^{\Delta\sigma}(a) = a_1 \left( \eta^{\Delta}(a) + \mu(a)\eta^{\Delta^2}(a) \right) = 0.$$

Now we turn to analyze what happens at $t = \rho^3(b)$. It is easy to see that if $b$ is left-dense, then the last terms of (10) vanish. So suppose that $b$ is left-scattered. Since $\sigma$ is delta differentiable, by Lemma 1 we cannot have points which are simultaneously left-dense and right-scattered. Hence, $\rho(b)$, $\rho^2(b)$ and $\rho^3(b)$ are right-scattered points. Now, by hypothesis $\eta^\Delta(\rho^2(b)) = 0$, hence we have by (3) that

$$\frac{\eta(\rho(b)) - \eta(\rho^2(b))}{\mu(\rho^2(b))} = 0.$$

But $\eta(\rho^2(b)) = 0$, hence $\eta(\rho(b)) = 0$. Analogously, we have

$$\eta^{\Delta^2}(\rho^2(b)) = 0 \Leftrightarrow \frac{\eta^\Delta(\rho(b)) - \eta^\Delta(\rho^2(b))}{\mu(\rho^2(b))} = 0,$$

from what follows that $\eta^\Delta(\rho(b)) = 0$. This last equality implies $\eta(b) = 0$. Applying previous expressions to the last terms of (10), we obtain:

$$\eta^{\sigma^2}(\rho^3(b)) = \eta(\rho(b)) = 0,$$

$$\eta^{\sigma\Delta}(\rho^3(b)) = \frac{\eta^{\sigma^2}(\rho^3(b)) - \eta^\sigma(\rho^3(b))}{\mu(\rho^3(b))} = 0,$$

$$\eta^{\sigma^3}(\rho^3(b)) = \eta(b) = 0,$$

$$\eta^{\sigma^2\Delta}(\rho^3(b)) = \frac{\eta^{\sigma^3}(\rho^3(b)) - \eta^{\sigma^2}(\rho^3(b))}{\mu(\rho^3(b))} = 0,$$

$$\eta^{\sigma\Delta^2}(\rho^3(b)) = \frac{\eta^{\sigma\Delta}(\rho^2(b)) - \eta^{\sigma\Delta}(\rho^3(b))}{\mu(\rho^3(b))}$$

$$= \frac{\frac{\eta^\sigma(\rho(b)) - \eta^\sigma(\rho^2(b))}{\mu(\rho^2(b))} - \frac{\eta^\sigma(\rho^2(b)) - \eta^\sigma(\rho^3(b))}{\mu(\rho^3(b))}}{\mu(\rho^3(b))}$$

$$= 0.$$

In view of our previous calculations,

$$\left[L_{u_1}(\cdot)\eta^{\sigma^2}(t)\right]_{t=a}^{t=\rho^3(b)} + \left[L_{u_2}(\cdot)\eta^{\sigma\Delta}(t)\right]_{t=a}^{t=\rho^3(b)} + \left[L_{u_3}(\cdot)\eta^{\Delta^2}(t)\right]_{t=a}^{t=\rho^3(b)}$$

$$+ \mu(\rho^3(b))\left\{L_{u_0}\eta^{\sigma^3} + L_{u_1}\eta^{\sigma^2\Delta} + L_{u_2}\eta^{\sigma\Delta^2} + L_{u_3}\eta^{\Delta^3}\right\}(\rho^3(b))$$

is reduced to[1]

$$L_{u_3}(\rho^3(b))\eta^{\Delta^2}(\rho^3(b)) + \mu(\rho^3(b))L_{u_3}(\rho^3(b))\eta^{\Delta^3}(\rho^3(b)). \tag{12}$$

Now note that

$$\eta^{\Delta^2\sigma}(\rho^3(b)) = \eta^{\Delta^2}(\rho^3(b)) + \mu(\rho^3(b))\eta^{\Delta^3}(\rho^3(b))$$

---

[1] In what follows there is some abuse of notation: $L_{u_3}(\rho^3(b))$ denotes $L_{u_3}(\cdot)|_{t=\rho^3(b)}$, that is, we substitute $t$ in $(\cdot) = (t, y_*^{\sigma^3}(t), y_*^{\sigma^2\Delta}(t), y_*^{\sigma\Delta^2}(t), y_*^{\Delta^3}(t))$ by $\rho^3(b)$.

and by hypothesis $\eta^{\Delta^2\sigma}(\rho^3(b)) = \eta^{\Delta^2}(\rho^2(b)) = 0$. Therefore,

$$\mu(\rho^3(b))\eta^{\Delta^3}(\rho^3(b)) = -\eta^{\Delta^2}(\rho^3(b)),$$

from which follows that (12) must be zero. We have just simplified (10) to

$$0 = \int_a^{\rho^3(b)} \left\{ L_{u_0}(\cdot)\eta^{\sigma^3}(t) - L_{u_1}^{\Delta}(\cdot)\eta^{\sigma^3}(t) \right.$$
$$\left. - L_{u_2}^{\Delta}(\cdot)\eta^{\sigma\Delta\sigma}(t) - L_{u_3}^{\Delta}(\cdot)\eta^{\Delta^2\sigma}(t) \right\} \Delta t. \quad (13)$$

In order to apply again the integration by parts formula, we must first make some transformations in $\eta^{\sigma\Delta\sigma}$ and $\eta^{\Delta^2\sigma}$. By (6) we have

$$\eta^{\sigma\Delta\sigma}(t) = \frac{1}{a_1}\eta^{\sigma^2\Delta}(t) \quad (14)$$

and

$$\eta^{\Delta^2\sigma}(t) = \frac{1}{a_1^2}\eta^{\sigma\Delta^2}(t). \quad (15)$$

Hence, (13) becomes

$$0 = \int_a^{\rho^3(b)} \left\{ L_{u_0}(\cdot)\eta^{\sigma^3}(t) - L_{u_1}^{\Delta}(\cdot)\eta^{\sigma^3}(t) \right.$$
$$\left. - \frac{1}{a_1}L_{u_2}^{\Delta}(\cdot)\eta^{\sigma^2\Delta}(t) - \frac{1}{a_1^2}L_{u_3}^{\Delta}(\cdot)\eta^{\sigma\Delta^2}(t) \right\} \Delta t. \quad (16)$$

By the same reasoning as before, (16) is equivalent to

$$0 = \int_a^{\rho^4(b)} \left\{ L_{u_0}(\cdot)\eta^{\sigma^3}(t) - L_{u_1}^{\Delta}(\cdot)\eta^{\sigma^3}(t) \right.$$
$$\left. - \frac{1}{a_1}L_{u_2}^{\Delta}(\cdot)\eta^{\sigma^2\Delta}(t) - \frac{1}{a_1^2}L_{u_3}^{\Delta}(\cdot)\eta^{\sigma\Delta^2}(t) \right\} \Delta t$$
$$+ \mu(\rho^4(b)) \left\{ L_{u_0}\eta^{\sigma^3} - L_{u_1}^{\Delta}\eta^{\sigma^3} - \frac{1}{a_1}L_{u_2}^{\Delta}\eta^{\sigma^2\Delta} - \frac{1}{a_1^2}L_{u_3}^{\Delta}\eta^{\sigma\Delta^2} \right\} (\rho^4(b))$$

and integrating by parts we obtain

$$0 = \int_a^{\rho^4(b)} \left\{ L_{u_0}(\cdot)\eta^{\sigma^3}(t) - L_{u_1}^{\Delta}(\cdot)\eta^{\sigma^3}(t) \right.$$
$$\left. + \frac{1}{a_1}L_{u_2}^{\Delta^2}(\cdot)\eta^{\sigma^3}(t) + \frac{1}{a_1^2}L_{u_3}^{\Delta^2}(\cdot)\eta^{\sigma\Delta\sigma}(t) \right\} \Delta t$$
$$- \left[ \frac{1}{a_1}L_{u_2}^{\Delta}(\cdot)\eta^{\sigma^2}(t) \right]_{t=a}^{t=\rho^4(b)} - \left[ \frac{1}{a_1^2}L_{u_3}^{\Delta}(\cdot)\eta^{\sigma\Delta}(t) \right]_{t=a}^{t=\rho^4(b)} \quad (17)$$
$$+ \mu(\rho^4(b)) \left\{ L_{u_0}\eta^{\sigma^3} - L_{u_1}^{\Delta}\eta^{\sigma^3} - \frac{1}{a_1}L_{u_2}^{\Delta}\eta^{\sigma^2\Delta} - \frac{1}{a_1^2}L_{u_3}^{\Delta}\eta^{\sigma\Delta^2} \right\} (\rho^4(b)).$$

Using analogous arguments to those above, we simplify (17) to

$$\int_a^{\rho^4(b)} \left\{ L_{u_0}(\cdot)\eta^{\sigma^3}(t) - L_{u_1}^{\Delta}(\cdot)\eta^{\sigma^3}(t) \right.$$
$$\left. + \frac{1}{a_1} L_{u_2}^{\Delta^2}(\cdot)\eta^{\sigma^2\Delta}(t) + \frac{1}{a_1^3} L_{u_3}^{\Delta^2}(\cdot)\eta^{\sigma^2\Delta}(t) \right\} \Delta t = 0.$$

Calculations as done before lead us to the final expression

$$\int_a^{\rho^5(b)} \left\{ L_{u_0}(\cdot)\eta^{\sigma^3}(t) - L_{u_1}^{\Delta}(\cdot)\eta^{\sigma^3}(t) \right.$$
$$\left. + \frac{1}{a_1} L_{u_2}^{\Delta^2}(\cdot)\eta^{\sigma^3}(t) - \frac{1}{a_1^3} L_{u_3}^{\Delta^3}(\cdot)\eta^{\sigma^3}(t) \right\} \Delta t = 0,$$

which is equivalent to

$$\int_a^{\rho^5(b)} \left\{ L_{u_0}(\cdot) - L_{u_1}^{\Delta}(\cdot) + \frac{1}{a_1} L_{u_2}^{\Delta^2}(\cdot) - \frac{1}{a_1^3} L_{u_3}^{\Delta^3}(\cdot) \right\} \eta^{\sigma^3}(t)\Delta t = 0. \qquad (18)$$

Applying Lemma 4 to (18), we obtain the Euler-Lagrange equation

$$L_{u_0}(\cdot) - L_{u_1}^{\Delta}(\cdot) + \frac{1}{a_1} L_{u_2}^{\Delta^2}(\cdot) - \frac{1}{a_1^3} L_{u_3}^{\Delta^3}(\cdot) = 0, \quad t \in [a, \rho^6(b)].$$

$\square$

Following exactly the same steps in the proofs of Lemma 4 and Theorem 2 for an arbitrary $r \in \mathbb{N}$, one easily obtains the Euler-Lagrange equation for problem (2).

**Theorem 3.** *(Necessary optimality condition for problems of the calculus of variations with higher-order delta derivatives) On a time scale $\mathbb{T}$ satisfying hypothesis (H), if $y_*$ is a weak local minimum for problem (2), then $y_*$ satisfies the Euler-Lagrange equation*

$$\sum_{i=0}^r (-1)^i \left(\frac{1}{a_1}\right)^{\frac{(i-1)i}{2}} L_{u_i}^{\Delta^i} \left( t, y_*^{\sigma^r}(t), y_*^{\sigma^{r-1}\Delta}(t), \ldots, y_*^{\sigma\Delta^{r-1}}(t), y_*^{\Delta^r}(t) \right) = 0,$$

$$(19)$$

$t \in [a, \rho^{2r}(b)]$.

*Remark 4.* The factor $\left(\frac{1}{a_1}\right)^{\frac{(i-1)i}{2}}$ in (19) comes from the fact that, after each time we apply the integration by parts formula, we commute successively $\sigma$ with $\Delta$ using (6) (see formulas (14) and (15)), doing this $\sum_{j=1}^{i-1} j = \frac{(i-1)i}{2}$ times for each of the parcels within the integral.

**Acknowledgments**

This work is part of the first author's PhD project and was partially supported by the Control Theory Group (cotg) of the Center for Research on Optimization and Control (CEOC), through the Portuguese Foundation for Science and Technology (FCT), cofinanced by the European Community Fund FEDER/POCI 2010. The authors are grateful to Dorota Mozyrska and Ewa Pawluszewicz for inspiring discussions during the Workshop on Mathematical Control Theory and Finance, Lisbon, 10-14 April 2007, where some preliminary results were presented; to an anonymous referee for helpful comments.

# References

1. Atici FM, Biles DC, Lebedinsky A (2006) An application of time scales to economics. Math. Comput. Modelling 43(7-8):718–726.
2. Bohner M (2004) Calculus of variations on time scales. Dyn. Sys. and Appl. 272(13):339–349.
3. Bohner M, Peterson A (2001) Dynamic equations on time scales: an introduction with applications. Birkhäuser, Boston
4. Hilscher R, Zeidan V (2004) Calculus of variations on time scales: weak local piecewise $C^1_{rd}$ solutions with variable endpoints. J. Math. Anal. Appl. 289(1):143–166.

# Finding Invariants of Group Actions on Function Spaces, a General Methodology from Non-Abelian Harmonic Analysis

Jean-Paul Gauthier, Fethi Smach, Cedric Lemaître, and Johel Miteran

LE2I, UMR CNRS 5158, Université de Bourgogne,
Bat. Mirande, BP 47870, 21078, Dijon CEDEX, France.
gauthier@u-bourgogne.fr, smach_fethi@yahoo.fr,
Cedric.Lemaitre@u-bourgogne.fr, miteranj@u-bourgogne.fr

**Summary.** In this paper, we describe a general method using the abstract non-Abelian Fourier transform to construct "rich" invariants of group actions on functional spaces.

In fact, this method is inspired of a classical method from image analysis: the method of Fourier descriptors, for discrimination among "contours" of objects. This is the case of the Abelian circle group, but the method can be extended to general non-Abelian cases.

Here, we improve on some of our previous developments on this subject, in particular in the case of compact groups and motion groups. The last point (motion groups) is in the perspective of invariant image analysis. But our method can be applied to many practical problems of discrimination, or detection, or recognition.

## 1 Introduction

In the paper, we consider the very general problem of finding "rich" invariants of the action of a (locally compact) group $G$ on functions over $G$ or over one of its homogeneous spaces. We start from a very old idea coming from a classical engineer's technique for invariant objects recognition: the Fourier-descriptors method. Invariant objects recognition is a critical problem in image processing. To solve it, numerous approaches have been proposed in the literature, often based on the computation of invariants followed by a classification method. Considering the group of motions of the plane, Gauthier and al. [9], [12], proposed a family of invariants, called motion descriptors, which are invariants in translation, rotations, scale and reflections. H. Fonga [7] applied them to grey level images. A recent survey on this question can be found in [20]. Another interesting paper closely connected to this work is [16].

In this paper, we develop and we give final results of a general theory of "Fourier descriptors". The paper contains really new results that justify the practical use of these "generalized Fourier descriptors".

The paper deals mostly with two cases: first, the case of compact groups, and second, the case of certain "motion groups", for the purpose of image analysis.

In a forthcoming paper [18], we show application of our results to several problems of pattern recognition (in particular, human-face recognition). In this last paper, a main point is that we apply 2D-invariant motion-descriptors for 3D recognition. The justification is clear: in practice we get a number of 2D images of the same object under several points of view. The motion descriptors being motion-invariants, we need a single picture for each point of view, independently of the position of the object. Also, in this paper, we use the invariants in the context of a "learning-machine" of Support-Vector-Machine (SVM) type [21].

However, in another practical context (3D data for instance) we could apply our methodology to the action of the group $SO_3 \ltimes \mathbb{R}^3$ of 3D-motions. Generalized motion descriptors for this group action can be computed easily using our theory.

We obtained a long time ago the results presented here in the case of compact groups. But proofs of them were never published. We give these proofs here (Theorem 5). Our final (original) result (in the case of the discrete 2-D motion groups acting on the plane) is stated in Theorem 8, and we sketch the proof.

Along the paper, we use the terminologies "Fourier descriptors", or "generalized Fourier descriptors" for general groups. When we want to focus on pattern recognition and motion groups, we use the terminology "motion descriptors".

Many technical details are omitted here. They will appear in [18].

## 2 Preliminaries

Let us start with a few preliminaries about:

- The classical Fourier descriptors for contours.
- The main facts about the abstract Fourier transform from group harmonic analysis. The example of the group $M_2$ of motions of the plane is treated explicitly.
- The generalization of Fourier descriptors for contours to Fourier descriptors in the large.

## 2.1 Classical Fourier descriptors for contours (the circle group)

The Fourier descriptors method is a very old method used for pattern analysis from the old days on. The oldest reference we were able to find is [17]. Basically, the method uses the good properties of standard Fourier series with respect to translations. For the sake of completeness, let us recall this basic idea, that has been used successfully several times for pattern recognition. For details, see for instance [17].

The method applies to the problem of discrimination of 2D-patterns by their **exterior** contour. Let the exterior contour be well defined, and regular enough (piecewise smooth, say). Assume that it is represented as a closed curve, arclength parameterized and denoted by $s(\theta)$. The variable $\theta$ is the arclength, from some arbitrary reference point $\theta_0$ on the contour, and $s(\theta)$ denotes the value of the angle between the tangent to the contour at $\theta$ and some privileged direction (the $x$-axis, say). By construction, the function $s(\theta)$ is obviously invariant under 2D translation of the pattern. Let now $\hat{s}_n$ denote the Fourier series of the periodic function $s(\theta)$. The only arbitrary object that makes the function $s$ non-invariant under motions (translations plus rotations) of the pattern, is the choice of the initial point $\theta_0$. As it is well known, a translation of $\theta_0$ by $a$, $\theta_0 := a + \theta_0$, changes $\hat{s}_n$ for $e^{ian}\hat{s}_n$, where $i = \sqrt{-1}$. (Here, the total arclength is normalized to $2\pi$). Set $\hat{s}_n = \rho_n e^{i\varphi_n}$. Let us define the "shifts of phases" $R_{n,m} = \frac{\varphi_n}{n} - \frac{\varphi_m}{m}$. Then, it is easy to check that the "discrete power spectral densities" $Pn = |\hat{s}_n|^2$ and the "shifts of phases" $R_{n,m}$ **form a complete set of invariants** of exterior contours, under motions of the plane. They are also homotetic-invariants as soon as the total arclength is normalized.

This result is extremely efficient for shape discrimination, it has been used an incredible number of times in many areas. It is very robust and physically interesting for several reasons (in particular the fact that the $P_n$ are just discrete "power spectral densities", and that both $P_n$ and $R_{n,m}$ can be computed very quickly using FFT algorithms). Also, the extraction of the "exterior-contour" is more or less a standard procedure in image processing.

**The main default** of the method is that it doesn't take any account of the "texture" of the pattern: two objects with similar exterior contours have similar "Fourier descriptors" $P_n$ and $R_{n,m}$.

This apparently naive method is in fact conceptually very important: as soon as one knows a bit about abstract harmonic analysis, one immediately thinks about possible abstract generalizations of this method. The first paper that we know in which this idea of "abstract generalization" of the method appears is the paper [4]. One of the authors here in worked on the subject, with several co-workers ([9], [11], [7], [12]). In particular, there is a lot of very interesting results in the theses [11] and [7]. A recent reference is [20].

Unfortunately, our results being very incomplete, they were never completely published. We would like here to give a series of more or less final result, not yet completely satisfactory, but very interesting and convincing.

They lead to the "**generalized Fourier descriptors**" that we use in the paper [18], and that look extremely efficient for objects discrimination, in addition to a standard Support-Vector-Machine technique. Moreover, at the end, they are computed in practice with standard Fourier integrals, then with FFT algorithms, and hence the algorithms are "fast".

## 2.2 The Fourier Transform on locally compact unimodular groups.

Classical Fourier descriptors for exterior contours will just correspond to the case of the "circle" group as the reader can check, i.e. the group of rotations $e^{i\theta}$ of the complex plane.

By a famous theorem of Weil, a locally compact group possesses a (almost unique) Haar-measure ([24]), i.e. a measure which is invariant under (left or right) translations. For instance the Haar measure of the circle group is $d\theta$ since $d(\theta + a) = d\theta$. A group is said unimodular if its left and right Haar measures can be taken equal (that is, the Haar measure associated with left or right translations). An abelian group is obviously automatically unimodular. A less obvious result is that a compact group is automatically unimodular.

The most pertinent examples for pattern recognition are of course the following:

1. The circle group $C$.
2. The group of motions of the plane $M_2$. It is the group of rotations and translations $(\theta, x, y)$ of the plane. As one can check, the product law on $M_2$ is

$$(\theta_1, x_1, y_1).(\theta_2, x_2, y_2) = \qquad (1)$$
$$(\theta_1 + \theta_2, \cos(\theta_1)x_2 - \sin(\theta_1)y_2 + x_1, \sin(\theta_1)x_2 + \cos(\theta_1)y_2 + y_1).$$

It represents the geometric composition of two motions. The main difference with the circle group is that it is not Abelian (commutative). This expresses the fact that rotations and translations of the plane do not commute. However, it is unimodular: the measure $d\theta dx dy$ is simultaneously left and right invariant.

3. The group of $y-$homotheties and $x-$translations of the upper two dimensional half plane: $(y_1, x_1).(y_2, x_2) = (y_1 y_2, x_1 + x_2)$. Here, the $y_i's$ are positive real numbers. Left and right Haar measure is $dx\frac{dy}{y}$ since $dx\frac{dy}{y} = d(x + a)\frac{d(by)}{by}$.
This abelian group is related to the classical Fourier-Mellin transform. A similar group of interest is the (abelian) group of $\theta$-rotations and $\lambda$ homotheties of the complex or two dimensional plane: $(\theta_1, \lambda_1).(\theta_2, \lambda_2) =$

$(\theta_1 + \theta_2, \lambda_1\lambda_2)$. Here again, the $\lambda_i$'s are positive real numbers but the $\theta_i$'s belong to the circle group. Of course, if one takes an image centered around it's gravity center, then, the effect of translations is eliminated, and it remains only the action of rotations and homotheties. Applying the theory developed in the second part of this paper to the case of this group leads to complete invariants with respect to motions and homotheties. This is related with the nice work of [10].

Unfortunately, in this case, the computation of all the invariants is based upon a preliminary estimation of the gravity center of the image. Hence, the invariants are simultaneously very sensitive to this preliminary estimation.

4. The group of translations, rotations and homotheties of the 2D plane itself (we don't write the multiplication but it is obvious) is unfortunately not unimodular. Hence the theory in this Section does not apply. It is why one has to go back to the previous group $M_2$.

5. The group $SO_3$ of rotations of $\mathbb{R}^3$. It is related to the human mechanisms of vision (see the paper [4]).

6. Certain rather **unusual groups** play a fundamental role in our theory below: the groups $M_{2,N}$ of motions, the rotation component of which is an integer multiple of $\frac{2\pi}{N}$. They are subgroups of $M_2$, and if $N$ is large, $M_{2,N}$ could be reasonably called the "group of translations and sufficiently small rotations". In some precise mathematical sense, $M_2$ is the limit when $N$ tends to infinity of the groups $M_{2,N}$.

For standard Fourier series and Fourier transforms, there are several general ingredients. Fourier series correspond to the circle group, Fourier transforms to the $\mathbb{R}$ (or more generally $\mathbb{R}^p$) group. In both cases, we have the formulas:

$$\hat{s}_n = \int_G s(\theta) \; e^{-in\theta} \; d\theta \tag{2}$$

$$\hat{f}(\lambda) = \int_G f(x) \; e^{-i\lambda x} \; dx.$$

Formally, in these two formulas appear an integration over the group $G$ with respect to the Haar measure (respectively $d\theta, dx$) of the function (respectively $s, f$) times (the inverse of) the "mysterious" term $e^{in\theta}$ (resp. $e^{i\lambda x}$). This term is the "character" term. It has to be interpreted as follows: For each $n$ (resp. $\lambda$), the map $\mathbb{C} \to \mathbb{C}$, $z \to e^{in\theta}z$ (resp. the map $z \to e^{i\lambda x}z$) is a unitary map (i.e. preserving the norm over $\mathbb{C}$), and the map $\theta \to e^{in\theta}$ (resp. $x \to e^{i\lambda x}$) is a continuous[1] group-homomorphism to the group of unitary linear transformations of $\mathbb{C}$. For a general topological group $G$, such a mapping is called a "character" of $G$.

---

[1] Along the paper, the topology over unitary operators on a Hilbert or Euclidian space is not the normic, but the strong topology.

The main basic result is the Pontryagin's duality theorem, that claims the following:

**Theorem 1.** *(Pontryagin's duality Theorem) The set of characters of an Abelian locally-compact group G is a locally-compact abelian group (under natural multiplication of characters), denoted by $G\hat{\ }$, and called the dual group of G. The dual group $(G\hat{\ })\hat{\ }$ of $G\hat{\ }$ is isomorphic to G.*

Then, the Fourier transform over $G$ is defined like that: it is a mapping from $\mathbb{L}^2(G,dg)$ (space of square integrable functions over $G$, with respect to the Haar measure $dg$), to the space $\mathbb{L}^2(G\hat{\ },d\hat{g})$, where $d\hat{g}$ is the haar measure over $G\hat{\ }$ :

$$f \rightarrow \hat{f}, \tag{3}$$

$$\hat{f}(\hat{g}) = \int_G f(g)\chi_{\hat{g}}(g^{-1})dg.$$

Here, $\hat{g} \in G\hat{\ }$ and $\chi_{\hat{g}}(g)$ is the value of the character $\chi_{\hat{g}}$ on the element $g \in G$.

As soon as one knows that the dual group of $\mathbb{R}$ is $\mathbb{R}$ itself, and the dual group of the circle group is the discrete additive group $\mathbb{Z}$ of integer numbers, it is clear that Formulas 2 are particular cases of Formula 3.

It happens that there is a generalization of the usual **Plancherel's Theorem**: The Fourier Transform[2] is an isometry from $\mathbb{L}^2(G, dg)$ to $\mathbb{L}^2(G\hat{\ }, d\hat{g})$. The general form of the inversion formula follows:

$$f(g) = \int_G f(\hat{g})\chi_{\hat{g}}(g)d\hat{g}. \tag{4}$$

In our cases $(\mathbb{R},C)$, this gives of course the usual formulas.

In the case of nonabelian groups, the generalization starts to be less straightforward. To define a reasonable Fourier transform, one cannot consider only characters (this is not enough for a good theory, leading to Plancherel's Theorem). One has to consider more general objects than characters, namely, unitary irreducible representations of $G$. A (continuous) unitary representation of $G$ consists of replacing $\mathbb{C}$ by a general complex Hilbert[3] space $H$, and the characters $\chi_{\hat{g}}$ by unitary linear operators $\chi_{\hat{g}}(g) : H \rightarrow H$, such that the mapping $g \rightarrow \chi_{\hat{g}}(g)$ is a continuous[4] homomorphism. Irreducible means that there is no nontrivial closed subspace of $H$ which is invariant under all the operators $\chi_{\hat{g}}(g)$, $g \in G$. Clearly, characters are very special cases of continuous unitary irreducible representations. The main fact is that, for locally compact nonabelian groups, to get Plancherel's formula, it is enough to replace characters by these representations.

---

[2] Precisely, Haar measures can be normalized so that Fourier transform is isometric.
[3] In the paper, all Hilbert spaces are assumed separable.
[4] For the strong topology of the unitary group $U(H)$.

**Definition 1.** *Two representations* $\chi_1, \chi_2$ *of* $G$, *with respective underlaying Hilbert spaces* $H_1, H_2$ *are said equivalent if there is a linear invertible operator* $A : H_1 \rightarrow H_2$, *such that, for all* $g \in G$ :

$$\chi_2(g) \circ A = A \circ \chi_1(g). \tag{5}$$

*More generally, a linear operator* $A$, *eventually noninvertible, meeting condition 5, is called an* **intertwining operator** *between the representations* $\chi_1, \chi_2$.

*The set of equivalence classes of unitary irreducible representations of* $G$ *is called the dual set of* $G$, *and is denoted by* $G^{\hat{}}$ ·

One of the main differences with the abelian case is that $G^{\hat{}}$ has in general no group structure. However, in this very general setting, Plancherel's Theorem holds:

**Theorem 2.** *Let* $G$ *be a locally compact unimodular group with Haar measure* $dg$. *Let* $G^{\hat{}}$ *be the dual of* $G$. *There is a measure over* $G^{\hat{}}$ *(called the Plancherel's measure, and denoted by* $d\hat{g}$*), such that, if we define the Fourier transform over* $G$ *as the mapping:*

$$\mathbb{L}^2(G, dg) \rightarrow \mathbb{L}^2(G^{\hat{}}, d\hat{g}), \tag{6}$$

$$f \rightarrow \hat{f},$$

$$\hat{f}(\hat{g}) = \int_G f(g) \chi_{\hat{g}}(g^{-1}) dg,$$

*then,* $\hat{f}(\hat{g})$ *is a Hilbert-Schmidt operator over the underlaying space* $H_{\hat{g}}$, *and the Fourier transform is an isometry.*

*As a consequence, the following inverse formula holds:*

$$f(x) = \int_{G^{\hat{}}} Trace[\hat{f}(\hat{g}) \chi_{\hat{g}}(g)] d\hat{g}. \tag{7}$$

More generally, if $\chi$ is a unitary representation of $G$ -**not necessarily irreducible**- one can define the Fourier transform $\hat{f}(\chi)$ by the same formula 6.

All this could look rather complicated. In fact, it is not at all, and we shall immediately make it explicit in the case of main interest for our applications to pattern recognition, namely the group of motions $M_2$.

In the following, for the group $M_2$, (and later on $M_{2,N}$), we take up the notations below:

**Notation 1** *Elements of the group are denoted indifferently by* $g = (\theta, x, y) = (\theta, X)$, *where* $X = (x, y) \in \mathbb{R}^2$. *The usual scalar product over* $\mathbb{R}^2$ *is denoted by* $< ., . >_{\mathbb{R}^2}$, *or simply* $< ., . >$ *if no confusion is possible. Then, the product over* $M_2$ *(resp.* $M_{2,N}$ *writes* $(\theta, X).(\alpha, Y) = (\theta + \alpha, R_\theta Y + X)$, *where* $R_\theta$ *is the rotation operator of angle* $\theta$.

*Example 1.* Group $M_2$ of motions of the plane.

-In that case, the unitary irreducible representations fall in two classes: 1. characters (one dimensional Hilbert space of the representation), 2. The other irreducible representations have infinite dimensional underlaying Hilbert space $H = \mathbb{L}^2(C, d\theta)$ where $C$ is the circle group $\mathbb{R}/2\pi\mathbb{Z}$, and $d\theta$ is the Lebesgue measure over $C$. These representations are parameterized by any ray $R$ from the origin in $\mathbb{R}^2$, $R = \{\alpha V, V$ some fixed nonzero vector in $\mathbb{R}^2$, $\alpha$ a real number, $\alpha > 0\}$. For $r \in R$ (the ray), the representation $\chi_r$ expresses as follows, for $\varphi(.) \in H$:

$$[\chi_r(\theta, X).\varphi](u) = e^{i<r, R_u X>}\varphi(u + \theta). \tag{8}$$

The Plancherel's measure has support the second class of representations, i.e. characters play no role in that case.

It is easily computed that the Fourier transform of $f \in \mathbb{L}^2(M_2, \text{Haar})$ writes, with $X = (x, y)$:

$$[\hat{f}(r).\varphi](u) = \int\int\int_{M_2} f(\theta, x, y)e^{-i<r, R_{u-\theta}X>} \times \varphi(u - \theta)d\theta dx dy. \tag{9}$$

**The main property of the general Fourier transform** that we will use in the paper concerns obviously its behavior with respect to translations of the group. Let $f \in \mathbb{L}^2(G, dg)$ and set $f_a(g) = f(ag)$. Due to the invariance of the Haar measure w.r.t. translations of $G$, we get the **crucial** generalization of a well known formula:

$$\hat{f}(\hat{g}) \circ \chi_{\hat{g}}(a) = \widehat{f_a}(\hat{g}). \tag{10}$$

## 2.3 General definition of the generalized Fourier descriptors, from those over the circle group

In the case of exterior contours of 2D patterns, the group under consideration is the circle group $C$. The set of invariants $P_n, R_{m,n}$ has first to be replaced by the (**almost equivalent**) set of invariants, $P_n, \tilde{R}_{m,n}$, where the new "phase invariants" $\tilde{R}_{m,n}$ are defined by:

$$\tilde{R}_{m,n} = \hat{s}_n \hat{s}_m \overline{\hat{s}_{n+m}}. \tag{11}$$

The justification of this definition is the following: it is easily checked that at least **on a residual subset** of $\mathbb{L}^2(C)$ **these sets of invariants are equivalent**. This is enough for our practical purposes.

*Remark 1.* 1. There is a counterexample in [12] showing that the second set of invariants is weaker (does not discriminate among all functions).

But in practice, discriminating over a very big dense subset of functions is enough. Moreover, it is unexpected to be able to do more, in general.

2. Nevertheless, in the case of the additive groups $\mathbb{R}^n$, these second invariants discriminate completely. This is shown in [11].

3. For complete invariants over $\mathbb{L}^2(G)$ in **the general abelian case**, generalizing those, see [12], [11], [7].

Now, an important fact has to be pointed out. There is a natural interpretation and generalization of the "phase-invariants" $\tilde{R}_{m,n}$ in terms of representations.

We are given an arbitrary unimodular group $G$, with Haar measure $dg$. We define the Fourier transform $\hat{f}$ of $f$, as the map from the set of (equivalence classes of) unitary irreducible representations of $G$, given by formula 6.

Let us state now a crucial definition, and a crucial theorem.

**Definition 2.** *The following sets $I_1, I_2$, are called respectively the first and second Fourier descriptors (or motion descriptors) of a map $f \in \mathbb{L}^2(G)$. For $\hat{g}, \hat{g}_1, \hat{g}_2 \in \hat{G}$,*

$$I_1^{\hat{g}}(f) = \hat{f}(\hat{g}) \circ \hat{f}(\hat{g})^*, \tag{12}$$

$$I_2^{\hat{g}_1,\hat{g}_2}(f) = \hat{f}(\hat{g}_1)\hat{\otimes}\hat{f}(\hat{g}_2) \circ \hat{f}(\hat{g}_1\hat{\otimes}\hat{g}_2)^*,$$

where $\hat{f}(\hat{g})^*$ denotes the adjoint of $\hat{f}(\hat{g})$, and where $\hat{g}_1\hat{\otimes}\hat{g}_2$ denotes the (equivalence class of) (Kronecker) Hilbert tensor product of the representations $\hat{g}_1$ and $\hat{g}_2$, and $\hat{f}(\hat{g}_1)\hat{\otimes}\hat{f}(\hat{g}_2)$ is the Hilbert tensor product of the Hilbert-Schmidt operators $\hat{f}(\hat{g}_1)$ and $\hat{f}(\hat{g}_2)$.

Then, clearly, in the particular case of the circle group, **these formulas coincide** with those defining $P_n, \tilde{R}_{m,n}$.

Let us temporarily say that a (grey-level) image $f$ on $G$ is a compactly supported real nonzero function over $G$, with positive values (the grey levels).

**Theorem 3.** *The quantity $I_1(f)$ is determined by $I_2(f)$ (by abuse, we write $I_1(f) \subset I_2(f)$) and $I_1(f), I_2(f)$ are invariant under translations of $f$ by elements of $G$.*

*Proof.* That $I_1(f)$ is determined by $I_2(f)$ comes from the fact that, $f$ being an image, taking for $\hat{g}_2$ the trivial character $c_0$ of $G$, we get that $I_2^{\hat{g}_1,\hat{g}_2}(f) = av(f)I_1^{\hat{g}_1}(f)$, where the "mean value" of $f$, $av(f) = \int_G f(g)dg > 0$, $av(f) = (I_2^{c_0,c_0})^{1/3}$. That $I_1^{\hat{g}}(f_a) = I_1^{\hat{g}}(f)$ (where $f_a(g) = f(ag)$), the translate of $f$ by $a$) comes from the classical property 10 of Fourier transforms. That $I_2^{\hat{g}_1,\hat{g}_2}(f) = I_2^{\hat{g}_1,\hat{g}_2}(f_a)$, comes from the other trivial fact, just a consequence of the definition,

$$\hat{f}_a(\hat{g}_1\hat{\otimes}\hat{g}_2) = \hat{f}(\hat{g}_1\hat{\otimes}\hat{g}_2) \circ (\chi_{\hat{g}_1}(a)\hat{\otimes}\chi_{\hat{g}_2}(a)),$$

and from the unitarity of the representations.

**Our purpose in the remaining of the paper** is to compute these invariants and to investigate about their completeness (at least on a big subset of $\mathbb{L}^2(G)$) and their pertinence. We will mostly consider either an Abelian or compact group $G$, or one of our motion groups $M_2$ and $M_{2,N}$.

# 3 The generalized Fourier descriptors for the motion group $M_2$

Here, using the results stated in Example 1, let us compute the generalized Fourier descriptors from the definition 2 for the group of motions $M_2$.

The following series of formulas comes from straightforward computations, using the results and notations stated in Example 1.

For $r_1, r_2 \in \mathbb{R}^2$ the tensor product $\chi_{r_1} \hat{\otimes} \chi_{r_2}$, (denoted also by $\chi_{r_1 \hat{\otimes} r_2}$) of the representations $\chi_{r_1}$ and $\chi_{r_2}$ can be written, for $\varphi \in \mathbb{L}^2(C \times C) \sim \mathbb{L}^2(C) \hat{\otimes} \mathbb{L}^2(C)$,

$$[\chi_{r_1 \hat{\otimes} r_2}(\theta, X)\varphi](u_1, u_2) = e^{i<R_{-u_2}r_1 + R_{-u_1}r_2, X>} \times \varphi(u_1 + \theta, u_2 + \theta). \quad (13)$$

Therefore, we have the following expression for the adjoint operator:

$$[\chi_{r_1 \hat{\otimes} r_2}(\theta, X)^*\varphi](u_1, u_2) = e^{-i<R_{\theta-u_2}r_1 + R_{\theta-u_1}r_2, X>}$$
$$\times \varphi(u_1 - \theta, u_2 - \theta). \quad (14)$$

**A very important point**: we consider functions $f$ on the group of motions that are functions of $X = (x, y)$ only (they do not depend on $\theta$, i.e. they are the "trivial" lifts on the group $M_2$ of functions on the plane $\mathbb{R}^2$). For the Fourier transform, we get, with $\varphi \in \mathbb{L}^2(C)$, $r \in R$,

$$[\hat{f}(r)\varphi](u) = \int_C \tilde{f}(R_{\theta-u}r)\varphi(u - \theta)d\theta = <\varphi(\theta), \overline{\tilde{f}(R_{-\theta}r)} >_{\mathbb{L}^2(C,d\theta)}, \quad (15)$$

in which $\tilde{f}(V)$ denotes as before the usual Fourier transform over (the Abelian group) $\mathbb{R}^2$ :

$$\tilde{f}(V) = \int_{\mathbb{R}^2} f(X)e^{-i<V,X>_{\mathbb{R}^2}} dxdy. \quad (16)$$

The adjoint of the Fourier transform is given by:

$$[\hat{f}(r)^*\varphi](u) = \overline{\tilde{f}(R_{-u}r)} < \varphi, 1 >_{\mathbb{L}^2(C)}, \quad (17)$$

where 1 is the constant function over $C$, with value 1.

It follows that:

$$[\hat{f}(r_1)^* \hat{\otimes} \hat{f}(r_2)^*\varphi](u_1, u_2) = \overline{\tilde{f}(R_{-u_2}r_1)}\overline{\tilde{f}(R_{-u_1}r_2)} \times \int\int_{C \times C} \varphi(a, b)dadb \quad (18)$$

The final expression we need, to compute the generalized Fourier descriptors, follows easily from (13):

$$\hat{f}(r_1 \hat{\otimes} r_2)\varphi(u_1, u_2) = \int_C \tilde{f}(R_{\theta - u_2} r_1 + R_{\theta - u_1} r_2) \times \varphi(u_1 - \theta, u_2 - \theta) d\theta \quad (19)$$

Using all these formulas, it is not hard to get our result for the generalized Fourier descriptors:

$$[I_1^r(f)\varphi](u) = \int_C |\tilde{f}(R_\theta r)|^2 d\theta \ < \varphi, 1 >_{\mathrm{L}^2(C)}, \quad (20)$$

$$[I_2^{r_1,r_2}(f)\varphi](u_1, u_2) = \int_C \tilde{f}(R_\theta(\hat{r}_1 + \hat{r}_2)) \overline{\tilde{f}}(R_\theta \hat{r}_1) \overline{\tilde{f}}(R_\theta \hat{r}_2) d\theta$$

$$\int \int_{C \times C} \varphi(a, b) da db,$$

$$\text{with } \hat{r}_i = R_{-u_i} r_i, \quad i = 1, 2.$$

Clearly, these invariants are completely determined by the following ones, that we use in practice (in [18] for instance):

$$I_1^r(f) = \int_C |\tilde{f}(R_\theta r)|^2 d\theta, \ r \in R, \quad (21)$$

$$I_2^{\xi_1,\xi_2}(f) = \int_C \tilde{f}(R_\theta(\xi_1 + \xi_2)) \overline{\tilde{f}}(R_\theta \xi_1) \overline{\tilde{f}}(R_\theta \xi_2) d\theta, \qquad \text{for } \xi_1, \xi_2 \in \mathbb{R}^2.$$

*Remark 2.* The generalized Fourier descriptors are real quantities (This is not an obvious fact for the second type invariants, but it is easily checked).

Completeness of these invariants is still an open question. However in the remaining of the paper we will prove certain completeness results in other very close cases.

# 4 The case of compact (non-Abelian) groups

This is the most beautiful part of the theory, showing in a very convincing way that the formulas 12 are really pertinent: in the compact case, (including the classical Abelian case of exterior contours), the generalized Fourier descriptors are weakly complete. This is due to the Tannaka-Krein duality theory. (See [14], [15]).

## 4.1 Chu and Tannaka categories, Chu and Tannaka dualities

Tannaka Theory is the generalization to compact groups of Pontryagin's duality theory.

The following facts are standard: The dual of a compact group is a **discrete set**, and all its unitary irreducible representations are **finite dimensional**.

The main lines of Tannaka theory is like that: we start with a compact group $G$.

1. There is the notion of a Tannaka category $\mathcal{T}_G$, that describes the structure of the finite dimensional unitary representations of $G$;
2. There is the notion of a quasi representation $\mathcal{Q}$ of a Tannaka category $\mathcal{T}_G$;
3. The set $rep(G)\hat{}$ of quasi representations of the Tannaka category $\mathcal{T}_G$ has the structure of a topological group;
4. The groups $rep(G)\hat{}$ and $G$ are naturally isomorphic. (Tannaka duality).

This scheme completely generalizes the scheme of Pontryagin's duality to the case of compact groups.

In fact, Tannaka duality theory is just a particular case of Chu duality, which will be **the crucial form of duality needed** for our purposes. Hence, let us introduce precisely Chu duality ([14], [5]), and Tannaka duality will just be **the particular case of compact groups**.

Let temporarily $G$ be an arbitrary topological group.

For all $n \in \mathbb{N}$ the set $rep_n(G)$ denotes the set of continuous unitary representations of $G$ over $\mathbb{C}^n$. $rep_n(G)$ is endowed with the following topology: a basis of open neighborhoods of $T \in rep_n(G)$ is given by the sets $W(K, T, \varepsilon)$, $\varepsilon > 0$, and $K \subset G$, a compact subset,

$$W(K, T, \varepsilon) = \{\tau \in rep_n(G) |\ ||T(g) - \tau(g)|| < \varepsilon, \forall g \in K\},$$

where the norm of operators is the usual Hilbert-Schmidt norm. If $G$ is locally compact, so is $rep_n(G)$.

**Definition 3.** *The Chu-Category of $G$ is the category $\pi(G)$, the objects of which are the finite dimensional unitary representations of $G$, and the morphisms are the intertwining operators.*

**Definition 4.** *A quasi-representation of the category $\pi(G)$ is a function $Q$ over $ob(\pi(G))$ such that $Q(\chi)$ belongs to $U(H_\chi)$, the unitary group over the underlaying space $H_\chi$ of the representation $\chi$, with the following properties:*

*0. $Q$ commutes with Hilbert direct-sum: $Q(\chi_1 \oplus \chi_2) = Q(\chi_1) \oplus Q(\chi_2)$*
*1. $Q$ commutes with the Hilbert tensor product: $Q(\chi_1 \hat{\otimes} \chi_2) = Q(\chi_1) \hat{\otimes} Q(\chi_2)$,*
*2. $Q$ commutes with the equivalence operators: for an equivalence $A$ between $\chi_1$ and $\chi_2$, $A \circ Q(\chi_1) = Q(\chi_2) \circ A$,*
*3. the mappings, $rep_n(G) \to U(\mathbb{C}^n)$, $\chi \to Q(\chi)$ are continuous.*

*The set of quasi-representations of the category $\pi(G)$ is denoted by $rep(G)\hat{}$.*

There are "natural" quasi representations of $G$ : for each $g \in G$, the mapping $\Omega_g(\chi) = \chi(g)$ defines a quasi-representation of $\pi(G)$.

*Remark 3.* $rep(G)\hat{}$ is a group with the multiplication $Q_1.Q_2(\chi) = Q_1(\chi).Q_2(\chi)$. The neutral element is $E$, with $E(\chi) = \Omega_e(\chi) = \chi(e)$, for $e$ the neutral of $G$.

There is a topology over $rep(G)\hat{}$ such that it becomes a topological group. A fundamental system of neighborhoods of $E$ is given by the sets $W(\hat{K}_{n_1}, ..., \hat{K}_{np}, \varepsilon)$, $\varepsilon > 0$ and $\hat{K}_{n_i}$ is compact in $rep_{n_i}(G)$, with $W(\hat{K}_{n_1}, ..., \hat{K}_{np}, \varepsilon) = \{Q \in rep(G)\hat{} \mid \|Q(\chi) - E(\chi)\| < \varepsilon, \forall \chi \in \cup \hat{K}_{n_i}\}$.

The first main result is that, as soon as $G$ is locally compact, the mapping $\Omega : G \to rep(G)\hat{}$, $g \to \Omega_g$ is a continuous homomorphism.

**Definition 5.** *A locally compact $G$ has the **duality property if** $\Omega$ is a topological group isomorphism.*

The main result is:

**Theorem 4.** *If $G$ is locally compact, Abelian, then $G$ has the duality property. (This is no more than Pontryagin's duality).*

   *If $G$ is compact, $G$ has the duality property. (This is Tannaka-Krein theory).*

In the last section of the paper, for the purpose of pattern recognition, we will use crucially the fact that **another class of groups, namely the Moore groups, have also the duality property.**

## 4.2 Generalized Fourier descriptors over compact groups

Our result in this section is based upon Tannaka theory, and shows that the **weak-completeness** -i.e. completeness over a residual subset of $\mathbb{L}^2(G,dg)$- of the generalized Fourier descriptors (which holds on the circle group, and which is crucial for pattern recognition of "exterior contours") **generalizes to compact separable groups.**

   If $G$ is compact separable, then, we have the following crucial but obvious lemma:

**Lemma 1.** *The subset $R$ of functions $f \in \mathbb{L}^2(G,dg)$ such that $\hat{f}(\hat{g})$ is invertible for all $\chi = \hat{g} \in G\hat{}$ is residual in $\mathbb{L}^2(G,dg)$.*

*Proof.* It follows from [6] that if $G$ is compact separable, then $G\hat{}$ is countable. For a fixed $\hat{g}$, the set of $f$ such that $\hat{f}(\hat{g})$ is not invertible is clearly open, dense. Hence, $R$ is a countable intersection of open-dense sets, in a Hilbert space.

The main theorem is:

**Theorem 5.** *Let $G$ be a compact separable group. Let $R$ be the subset of elements of $\mathbb{L}^2(G,dg)$ on which the Fourier transform takes values in invertible operators. Then $R$ is residual in $\mathbb{L}^2(G,dg)$, and the generalized Fourier descriptors discriminate over $R$.*

*Proof.* Let us take two functions $f, h \in R$, such that the associated generalized Fourier descriptors are equal. The equality of the first-type Fourier descriptors gives $\hat{f}(\hat{g}) \circ \hat{f}(\hat{g})^* = \hat{h}(\hat{g}) \circ \hat{h}(\hat{g})^*$, for all $\hat{g} \in \hat{G}$. Since $\hat{f}(\hat{g})$ is invertible, we deduce that there is $u(\hat{g}) \in U(H_{\hat{g}})$, such that $\hat{f}(\hat{g}) = \hat{h}(\hat{g})\, u(\hat{g})$.

If $\chi$ is a reducible unitary representation, it is a finite direct sum of irreducible representations, and therefore, the equality $\hat{f}(\chi) \circ \hat{f}(\chi)^* = \hat{h}(\chi) \circ \hat{h}(\chi)^*$, for all $\hat{g}_i \in \hat{G}$ also defines an invertible $u(\chi) = \hat{h}(\chi)^{-1}\, \hat{f}(\chi)$. (By the finite sum decomposition, $\hat{h}(\chi) = \dot{\oplus}\hat{h}(\hat{g}_i)$, hence $\hat{h}(\chi)$ is invertible).

Moreover it is obvious that the mappings $rep_n(G) \to M(n, \mathbb{C})$, $\chi \to \hat{f}(\chi)$ are continuous therefore the mapping $\chi \to u(\chi) = \hat{h}(\chi)^{-1}\, \hat{f}(\chi)$ is also continuous.

Also, the equality of the (second) Fourier descriptors for the irreducible representations, [due to the finite decomposition of any representation in a direct sum of irreducible ones, plus the usual properties of Hilbert tensor product] implies the equality of Fourier descriptors for arbitrary (non-irreducible) unitary finite-dimensional representations, i.e., if $\chi, \chi'$ are such unitary representations, non necessarily irreducible, we have also:

$$\hat{f}(\chi)\hat{\otimes}\hat{f}(\chi') \circ \hat{f}(\chi\hat{\otimes}\chi')^* = \hat{h}(\chi)\hat{\otimes}\hat{h}(\chi') \circ \hat{h}(\chi\hat{\otimes}\chi')^*. \tag{22}$$

Replacing in this last equality $\hat{f}(\chi) = \hat{h}(\chi)\, u(\chi)$, and taking into account the fact that all the $\hat{f}(\chi), \hat{h}(\chi)$ are invertible, we get that:

$$u(\chi\hat{\otimes}\chi') = u(\chi)\hat{\otimes}u(\chi'), \tag{23}$$

for all finite dimensional unitary representations $\chi, \chi'$ of $G$.

Now, for such $\chi, \chi'$, and for $A$ intertwining $\chi$ and $\chi'$, we have also $A\hat{f}(\chi) = \int_G f(g)A\chi(g^{-1})dg = \int_G f(g)\chi'(g^{-1})Adg = \hat{f}(\chi')A$. It follows that $A\hat{h}(\chi)u(\chi) = \hat{h}(\chi')u(\chi')A$, hence, $\hat{h}(\chi')Au(\chi) = \hat{h}(\chi')u(\chi')A$, in which $\hat{h}(\chi')$ is invertible. Therefore, $Au(\chi) = u(\chi')A$. By Definition 4, $u$ is a quasi-representation of the category $\pi(G)$. By Theorem 4, $G$ has the duality property, and for all $\hat{g} \in \hat{G}$, $u(\hat{g}) = \chi_{\hat{g}}(g_0)$ for some $g_0 \in G$. Then:

$$\hat{f}(\hat{g}) = \hat{h}(\hat{g})\chi_{\hat{g}}(g_0),$$

and, by the main property 10 of Fourier transforms, $\hat{f} = \widehat{h_a}$, $f = h_a$ for some $a \in G$.

# 5 The case of the group of motions with small rotations $M_{2,N}$

This section contains our final results. We will consider the action on the plane of the group $M_{2,N}$ of translations and small rotations. In the case where $N$ is an odd number, we will be able to achieve a full theory and to get a weak-completion result. Several intermediate technical lemmas are not proven here in. We refer to the forthcoming paper [18]. However, these lemmas are pure technical details, the main ideas being explained here.

## 5.1 Moore groups and duality for Moore groups

For details, we refer to [14]. We already know that compact groups have all their unitary irreducible representations of finite dimension. But they are not the only ones.

**Definition 6.** *A Moore group is a locally-compact group, such that all its unitary irreducible representations have finite-dimensional underlaying Hilbert space.*

**Theorem 6.** *The groups $M_{2,N}$ are Moore groups.*

**Theorem 7.** *[14] (Chu duality) Moore groups (separable) have the duality property.*

Then, we will try to copy what has been done for compact groups to our Moore groups. There are several difficulties, due to the fact that the functions under consideration (the images) are very special functions over the group. In fact, they are functions over the homogeneous space $\mathbb{R}^2$ of $M_{2,N}$.

## 5.2 Representations, Fourier transform and generalized Fourier descriptors over $M_{2,N}$

In fact, considering "images", we will be interested only with functions on $M_{2,N}$ that are also functions on the plane $\mathbb{R}^2$. One of the main problems, as we shall see, is that there are several possible "lifts" of the functions of $\mathbb{L}^2(\mathbb{R}^2)$ on $\mathbb{L}^2(M_{2,N})$, and that the "trivial" lift is bad for our purposes.

Typical elements of $M_{2,N}$ are still denoted by $g = (\theta, x, y) = (\theta, X)$, $X = (x, y) \in \mathbb{R}^2$, but now, $\theta \in \check{N} = \{0, ..., N-1\}$. Each such $\theta$ represents a rotation of angle $\frac{2\theta\pi}{N}$, that we still denote by $R_\theta$.

The Haar measure is the tensor product of the uniform measure over $\check{N}$ and the Lebesgue measure over $\mathbb{R}^2$. The dual space $\hat{G}$ is the union of the finite set $\mathbb{Z}/N\mathbb{Z} = \check{N}$ (characters) with the "Slice of Cake" $\mathcal{S}$, corresponding to nonzero values of $r \in \mathbb{R}^2$ of angle $\alpha(r)$, $0 \leq \alpha(r) < \frac{2\pi}{N}$. The support of the Plancherel Measure is $\mathcal{S}$ (characters are of no use).

Here $\varphi \in \mathbb{C}^N$, i.e. $\varphi : \check{N} \to \mathbb{C}$. We have exactly the same formula as for $M_2$ :

$$[\chi_r(\theta, X).\varphi](u) = e^{i<r, R_u X>} \varphi(u + \theta), \qquad (24)$$

but $r \in \mathcal{S}$ and the map $l^2(\check{N}) \to l^2(\check{N})$, $\varphi(u) \to \varphi(u + \theta)$, is just the $\theta$−shift operator over $\mathbb{C}^N$.

The Fourier transform has a similar expression to formula 9:

$$[\hat{f}(r).\varphi](u) = \sum_{\check{N}} (\int \int_{\mathbb{R}^2} f(\theta, x, y)e^{-i<r, R_{u-\theta} X>} \times \varphi(u - \theta)dxdy) \qquad (25)$$

Similar computations to those of Section 3 lead to the final formula for the Fourier descriptors relative to the trivial lift of functions $f$ over $\mathbb{R}^2$ into functions over $M_{2,N}$ (not depending on $\theta$) :

$$I_1^r(f) = \sum_{\theta \in \check{N}} |\tilde{f}(R_\theta r)|^2 d\theta, \ r \in R, \qquad (26)$$

$$I_2^{\xi_1, \xi_2}(f) = \sum_{\theta \in \check{N}} \tilde{f}(R_\theta(\xi_1 + \xi_2)) \overline{\tilde{f}}(R_\theta \xi_1) \overline{\tilde{f}}(R_\theta \xi_2) d\theta,$$

for $\xi_1, \xi_2 \in \mathbb{R}^2$.

By Theorem 3, these **generalized Fourier descriptors are invariant under the action of** $M_{2,N}$ **on** $\mathbb{L}^2(\mathbb{R}^2)$. Let us explain the main problem that appears when we try to generalize the theorem 5 of Section 4.2.

For this, we have to consider the special expression of the Fourier transform of the "trivial lift" of a function on the plane. We have a similar expression to formula 15 in Section 3:

$$[\hat{f}(r)\varphi](u) = \sum_{\check{N}} \tilde{f}(R_{\theta-u}r)\varphi(u - \theta) = < \varphi(\theta), \overline{\tilde{f}}(R_{-\theta}r) >_{l^2(\check{N})} .$$

The **crucial** point in the proof of the main theorem 5 is that the operators $\hat{f}(r)$ are all invertible. But, here, it is not at all the case since the operators defined by the formula above are far from invertible: **they always have rank 1**.

To overcome this difficulty, **we have to chose another lift of functions on the plane to functions on** $M_{2,N}$, **the trivial lift being too rough**. This is what we do in the next section.

## 5.3 The cyclic-lift from $\mathbb{L}^2(\mathbb{R}^2)$ to $\mathbb{L}^2(M_{2,N})$

From now on, we consider functions on $\mathbb{R}^2$, that are square-summable, and that have their support contained in a translate of a given compact set $K$ (the "screen").

Given a compactly supported function in $\mathbb{L}^2(\mathbb{R}^2, \mathbb{R})$, we can define its average and its (weighted) centroid, as follows:

$$av(f) = \int_K f(x, y)dxdy,$$

$$centr(f) = (x_f, y_f) = X_f = (\int_K xf(x, y)dxdy, \int_K yf(x, y)dxdy).$$

**Definition 7.** *The cyclic-lift of a compactly supported* $f \in \mathbb{L}^2(\mathbb{R}^2, \mathbb{R})$, *with nonzero average, onto* $\mathbb{L}^2(M_{2,N})$ *is the function* $f^c(\theta, x, y) = f(R_\theta X + \frac{centr(f)}{av(f)})$.

Note that $\frac{centr(f)}{av(f)}$ is the "geometric center" of the image $f$ and that $f^c(0, X)$ is the "centered image".

The set of $K$-supported real valued functions is a closed subspace $\mathcal{H} = \mathbb{L}^2(K)$ of $\mathbb{L}^2(\mathbb{R}^2)$. The set $\mathcal{I}$ of elements of $\mathcal{H}$ with nonzero average is an open subset of $\mathcal{H}$, therefore it has the structure of a Hilbert manifold. This is important since we shall apply to this space the parametric transversality theorem of [1].

**Definition 8.** *from now on, a (grey level, or one-color) "image"* $f$ *is an element of* $\mathcal{I}$.

Notice that moreover, usual images have positive value. (grey or color levels vary between zero and 1). This will be of no importance here in.

It is not so hard to check that $f$ and $g$ **differ from a motion with angle** $\frac{4k\pi}{N}$ **if and only if** $f^c$ **and** $g^c$ **differ from a motion with angle equal to** $\frac{2k\pi}{N}$.

In this way we reduce the problem of equivalence with rotation of certain multiples of a small angle to **the problem of equivalence of the cyclic lifts over** $M_{2,N}$.

This problem will be treated now, with the same method as in Section 4 (case of compact groups). For crucial reasons that will appear clearly below, we will consider only the case of an **odd** $N = 2n + 1$. Note that if $N$ is odd, when $k$ varies in $\check{N}$, $2k \bmod N$ also varies in $\check{N}$.

### 5.4 Fourier transform, generalized Fourier descriptors of cyclic lifts over $M_{2,2n+1}$

Using the expression 24 of the unitary irreducible representations over $M_{2,N}$, easy computations give the following results:

For $r_1, r_2 \in S$,

$$[\chi_{r_1 \hat{\otimes} r_2}(\theta, V)\varphi](u1, u2) = e^{i<R_{-u_2}r_1 + R_{-u_1}r_2, V>} \times \varphi(u_1 + \theta, u_2 + \theta) \quad (27)$$

Notice that this expression is exactly the same as 13. As a consequence, again:

$$\chi_{r_1 \hat\otimes r_2}(\theta, X)^* \varphi](u_1, u_2) = e^{-i<R_\theta - u_2 r_1 + R_\theta - u_1 r_2, X>} \times \varphi(u_1 - \theta, u_2 - \theta). \quad (28)$$

For the Fourier transform of a cyclic lift $f^c$, we get:

$$[\widehat{f^c}(r)\Psi](u) = \sum_\alpha \tilde f(R_{2\alpha + u} r) e^{i<R_{2\alpha + u} r, \frac{1}{av(f)} X f>} \Psi(-\alpha) \quad (29)$$

$$= \sum_{\alpha \in \breve N} \tilde f(R_{u - 2\alpha} r) e^{i<R_{u - 2\alpha} r, \frac{1}{av(f)} X f>} \Psi(\alpha).$$

Here, as above, $\tilde f(V)$ denotes the usual 2-D Fourier transform of $f$ at $V$. We get also:

$$[\widehat{f^c}(r)^* \Psi](u) = \sum_{\alpha \in \breve N} \overline{\tilde f}(R_{\alpha - 2u} r) e^{-i<R_{\alpha - 2u} r, \frac{1}{av(f)} X f>} \Psi(\alpha). \quad (30)$$

The last expression we need is:

$$[\widehat{f^c}(r_1 \hat\otimes r_2)\varphi](u_1, u_2) = \sum_{\alpha \in \breve N} \tilde f(R_{2\alpha - u_2} r_1 + R_{2\alpha - u_1} r_2) \quad (31)$$

$$e^{i<R_{2\alpha - u_2} r_1 + R_{2\alpha - u_1} r_2, \frac{1}{av(f)} X f>} \varphi(u_1 - \alpha, u_2 - \alpha).$$

Formula 30 leads to:

$$[\widehat{f^c}(r_1)^* \hat\otimes \widehat{f^c}(r_2)^* \varphi(u_1, u_2) = \quad (32)$$

$$\sum_{(\alpha_1, \alpha_2) \in \breve N \times \breve N} \overline{\tilde f}(R_{\alpha_2 - 2u_2} r_1) \overline{\tilde f}(R_{\alpha_1 - 2u_1} r_2)$$

$$e^{-i<R_{\alpha_2 - 2u_2} r_1 + R_{\alpha_1 - 2u_1} r_2, \frac{1}{av(f)} X f>} \times \varphi(\alpha_1, \alpha_2).$$

Now, we can perform the computation of the generalized Fourier descriptors. After computations based upon the formulas just established, we get for the self adjoint matrix $I_1^r(f) = \hat f(r) \circ \hat f(r)^*$ :

$$I_1^r(f)_{l,k} = \sum_{j \in \breve N} \tilde f(R_{l - 2j} r) \overline{\tilde f}(R_{k - 2j} r) e^{i<(R_l - R_k) R_{-2j} r, \frac{1}{av(f)} X f>},$$

and for the phase invariants $I_2^{r_1, r_2}(f)$ :

$$[I_2^{r_1, r_2}(f)\Psi](u_1, u_2) =$$

$$\sum_{j \in \breve N} \sum_{(\omega_1, \omega_2) \in \breve N} \tilde f(R_{2j - u_2} r_1 + R_{2j - u_1} r_2) \overline{\tilde f}(R_{\omega_2 - 2u_2 + 2j} r_1)$$

$$\overline{\tilde f}(R_{\omega_1 - 2u_1 + 2j} r_2) \times$$

$$e^{i<(I - R_{\omega_2} - u_2) R_{2j - u_2} r_1 + (I - R_{\omega_1} - u_1) R_{2j - u_1} r_2, \frac{1}{av(f)} X f>} \times$$

$$\Psi(u_1, u_2).$$

Since $N$ is odd, setting $m = 2j$, we get:

$$I_1^r(f)_{l,k} = \sum_{m \in \check{N}} \tilde{f}(R_{l-m}r)\overline{\tilde{f}}(R_{k-m}r)e^{i<(R_l-R_k)R_{-m}r, \frac{1}{av(f)}Xf>}, \qquad (33)$$

and also, we see easily that $I_2^{r_1,r_2}(f)$ is completely determined by the quantities:

$$\widetilde{I_2^{r_1,r_2}}(f)(u_1, u_2, \omega_1, \omega_2) = \qquad (34)$$
$$\sum_{m \in \check{N}} \tilde{f}(R_{m-u_2}r_1 + R_{m-u_1}r_2)\overline{\tilde{f}}(R_{\omega_2-2u_2+m}r_1) \times$$
$$\overline{\tilde{f}}(R_{\omega_1-2u_1+m}r_2) \times$$
$$e^{i<(I-R_{\omega_2}-u_2)R_{m-u_2}r_1+(I-R_{\omega_1}-u_1)R_{m-u_1}r_2, \frac{1}{av(f)}Xf>}$$

Setting $u_2 = -l_2, \omega_2 - 2u_2 = k2, u_1 = -l_1, \omega_1 - 2u_1 = k_1$, we get:

$$\widetilde{I_2^{r_1,r_2}}(f)(l_1, l_2, k_1, k_2) = \qquad (35)$$
$$\sum_{m \in \check{N}} \tilde{f}(R_{m+l_2}r_1 + R_{m+l_1}r_2)\overline{\tilde{f}}(R_{k_2+m}r_1)$$
$$\overline{\tilde{f}}(R_{k_1+m}r_2) \times e^{i<(R_{l_2}-R_{k_2})R_m r_1+(R_{l_1}-R_{k_1})R_m r_2, \frac{1}{av(f)}Xf>}$$

*Remark 4.* Consider the particular case $l_2 = k_2, l_1 = k_1$, and set $\xi_1 = R_{k_2}r_1$, $\xi_2 = R_{k_1}r_2$, then, we get:

$$\widetilde{I_2^{\xi_1,\xi_2}}(f)(l_1, l_2) = \sum_{m \in \check{N}} \tilde{f}(R_m(\xi_1 + \xi_2))\overline{\tilde{f}}(R_m\xi_1)\overline{\tilde{f}}(R_m\xi_2). \qquad (36)$$

Note that this is just the **discrete version of the (continuous) invariants of type 2**, in Formula 21. Note also that, making the change of variables $\xi_1 = R_{k_2}r_1$, $\xi_2 = R_{k_1}r_2$, $\xi_3 = R_{l_2}r_1 + R_{l_1}r_2$, we get:

$$\widetilde{I_3^{\xi_1,\xi_2,\xi_3}}(f) = \sum_{m \in \check{N}} \tilde{f}(R_m\xi_3)\overline{\tilde{f}}(R_m\xi_1)\overline{\tilde{f}}(R_m\xi_2)e^{i<R_m(\xi_3-\xi_1-\xi_2), \frac{1}{av(f)}Xf>}.$$

which is the final (discrete) form of our invariants.

Therefore, at the end, we have 3 sets of generalized Fourier descriptors (type-1, type-2, type-3):

$$\widetilde{I_1^r(f)}_{l,k}, \quad \widetilde{I_2^{\xi_1,\xi_2}}(f) \subset \widetilde{I_3^{\xi_1,\xi_2,\xi_3}}(f)$$

$$\widetilde{I_1^r(f)}_{l,k} = \sum_{m \in \check{N}} \tilde{f}(R_{l-m}r)\overline{\tilde{f}}(R_{k-m}r)e^{i<(R_l-R_k)R_{-m}r,\frac{1}{av(f)}Xf>},$$

$$\widetilde{I_2^{\xi_1,\xi_2}}(f) = \sum_{m \in \check{N}} \tilde{f}(R_m(\xi_1+\xi_2))\overline{\tilde{f}}(R_m\xi_1)\overline{\tilde{f}}(R_m\xi_2),$$

$$\widetilde{I_3^{\xi_1,\xi_2,\xi_3}}(f) = \sum_{m \in \check{N}} \tilde{f}(R_m\xi_3)\overline{\tilde{f}}(R_m\xi_1)\overline{\tilde{f}}(R_m\xi_2)e^{i<R_m(\xi_3-\xi_1-\xi_2),\frac{1}{av(f)}Xf>}.$$

As we shall see these descriptors are weakly complete (i.e. they discriminate over a residual subset of the set of images under the action of motions of angle $\frac{4k\pi}{2n+1}$, i.e. $\frac{2k'\pi}{2n+1}$).

## 5.5 Completeness of the discrete generalized Fourier descriptors

This is a rather hard work. We try to follow the scheme of the proof of Theorem 5, and at several points, there are technical difficulties.

Here, as above, a compact $K \subset \mathbb{R}^2$ is fixed, containing a neighborhood of the origin ($K$ is the "screen"), and an image is an element of $\mathcal{I}$, from Definition 8.

Let us consider the subset $\mathcal{G} \subset \mathcal{I}$ of "generic images", defined as follows. For $f \in \mathcal{I}$, $\tilde{f}^t$ denotes the ordinary 2-D Fourier transform of $f^c(0, X)$ as an element of $\mathbb{L}^2(\mathbb{R}^2)$. Set as above $X = (x, y) \in \mathbb{R}^2$ (but here $X$ should be understood as a point of the frequency plane). The function $\tilde{f}^t(X)$ is a complex-valued function of $X$, analytic in $X$. (Paley-Wiener). For $r \in \mathbb{R}^2$, denote by $\omega_r \in \mathbb{C}^N$ the vector $\omega_r = (\tilde{f}^t(R_0r), ..., \tilde{f}^t(R_{\theta_i}r), ..., \tilde{f}^t((R_{\theta_{N-1}}r))$.

Denote also by $\Omega_r$ the circulant matrix associated to $\omega_r$. If $F_N$ denotes the usual DFT matrix of order $N$ (i.e. the $N \times N$ unitary matrix representing the Fourier Transform over the Abelian group $\mathbb{Z}/N\mathbb{Z}$), then the vector of eigenvalues $\delta_r$ of $\Omega_r$ meets $\delta_r = F_N\omega_r$.

**Definition 9.** *The generic set $\mathcal{G}$ is the subset of $\mathcal{I}$ of elements such that $\Omega_r$ is an invertible matrix for all $r \in \mathbb{R}^2$, $r \neq 0$, except for a (may be countable) set of isolated values of $r$, for which $\Omega_r$ has a zero eigenvalue with simple multiplicity.*

The next Lemma shows that if $N$ is an odd integer number, then $\mathcal{G}$ is very big. It is proven by using parametric transversality arguments from [1].

**Lemma 2.** *Assume that $N$ is odd. Then, $\mathcal{G}$ is residual.*

This lemma is false for even $N$, due to the special properties of usual Fourier transforms of real-valued functions.

Now let us take $f, g \in \mathcal{G}$, and assume that their discrete generalized Fourier descriptors from Section 5.4 are equal.

We can apply the reasoning of Section 4.2 to construct a quasi-representation of the category $\pi(M_{2,N})$ at points where $\Omega_r(f)$ and $\Omega_r(g)$ are invertible only.

Recall the formula 29 for our Fourier Transform in the case of $M_{2,N}$:

$$[\widehat{f^c}(r)\Psi](u) = \sum_{\alpha \in \check{N}} \tilde{f}(R_{u-2\alpha}r)e^{i<R_{u-2\alpha}r,\frac{1}{av(f)}X_f>}\Psi(\alpha)$$

$$= \sum_{\alpha \in \check{N}} \tilde{f}^t(R_{u-2\alpha}r)\Psi(\alpha),$$

$$\text{with } f^t(x) = f(x + \frac{X_f}{av(f)}) = f^c(0,x),$$

by the basic property of the usual 2D Fourier transform with respect to translations.

Since $N$ is odd (a crucial point again), it is also equal to:

$$[\hat{f}^c(r)\Psi](u) = \sum_{\alpha \in \check{N}} \tilde{f}^t(R_{u-\alpha}r)(C\Psi)(\alpha). \tag{37}$$

where $C$ is a certain universal unitary operator (permutation).

This formula can also be read, in a matrix setting, as:

$$\hat{f}^c(r) = \Omega_r(f)C. \tag{38}$$

Also, by the equality of the invariants, the points where $\Omega_r(f)$ and $\Omega_r(g)$ are non-invertible are the same.

Out of these isolated points, we can apply the same reasoning as in the compact case, Section 4.2. Hence, the equality of the first invariants gives:

$$\hat{f}^c(r)\hat{f}^c(r)^* = \Omega_r(f)\Omega_r(f)^* = \hat{g}^c(r)\hat{g}^c(r)^* = \Omega_r(g)\Omega_r(g)^*.$$

Since at nonsingular points $\Omega_r(f)$ and $\Omega_r(g)$ are invertible, this implies that there is a unitary matrix $U(r)$ such that $\hat{g}^c(r) = \hat{f}^c(r)U(r)$.

Let $I = \{r_i | \Omega_{r_i} \text{ is singular}\}$. Out of $I$, $U(r)$ is an analytic function of $r$, since $U(r) = [\hat{f}^c(r)]^{-1}\hat{g}^c(r)$.

Now, we will need some results about unitary representations, namely:

**R1.** Two finite dimensional unitary representations that are equivalent are unitarily equivalent,

and the more difficult one, that we state in our special case only, and which is a consequence of the "Induction-reduction" theorem of Barut [2] (however, once one knows the result, he can easily check it by direct computations in the special case).

**R2.** For $r_1, r_2 \in \mathbb{R}^2$, the representation $\chi_{r_1 \hat{\otimes} r_2}$ is equivalent (hence unitarily equivalent by **R1**) to the direct Hilbert sum of representations $\oplus_{k \in \check{N}} \chi_{r_1 + R_k r_2}$.

This means that , if we take $r_1, r_2$ out of $I$, but $r_1 + R_{k_0} r_2 \in I$, and $r_1 + R_k r_2 \notin I$ for $k \neq k_0$ (which is clearly possible), and if $A$ denotes the unitary equivalence between $\chi_{r_1 \hat{\otimes} r_2}$ and $\hat{\oplus}_{k \in \check{N}} \chi_{r_1 + R_k r_2}$, setting $\xi_k = r_1 + R_k r_2$, we can write that the block diagonal matrix $\Delta_f = diag(\hat{f}^c(\xi_0), ..., \hat{f}^c(\xi_{N-1}))$ satisfies:

$$\Delta_f = \Delta_g A U(r_1)^* \hat{\otimes} U(r_2)^* A^{-1}. \tag{39}$$

Indeed, this comes from the equality of the second-type descriptors:

$$\hat{f}^c(\chi_{r_1}) \hat{\otimes} \hat{f}^c(\chi_{r_2}) \circ \hat{f}^c(\chi_{r_1} \hat{\otimes} \chi_{r_2})^* = \tag{40}$$
$$\hat{g}^c(\chi_{r_1}) \hat{\otimes} \hat{g}^c((\chi_{r_2}) \circ \hat{g}^c(\chi_{r_1} \hat{\otimes} \chi_{r_2})^*,$$

and since $\hat{g}^c(\chi_{r_1}) \hat{\otimes} \hat{g}^c((\chi_{r_2}) = \hat{f}^c(\chi_{r_1}) \hat{\otimes} \hat{f}^c(\chi_{r_2}) \circ U(r_1) \hat{\otimes} U(r_2)$ and both are invertible operators, then, replacing in 40, we get:

$$\hat{f}^c(\chi_{r_1} \hat{\otimes} \chi_{r_2}) \circ \hat{f}^c(\chi_{r_1})^* \hat{\otimes} \hat{f}^c(\chi_{r_2})^* =$$
$$\hat{g}^c(\chi_{r_1} \hat{\otimes} \chi_{r_2}) \circ U(r_1)^* \hat{\otimes} U(r_2)^* \circ \hat{f}^c(\chi_{r_1})^* \hat{\otimes} \hat{f}^c(\chi_{r_2})^*,$$

which implies,

$$\hat{f}^c(\chi_{r_1} \hat{\otimes} \chi_{r_2}) = \hat{g}^c(\chi_{r_1} \hat{\otimes} \chi_{r_2}) \circ U(r_1)^* \hat{\otimes} U(r_2)^*.$$

Using the equivalence $A$, we get:

$$A \hat{f}^c(\chi_{r_1} \hat{\otimes} \chi_{r_2}) A^{-1} = A \hat{g}^c(\chi_{r_1} \hat{\otimes} \chi_{r_2}) A^{-1} A \circ U(r_1)^* \hat{\otimes} U(r_2)^* A^{-1}.$$

This last equality is exactly (39).

*Remark 5.* The following fact is important: the matrix $A$ is a constant. This comes again from the "Induction-Reduction" Theorem of [2] (or from direct computation): the equivalence $A : \mathbb{L}^2(\check{N}) \hat{\otimes} \mathbb{L}^2(\check{N}) \approx \mathbb{L}^2(\check{N} \times \check{N}) \rightarrow \hat{\oplus}_{k \in \check{N}} \mathbb{L}^2(\check{N})$, is given by $A\varphi = \hat{\oplus}_{k \in \check{N}} \varphi_k$, with $\varphi_k(l) = \varphi(l, l - k)$. Hence, its matrix is independent of $r_1, r_2$.

Let us rewrite (39) as $\Delta_f = \Delta_g H$, for some unitary matrix $H$. Since $N - 1$ corresponding blocks in $\Delta_f$ and $\Delta_g$ are invertible, it follows that $H$ is also block diagonal. Since it is unitary, all diagonal blocks are unitary. In particular, the $k_0^{th}$ block is unitary. Also, $H = A \circ U(r_1)^* \hat{\otimes} U(r_2)^* A^{-1}$ is an analytic function of $r_1, r_2$. Moving $r_1, r_2$ in a neighborhood moves $r_1 + R_{k_0} r_2$ in a neighborhood. If we read the $k_0^{th}$ line of the equality $\Delta_f = \Delta_g H$, we get $\Delta_f(\chi_{r_1 + R_{k_0} r_2}) = \Delta_g(\chi_{r_1 + R_{k_0} r_2}) H_{k_0}(r_1, r_2)$, where $H_{k_0}(r_1, r_2)$ is unitary, and analytic in $r_1, r_2$. It follows that, by analyticity outside $I$, that $U(r)$ prolongs analytically to all of $\mathbb{R}^2 \backslash \{0\}$, in a unique way. The equality $\hat{g}^c(r) = \hat{f}^c(r) U(r)$ holds over $\mathbb{R}^2 \backslash \{0\}$.

Now, for the characters $\hat{K}_n$, $n \in \mathbb{Z}/N\mathbb{Z}$, it is easily computed that $\hat{f}^c(\hat{K}_n) = av(f) \sum_k e^{2\pi i n k/N}$. In particular $\hat{f}^c(0) = Nav(f)$.

The equality of the second type invariants imply that $av(f) = av(g)$. Moreover, if $\hat{f}^c(\hat{K}_0) \neq 0$, $\hat{g}^c(\hat{K}_0) = \hat{f}^c(\hat{K}_0)$. This implies the choice $U(\hat{K}_0) = 1$

For $n \neq 0$, note that $\hat{f}^c(\hat{K}_n)$ and $\hat{g}^c(\hat{K}_n)$ are zero. Hence we cannot define $U(\hat{K}_n)$ in the same way. In fact, we will consider the representations $\chi_{n,r} \approx \chi_r$,

$$\chi_{n,r} = \hat{K}_n \otimes \chi_r. \tag{41}$$

The representation $\chi_{n,r}$ is equivalent to $\chi_r$, the equivalence being $A_n$,

$$A_n(u) = e^{\frac{2\pi i}{N} un} = \varepsilon^{un}. \tag{42}$$

Also, we set

$$U(\chi_{n,r}) = U(n,r) = [\hat{f}^c(\chi_{n,r})]^{-1} \hat{g}^c(\chi_{n,r}) = A_{-n} U(r) A_n. \tag{43}$$

It follows that, wherever $U(r)$ is defined, $U(n,r)$ is also defined. We set also:

$$U(\hat{K}_n)Id = U(r)^* A_{-n} U(r) A_n = U(r)^* U(n,r). \tag{44}$$

A-priori, $U(\hat{K}_n)$ is ill defined, for several reasons. The crucial lemma 3 below shows not only that it is actually well defined but also:

$$U(\hat{K}_n) = e^{in\theta_0}, \text{ for some } \theta_0 = \frac{2\pi k_0}{N}. \tag{45}$$

*Remark 6.* At this point, we could already conclude from 45 directly (but not so easily) our result, i.e. $\tilde{h}^t(R_\theta r)$ and $\tilde{f}^t(R_\theta r)$ differ from a rotation $R_{\theta_0}$. However it is rather easy to see that this is in fact just "Chu-duality".

Note that, to conclude 45, we need the lemma 3, which is the most complicated among the series of lemmas just below.

Let us define $U(\chi)$ for any $p$-dimensional representation $\chi$ ($p$ arbitrary).

As a unitary representation $\chi$ is unitarily equivalent to $\overset{p}{\underset{i=1}{\oplus}} \chi_{r_i} \overset{k}{\underset{i=1}{\oplus}} \hat{K}_{n_i} = \oplus \chi_i$, $r_i \in S$, i.e. $\chi = A\Delta\chi_i A^*$, where $A$ is some unitary matrix, and $\Delta\chi_i$ is a block diagonal of irreducible representations $\chi_i$.

We define $U(\chi) = A\Delta U\chi_i A^*$.

The proofs of the following lemmas 3, 7, 8 are easy but rather technical. We don't give them here. The reader is referred to [18].

**Lemma 3.** *$U$ is well defined.*

**Lemma 4.** *At a point $\chi = A(\chi_{r_1} \oplus \ldots \oplus \chi_{r_p} \oplus \hat{K}_{k1} \oplus \ldots \oplus \hat{K}_{kl})A^* = A(\chi_r \oplus \chi_{\hat{K}})A^*$, where $r_1, \ldots r_p \notin I$, we have:*

$$U(\chi) = A(\ldots \hat{f}^c(\chi_r)^{-1} \hat{g}^c(\chi_r) \oplus \ldots e^{ik\theta_0} \oplus \ldots)A^*.$$

**Lemma 5.** $U(\chi \dot\oplus \chi') = U(\chi) \dot\oplus U(\chi')$.

**Lemma 6.** *If* $A\chi = \chi'A$, $A$ *unitary, then:* $AU(\chi) = U(\chi')A$.

The Lemmas 5, 6 are just trivial consequences of the definition of $U(\chi)$.

**Lemma 7.** $U$ *is continuous.*

**Lemma 8.** $U(\chi \dot\otimes \chi') = U(\chi) \dot\otimes U(\chi')$.

Lemmas 3, 4, 5, 6, 7, show that $U$ is a quasi-representation of the category $\pi(M_{2,N})$.

Since $M_{2,N}$ has the duality property, $U(\chi) = \chi(g_0)$ for some $g_0 \in M_{2,N}$. Also, we have:

$$\hat{g}^c(\chi_r) = \hat{f}^c(\chi_r) U(\chi_r) = \hat{f}^c(\chi_r) \chi_r(g_0) = \hat{f}^c_{g_0}(\chi_r),$$

by the fundamental property of the Fourier transform.

The support of the Plancherel's measure being given by the (non-character) unitary irreducible representations $\chi_r$, by the inverse Fourier transform, we get $g^c = f^c_{g_0}$, for some $g_0 \in M_{2,N}$, which is what we needed to prove.

**Theorem 8.** *If the (Three types of) discrete generalized Fourier descriptors of two images $f, g \in \mathcal{G}$ are the same, and if $N$ is odd, then the two images differ from a motion, the rotation of which has angle $\frac{4k\pi}{N}$ (i.e. $\frac{2k'\pi}{N}$ since $N$ is odd) for some $k$. Remind that $\mathcal{G}$ is a residual subset of the set of images of size $K$.*


# 6 Conclusion

In this paper, we have developed a rather general theory of "motion descriptors", based upon the basic duality concepts of abstract harmonic analysis.

We have applied this theory to several motion groups, and to the general case of compact groups, completing previous results.

This theory leads to rather general families of invariants under group actions operating on functions (images). We have proved weak completeness -i.e. completeness over a large (residual) subset of the set of images- in the case of several special groups, including motion groups "with small basic rotation". These invariants are at most cubic expressions of the functions (images).

A number of interesting theoretical questions remain open (such as completeness for the usual group of motions $M_2$). Let us point out the following:

1. There is a final form of duality Theory, which is given by "Tatsuuma duality", see [14], [19]. This is a generalization of Chu duality, to general locally compact (type 1) groups. In particular, it works for $M_2$. Unfortunately, huge difficulties appear when trying to use it in our context. However this is a challenging subject.

2. Computation of the generalized Fourier descriptors reduces to usual FFT evaluations.
3. The first and second-type descriptors, that arise via the trivial or the cyclic lift have a very interesting practical feature: they don't depend on an estimation of the centroid of the image. This is a strong point in practice.
4. Otherwise, the variables that appear in the generalized Fourier descriptors have clear frequency interpretation. Hence, depending on the problem (a high or low frequency texture), one can chose the actual values of these frequency variables in certain adequate ranges.

# References

1. Abraham R, Robbin J (1967) Transversal Mappings and Flows. W.A. Benjamin Inc.
2. Barut AO, Raczka R (1986) Theory of Group representations and applications, second edition. World Scientific
3. Boser BE, Guyon IM, Vapnik V (1992) A Training Algorithm for Optimal Margin classifiers. In: proc. Fifth Ann. Workshop Computational Learning theory, 144–152. ACM Press
4. Chen S (1985) A new vision system and the Fourier Descriptors Method by Group Representations. IEEE CDC conference, 1985, Las Vegas, USA.
5. Chu H (1966) Compactification and duality of topological groups. Trans. American Math. Society, 123:310–324.
6. Dixmier J (1969) Les $C^*$-algèbres et leurs repréentations. Gauthier-Villars, Paris
7. Fonga H (1992) Analyse harmonique sur les groupes et reconnaissance de formes. PHD Tthesis, université de Grenoble
8. Fukunaga K (1990) Introduction to statistical Pattern Recognition, second ed. Academic Press
9. Gauthier JP, Bornard G, Zilbermann M (1991) Harmonic analysis on motion groups and their homogeneous spaces. IEEE Trans on Systems, Man and Cyb., 21.
10. Ghorbel F (1994) A complete invariant description for grey level images by the harmonic analysis approach. Pattern Recognition Letters, 15.
11. Gourd F (1990) Quelques Méthodes d'Analyse Harmonique en Théorie de la Détection et de la Reconnaissance de Formes. PHD thesis, Université de Grenoble
12. Gourd F, Gauthier JP, Younes H (1989) Une méthode d'invariants de l'analyse harmonique en reconnaissance de formes, Traitement du Signal, 6(3).
13. Hewitt E, Ross KA (1963) Abstract Harmonic analysis. Springer Verlag, Berlin
14. Heyer H (1970) Dualität über LokalKompakter Gruppen. Springer Verlag
15. Joyal A, Street R (1991) An Introduction to Tannaka Duality and Quantum Groups. Lecture notes in maths 1488, Springer, pp. 411–492.
16. Kyatkin AB, Chirikjan GS (1999) Pattern matching as a correlation on the discrete Motion Group. Computer Vision and Image Understanding, 74(1)22–35.
17. Persoon, King Sun Fu (1977) Shape discrimination using Fourier descriptors. IEEE systems man and cybernetics, SMC 7(3).

18. Smach F, Lemaître C, Gauthier JP, Miteran J (2006) Generalized Fourier Descriptors with Applications to object recognition in SVM Context. Submitted
19. Tatsuuma N (1967) A duality Theorem for locally compact groups. J. Math. Kyoto Univ 6:187–293.
20. Turski J (2004) Geometric Fourier Analysis of the conformal camera for active vision. SIAM review, 46(1).
21. Vapnik VN (1998) The statistical Learning Theory. Springer
22. Vilenkin N (1969) Fonctions spéciales et théorie de la représentation des groupes. Dunod, Paris
23. Warner G (1972) Harmonic Analysis on semi-simple Lie Groups. Springer Verlag
24. Weil A (1965) L'intégration dans les groupes topologiques et ses applications. Hermann, Paris

# Nonholonomic Interpolation for Kinematic Problems, Entropy and Complexity

Jean-Paul Gauthier[1] and Vladimir Zakalyukin[2]

[1] LE2I, UMR CNRS 5158, Université de Bourgogne,
   Bat. Mirande, BP 47870, 21078, Dijon CEDEX France.
   `gauthier@u-bourgogne.fr`
[2] Moscow State University, 119 993, Leninski Gori, 1, Moscow, Russia.
   `zakalyu@mail.ru`

**Summary.** Here we present the main lines of a theory we developed in a series of previous papers, about the motion planning problem in robotics. We illustrate the theory with a few academic examples.

   Our theory, although at its starting point, looks promising even from the "constructive" point of view. It does not mean that we have precise general algorithms, but the theory contains this potentiality.

   The robot is given under the guise of a set of linear kinematic constraints (a distribution). The cost is specified by a riemannian metric on the distribution. Given a non-admissible path for the robot, i.e. a path that does not satisfy the kinematic constraints), our theory allows to evaluate precisely and constructively the "metric complexity" and the "entropy" of the problem. This estimation of metric complexity provides methods for **approximation** of nonadmissible paths by admissible ones, while the estimation of entropy provides methods for **interpolation** of the nonadmissible path by admissible.

## 1 Introduction, examples, position of the problems

What we call the "motion planning problem" here is a particular case of the very general problem of outstanding importance in control theory: given a control system, and given a non-admissible trajectory, i.e. any curve in the phase space, which is not a trajectory of the system, try to approximate it by an admissible one. Do it in some optimal way.

   The particular case we consider here is the case of a kinematic system defined by a linear set of nonholonomic constraints (i.e. a nonintegrable distribution). We approximate paths in the "Subriemannian sense", a concept which will be made clear later on.

   The paper refers constantly to the series of works of the F. Jean [13, 14, 15] and of the authors [6, 7, 8, 9, 10, 11].

All along the paper, there is a small parameter $\varepsilon$ (we want to approximate up to $\varepsilon$) and certain quantities $f(\varepsilon), g(\varepsilon)$ go to $+\infty$ when $\varepsilon$ tends to zero. We say that such quantities are equivalent ($f \simeq g$) if $\lim_{\varepsilon \to 0} \frac{f(\varepsilon)}{g(\varepsilon)} = 1$.

## 1.1 A few academic examples of kinematic systems subject to motion planning

In the paper we provide a generic classification of motion planning problems for low values of the corank $k$ of the distribution. The three examples below certainly belong to the "folklore" but they describe certain generic exceptions that appear naturally in the classification.

*Example 1.* (The "car" robot)

$$\dot{x} = \cos(\theta)u \tag{1}$$
$$\dot{y} = \sin(\theta)u,$$
$$\dot{\theta} = v.$$

In this example, $(x, y)$ is the position of the car on the plane, and $\theta$ is the angle between the driving wheel and the $x$-axis in the $(x, y)$ plane. The two controls, $u, v$, correspond to turning the wheel (with $v$) and pushing in the direction of the wheel (with $u$).

*Example 2.* (The car with a trailer)

$$\dot{x} = \cos(\theta)u, \tag{2}$$
$$\dot{y} = \sin(\theta)u,$$
$$\dot{\theta} = v,$$
$$\dot{\varphi} = v - \frac{\sin(\varphi)}{l}u.$$

Here, $l$ is the length of the "attach" between the car and the trailer, and $\varphi$ is the angle between the axes of the car and the trailer.

*Example 3.* (The ball rolling on a plate)

$$\dot{x} = u, \quad \dot{y} = v,$$
$$\dot{R} = \begin{pmatrix} 0 & 0 & u \\ 0 & 0 & v \\ -u & -v & 0 \end{pmatrix} R,$$

where $R \in SO(3, \mathbb{R})$. Here, $x, y$ are the coordinates of the contact point between the sphere and the plane on which it is rolling.

In the example 2, let us consider the following non admissible trajectory $\Gamma : [0,1] \to R^2 \times (S_1)^2$, where $S_1 = \mathbb{R}/2\pi\mathbb{Z}$,

$$\Gamma(s) = (x(s), y(s), \theta(s), \varphi(s)) = (s, 0, \frac{\pi}{2}, 0).$$

It describes a sort of "parking problem" in which we want both the car and the trailer to move along the $x$-axis, while remaining perpendicular to the $x$-axis.

In the example 3, we may consider a trajectory $\Gamma(s)$ of the type: $\Gamma(s) = (x(s), y(s), R_0)$, where $(x(s), y(s))$ is any curve on the plane (later we will chose a line segment) and $R_0$ is a constant frame. It means that a cat is standing quietly on the ball, remaining at the top and looking in the same direction while the ball moves along the curve $(x(s), y(s))$.

## 1.2 The subriemannian cost, the metric complexity and the interpolation-entropy

We want to approximate nonadmissible curves in some optimal sense, but in these purely kinematic problems, we would like the result (admissible curve) to be "optimal" independently of its parametrization. A natural way to do this is to consider the subriemannian length of admissible curves. If $\Delta$ denotes the distribution (specifying the nonholonomic constraints) then we take a riemannian metric $g$ over $\Delta$, that allows to measure the length of tangent vectors to admissible curves, and therefore the length of admissible curves.

The couple $(\Delta, g)$ is called a subriemannian metric, and the subriemannian length of any admissible curve is actually independent of the parametrization. Also, if $\Delta$ is completely non-integrable, any two points $x, y$ in phase space can be connected by some admissible curve, and the minimum of these lengths is the subriemannian distance $d(x, y)$.

The motion planning problem is local around the compact curve $\Gamma$. Hence we can always consider that the phase space is some open set in $\mathbb{R}^n$. The rank of the distribution will be denoted by $p$, and we can always assume the existence of a global orthonormal frame $\mathcal{F}$ for the metric over $\Delta$, $\mathcal{F} = (F_1, ... F_p)$ where $k = n - p$ is the codimension (or corank) of $\Delta$.

Then, we consider systems in "control system form" (which is the case in the 3 examples above):

$$\dot{x} = \sum_{i=1}^{p} F_i(x) u_i. \tag{3}$$

The $F_i$, $i = 1, ..., p$, form an orthonormal frame for the metric, therefore the distribution $\Delta$ is just $span(F_1, ..., F_p)$ and the length of an admissible curve $\gamma : [0, T] \to \mathbb{R}^n$ corresponding to a control $u(t)$, $t \in [0, T]$ is just:

$$l(\gamma) = \int_0^T \sqrt{\sum_{i=1}^{p} (u_i)^2}.$$

**Definition 1.** *A motion planning problem $\mathcal{P}$ is a triple $\mathcal{P} = (\Delta, g, \Gamma)$, where $(\Delta, g)$ is a (rank $p$) subriemannian metric over $\mathbb{R}^n$ and $\Gamma : [0, T] \to \mathbb{R}^n$ is a smooth curve. The set of motion planning problems is endowed with the $C^\infty$ topology over compact sets.*

We are interested with **generic** motion planning problems only. In particular, $\Gamma$ and $\Delta$ are transversal except maybe at isolated points. Also some singularities of $\Delta$ may appear along $\Gamma$. In that case they are always isolated in $\Gamma$.

Given a motion planning $\mathcal{P}$ we consider the subriemannian tube $T_\varepsilon$ and the subriemannian cylinder $C_\varepsilon$ around $\Gamma$, with respect to the subriemannian distance, where $\varepsilon$ is the small parameter.

**Definition 2.** *The metric complexity $MC(\varepsilon)$ (resp. the interpolation entropy $E(\varepsilon)$) of the problem $\mathcal{P}$ is $\frac{1}{\varepsilon}$ times the minimum length of an admissible curve connecting the endpoints $\Gamma(0), \Gamma(T)$ of $\Gamma$, and contained in the tube $T_\varepsilon$ (resp. $\varepsilon$-interpolating $\Gamma$, that is, in any segment of length $\geq \varepsilon$, there is a point of $\Gamma$). These quantities $MC(\varepsilon)$, $E(\varepsilon)$ are functions of $\varepsilon$ that tend to $+\infty$ when $\varepsilon$ tends to zero. They are of course considered up to equivalence.*

**Definition 3.** *An **asymptotic optimal synthesis** for complexity (resp. for entropy) is a one-parameter family $\gamma_\varepsilon : [0, T\gamma_\varepsilon] \to \mathbb{R}^n$, (that we may take arclength parameterized, i.e. $\sum_{i=1}^{p}(u_i(t))^2 = 1$) of admissible curves that realize an equivalent of the metric complexity (resp. the entropy), i.e:*

1. *$\gamma_\varepsilon(0) = \Gamma(0), \quad \gamma_\varepsilon(T\gamma_\varepsilon) = \Gamma(T)$,*
2. *$\gamma_\varepsilon([0, T\gamma_\varepsilon]) \subset T_\varepsilon$,*
3. *(for entropy only) $\gamma_\varepsilon$ is $\varepsilon$-interpolating, i.e. $\gamma_\varepsilon$ connects (a finite number of) points of $\Gamma$ by pieces of length less than or equal to $\varepsilon$.*
4. *$MC(\varepsilon) \simeq \frac{l(\gamma_\varepsilon)}{\varepsilon}$ (resp. $E(\varepsilon) \simeq \frac{l(\gamma_\varepsilon)}{\varepsilon}$).*

Asymptotic optimal syntheses always do exist, and our theory **is constructive** in the sense that not only we compute the metric complexity and entropy, but also we exhibit (more or less explicitly) the corresponding asymptotic optimal syntheses.

*Remark 1.* In robotics, usually, robots have to move from a source point to a target point in the phase space, avoiding obstacles. The problem is solved in two steps:

First, find a curve (nonadmissible) connecting source to target and avoiding the obstacles. There are more or less standard procedures to do this, and we don't address this question here in.

Second, approximate this curve by an admissible one, up to sufficiently small $\varepsilon$. This is the problem under discussion here. Moreover, $\varepsilon$ very small corresponds to the case of a phase space with a lot of obstacles.

The asymptotics of the number of (subriemannian) balls of radius $\varepsilon$ to cover the curve $\Gamma$ is the standard Kolmogorov's entropy $E_K(\varepsilon)$. Clearly we always have $E(2\varepsilon) \leq E_K(\varepsilon)$. Some example is known where the equality is not reached. But, it is a highly nongeneric situation.

**Theorem 1.** *In all the (generic) situations we study, the following equality holds: $E(2\varepsilon) \simeq E_K(\varepsilon)$.*

*Proof.* It follows from our papers that in all the situations under consideration, we produce asymptotic optimal syntheses $\gamma_\varepsilon$ for entropy, with the following property. Let $\gamma_\varepsilon$ be arclength parameterized. Consider any interpolating interval of $\gamma_\varepsilon$ of the form $[t, t + a]$, $\gamma_\varepsilon(t) \in \Gamma$, $\gamma_\varepsilon(t + a) \in \Gamma$, $a \leq \varepsilon$. Then, the subriemannian ball of radius $\frac{\varepsilon}{2}$ centered at $\gamma_\varepsilon(t + \frac{a}{2})$ covers entirely the piece segment of $\Gamma$ between $\gamma_\varepsilon(t)$ and $\gamma_\varepsilon(t + a)$. It follows that $E_K(\frac{\varepsilon}{2}) \leq E(\varepsilon)$.

In other terms, generically there are no "metric cusps".

## 1.3 The results for examples 2 and 3

We don't deal with Example 1, which is of the same kind but simpler than example 2.

The distribution $\Delta$ in Example 2 is the so called Engel's distribution. There is an "abnormal flow" for the distribution. That is, there is a vector field the trajectories of which are extremals of the Pontryagin's maximum principle, corresponding to zero Hamiltonian. We chose the vector field $F_1$ to generate this abnormal flow. Then $F_2$ is the unit orthogonal vector field to $F_1$ in $\Delta$. We set $F_3 = [F_1, F_2]$, and this choice defines an orthonormal frame, hence in particular a subriemannian metric $g'$ in the derived distribution $\Delta'$. We take a one form $\omega$ which vanishes on $\Delta'$ and which has value 1 on the vectors $\dot{\Gamma}(t)$. This form $\omega$ is unique up to multiplication by a smooth function which is 1 on $\Gamma$. Then we define a field $A(t)$ of skew-symmetric endomorphisms of $\Delta$ along $\Gamma$, as follows:

$$< A(t)X, Y >_{g'} = d\omega(X, Y), \ \forall X, Y \in \Delta_{\Gamma(t)}.$$

We set $\delta(t) = ||A(t)||_{g'}$. The invariant $\delta(t)$ does not depend on the choice of $\omega$, and there is a universal constant $\hat{\sigma} \sim 0.00580305$ such that:

$$E(\varepsilon) \simeq \frac{3}{2\hat{\sigma}\varepsilon^3} \int_\Gamma \frac{dt}{\delta(t)}. \tag{4}$$

Moreover asymptotic optimal syntheses can be computed explicitly in terms of the Jacobi elliptic functions. In particular, the solution of the parking problem set in Section 1.1 is depicted on Figure 1. On this figure, the $x$ axis along which the system is expected to move transversely, has a magnified scale, which makes some angle look not small. Optimal controls express in terms of the Jacobi elliptic functions as follows:

$$v(t) = 1 - dn(K(1 + \frac{4t}{\varepsilon}))^2, \tag{5}$$

$$u(t) = -2dn(K(1 + \frac{4t}{\varepsilon}))sn(K(1 + \frac{4t}{\varepsilon})) \sin \frac{\varphi_0}{2},$$

where $2\ Eam(K) = K$ and $K(k)$ is the quarter period of the Jacobi elliptic functions of modulus $k = \frac{\sin(\varphi_0)}{2}$, $\varphi_0 \sim 130°$ (See Love, [17]).

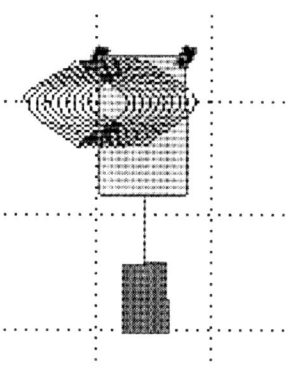

**Fig. 1.** parking of a car with a trailer.

For Example 3 there is no abnormal flow but through each point of $\Gamma$ there is some abnormal trajectory transversal to $\Gamma$ (in the sense of Pontryagin's maximum principle). We chose the unit vector field $F_1$ to generate these trajectories. The following construction will be independent of this choice. We chose $F_2, F_3$ as in the Engel's case, and we set $F_4 = [F_1, F_3]$, $F_5 = [F_2, F_3]$. We denote by $\omega$ a one form which vanishes on $\Delta'$ and such that $1 = \omega(\Gamma) = \omega(F_5)$. Again the results will be independent of the choice of $\omega$. Set $\gamma(t) = \omega(F_4)$. Then, the entropy is given by the same formula as 4, namely:

$$E(\varepsilon) \simeq \frac{3}{2\hat{\sigma}\varepsilon^3} \int_\Gamma \frac{dt}{\gamma(t)},$$

and the controls corresponding to an asymptotic optimal synthesis are still given by the formula 5.

The trajectory of the contact point of the rolling ball with the plane is shown on Figure 2. It is some Euler's elastica (Love, [17]). The cross on the ball figures the position of the cat.

## 1.4 Content of the paper

In the section 2, we introduce several basic concepts and tools. The first are the notions of normal coordinates and normal form, generalizing the normal

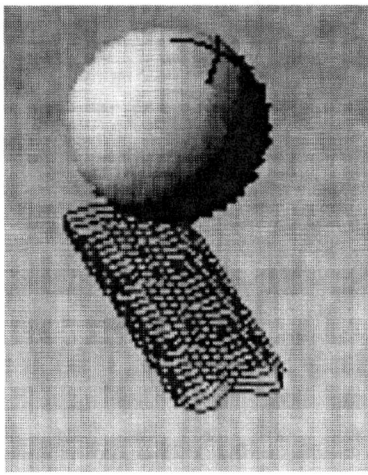

**Fig. 2.** The cat on a ball on a plane.

coordinates of riemannian geometry. These tools allow to define some adequate notions of "Nilpotent approximation along $\Gamma$". We give also two crucial technical lemmas: 1. Reduction to Nilpotent approximation, 2. the **logarithmic Lemma**, allowing to manage with the general situation where the curve $\Gamma$ crosses transversally a singularity of codimension 1 of the distribution $\Delta$.

Normal coordinates come from all our previous papers, and the logarithmic lemma is stated in [11]. The reduction to nilpotent approximation is shown in [10].

Section 3 deals with the codimension one case. Further results and details for this codimension one case can be found in [6, 7].

Section 4 deals with all generic cases where the corank $k$ of $\Delta$ is three at most. It is a remarkable fact that, up to corank 3, in almost all cases we have the following relation between entropy and metric complexity.

**Theorem 2.** *If $k \leq 3$ then except in the two cases described in the previous section, i.e. Engel $(p = 2, k = 2)$ and rolling ball $(p = 2, k = 3)$ (in which cases we don't know the expression of the metric complexity), we have the following relation, generically: $E(\varepsilon) \simeq 2\pi MC(\varepsilon)$.*

This last theorem comes from [10].

Section 5 states what we know in the cases where $k =$ codim$(\Delta) \geq 4$. We provide an almost complete generic classification up to corank $k = 8$, and some extra results up to corank $k = 10$. These results come mostly from the papers [9, 10, 11].

In our conclusion 6, we want in particular to point out a presumably very important "robustness property" of our approach. This property is a consequence of all these results: in fact, very often, the optimal syntheses (after certain natural reparameterization of $\Gamma$) depend neither on the system nor on the metric or on the curve $\Gamma$: they depend only on the structure of the distribution (its growth vector).

## 2 A few useful tools

### 2.1 Normal coordinates and normal form

The normal coordinates we introduce follow the old idea of "normal coordinates" in riemannian geometry. They are coordinates in which $\Gamma$ and certain subriemannian geodesics are normalized in a convenient way. Several versions of these coordinates were previously introduced, first in [4] in a formal way completely forgetting about geometric ideas. This paper [4] is rather important since it provides certain power series expansion of the normal coordinates. This expansion could be generalized here in order to approximate effectively our normal coordinates and normal forms.

After the paper [4], the normal coordinates (and the normal form) were interpreted in geometric terms, and extended to different other situations (see the papers [1], [2], [3]).

Consider a motion planning problem $\mathcal{P} = (\Delta, g, \Gamma)$ not necessarily one-step-bracket-generating. Take a (germ along $\Gamma$ of) **parameterized $k$-dimensional surface** $S$, transversal to $\Delta$,

$$S = \{q(s_1, ..., s_{p-1}, t) \in \mathbb{R}^n\}, \text{with } q(0, ..., 0, t) = \Gamma(t).$$

Such a germ does exist if $\Gamma$ is not tangent to $\Delta$. The exclusion of a neighborhood of an isolated point where $\Gamma$ is tangent to $\Delta$, that is $\Gamma$ becomes "almost admissible", does not affect the estimations in this paper. Moreover, around such a point, motion planning is trivial.

We denote by $\mathcal{C}_\varepsilon^S = \{\xi; \ d(S, \xi) = \varepsilon\}$ the subriemannian cylinder of radius $\varepsilon$ around $S$, and by $\mathcal{T}_\varepsilon^S$ the corresponding tube.

**Lemma 1.** *(Normal coordinates with respect to $S$). There are mappings $x : \mathbb{R}^n \to \mathbb{R}^p$, $y : \mathbb{R}^n \to \mathbb{R}^{k-1}$, $w : \mathbb{R}^n \to \mathbb{R}$, such that $\xi = (x, y, w)$ is a coordinate system on some neighborhood of $S$ in $\mathbb{R}^n$, such that:*
   *0. $S(y, w) = (0, y, w)$, $\Gamma = \{(0, 0, w)\}$,*
   *1. $\Delta_{|S} = \ker dw \cap_{i=1,...k-1} \ker dy_i$, $g_{|S} = \sum_{i=1}^p (dx_i)^2$,*
   *2. $\mathcal{C}_\varepsilon^S = \{\xi| \sum_{i=1}^p x_i^2 = \varepsilon^2\}$,*
   *3. geodesics of the Pontryagin's maximum principle ([18]) meeting the transversality conditions w.r.t. $S$ are straight lines through $S$, contained in the planes $P_{y_0, w_0} = \{\xi|(y, w) = (y_0, w_0)\}$. Hence, they are orthogonal to $S$.*

*These normal coordinates are unique up to changes of coordinates of the form*

$$\tilde{x} = T(y, w)x, (\tilde{y}, \tilde{w}) = (y, w), \tag{6}$$

*where $T(y, w) \in O(p)$, the $p \times p$ orthogonal group.*

## Frames

A motion planning problem can be specified by a couple $(\Gamma, F)$, where $F = (F_1, ..., F_p)$ is a $g$-orthonormal frame of vector fields generating $\Delta$. Hence, we will also write $\mathcal{P} = (\Gamma, F)$. If a global coordinate system $(x, y, w)$, not necessarily normal, is given on a neighborhood of $\Gamma$ in $\mathbb{R}^n$, with $x \in \mathbb{R}^p$, $y \in \mathbb{R}^{k-1}$, $w \in \mathbb{R}$, then we write:

$$F_j = \sum_{i=1}^{p} \mathcal{Q}_{i,j}(x, y, w) \frac{\partial}{\partial x_i} + \sum_{i=1}^{k-1} \mathcal{L}_{i,j}(x, y, w) \frac{\partial}{\partial y_i} + \mathcal{M}_j(x, y, w) \frac{\partial}{\partial w}, \quad (7)$$

$$j = 1, ..., p.$$

Hence, the SR metric is specified by the triple $(\mathcal{Q}, \mathcal{L}, \mathcal{M})$ of smooth $x, y, w$-dependent matrices.

## The general normal form

Fix a surface $S$ as in Section 2.1 and a normal coordinate system $\xi = (x, y, w)$.

**Theorem 3.** *(Normal form) There is a unique orthonormal frame $F = (\mathcal{Q}, \mathcal{L}, \mathcal{M})$ for $(\Delta, g)$ with the following properties:*
  1. *$\mathcal{Q}(x, y, w)$ is symmetric, $\mathcal{Q}(0, y, w) = Id$ (the identity matrix),*
  2. *$\mathcal{Q}(x, y, w)x = x$,*
  3. *$\mathcal{L}(x, y, w)x = 0$, $\mathcal{M}(x, y, w)x = 0$.*
  *Conversely if $\xi = (x, y, w)$ is a coordinate system satisfying conditions 1, 2, 3 above, then $\xi$ is a normal coordinate system for the SR metric defined by the orthonormal frame $F$ with respect to the parameterized surface $\{(0, y, w)\}$.*

Clearly, this normal form is invariant under the changes of normal coordinates of the form (6).

Let us write:

$$\mathcal{Q}(x, y, w) = Id + Q_1(x, y, w) + Q_2(x, y, w) + ...,$$
$$\mathcal{L}(x, y, w) = 0 + L_1(x, y, w) + L_2(x, y, w) + ...,$$
$$\mathcal{M}(x, y, w) = 0 + M_1(x, y, w) + M_2(x, y, w) + ...,$$

where $Q_j, L_j, M_j$ are matrices depending on $\xi$, the coefficients of which have order $j$ w.r.t. $x$ (i.e. they are in the $j^{th}$ power of the ideal of $C^\infty(x, y, w)$ generated by the functions $x_r$, $r = 1, ..., p$). In particular, $Q_1$ is linear in $x$, $Q_2$ is quadratic, etc... Set $u = (u_1, ..., u_p) \in \mathbb{R}^p$. Then $\sum_{j=1}^{k-1} L_{1_j}(x, y, w)u_j = L_{1,y,w}(x, u)$ is bilinear in $(x, u)$, and $\mathbb{R}^{p-1}$-valued. Its $i^{th}$ component is the bilinear expression denoted by $L_{1,i,y,w}(x, u)$. Similarly $\sum_{j=1}^{k-1} M_{1_j}(x, y, w)u_j = M_{1,y,w}(x, u)$ is a quadratic form in $(x, u)$. The corresponding matrices are denoted by $L_{1,i,y,w}$, $i = 1, ..., k-1$, and $M_{1,y,w}$.

The following was proved in [2], [3] for corank 1, but holds in general.

**Proposition 1.** *1. $Q_1 = 0$,*
  *2. $L_{1,i,y,w}$, $i = 1, ..., k-1$, and $M_{1,y,w}$ are skew symmetric matrices.*

## 2.2 Nilpotent approximations along $\Gamma$

The generic cases we care about don't use brackets of order more than 3. Brackets of order 3 are used at isolated points of $\Gamma$ only. This is due to the fact that either we consider the generic cases of corank $\leq 3$ treated in Sections 3, 4 or only certain cases of corank more than 3 from Section 5.

First we have to chose the surface $S$. In the one-step-bracket-generating case, we chose the surface $S$ arbitrary (but transverse to $\Delta$). The nilpotent approximation will not depend on this choice. In the Engel case, we show below (Example 7) how to chose $S$. In the cases $p = 2$, $k = 3$, we chose $S$ as explained in Example 8. Isolated points where third brackets are necessary may appear.

After this choice, we assign certain weights to the variables $x_i, y_j, w$ and the operators $\frac{\partial}{\partial x_i}, \frac{\partial}{\partial y_j}, \frac{\partial}{\partial w}$ in agreement with the order of the variables w.r.t. the small parameter $\varepsilon$ inside the tube $T_\varepsilon$, considering admissible trajectories starting from $\Gamma$ at time 0. For instance the weight of $x_i, i = 1, ..., p$ is 1 since $||\dot{x}||_{\mathbb{R}^p} \leq 1$ (admissible curves are arclength parameterized). Then, the weight of the operators $\frac{\partial}{\partial x_i}$ is -1 since the Lie derivative of monomials with respect to $\frac{\partial}{\partial x_i}$ makes them decrease of weight one. The weight of $y_j, j = 1, ..., k$ may be 2 or 3: for instance in the one step bracket generating case, the expression of $\dot{y}_j$ starts with a linear term in $x$, then inside $T_\varepsilon$, $|y_j| \leq k\varepsilon^2$ for a certain constant $k$. Therefore the weight of $y_j$ is 2. It follows that the weight of $\frac{\partial}{\partial y_j}$ is $-2$. If the expression of $\dot{y}_j$ starts with a quadratic term, then $y_j$ has weight 3 and $\frac{\partial}{\partial y_j}$ has weight $-3$. After assigning in the same way a weight to $\frac{\partial}{\partial w}$, **we finally set that the weight of $w$ is zero** ($w$ is just the parameter of $\Gamma$).

Note that the weights are constant, except may be at isolated points of $\Gamma$.

**Definition 4.** *The nilpotent approximation $\hat{\mathcal{P}}$ of $\mathcal{P}$ along $\Gamma$ is defined as follows.*

*1. On a piece segment of $\Gamma$ where there is no isolated point corresponding to a codimension one singularity of the distribution, we keep only the terms of order -1 in the normal form (i.e. an orthonormal frame for $\hat{\mathcal{P}}$ is the normal frame of $\mathcal{P}$ truncated at order -1),*

*2. In a neighborhood of a singular isolated point, we keep also the terms of order $\leq$ -1, but taking into account the weights at the isolated point.*

*Example 4.* The nilpotent approximation in the one-step-bracket generating has the following form, (using control system notations):

$$\dot{x} = u; \tag{8}$$

$$\dot{y}_i = \frac{1}{2}x'L_i(w)u; \quad i = 1, ..., k - 1;$$

$$\dot{w} = \frac{1}{2}x'M(w)u.$$

Here $x'$ is the transpose of $x$ and $w$ is the coordinate along $\Gamma$. The matrices $L_i, M$, depending on $w$, are skew symmetric.

The $y, w$ space identifies with the surface $S$. It can be also identified to the quotient $T_{\Gamma(w)}\mathbb{R}^n/\Delta_{\Gamma(w)}$, and the mapping $(x, u) \rightarrow (x'L^1(w)u, ..., x'L^{p-1}(w)u, x'M(w)u)$ from $\Delta_{\Gamma(w)} \times \Delta_{\Gamma(w)}$ to $T_{\Gamma(w)}\mathbb{R}^n/\Delta_{\Gamma(w)}$ is just the coordinate form of the bracket mapping modulo distribution, $[.,.]/\Delta$, which is a tensor, as is well known.

*Example 5.* (**Nilpotent approximation in 3-dimensional contact case**). In the 3 dimensional case the (parameterized) surface $S$ reduces to the (parameterized) curve $\Gamma$. There is no $y$-variable. By [4] the normal form along $\Gamma$ can be written as:

$$F_1 = \frac{\partial}{\partial x_1} - x_2\beta(x_1\frac{\partial}{\partial x_2} - x_2\frac{\partial}{\partial x_1}) + \frac{x_2}{2}\gamma\frac{\partial}{\partial w}, \tag{9}$$

$$F_2 = \frac{\partial}{\partial x_2} + x_1\beta(x_1\frac{\partial}{\partial x_2} - x_2\frac{\partial}{\partial x_2}) - \frac{x_1}{2}\gamma\frac{\partial}{\partial w},$$

where $\beta, \gamma$ are smooth functions of all variables $x_1, x_2, w$.

The distribution $\Delta$ is contact iff $\gamma(0, 0, w) \neq 0$. Then the nilpotent approximation along $\Gamma$ is:

$$F_1 = \frac{\partial}{\partial x_1} + \frac{x_2}{2}\gamma(0, 0, w)\frac{\partial}{\partial w}, \tag{10}$$

$$F_2 = \frac{\partial}{\partial x_2} + -\frac{x_1}{2}\gamma(0, 0, w)\frac{\partial}{\partial w},$$

*Example 6.* (**Nilpotent approximation in Martinet case**). We consider the case of a singular isolated point $w_0 = 0$. If $\gamma(0, 0, 0) = 0$ , then the distribution is generically of "Martinet type" (i.e. second brackets allow to fill in the tangent space). In that case, in a neighborhood of such a point, the nilpotent approximation is (since $\frac{\partial}{\partial w}$ has order $-3$):

$$F_1 = \frac{\partial}{\partial x_1} + \frac{x_2}{2}(\gamma_0(w) + \gamma_1(w)x_1 + \gamma_2(w)x_2)\frac{\partial}{\partial w}, \tag{11}$$

$$F_2 = \frac{\partial}{\partial x_2} + -\frac{x_1}{2}(\gamma_0(w) + \gamma_1(w)x_1 + \gamma_2(w)x_2)\frac{\partial}{\partial w},$$

where $\gamma_0(0) = 0$. What will be important in that case is that $\frac{\partial\gamma_0}{\partial w}(0) = \sigma \neq 0$.

In the two following examples, we keep the notations of Section 1.3.

*Example 7.* (**The Engel's case of Example 2**). In that case we chose the parameterized surface $S$ as follows: $S(y, w) = \exp(yF_3)(\Gamma(w))$. We chose also $T(y, w)$ for $F_1 = \frac{\partial}{\partial x_1}$, $F_2 = \frac{\partial}{\partial x_2}$ along $\Gamma$. Then, the nilpotent approximation is:

$$F_1 = \frac{\partial}{\partial x_1} + \frac{x_2}{2}\frac{\partial}{\partial y} + \frac{1}{2}x_1 x_2 \delta(w)\frac{\partial}{\partial w},$$

$$F_2 = \frac{\partial}{\partial x_2} - \frac{x_1}{2}\frac{\partial}{\partial y} - \frac{1}{2}(x_1)^2 \delta(w)\frac{\partial}{\partial w}.$$

Note that $\delta(w) = ||A(w)||_{g'}$, where $A(.)$ has been defined in Section 1.3 and then for a generic motion planning problem $\mathcal{P}$, $\delta(w)$ never vanishes.

*Example 8.* (**The rolling ball case of Example 3**). We chose:

$$S(y_1, y_2, w) = \exp(y_2 F_5)^\circ \exp(y_1 F_3)(\Gamma(w)),$$

and $T(y, w)$ for the abnormals meeting the transversality conditions of Pontryagin's maximum principle with respect to $\Gamma$ be trajectories of $F_1$. We get the nilpotent approximation:

$$F_1 = \frac{\partial}{\partial x_1} + \frac{x_2}{2}\frac{\partial}{\partial y_1} - \frac{(x_2)^2}{3}\frac{\partial}{\partial y_2} + \frac{1}{2}x_1 x_2 \gamma(w)\frac{\partial}{\partial w},$$

$$F_2 = \frac{\partial}{\partial x_2} - \frac{x_1}{2}\frac{\partial}{\partial y_1} + \frac{x_1 x_2}{3}\frac{\partial}{\partial y_2} - \frac{1}{2}(x_1)^2 \gamma(w)\frac{\partial}{\partial w}.$$

Note that at some isolated points $\gamma$ may be zero. In that case we proceed as in Example 6.

## 2.3 Two crucial lemmas

Nilpotent approximations dominate along $\Gamma$, and they give the entropy and metric complexity of the problem as stated in the next lemma.

**Lemma 2.** (*Reduction to nilpotent approximation*) *In all the cases under consideration here, the entropy and metric complexity of the problem $\mathcal{P}$ and its nilpotent approximation $\mathcal{NP}$ are equal.*

This lemma is proven in [10] (for entropy only but it holds also for the metric complexity).

In the case of generic codimension 1 singularities of the distribution $\Delta$, we have also the very important following lemma, proven in [11], for the specific cases we need in this paper (i.e. brackets of maximum order 3 at generic isolated singularities). However a similar lemma should be true in a more general setting.

Assume that the brackets of higher order at singular isolated points have order $r + 1$ (then it is $r$ at generic points). Consider the distribution $\Delta^{(r-1)}$ (i.e. $\Delta$ bracketed $r - 1$ times by itself). Of course, $\Delta^{(0)} = \Delta$. Then consider the forms $\omega$ which are zero on $\Delta^{(r-1)}$ and which take value 1 on $\dot{\Gamma}$. We get an affine family $\omega_\lambda$ of these.

In the cases we consider, we have either $r = 1$ or $r = 2$, that is at isolated points we need at most the third Bracket. If $r = 2$, then $k = 3$ (if $k = 2$, i.e.

Engel's case, there is generically no singular point along $\Gamma$). Then we have a canonical metric on $\Delta'/\Delta$ : unit vector for this metric is given by the bracket of any two unit orthogonal vectors in $\Delta$. Via of our nilpotent approximation along $\Gamma$, it provides a metric $g'$ on $\Delta'$ (along $\Gamma$). Then, in all cases, we have a riemannian metric $g'$ on $\Delta^{(r-1)}$. If $r = 1$ then $g' = g$. Let us define along $\Gamma$ an affine family of operators $A(w) : \Delta^{(r-1)} \rightarrow \Delta^{(r-1)}$, skew symmetric w.r.t $g'$ by:

$$d\omega_\lambda(X, Y) = < A_\lambda(w)X, Y >, \text{ for all } X, Y \text{ in } \Delta^{(r-1)}_{\Gamma(w)}. \qquad (12)$$

Set:

$$\chi(w) = \inf_\lambda ||A_\lambda(w)||. \qquad (13)$$

**Lemma 3.** *(**Logarithmic Lemma**) In the cases under consideration in the paper we have, at generic points of $\Gamma$, formulas for the entropy of the type*

$$E(\varepsilon) \simeq \frac{1}{\varepsilon^p} \int_\Gamma \frac{\alpha(w)dw}{\chi(w)}, \text{ where } \alpha \text{ is a certain other invariant, (this formula is valid}$$

*on pieces of $\Gamma$ containing no singular point). Assume that there are codimension 1 singularities of $\Delta$ at points $w_1, ..., w_l$. Then $\chi$ has a nonzero derivative $\chi'$ at these points, and the entropy expresses as:* $E(\varepsilon) \simeq -\frac{2\ln(\varepsilon)}{\varepsilon^p} \sum_{i=1}^{l} \frac{a(w_i)}{|\chi'(w_i)|}.$

*Remark 2.* Note for instance that it is the case in Example 8, where $\gamma(w)$ there is equal to $\chi(w)$ here. It is also true in Martinet case, Example 6, where $\sigma = \chi'(w)$.

There is one more general fact in this study that we use everywhere without stating it: there are other generic isolated points where the curve $\Gamma$ is tangent to a certain distribution at some unavoidable points (for instance if $n = 3$, $p = 2$ then $\Gamma$ may be tangent to $\Delta$; another case is $n = 4$, $p = 2$ where $\Gamma$ may be tangent to $\Delta'$). In such situations, in the denominator of some expression of the form $\frac{1}{\varepsilon^p} \int_\Gamma \frac{\alpha(w)dw}{\chi(w)}$ the invariant $\chi$ tends to infinity. However, this expression still makes sense. In these cases, always, expressions of entropy or metric complexity **are still valid.**

## 3 The codimension one case

In the situation where $\Delta$ has corank one, **the normal form gives everything** (metric complexity, entropy and explicit asymptotic optimal synthesis) **even without using any variational principle:** all the results are directly readable on the normal form. Details about this corank one case can be found in [6, 7]. Let us summarize the main results and sketch some proofs.

The cylinder $C_\varepsilon$ is smooth (which is not true in higher corank) except in a neighborhood of unavoidable points where $\Gamma$ is tangent to $\Delta$. But as we

said, at these points, asymptotic optimal synthesis is trivial, and expression
of complexity and entropy is still valid. Therefore in this section we assume
that $\Gamma$ is transversal to $\Delta$.

In the normal form of Theorem 3 and inside $T_c$ let us read the equation
of the $w$ variable:

$$\dot{w} = \frac{1}{2}x'M(w)u + O(\varepsilon^2).$$

This implies, since we take the admissible curves arclength parameterized
that $\dot{w} \leq \frac{1}{2}\varepsilon\|M(w)\| \; \|u\| + O(\varepsilon^2) = \frac{1}{2}\varepsilon\|M(w)\| + O(\varepsilon^2)$, then $dt \geq$
$\frac{2}{\varepsilon}\frac{dw}{\|M(w)\|(1+O(\varepsilon))}$. Hence obviously, $MC(\varepsilon) \geq \frac{2}{\varepsilon^2}\int_\Gamma \frac{dw}{\|M(w)\|}$. We can chose the

matrices $T(w)$ in 6 for $M(w)$ is in $2 \times 2$ block diagonal form, and the first

block $\begin{pmatrix} 0 & \chi(w) \\ -\chi(w) & 0 \end{pmatrix}$ is such that $\|M(w)\| = \chi(w)$. Then we consider the

two dimensional cylinder $C^1(\varepsilon) = \{(x_1)^2 + (x_2)^2 = \varepsilon^2, x_3 = \ldots = x_p = 0\}$.
The distribution $\Delta$ being transversal to $\Gamma$, it is also transversal to $C^1(\varepsilon)$ for $\varepsilon$
small enough. The intersection of the distribution and the tangent bundle to
$C^1(\varepsilon)$ defines a vector field $X_\varepsilon$ on $C^1(\varepsilon)$. We take a trajectory $\gamma_\varepsilon$ of this vector
field. We complete it with two horizontal (i.e. corresponding to $w$=constant)
geodesics from Lemma 1 (point 3) that connect endpoints of $\Gamma$ to $C^1(\varepsilon)$ at the
price $2\varepsilon$ (without any effect on the final estimation of the metric complexity).
The reader can compute directly with the normal form that the cost of this
strategy is equivalent to $\frac{2}{\varepsilon^2}\int_\Gamma \frac{dw}{\|M(w)\|}$. Therefore,

$$MC(\varepsilon) \simeq \frac{2}{\varepsilon^2}\int_\Gamma \frac{dw}{\|M(w)\|},$$

and this strategy provides an asymptotic optimal synthesis. It is given by
the two pieces of horizontal geodesics just described, plus a trajectory of the
vector field $X_\varepsilon$. Then in fact, everything is reduced to the case $n = 3$, for
which the asymptotic optimal synthesis is described on Figure 3 (a).

The proof of theorem 2 in that case follows form the normal form and the
usual Euclidian isoperimetric inequality.

There are other cylinders $C'_\varepsilon$, that are tangent to $\Gamma$ and transversal to $\Delta$,
of perimeter $\varepsilon$, also in the plane $x_3 = \ldots = x_p = 0$, such that the asymptotic
optimal synthesis for entropy is given by the vector field $X'_\varepsilon$ obtained by
intersecting $\Delta$ with the tangent bundle to $C'_\varepsilon$. This is shown on Figure 3 (b).

The only remaining generic situation in corank 1 is the Martinet case when
$n = 3$. What happens is depicted on Figure 4 for the metric complexity (there
is a similar picture for entropy): there is a limit cycle on the cylinder $C_\varepsilon = C^1_\varepsilon$.

The asymptotic optimal synthesis is like that: 1. Connect the endpoints of
$\Gamma$ to $C_\varepsilon$ by horizontal geodesics. 2. Follow a trajectory of $X_\varepsilon$ as long as the
variable $w$ varies in a monotonic way. When the derivative of $w$ changes sign,

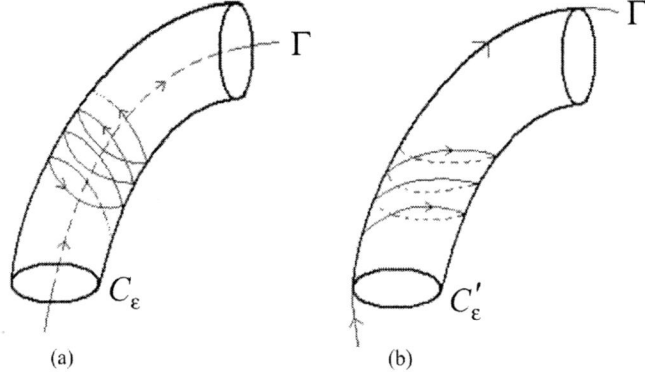

**Fig. 3.** Contact case.

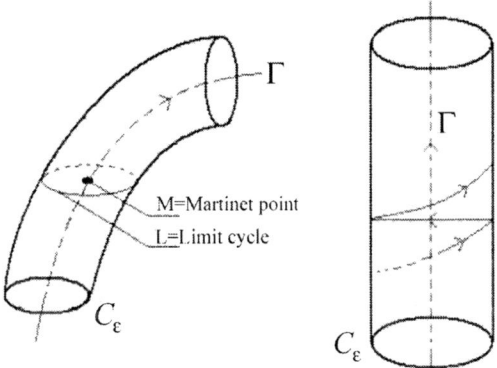

**Fig. 4.** Martinet case.

cross the cylinder by using a horizontal geodesic (given by Lemma 1 point 3). This allows to cross the singularity. 3. Continue in a symmetric way by following a trajectory of the vector field $-X_\varepsilon$.

The proof of the fact that this strategy is an asymptotic optimal synthesis for the metric complexity follows just from explicit estimations from the normal form 11.

It is remarkable that the same type of direct estimate for Entropy shows the relation $E(\varepsilon) \simeq 2\pi MC(\varepsilon)$.

To summarize, let $\omega$ be the (unique up to multiplication by a function which is unit on $\Gamma$) one form such that $\omega(\dot{\Gamma}) = 1, \omega(\Delta) = 0$. Let $A(t)$ be the one parameter family of (skew symmetric w.r.t. $g$) endomorphisms of $\Delta$ along $\Gamma$, defined by $d\omega(X,Y) = <A(t)X,Y>_g$, for all $X,Y$ in $\Delta_{\Gamma(t)}$. Set $\chi(t) = ||A(t)||_g$.

**Theorem 4.** *Codimension $k = 1$. The metric complexity and entropy satisfy, for generic problem $\mathcal{P}$ :*

$$E(\varepsilon) \simeq 2\pi MC(\varepsilon),$$

$$MC(\varepsilon) \simeq \frac{2}{\varepsilon^2} \int_\Gamma \frac{dt}{\chi(t)}, \quad \text{if } n > 3 \text{ or there is no Martinet point on } \Gamma,$$

$$MC(\varepsilon) \simeq \frac{-\ln(\varepsilon)}{\varepsilon^2} \sum_{\text{Martinet points } t_i} \frac{4}{|\chi'(t_i)|}, \quad \begin{matrix} \text{if } n = 3 \text{ and there are Martinet} \\ \text{points.} \end{matrix}$$

*Moreover, the asymptotic optimal synthesis is explicit, as described above in this section.*

There are complementary results about nongeneric cases with $k = 1$ in the paper [7].

## 4 Codimension smaller or equal to three

### 4.1 One step bracket generating case

We use the reduction to nilpotent approximation (Lemma 2) and the normal form for the nilpotent approximation (8).

Consider again the bracket tensor mapping $\mathcal{B}_w = [.,.]/\Delta : \Delta_{\Gamma(w)} \times \Delta_{\Gamma(w)} \to T_{\Gamma(w)}\mathbb{R}^n/\Delta_{\Gamma(w)}$. Set

$$B_{1,1}(w) = \{(X,Y) \in \Delta_{\Gamma(w)} \times \Delta_{\Gamma(w)}; ||X||_g, ||Y||_g \leq 1\},$$

i.e. the product of two unit balls $B_1(w)$ in $\Delta_{\Gamma(w)}$. Set $\mathcal{K}(w) = \mathcal{B}_w(B_{1,1}(w))$.

**Definition 5.** *The set $\mathcal{K}(w)$ is strictly convex in the direction of $\dot{\Gamma}(w)$ if one of the two equivalent following statements holds:*

$(S_1)$ *There is $x^* = \lambda\dot{\Gamma}(w) \in \mathcal{K}(w)$, $\lambda > 0$, and $\omega \in (T_{\Gamma(w)}\mathbb{R}^n/\Delta_{\Gamma(w)})^* \approx (\mathbb{R}^p)^*$ (dual space of $T_{\Gamma(w)}\mathbb{R}^n/\Delta_{\Gamma(w)}$) such that, for all $z \in \mathcal{K}(w)$,*

$$\omega(x^*) \geq \omega(z); \tag{14}$$

$(S_2)$ *If $V^* = \{\omega \in (\mathbb{R}^p)^*, \omega(\dot{\Gamma}(w)) = 1\}$, then there exists $\omega^* \in V^*$, $x^* = \lambda\dot{\Gamma}(w) \in \mathcal{K}(w)$, $\lambda > 0$, with:*

$$\omega^*(x^*) = \sup_{x \in \mathcal{K}(w)} \omega^*(x) = \inf_{w \in V^*} \sup_{x \in \mathcal{K}(w)} \omega(x).$$

Then, the key observation is given by the following Lemma.

**Lemma 4.** *(Corank 1,2,3). For a generic (open dense) motion planning problem $\mathcal{P}$, for all $t \in [0, T]$, $\mathcal{K}(w)$ is strictly convex in the direction of $\dot{\Gamma}(w)$.*

*Proof.* We just sketch the proof. Technical details can be found in [8]. First, for these values of corank, all the matrices $A_\lambda(w)$ from Formula 12 have simple nonzero eigenvalues (except at isolated points that have no influence on the final result, if $k = 3$). Then we can chose the changes of normal coordinates (6) and $\Gamma$-preserving changes of coordinates in the surface $S$, for the matrix $M(w)$ in the expression (8) of the Nilpotent approximation verifies: $A_0(w) = M(w)$, $\chi(w) = ||M(w)||$ (where $\chi(w)$,the main invariant, has been defined in 13). Also, we can put $M(w)$ in block-diagonal $2 \times 2$ form, such that the first block (corresponding to normal coordinates $x_1, x_2$ is $\begin{pmatrix} 0 & \chi(w) \\ -\chi(w) & 0 \end{pmatrix}$.

Then convexity in the direction of $\dot{\Gamma}$ just means that the skew-symmetric matrices $L_i(w)$, $i = 1, ..., p-1$, verify:

$$L_i(w)_{1,2} = L_i(w)_{2,1} = 0. \tag{15}$$

Indeed, $L_i(w)_{1,2} = -L_i(w)_{2,1}$ are the $y$-coordinates of the image of the bracket mapping $[.,.]/\Delta$ and $\dot{\Gamma} \bmod \Delta$ has coordinates $y = 0$, $w = w^*$. Then (15) is necessary to get Property (14).

Now, assume this property 15 does not hold. Set $P(\lambda) = \sup_{||x||, ||z|| \leq 1} x'A_\lambda z = ||A_\lambda||^2$. This sup is attained for certain unit vectors $x(\lambda)$, $z(\lambda)$, that are moreover orthogonal. Then, the $\inf_\lambda P(\lambda)$ is attained at $\lambda^* = 0$. Therefore, $\frac{dP(\lambda)}{d\lambda} = 0$ for $\lambda = 0$. We rewrite $P(\lambda) = x'(\lambda)(M + \sum_{i=1}^{k-1} \lambda_i L_i) z(\lambda)$. We get $0 = x'(0)L_i z(0) + \frac{dx'}{d\lambda_i}(0) M z(0) + x'(0) M \frac{dz}{d\lambda_i}(0)$. But both terms $I = \frac{dx'}{d\lambda_i}(0) M z(0)$ and $II = x'(0) M \frac{dz}{d\lambda_i}(0)$ vanish: indeed $M z(0) = \chi(w) x(0)$, then $I = \chi(w) \frac{dx'}{d\lambda_i}(0) x(0)$. But $< x(\lambda), x(\lambda) > = 1$. It follows that $0 = x'(0)L_i z(0)$, which is what we need.

*Remark 3.* This proof shows also that, in future situations where $\mathcal{K}(w)$ will be not strictly convex in the direction of $\dot{\Gamma}(w)$, we can chose normal coordinates and $\Gamma$-preserving coordinates in $S$ such that: $\inf_\lambda ||A_\lambda(w)||$ is attained at $\lambda = 0$, and $A_0(w) = M(w)$ has **double eigenvalue of maximum modulus.** This will be a main fact in the next Section 5.

The lemma 4 means in fact that, as in Section 3, everything can be reduced to the three dimensional contact case, at least for the nilpotent approximation.

Actually, with the same argument as in corank 1, we have $MC(\varepsilon) \geq \frac{2}{\varepsilon^2} \int_\Gamma \frac{dw}{||M(w)||}$. We consider the normal coordinates, as in the proof of Lemma 4, where $M(w)$ is block diagonal, with first block equal to $\begin{pmatrix} 0 & \chi(w) \\ -\chi(w) & 0 \end{pmatrix}$. We consider the cylinder $C^1(\varepsilon) = \{(x_1)^2 + (x_2)^2 = \varepsilon^2, x_3 = \ldots = x_p = 0\}$. Provided that $\varepsilon$ is small enough $S$ and $C^1(\varepsilon)$ are transversal since $S$ is assumed transversal to $\Delta$. Again the intersection of $\Delta$ with the tangent bundle to $C^1(\varepsilon)$ provides a unit vector field denoted by $X_\varepsilon$ (two such opposite vector fields in fact). The key point is the convexity that implies the property (15). From this property, it follows that the vector field $X_\varepsilon$ has the form:

$$X_\varepsilon = \sum_{i=1}^p X_i \frac{\partial}{\partial x_i} + X_n \frac{\partial}{\partial w}.$$

Therefore, If we denote by $C^2(\varepsilon) = \{(x_1)^2 + (x_2)^2 = \varepsilon^2, x_3 = \ldots = x_p = 0, y = 0\}$ the 2-dimensional sub-cylinder corresponding to $y = 0$, $X_\varepsilon$ is a vector field on $C^2(\varepsilon)$. It is easy to check that trajectories of this vector field, plus two horizontal pieces of geodesics of length $\varepsilon$, (to connect the endpoints of $\Gamma$ to $C^2(\varepsilon)$) provide admissible paths with length exactly equal to $\frac{2}{\varepsilon^2} \int_\Gamma \frac{dw}{||M(w)||}$.

*Remark 4.* In fact, for the nilpotent approximation in adequate normal coordinates, using property 15, and restricting to the cylinder $C^2(\varepsilon)$ we get exactly the 3-dimensional contact case.

We conclude:

**Theorem 5.** *(*$k \leq 3$*, one step bracket generating). Entropy is still given by the formula $E(\varepsilon) \simeq \frac{2}{\varepsilon^2} \int_\Gamma \frac{dw}{\chi(w)}$. In certain normal coordinates, the problem $\hat{\mathcal{P}}$ (motion planning for nilpotent approximation) reduces to the 3-dimensional contact case by restriction to some subspace. Moreover again, $E(\varepsilon) \simeq 2\pi MC(\varepsilon)$.*

## 4.2 Remaining cases

All cases corresponding to $k \leq 3$ are covered (except at logarithmic isolated points) by the one step bracket generating case plus certain exceptions. The logarithmic lemma may be applied at the isolated points in both the one step bracket generating case and the exceptions.

The (generic) exceptions are:

- $p = 2, k = 2$, Engel's case, treated in Section 1.3,
- $p = 2, k = 3$, the cat on the ball on the plane, treated also in Section 1.3.

## 5 Codimension more than three

Most of the results in this Section come from [10, 11].

If $\mathcal{P}$ is one step bracket generating, the situation becomes very different from the case $k \leq 3$ since the sets $\mathcal{K}(w)$ may be not strictly convex in the direction of $\dot{\Gamma}(w)$. The first really wild case is the case $p = 4$, $k = 6$ (i.e. a 4-distribution in $\mathbb{R}^{10}$). In that case, consider the projectivisation $PB$ of the bracket mapping,

$$PB = [.,.]/\Delta : G_{2,4} \to P\mathbb{R}^6,$$

where $G_{2,4}$ is the grassmannian of 2-planes in $\mathbb{R}^4$, and $P\mathbb{R}^6$ is the 5-dimensional projective plane. The dimension $\dim(G_{2,4})$ is 4. Hence in generic situation $\mathcal{K}(w)$ never intersects the direction of $\dot{\Gamma}(w) \bmod \Delta$ (except maybe at isolated points). In particular, $\mathcal{K}(w)$ is never convex in the direction of $\dot{\Gamma}(w)$.

Also, due to Remark 3, we may consider that the problem $\mathcal{P}$ has nilpotent approximation in normal form (8) where the skew symmetric matrices $M(w)$ have double eigenvalue, and

$$\chi(w) = ||M(w)|| = \inf_{\lambda} ||A_\lambda(w)|| =$$

$$= \inf_{\lambda} ||M(w) + \sum_{i=1}^{k-1} \lambda_i L_i(w)|| = ||A_0(w)||. \tag{16}$$

Then for fixed $w$ we have an affine pencil of skew symmetric matrices $M + \sum_{i=1}^{k-1} \lambda_i L_i$, that we denote in abbreviated notations by $M + \lambda L$. It is natural, considering double eigenvalue matrices in the Lie algebra $so(4, \mathbb{R})$, to consider the decomposition in pure quaternions and pure skew quaternions:

$$so(4, \mathbb{R}) = P \oplus \hat{P},$$

where $P$ is the vector space of pure quaternions, generated by $i, j, k$ :

$$i = \begin{pmatrix} 0 & -1 & 0 & 0 \\ 1 & 0 & 0 & 0 \\ 0 & 0 & 0 & -1 \\ 0 & 0 & 1 & 0 \end{pmatrix}, \quad j = \begin{pmatrix} 0 & 0 & -1 & 0 \\ 0 & 0 & 0 & 1 \\ 1 & 0 & 0 & 0 \\ 0 & -1 & 0 & 0 \end{pmatrix}, \quad k = \begin{pmatrix} 0 & 0 & 0 & -1 \\ 0 & 0 & -1 & 0 \\ 0 & 1 & 0 & 0 \\ 1 & 0 & 0 & 0 \end{pmatrix},$$

while $\hat{P}$ is generated by $\hat{i}, \hat{j}, \hat{k}$, with:

$$\hat{i} = \begin{pmatrix} 0 & -1 & 0 & 0 \\ 1 & 0 & 0 & 0 \\ 0 & 0 & 0 & 1 \\ 0 & 0 & -1 & 0 \end{pmatrix}, \quad \hat{j} = \begin{pmatrix} 0 & 0 & 1 & 0 \\ 0 & 0 & 0 & 1 \\ -1 & 0 & 0 & 0 \\ 0 & -1 & 0 & 0 \end{pmatrix}, \quad \hat{k} = \begin{pmatrix} 0 & 0 & 0 & -1 \\ 0 & 0 & 1 & 0 \\ 0 & -1 & 0 & 0 \\ 1 & 0 & 0 & 0 \end{pmatrix}.$$

As we already noticed two degrees of freedom remain in the normal form of the nilpotent approximation:

- we may consider $\Gamma$-preserving (linear) changes of coordinates in the surface $S$.
- We may also consider changes of normal coordinates 6, that act at the level of the nilpotent approximation via conjugation by elements of $SO(4, \mathbb{R})$ on the pencil of matrices $(M + \lambda L)$.

We have the following theorem:

**Theorem 6.** *Outside an arbitrarily small neighborhood of a finite subset of $\Gamma$, there is a choice of the normal coordinates, and a parametrization of $S$ (preserving the parametrization of $\Gamma$), such that the nilpotent approximation takes the form:*

$$\dot{x} = u, \tag{17}$$

$$\dot{y}_1 = \frac{1}{2}x'(\hat{\imath} + \varrho(w)i)u,$$

$$\dot{y}_2 = \frac{1}{2}x'ju, \ \dot{y}_3 = \frac{1}{2}x'ku, \ \dot{y}_4 = \frac{1}{2}x'\hat{\jmath}u, \ \dot{y}_5 = \frac{1}{2}x'\hat{k}u,$$

$$\dot{w} = \frac{1}{2}\chi(w)x'iu.$$

*Here $x'$ is the transpose of $x$ and $\varrho(w)$ is a certain invariant of the motion planning problem, $-1 \leq \varrho(w) \leq 1$. The value $\varrho(w) = \pm 1$ when $\mathcal{K}(w)$ is strictly convex in the direction of $\Gamma$. Otherwise, the direction of $\dot{\Gamma}_w$ avoids $\mathcal{K}(w)$.*

Then, the invariant $1 - |\varrho(w)|$ measures the "distance to convexity" of $\mathcal{K}(w)$.

Using this normal form, the Pontryagin's maximum principle, the reduction to nilpotent approximation and using deeply the structure of quaternions and skew quaternions, we can explicitly construct an asymptotic optimal synthesis for the nilpotent approximation and prove the following theorem:

**Theorem 7.** *($p = 4, k = 6$) Generically $\chi$ vanishes at isolated points (where $\Delta$ is only 2 steps bracket generating). On pieces of $\Gamma$ containing no such point, we have:*

$$E(\varepsilon) \simeq \frac{2\pi}{\varepsilon^2} \int_{\Gamma} \frac{(3 - |\varrho(w)|)}{\chi(w)} dw, \tag{18}$$

*otherwise, applying the logarithmic lemma to the $r$ singular points*

$$E(\varepsilon) \simeq -\frac{4\pi \ln(\varepsilon)}{\varepsilon^2} \sum_{i=1}^{r} \frac{(3 - |\varrho(w_i)|)}{\chi'(w_i)}.$$

Note that, when $|\varrho(.)| \equiv 1$, i.e. the problem is strictly convex, we get the formulas of Section 4.

In the cases where $p = 4$, but $k = 4$ or $5$, the situation is a bit more complicated: we consider the "body" $\mathcal{K}(w)$ moving along the curve $\Gamma$. It may happen generically that on some open intervals $\mathcal{K}(w)$ is convex in the direction of $\dot{\Gamma}$, or it may happen that it is not. On the intervals where it is convex, we still have the usual estimation $E(\varepsilon) \simeq \frac{4\pi}{\varepsilon^2} \int_\Gamma \frac{dw}{\chi(w)}$. On the pieces where it is not, we have:

$$\frac{4\pi}{\varepsilon^2} \int_\Gamma \frac{dw}{\chi(w)} \leq E(\varepsilon) \leq \frac{6\pi}{\varepsilon^2} \int_\Gamma \frac{dw}{\chi(w)}, \tag{19}$$

To get these estimates, we use the same method as in the $(4, 10)$ case. For this we need invariants and normal forms for $(k-1)$dimensional pencils of $4 \times 4$ skew symmetric matrices. We get:

- Case $k = 4$ : The pencil $L + \lambda M$ can be reduced to the form: $M = \alpha\hat{i} + \beta\hat{j} + \gamma\hat{k}$, $L = \{i + \rho_1\hat{i}, j + \rho_2\hat{j}, k + \rho_3\hat{k}\}$, where $\alpha, \beta, \gamma, \varrho_1, \varrho_2, \varrho_3$ are real invariants.
- case $k = 5$ : The pencil $L + \lambda M$ can be reduced to the form: $M = \alpha(\cos(\theta)\hat{i} + \sin(\theta)\hat{j})$, $L = \{i + \rho_1\hat{i}, j + \rho_2\hat{j}, k, \hat{k}\}$, where $\alpha > 0$ and $\theta, \varrho_1, \varrho_2$ are the invariants. The problem is strictly convex iff $\frac{\varrho_1^2 - 1}{1 - \varrho_2^2} > 0$. In that case, $\chi = \alpha|\cos(\theta + \zeta)|$, with $\tan(\zeta) = (\frac{\varrho_1^2 - 1}{1 - \varrho_2^2})^{\frac{1}{2}}$. It is a case where convex and nonconvex situation generically coexist on different open subsets of $\Gamma$.

*Remark 5.* In the one step bracket generating situation, there is another principle, proven in our papers: the cases where $p$ is odd always reduce to $p = p-1$.

For $p = 5$ and $4 \leq k \leq 10$ as well as for $p > 5$ and $4 \leq k \leq 8$ problems are generically one-step bracket generating (with possible logarithmic points). In these cases $M(w)$ has still double (and not triple) eigenvalue and the quaternions remain in the picture. The estimation (19) still holds or, for $r$ logarithmic points:

$$-\frac{8\pi \ln(\varepsilon)}{\varepsilon^2} \sum_{i=1}^r \frac{dw}{\chi(w)} \leq E(\varepsilon) \leq -\frac{12\pi \ln(\varepsilon)}{\varepsilon^2} \sum_{i=1}^r \frac{dw}{\chi(w)}.$$

Finally, the last exception for $k \leq 4$ is $k = 4$, $p = 3$. For a generic problem $\mathcal{P}$ the distribution $\Delta''$ spans the ambient space, without generic singular point, and in that case, we get an estimation of the type:

$$\frac{2}{3\hat{\sigma}\varepsilon^3} \int_\Gamma \frac{dw}{\varphi(w)} \leq E(\varepsilon) \leq \frac{2}{\hat{\sigma}\varepsilon^3} \int_\Gamma \frac{dw}{\varphi(w)},$$

where $\varphi$ is another invariant (see [11] for details).

# 6 Conclusion

Of course it remains a lot of work to apply fully the results of our theory to real problems. However, we started to realize some "academic applications" of the theory, both in simulation and in practice, and it looks extremely efficient.

At the step where we are since everything is constructive a precise algorithm could be obtained for low corank $k \leq 10$. For this, the strategy is more or less straightforward:

- we need an algorithm to estimate nilpotent approximations along $\Gamma$. This is just computations in formal power series, as were developed in [4] for $n = 3$,
- we need to put some "feedback" inside the strategy. This can be done in several natural ways.

We were able check in the applications that the methodology appears extremely robust. To justify this theoretically, let us point out the following fact valid up to corank $k = 3$, which is not negligible for practical problems:

The (practical) robustness just reflects a certain (mathematical) stability property:

On pieces of $\Gamma$ without logarithmic points, the entropy (and metric complexity) are of the form $\frac{a}{\varepsilon^l} \int_\Gamma \frac{dw}{\varkappa(w)}$. This holds for the one step bracket generating but also Engel's case and rolling ball case, that is, all generic cases. If we reparameterize $\Gamma$ by setting $\frac{dw}{\varkappa(w)} = d\tilde{w}$, we get the same expression but moreover: the asymptotic optimal synthesis for nilpotent approximation is the same whatever the system (depends only on the growth vector of the distribution). In other terms there is no invariant at all independently of the kinematic system $\Delta$, the metric $g$ and the curve $\Gamma$ :

- The one step bracket generating case reduces always to the contact 3-dimensional Heisenberg case, with normal form:

$$
\begin{aligned}
F_1 &= \frac{\partial}{\partial x_1} + \frac{x_2}{2} \frac{\partial}{\partial w}, \\
F_2 &= \frac{\partial}{\partial x_2} + -\frac{x_1}{2} \frac{\partial}{\partial w},
\end{aligned}
\tag{20}
$$

- The Engel's case $p = 2$, $k = 2$ reduces to:

$$
\dot{x}_1 = u_1, \ \dot{x}_2 = u_2,
$$
$$
\dot{y} = \frac{1}{2}(x_2 u_1 - x_1 u_2),
$$
$$
\dot{w} = \frac{1}{2} x_1 (x_2 u_1 - x_1 u_2),
$$

- Finally, the case $p = 2$, $k = 3$ also reduces to:

$$\dot{x}_1 = u_1, \; \dot{x}_2 = u_2,$$

$$\dot{y}_1 = \frac{1}{2}(x_2 u_1 - x_1 u_2),$$

$$\dot{y}_2 = -\frac{2}{3}x_2(x_2 u_1 - x_1 u_2),$$

$$\dot{w} = \frac{1}{2}x_1(x_2 u_1 - x_1 u_2).$$

Similarly all logarithmic situations can be renormalized at the level of the nilpotent approximation, no invariant playing any role but the growth vector of the distribution.

# References

1. Agrachev AA, Chakir HEA, Gauthier JP (1998) Subriemannian Metrics on $R^3$, in Geometric Control and Nonholonomic Mechanics. In: Mexico City 1996, 29–76. Proc. Can. Math. Soc. 25.
2. Agrachev AA, Gauthier JP (1999) Subriemannian Metrics and Isoperimetric Problems in the Contact Case. In: in honor L. Pontriaguin, 90th birthday commemoration. Contemporary Maths, 64:5–48 (Russian). English version: journal of Mathematical sciences, 103(6):639–663.
3. Charlot G (2002) Quasi Contact SR Metrics: Normal Form in $R^{2n}$, Wave Front and Caustic in $R^4$; Acta Appl. Math., 74(3):217–263.
4. Chakir HEA, Gauthier JP, Kupka IAK (1996) Small Subriemannian Balls on $R^3$, Journal of Dynamical and Control Systems, 2(3):359–421.
5. Clarke FH (1983) Optimization and nonsmooth analysis. John Wiley & Sons
6. Gauthier JP, Monroy-Perez F, Romero-Melendez C (2004) On complexity and Motion Planning for Corank one SR Metrics. COCV 10:634–655.
7. Gauthier JP, Zakalyukin V(2005) On the codimension one Motion Planning Problem. JDCS, 11(1):73–89.
8. Gauthier JP, Zakalyukin V (2005) On the One-Step-Bracket-Generating Motion Planning Problem. JDCS, 11(2)215–235.
9. Gauthier JP, Zakalyukin V (2005) Robot Motion Planning, a wild case. Proceedings of the Steklov Institute of Mathematics, 250:56–69.
10. Gauthier JP, Zakalyukin V (2006) On the motion planning problem, complexity, entropy, and nonholonomic interpolation. Journal of dynamical and control systems, 12(3).
11. Gauthier JP, Zakalyukin V (2007) Entropy estimations for motion planning problems in robotics. In: Volume In honor of Dmitry Victorovich Anosov, Proceedings of the Steklov Institute of Mathematics, 256(1):62–79.
12. Gromov M (1996) Carnot Caratheodory Spaces Seen from Within. In: Bellaiche A, Risler J.J, 79–323. Birkhauser
13. Jean F (2001) Complexity of Nonholonomic Motion Planning. International Journal on Control, 74(8):776–782.

14. Jean F (2003) Entropy and Complexity of a Path in SR Geometry. COCV, 9:485–506.
15. Jean F, Falbel E (2003) Measures and transverse paths in SR geometry. Journal d'Analyse Mathématique, 91:231–246.
16. Laumond JP (ed) (1998) Robot Motion Planning and Control, Lecture notes in Control and Information Sciences 229, Springer Verlag
17. Love AEH (1944) A Treatise on the Mathematical Theory of Elasticity, forth edition. Dover, New-York
18. Pontryagin L, Boltyanski V, Gamkelidze R, Michenko E (1962) The Mathematical theory of optimal processes. Wiley, New-York

# Instalment Options: A Closed-Form Solution and the Limiting Case

Susanne Griebsch[1], Christoph Kühn[2], and Uwe Wystup[1]

[1] Frankfurt School of Finance & Management,
Sonnemannstrasse 9-11, 60314 Frankfurt am Main, Germany.
s.griebsch@frankfurt-school.de, uwe.wystup@mathfinance.com
[2] Frankfurt MathFinance Institute, Goethe-University Frankfurt,
Robert-Mayer-Straße 10, 60054 Frankfurt am Main, Germany.
ckuehn@math.uni-frankfurt.de

**Summary.** In Foreign Exchange Markets Compound options (options on options) are traded frequently. Instalment options generalize the concept of Compound options as they allow the holder to prolong a Vanilla Call or Put option by paying instalments of a discrete payment plan. We derive a closed-form solution to the value of such an option in the Black-Scholes model and prove that the limiting case of an Instalment option with a continuous payment plan is equivalent to a portfolio consisting of a European Vanilla option and an American Put on this Vanilla option with a time-dependent strike.

**Keywords**: Exotic Options
**JEL classification**: C15, G12

## 1 Introduction

An Instalment Call or Put option works similar like a Compound Call or Put respectviely, but allows the holder to pay the premium of the option in instalments spread over time. A first payment is made at inception of the trade. The buyer receives the *mother option*. On the following payment days the holder of the Instalment option can decide to prolong the contract and obtain the *daughter option*, in which case he has to pay the second instalment of the premium, or to terminate the contract by simply not paying any more. After the last instalment payment the contract turns into a plain Vanilla Call or Put option. For an Instalment Put option we illustrate two scenarios in Figure 1.

### 1.1 Example

Instalment options are typically traded in Foreign Exchange markets between banks and corporates. For example, a company in the EUR-zone wants to

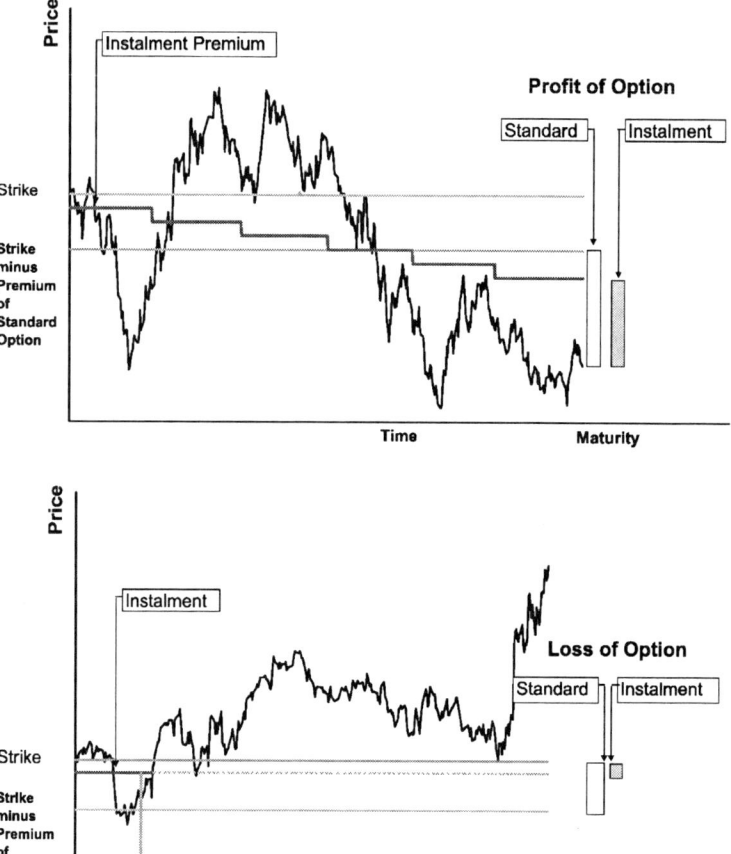

**Fig. 1.** Comparison of two scenarios of an Instalment option. The first diagram shows a continuation of all instalment payments until expiration. The second one shows a scenario where the Instalment option is terminated after the first decision date.

hedge receivables from an export transaction in USD due in 12 months time. It expects a stronger EUR/weaker USD. The company wishes to be able to buy EUR at a lower spot rate if EUR becomes weaker on the one hand, but on the other hand be fully protected against a stronger EUR. The future income in USD is yet uncertain but will be under review at the end of each quarter.

In this case a possible form of protection that the company can use is to buy a EUR Instalment Call option with 4 equal quarterly premium payments as for example illustrated in Table 1.

| Spot reference | 1.1500 EUR-USD |
|---|---|
| Maturity | 1 year |
| Notional | USD 1,000,000 |
| Company buys | EUR Call USD Put strike 1.1500 |
| Premium per quarter of the Instalment | USD 12,500.00 |
| Premium of the corresponding Vanilla Call | USD 40,000.00 |

**Table 1.** Example of an Instalment Call. Four times the instalment rate sums up to USD 50,000, which is more than buying the corresponding plain Vanilla for USD 40,000.

The company pays 12,500 USD on the trade date. After one quarter, the company has the right to prolong the Instalment contract. To do this the company must pay another 12,500 USD. After 6 months, the company has the right to prolong the contract and must pay 12,500 USD in order to do so. After 9 months the same decision has to be taken. If at one of these three decision days the company does not pay, then the contract terminates. If all premium payments are made, then in 9 months the contract turns into a plain Vanilla EUR Call.

Of course, besides not paying the premium, another way to terminate the contract is always to sell it in the market. So if the option is not needed, but deep in the money, the company can take profit from paying the premium to prolong the contract and then selling it.

If the EUR-USD exchange rate is above the strike at maturity, then the company can buy EUR at maturity at a rate of 1.1500.

If the EUR-USD exchange rate is below the strike at maturity the option expires worthless. However, the company would benefit from being able to buy EUR at a lower rate in the market.

Compound options can be viewed as a special case of Instalment options, and the possible variations of Compound options such as early exercise rights or deferred delivery apply analogously to Instalment options.

## 1.2 Reasons for Trading Compound and Instalment Options

We observe that Compound and Instalment options are always more expensive than buying the corresponding Vanilla option, sometimes substantially more expensive. So why are people buying them? One reason may be the situation that a treasurer has a budget constraint, i.e. limited funds to spend for foreign exchange risk management. With an Instalment he can then split the premium over time. This would be inefficient accounting, but a situation like this is not uncommon in practice. However, the essential motivation for a treasurer dealing with an uncertain cash-flow is the situation where he buys a Vanilla instead of an Instalment, and then is faced with a far out of the money Vanilla at time $t_1$, then selling the Vanilla does not give him as much as the savings between the Vanilla and the sum of the instalment payments. With an

Instalment, the budget to spend for FX risk can be planned and controlled. This additional optionality comes at a cost beyond the vanilla price.

From a trader's viewpoint, an instalment is a bet on the future change of the term structure of volatility. For instance, if the forward volatility (12) is higher than a trader's belief of the later materializing volatility, then he would go short an instalment. Some volatility arbitrage focussed hedge funds are trying to identify situations like this.

### 1.3 Literature on Instalment Options

There is not much literature available on the valuation of Instalment options. Here we mention the papers we know about this topic which were published in the past years.

In the paper of Davis, Tompkins and Schachermayer [6] no-arbitrage bounds on the price of Instalment options are derived, which are used to set up static hedges and to compare them to dynamic hedging strategies. Ben-Ameur, Breton and François [2] develop a dynamic programming procedure to compute the value of Instalment options and investigate the properties of Instalment options through theoretical and numerical analysis. Recently Ciurlia and Roko [3] construct a dynamic hedging portfolio and derive a Black-Scholes partial differential equation for the initial value of an American continuous Instalment option. In [11] Kimura and Kikuchi develop a Laplace transform based valuation of Instalment options. The valuation and risk management of Instalment options is related to Bermudan options as in both cases there is a discrete time scale with time points requiring decisions. For details on Bermudan contracts see, e.g., Baviera and Giada [1], Henrard [9] and Pietersz and Pelsser [12].

In the next section we discuss the valuation of Instalment options in the Black-Scholes model in closed-form. In Section 3 we examine the limiting case of an Instalment option, where instalment rates are paid continuously over the lifetime of the option. We will see, that this limiting case can be expressed *model-independently* as a portfolio of other options. In Section 4 we analyze the performance and convergence of our results numerically. Concluding remarks are given in Section 5.

## 2 Valuation in the Black-Scholes Model

The goal of this section is to obtain a closed-form formula for the $n$-variate Instalment option in the Black-Scholes model. For the cases $n = 1$ and $n = 2$ the Black-Scholes formula and Geske's Compound option formula (see [8]) are already well known.

We consider an exchange rate process $S_t$, whose evolution is modeled by a geometric Brownian motion

$$\frac{dS_t}{S_t} = (r_d - r_f)dt + \sigma dW_t, \tag{1}$$

where $W$ is a standard Brownian motion, the volatility is denoted by $\sigma$ and the domestic and foreign interest rates are denoted by $r_d$ and $r_f$ respectively. This means

$$S_T = S_0 \exp\left(\left(r_d - r_f - \frac{\sigma^2}{2}\right)T + \sigma\sqrt{T}Z\right), \tag{2}$$

where $S_0$ is the current exchange rate, $Z$ is a standard normal random variable and $T$ the time to maturity of the option.

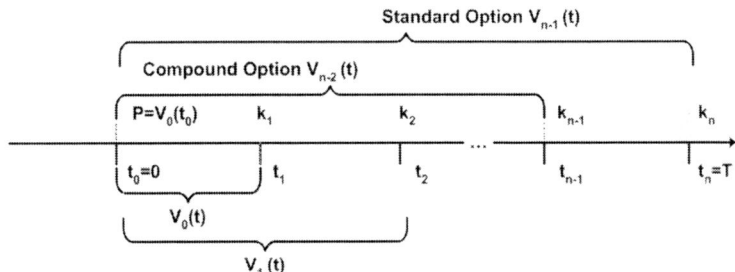

**Fig. 2.** Lifetime of the options with value $V_i$

As illustrated in Figure 2 we let $t_0 = 0$ be the Instalment option inception date and $t_0 < t_1 < t_2 < \ldots < t_n = T$ a schedule of decision dates in the contract on which the option holder has to decide whether to continue to pay the premiums $k_1, k_2, \ldots, k_{n-1}$ to keep the option alive. These premiums can be chosen to be all equal or to have different values. However, the first premium $V_0$ of the Instalment option is determined dependent on the other premiums. To compute the value of the Instalment option, which is the up front payment $V_0$ at $t_0$ to enter the contract, we begin with the option payoff at maturity $T$

$$V_n(s) := [\phi_n(s - k_n)]^+ := \max(\phi_n(s - k_n), 0), \tag{3}$$

where $s = S_T$ is the price of the underlying currency at $T$, $k_n$ the strike price and as usual $\phi_n = +1$ for the underlying standard Call option, $\phi_n = -1$ for a Put option. $V_n$ is the value of the underlying option at time $t_n$, whose value at time $t_{n-1}$ is given by its discounted expectation. And in turn we can again define a payoff function on this value, which would correspond to the payoff of a Compound option.

Generally, at time $t_i$ the option holder can either terminate the contract or pay $k_i$ to continue. Therefore by the risk-neutral pricing theory, the time-$t_i$-value is given by the backward recursion

$$V_i(s) = \left[ e^{-r_d(t_{i+1}-t_i)} \mathbb{E}[V_{i+1}(S_{t_{i+1}}) \mid S_{t_i} = s] - k_i \right]^+$$

$$\text{for } i = 1, \ldots, n-1, \qquad (4)$$

where $V_n(s)$ is given by (3). Following this principal the unique arbitrage-free initial premium of the Instalment option – given $k_1, \ldots, k_n$ – is

$$k_0 := V_0(s) = e^{-r_d(t_1-t_0)} \mathbb{E}[V_1(S_{t_1}) \mid S_{t_0} = s]. \qquad (5)$$

In practice, we normally want to have

$$k_0 = k_1 = \cdots = k_{n-1}. \qquad (6)$$

We notice that one way to determine the value of this Instalment option is to evaluate the nested expectations in Equation (5) through multiple numerical integration of the payoff functions via backward iteration. Another numerical procedure by Ben-Ameur, Breton and François is presented in [2]. In this paper the recursive structure of the value in Equation (5), which is illustrated in Figure 2, is used to develop a dynamic programming procedure to price Instalment options. Thirdly, it is possible to compute the value in closed-form, which is one of the results of this paper.

## 2.1 The Curnow and Dunnett Integral Reduction Technique

For the derivation of the closed-form pricing formula of an Instalment option, we see from Equations (2) and (4) that we need to compute integrals of the form

$$\int_{\mathbb{R}} [\text{option value}(y) - \text{strike}]^+ n(y) dy$$

with respect to the standard normal density $n(\cdot)$. This essentially means to compute integrals of the form

$$\int_{-\infty}^{h} N_i(f(y)) n(y)\, dy,$$

where $N_i(\cdot)$ is the $i$-dimensional cumulative normal distribution function, $f$ some vector-valued function and $h$ some boundary. The following result provides this relationship.

Denote the $n$-dimensional multivariate normal distribution function with upper limits $h_1, \ldots, h_n$ and correlation matrix $R_n := (\rho_{ij})_{i,j=1,\ldots,n}$ by $N_n(h_1, \ldots, h_n; R_n)$, and the univariate standard normal density function by $n(\cdot)$. Let the correlation matrix be non-singular and $\rho_{11} = 1$.

Under these conditions Curnow and Dunnett [4] derive the following *reduction formula*

$$N_n(h_1, \cdots, h_n; R_n) = \int_{-\infty}^{h_1} N_{n-1}\left(\frac{h_2 - \rho_{21}y}{(1 - \rho_{21}^2)^{1/2}}, \cdots, \frac{h_n - \rho_{n1}y}{(1 - \rho_{n1}^2)^{1/2}}; R_{n-1}^*\right) n(y) dy,$$

where the $n - 1$-dimensional correlation matrix $R^*$ is given by

$$R_{n-1}^* := (\rho_{ij}^*)_{i,j=2,\ldots,n}, \qquad \rho_{ij}^* := \frac{\rho_{ij} - \rho_{i1}\rho_{j1}}{(1 - \rho_{i1}^2)^{1/2}(1 - \rho_{j1}^2)^{1/2}}. \tag{7}$$

## 2.2 A Closed-Form Solution for the Value of an Instalment Option

An application of the above result of Curnow and Dunnett yields the derived closed-form pricing formula for Instalment options given in Theorem 1. Before stating the result, we will make an observation about its structure.

The formula in Theorem 1 below has a similar structure as the Black-Scholes formula for Basket options, namely $S_0 N_n(\cdot) - k_n N_n(\cdot)$ minus the later premium payments $k_i N_i(\cdot)$ ($i = 1, \ldots, n-1$). This structure is a result of the integration of the Vanilla option payoff,

$$\int_{\mathbb{R}} [S_T(y) - \text{strike}]^+ n(y) dy$$

which is again integrated after substracting the next instalment,

$$\int_{\mathbb{R}} [\text{Vanilla option value}(y) - \text{strike}]^+ n(y) dy$$

which in turn is integrated with the following instalment and so forth. By this iteration the Vanilla payoff is integrated with respect to the normal density function $n$ times and the $i$-th payment is integrated $i$ times for $i = 1, \ldots, n-1$.

**Theorem 1.** *Let* $\mathbf{k} = (k_1, \ldots, k_n)$ *be the strike price vector,* $\mathbf{t} = (t_1, \ldots, t_n)$ *the vector of the exercise dates of an n-variate Instalment option and* $\boldsymbol{\phi} = (\phi_1, \ldots, \phi_n)$ *the vector of the Put/Call-indicators of these n options. The value function of an n-variate Instalment option is given by*

$$V_0(S_0, \mathbf{k}, \mathbf{t}, \boldsymbol{\phi}) =$$

$$= e^{-r_f t_n} S_0 \phi_1 \cdot \ldots \cdot \phi_n \times$$

$$\times N_n\left[\frac{\ln \frac{S_0}{S_1^*} + \mu^{(+)}t_1}{\sigma\sqrt{t_1}}, \frac{\ln \frac{S_0}{S_2^*} + \mu^{(+)}t_2}{\sigma\sqrt{t_2}}, \ldots, \frac{\ln \frac{S_0}{S_n^*} + \mu^{(+)}t_n}{\sigma\sqrt{t_n}}; R_n\right] -$$

$$- e^{-r_d t_n} k_n \phi_1 \cdot \ldots \cdot \phi_n \times$$

$$\times N_n\left[\frac{\ln \frac{S_0}{S_1^*} + \mu^{(-)}t_1}{\sigma\sqrt{t_1}}, \frac{\ln \frac{S_0}{S_2^*} + \mu^{(-)}t_2}{\sigma\sqrt{t_2}}, \ldots, \frac{\ln \frac{S_0}{S_n^*} + \mu^{(-)}t_n}{\sigma\sqrt{t_n}}; R_n\right] -$$

$$- e^{-r_d t_{n-1}} k_{n-1} \phi_1 \cdot \ldots \cdot \phi_{n-1} \times$$

$$\times N_{n-1}\left[\frac{\ln \frac{S_0}{S_1^*} + \mu^{(-)}t_1}{\sigma\sqrt{t_1}}, \frac{\ln \frac{S_0}{S_2^*} + \mu^{(-)}t_2}{\sigma\sqrt{t_2}}, \ldots, \frac{\ln \frac{S_0}{S_{n-1}^*} + \mu^{(-)}t_{n-1}}{\sigma\sqrt{t_{n-1}}}; R_{n-1}\right] -$$

$$\vdots$$

$$- e^{-r_d t_2} k_2 \phi_1 \phi_2 N_2 \left[ \frac{\ln \frac{S_0}{S_1^*} + \mu^{(-)} t_1}{\sigma \sqrt{t_1}}, \frac{\ln \frac{S_0}{S_2^*} + \mu^{(-)} t_2}{\sigma \sqrt{t_2}}; \rho_{12} \right]$$

$$- e^{-r_d t_1} k_1 \phi_1 N \left[ \frac{\ln \frac{S_0}{S_1^*} + \mu^{(-)} t_1}{\sigma \sqrt{t_1}} \right] = \tag{8}$$

$$= e^{-r_f t_n} S_0 \prod_{i=1}^{n} \Phi_i N_n \left[ \left( \frac{\ln \frac{S_0}{S_m^*} + \mu^{(+)} t_m}{\sigma \sqrt{t_m}} \right)_{1,\ldots,n} \right] -$$

$$- \sum_{i=1}^{n} e^{-r_d t_i} k_i \prod_{j=1}^{i} \Phi_j N_i \left[ \left( \frac{\ln \frac{S_0}{S_m^*} + \mu^{(-)} t_m}{\sigma \sqrt{t_m}} \right)_{1,\ldots,i} \right], \tag{9}$$

where $\mu^{(\pm)}$ is defined as $r_d - r_f \pm \frac{1}{2}\sigma^2$.

The correlation coefficients in $R_i$ of the $i$-variate normal distribution function can be expressed through the exercise times $t_i$,

$$\rho_{ij} = \sqrt{t_i/t_j} \quad for \quad i,j = 1,\ldots,n \quad and \quad i < j. \tag{10}$$

$S_i^*$ $(i = 1,\ldots,n)$ denotes the price of the underlying at time $t_i$ for which the price of the underlying option is equal to $k_i$,

$$V_i(S_i^*) \overset{!}{=} k_i.$$

Remark 1. $S_i^*$ $(i = 1,\ldots,n)$ is determined as the largest resp. smallest spot price $S_t$ for which the initial price of the corresponding renewed $i$-th-Instalment option $(i = 1,\ldots,n)$ is equal to zero. In the case of calls $S_i^*$ is the largest underlying price at which the renewed Instalment option becomes worthless. This problem can be solved by a root finding procedure, e.g. Newton-Raphson. For a Vanilla Call the root $S_{n-1}^*$ always exists and is unique as the Black-Scholes price of a Vanilla Call is a bijection in the starting price of the underlying. Even for a simple Vanilla Put the root $S_{n-1}^*$ may not exist, because the price of a Put is bounded above. In general the existence of the $S_i^*$ can't be guaranteed, but has to be checked on an individual basis. If one of the $S_i^*$ does not exist, then the pricing formula cannot be applied. It means that the strikes are chosen too large. In such a case the strike $k_i$ has to be lowered. In particular, if $\phi_n = -1$ we need to ensure that $\sum_{i=0}^{n-1} k_i < k_n$. This means, that because the price of a vanilla put is bounded above by the strike price, the sum of the future payments must not exceed the upper bound. In practice, arbitrary mixes of calls and puts do not occur. The standard case is a series of calls on a final vanilla product.

Remark 2. The correlation coefficients $\rho_{ij}$ of these normal distribution functions contained in the formula arise from the overlapping increments of the

Brownian motion, which models the price process of the underlying $S_t$, at the particular exercise times $t_i$ and $t_j$.

*Proof.* The proof is established with Equation (7)[3].

Obviously Equation (8) readily extends to a term structure of interest rates and volatility. Therefore we will now deal with how to compute the necessary forward volatilities.

## 2.3 Forward Volatility

The *daughter option* of the Compound option requires knowing the volatility for its lifetime, which starts on the exercise date $t_1$ of the *mother option* and ends on the maturity date $t_2$ of the daughter option. This volatility is not known at inception of the trade, so the only proxy traders can take is the *forward volatility* $\sigma_f(t_1, t_2)$ for this time interval. In the Black-Scholes model the *consistency equation* for the forward volatility is given by

$$\sigma^2(t_1)(t_1 - t_0) + \sigma_f^2(t_1, t_2)(t_2 - t_1) = \sigma^2(t_2)(t_2 - t_0), \qquad (11)$$

where $t_0 < t_1 < t_2$ and $\sigma(t)$ denotes the at-the-money volatility up to time $t$. We extract the forward volatility via

$$\sigma_f(t_1, t_2) = \sqrt{\frac{\sigma^2(t_2)(t_2 - t_0) - \sigma^2(t_1)(t_1 - t_0)}{t_2 - t_1}}. \qquad (12)$$

## 2.4 Forward Volatility Smile

The more realistic way to look at this unknown forward volatility is that the fairly liquid market of Vanilla Compound options could be taken to back out the forward volatilities since this is the only unknown. These should in turn be consistent with other forward volatility sensible products like forward start options, window barrier options or faders.

In a market with smile the payoff of a Compound option can be approximated by a linear combination of Vanillas, whose market prices are known. For the payoff of the Compound option itself we can take the forward volatility as in Equation (12) for the at-the-money value and the smile of today as a proxy. More details on this can be found, e.g. in Schilling [14]. The actual forward volatility, however, is a trader's view and can only be taken from market prices. More details on how to include weekend and holiday effects into the forward volatility computation can be found in Wystup [16].

---

[3] A variation of Formula (8) has been independently derived by Thomassen and Wouve in [15].

## 3 Instalment Options with a Continuous Payment Plan

We will now examine what happens if we make the difference between the instalment payment dates $t_i$ smaller. This will also cause the prolongation payments $k_i$ to become smaller. In the limiting case the holder of the continuous instalment plan keeps paying at a rate $p$ per time unit until she decides to terminate the contract. It is intuitively clear that the above procedure converges as the sum of the strikes increases and is bounded above by the price of the underlying (a call option will never cost more than the underlying). In the limiting case it appears also intuitively obvious that the instalment plan is equivalent to the corresponding Vanilla plus a right to return it any time at a pre-specified rate, which is equivalent to the somehow discounted cumulative prolongation payment which one would have to pay for the time after termination. We will now formalize this intuitive idea.

Let $g = (g_t)_{t\in[0,T]}$ be the stochastic process describing the discounted net payoff of an Instalment option expressed as multiples of the domestic currency. If the holder stops paying the premium at time $t$, the difference between the option payoff and premium payments (all discounted to time 0) amounts to

$$g(t) = \begin{cases} e^{-r_d T}(S_T - K)^+ \mathbf{1}_{(t=T)} - \frac{p}{r_d}(1 - e^{-r_d t}) & \text{if } r_d \neq 0 \\ (S_T - K)^+ \mathbf{1}_{(t=T)} - pt & \text{if } r_d = 0 \end{cases}, \quad (13)$$

where $K$ is the strike. Given the premium rate $p$, the Instalment option can be taken as an American contingent claim with a payoff which may become negative. Thus, the unique no-arbitrage premium $P_0$ to be paid at time 0 (supplementary to the rate $p$) is given by

$$P_0 = \sup_{\tau \in \mathcal{T}_{0,T}} \mathbb{E}_Q(g_\tau), \quad (14)$$

where $Q$ denotes the risk-neutral measure and $\mathcal{T}_{0,T}$ denotes the set of stopping times with values in $[0, T]$. Ideally, $p$ is chosen as the *minimal* rate such that

$$P_0 = 0. \quad (15)$$

Note that $P_0$ from Equation (14) can never become negative as it is always possible to stop payments immediately. Thus, besides (15), we need a minimality assumption to obtain a unique rate. We want to compare the Instalment option with the American contingent claim $f = (f_t)_{t\in[0,T]}$ given by

$$f_t = e^{-r_d t}(K_t - C_E(T - t, S_t))^+, \quad t \in [0, T], \quad (16)$$

where $K_t = \frac{p}{r_d}\left(1 - e^{-r_d(T-t)}\right)$ for $r_d \neq 0$ and $K_t = p(T-t)$ when $r_d = 0$. $C_E$ is the value of a standard European Call. Equation (16) represents the payoff of an American Put on a European Call where the variable strike $K_t$ of the Put equals the part of the instalments *not* to be paid if the holder decides to terminate the contract at time $t$. Define by $\tilde{f} = (\tilde{f}_t)_{t \in [0,T]}$ a similar American contingent claim with

$$\tilde{f}(t) = e^{-r_d t} \left[ (K_t - C_E(T - t, S_t))^+ + C_E(T - t, S_t) \right], \quad t \in [0, T]. \quad (17)$$

As the process $t \mapsto e^{-r_d t} C_E(T - t, S_t)$ is a $Q$-martingale we obtain that

$$\sup_{\tau \in \mathcal{T}_{0,T}} \mathbb{E}_Q(\tilde{f}_\tau) = C_E(T, s_0) + \sup_{\tau \in \mathcal{T}_{0,T}} \mathbb{E}_Q(f_\tau). \quad (18)$$

**Theorem 2.** *An Instalment Call option with continuous payments is the sum of a European Call plus an American Put on this European Call, i.e.*

$$\underbrace{P_0 + p \int_0^T e^{-r_d s} \, ds}_{total\ premium\ payments} = C_E(T, s_0) + \sup_{\tau \in \mathcal{T}_{0,T}} \mathbb{E}_Q(f_\tau),$$

*where $P_0$ is the Instalment option price and $\sup_{\tau \in \mathcal{T}_{0,T}} \mathbb{E}_Q(f_\tau)$ is the price of an American put with a time-dependent strike.*

*Proof.* Define a new claim $\tilde{g} = (\tilde{g}_t)_{t \in [0,T]}$ differing from $g$ only by a constant, namely $\tilde{g}(t) = g(t) + p \int_0^T e^{-r_d s} \, ds$. In view of (18) we have to show that

$$\sup_{\tau \in \mathcal{T}_{0,T}} \mathbb{E}_Q[\tilde{g}(\tau)] = \sup_{\tau \in \mathcal{T}_{0,T}} \mathbb{E}_Q[\tilde{f}(\tau)]. \quad (19)$$

The inequality with $\leq$ in (19) is obvious as we have $\tilde{g} \leq \tilde{f}$ pointwise. Let us show the other direction. Denote by $V = (V_t)_{t \in [0,T]}$ the Snell envelope of the potentially larger process $\tilde{f}$, i.e. $V$ is a càdlàg process (right continuous paths with left limits) with

$$V_t = \text{ess.sup}_{\tau \in \mathcal{T}_{t,T}} \mathbb{E}_Q[\tilde{f}(\tau) \mid \mathcal{F}_t], \quad P\text{-a.s.}, \quad t \in [0, T],$$

where $(\mathcal{F}_t)_{t \in [0,T]}$ is the canonical filtration of the process $S$. Define by $h = h(u, s)$ the value of the Call plus the Put on the Call, if the initial price of the underlying is $s \in \mathbb{R}_+$ and time to maturity of the contract is $u \in \mathbb{R}_+$, i.e.

$$h(u, s) = \sup_{\tau \in \mathcal{T}_{0,u}} \mathbb{E}_s \left[ e^{-r_d \tau} \left[ \left[ \frac{p}{r_d}(1 - e^{-r_d(u-\tau)}) - C_E(u - \tau, \tilde{S}_\tau) \right]^+ \right. \right.$$
$$\left. \left. + C_E(u - \tau, \tilde{S}_\tau) \right] \right],$$

where $\tilde{S}$ is again a geometric Brownian motion with the same probabilistic characteristics as $S$. Using the Markov property of $S$ we can apply Theorem 3.4 in El Karoui/Lepeltier/Millet (1992) and obtain

$$V_t = \text{ess.sup}_{\tau \in \mathcal{T}_{t,T}} \mathbb{E}_Q \left[ \widetilde{f}(\tau) \mid \mathbb{F}_t \right] = \text{ess.sup}_{\tau \in \mathcal{T}_{t,T}} \mathbb{E}_Q \left[ \widetilde{f}(\tau) \mid S_t \right] =$$
$$= e^{-r_d t} h(T - t, S_t).$$

As $\widetilde{f}$ has continuous paths the optimal exercise time is given by

$$\widehat{\tau} = \inf\{t \in [0, T] \mid V_t = \widetilde{f}(t)\} =$$
$$= \inf\{t \in [0, T] \mid e^{-r_d t} h(T - t, S_t) = \widetilde{f}(t)\}. \tag{20}$$

Keeping this in mind, we want to show that

$$h(u, s) > C_E(u, s) \quad \text{for all } u > 0, s > 0. \tag{21}$$

As the process $t \mapsto e^{-r_d t} C_E(T - t, S_t)$ is a martingale we can pull it out of the optimal stopping problem and obtain

$$h(u, s) = C_E(u, s) + \sup_{\tau \in \mathcal{T}_{0,u}} \mathbb{E}_s \left[ e^{-r_d \tau} \left[ \frac{p}{r_d}(1 - e^{-r_d(u - \tau)}) - C_E(u - \tau, \widetilde{S}_\tau) \right]^+ \right],$$

and thus $h(u, s) > C_E(u, s)$, for all $u > 0, s > 0$, as the underlying Call $C_E$ can always get into the money with positive probability as long as $u > 0$. Therefore, we obtain for $t \in [0, T)$ and $s \in (0, \infty)$ the following implication

$$h(T - t, s) = (K_t - C_E(T - t, s))^+ + C_E(T - t, s) \Rightarrow K_t > C_E(T - t, s). \tag{22}$$

This means that by (20) $\widetilde{f}$ is only exercised prematurely when $K_t > C_E(T - t, S_t)$. But, in this case we have $\widetilde{f}(t) = \widetilde{g}(t)$. As at maturity the payoffs of $\widetilde{f}$ and $\widetilde{g}$ coincide anyway, we have for the optimal exercise time $\widehat{\tau}$ of the process $\widetilde{f}$

$$\widetilde{f}(\widehat{\tau}) = \widetilde{g}(\widehat{\tau}), \quad P\text{-a.s.}$$

Therefore we arrive at (19) and the assertion of the theorem follows.

*Remark 3.* We could use the same argument to prove that an Instalment put option is the sum of a European Put plus an American Put on this European Put.

## 4 Numerical Results

### 4.1 Implementational Aspects

In the appendix we give a sample implementation for the discrete case of an Instalment option in both

- Mathematica, which solves the nested integration for the value recursively as mentioned in Section 2, and
- R, which computes the value using Equation (8) in Theorem 1.

Both implementations are used to investigate the performance and convergence of Instalment option values.

## 4.2 Performance

To compare the various methods to determine the value of Instalment options – to calculate the initial premium at time 0 dependent on the remaining strikes – we compare values of a specific trivariate Instalment option. We implement on the same machine

1. a binomial tree method in C++,
2. the closed-form formula in the statistical language R (see [13]),
3. the dynamic programming algorithm of Breton et al. [2],
4. a numerical integration using Gauß quadrature methods with 50, 000 supporting points, and
5. a recursive algorithm implemented in Mathematica for the calculation of the value of an $n$-variate Instalment option.

In Table 2 the results and computation times of all these five methods are shown for two representative examples of a 3-variate-Instalment option. The computational times are given in seconds and lie close together for most of the applied techniques.

| Numerical Method | Value of $V_3$ | | CPU Time |
|---|---|---|---|
| Binomial trees for $n = 4000$ | 1.69053 | 0.0137335 | 1109 |
| Closed-form formula for $n = 3$ | 1.69092 | 0.0137339 | < 1 |
| Algorithm based on [2] with $p = 4000$ | 1.69084 | 0.0137332 | 168 |
| Numerical integration (50000-point Gauß-Legendre) | 1.69087 | 0.0137339 | 176 |
| Numerical integration with Mathematica | 1.69091 | 0.0137299 | 47 |

**Table 2.** Performance comparison of Instalment valuation algorithms. We use $S_0 = 100$, $k_1 = 100$, $k_{2,3} = 3$, $\sigma = 20\%$, $r_d = 10\%$, $r_f = 15\%$, $T = 1$, $\Delta t = 1/3$, $\phi_{1,2,3} = 1$ and $S_0 = 1.15$, $k_1 = 1.15$, $k_{2,3} = 0.02$, $\sigma = 10\%$, $r_d = 1\%$, $r_f = 2\%$, $T = 1, t = 1/3$, $\phi_{1,2,3} = 1$.

Our experiences with the application of these methods show that

- The results using binomial tree methods oscillate heavily even with a large depth of the tree. Our examples show variations in the fourth digit of the value by using binomial trees with a depth of the tree from up to 7000.
- The trivariate formula is the fastest of all compared methods. Its accuracy and computation time essentially depend on the quality of the root finding procedure and on the calculation of the multivariate normal distribution function.
- The techniques, which are based on numerical integration as well as the dynamic programming approach of Breton et al. [2] lie in the middle field of all observed computation times.

## 4.3 Convergence

We illustrate the convergence of the overall Instalment premium to the limiting case in Figure 3.

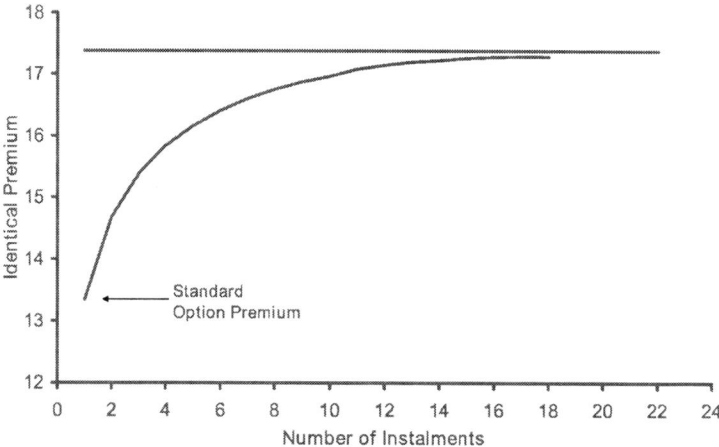

**Fig. 3.** Convergence of uniform premium in discrete case to continuous premium. We have used the data $S_0 = 100$, $K = 95$, $\sigma = 0.2$, $r_d = 0.05$, $r_f = 0$, $T = 1$.

Here we investigate our result in Theorem 2 for a practical example, where a number of identical premiums of their corresponding $n$-variate-Instalment option for $n = 1, ..., 18$ is evaluated. The identical premium of a 1-variate-Instalment option is obviously the value of a standard Call option at time 0. All other identical premiums are calculated by a root finding procedure with respect to the strike price of the function

$$V_0(k) - k = 0,$$

which is the value of the particular Instalment option at time 0 minus the strike price. Here we use the closed-form Equation (8). It is implemented in the statistical language R (see [13]) as it contains the multivariate normal distribution function. The source code is listed in the appendix.

This calculation requires a high degree of accuracy and therefore takes a long time to compute. The identical premium for a 18-variate-Instalment option is 17.28. The limit $U$ is calculated following Theorem 2 using the value of a European Call plus an American Put on this Call. The Black-Scholes formula is used to determine the value of the European Call, and for the calculation of the American product of the portfolio we use binomial tree methods. The limit $U$ lies approximately at 17.51 for the parameter set in Figure 3 and is approached here from below.

# 5 Summary

We have presented a closed-form solution for Instalment Call and Put options in the Black-Scholes model, discussed its application and numerical implementation. We proved the equivalence of the limiting case of a continuous instalment plan with a portfolio of the corresponding Vanilla and an American Put on that claim with a time dependent strike.

Further research could be done to explore closed-form valuation of Instalment options in models beyond Black-Scholes, such as stochastic volatility models or behavior in interest rate models. The case of Compound options ($n = 2$) has been examined in stochastic volatility models by Fouque and Han [5].

Another approach would be to analyze Instalment options with a more generalized payoff function at maturity, so that the Instalment plan's final option is a more exotic product than a Vanilla.

## Acknowledgments

We would like to thank Michèle Vanmaele, Robert G. Tompkins, Manuel Guerra and an anonymous referee for their help and comments.

# A Mathematica Code

## A.1 The Package instalment.m

```
BeginPackage["Options Instalment"]

Instalment::usage = "Instalment[S,K,T,vol,rd,rf,phi,N] \n
Black-Scholes value for European Instalment options\n
S: spot\n
K: strike list of individual options\n
T: time differences to maturity in years between individual options\n
beginning with Vanilla option maturity\n
vol: volatility\n
rd: domestic risk free rate: discounting is done as Exp[-T[[i]]*rd] \n
rf: foreign risk free rate: discounting is done as Exp[-T[[i]]*rf]\n
phi: list of +1 for Calls, -1 for Puts\n
N: number of options in Instalment option"

Begin["Private"]

ncum[x_]:=1/2*(Erf[x/Sqrt[2]]+1); (*cumulative standard normal*)
ndf[x_]:=Evaluate[D[ncum[x],x]]; (*standard normal density*)

Vanilla[x_,K_,vol_,r_,rf_,T_,fi_]:=Block[dp,dm,
```

```
dp=(Log[x/K]+(r-rf+0.5*vol*vol)*T)/(vol*Sqrt[T]);
dm=(Log[x/K]+(r-rf-0.5*vol*vol)*T)/(vol*Sqrt[T]);
fi*(Exp[-rf*T]*x*ncum[fi*dp]-Exp[-r*T]*K*ncum[fi*dm])];

Instalment[S_,K_,T_,vol_,rd_,rf_,phi_,N_]:=Block[mu,
    mu=rd-rf-0.5*vol*vol;
  If[N==1,Vanilla[S,K[[1]],vol,rd,rf,T[[1]],phi[[1]]],
    Exp[-T[[N]]*rd]*
  NIntegrate[
    Max[0,phi[[N]]*(Instalment[S*Exp[vol*Sqrt[T[[N]]]*z+mu*T[[N]]],
       K,T,vol,rd,rf,phi,N-1]-
    K[[N]])]*ndf[z],z,-10,10]]];

End[]
EndPackage[]
```

## A.2 The Testing Environment instalment_testenv.nb

```
spot = 100
vol = 0.2
tau = {1/3, 1/3, 1/3}
rd = 0.10
rf = 0.15
strike = {100, 3, 3}
phi = {1, 1, 1}

Instalment[spot, strike, tau, vol, rd, rf, phi, 3]

1.69085
```

# B R Code

## B.1 The R Functions

```
installments←function(spot,strikes,times,phis,rd,rf,sigma,interval){
n←length(times)
s←1:(n-1)
roots←rep(0,n)
roots[n]←strikes[n]

for(i in s){
tau←rep(0,i)
for(j in (1:i))
tau[j]←times[n-j+1]-times[n-i]

f←function(x){recur(x,strikes[(n+1-i):n],rev(tau),phis[(n+1-i):n],rd,
```

```
rf,sigma,roots[(n-i+1):n])-strikes[i]    }
roots[n-i]←uniroot(f,interval)[1]    }

result←recur(spot,strikes,times,phis,rd,rf,sigma,roots)
return(result)    }

recur←function(spot,k,t,phis,rd,rf,sigma,roots){

library(mnormt)

n←length(t)
s←1:n
args1←rep(0,n)
args2←rep(0,n)
multi←rep(0,n)
rho←matrix(rep(0,n^2),nrow=n,ncol=n)

for(i in s){
for(j in i:n){
rho[i,j]←sqrt(t[i]/t[j])
rho[j,i]←rho[i,j]    }    }

muplus←rd-rf+0.5*sigma^2
muminus←rd-rf-0.5*sigma^2

for(i in s){
args1[i]←(log(spot/roots[[i]])+muplus*t[i])/(sigma*sqrt(t[i]))
args2[i]←(log(spot/roots[[i]])+muminus*t[i])/(sigma*sqrt(t[i]))    }

for(i in s){
if(i==1)
multi[i]←prod(phis[1:i]) *pmnorm(x=args2[1:i],mean=0,varcov=1,abseps=
1e-10)[1]
else
multi[i]←prod(phis[1:i]) *pmnorm(x=args2[1:i],mean=rep(0,i),varcov=
rho[1:i,1:i])[1]    }

if(n==1)
part1←exp(-rf*t[n])*spot*prod(phis)
*pmnorm(x=args1,mean=0,varcov=1,abseps=1e-10)[1]
else
part1←exp(-rf*t[n])*spot*prod(phis)
*pmnorm(x=args1,mean=rep(0,i),varcov=rho[1:i,1:i])[1]

part2← sum(exp(-rd*t)*k*multi)return(part1-part2)    }
```

## B.2 The R Testing Environment

```
interval←c(0,150)
strikes←c(3,3,100)
times←c(1/3,2/3,1.0)
phis←c(1,1,1)
rd=0.1
rf=0.15
sigma=0.2
spot=100
installments(spot,strikes,times,phis,rd,rf,sigma,interval)

interval←c(0,10)
strikes←c(0.02,0.02,1.15)
times←c(1/3,2/3,1.0)
phis←c(1,1,1)
rd=0.01
rf=0.02
sigma=0.1
spot=1.15
installments(spot,strikes,times,phis,rd,rf,sigma,interval)
```

# References

1. Baviera R, Giada L (2006) Perturbative Approach to Bermundan Option Pricing. Working Paper 941318, SSRN.
2. Ben-Hameur H, Breton M, François P (2006) A Dynamic Programming Approach to Price Installment Options. European Journal of Operational Research 169:667–676.
3. Ciurlia P, Roko I (2005) Valuation of American Continuous-Installment Options. Computational Economics 1(2):143–165.
4. Curnow RN, Dunnett CW (1962) The Numerical Evaluation of Certain Multivariate Normal Integrals. Ann. Math. Statist 33:571–579.
5. Fouque J, Han C (2005) Evaluation of Compound Options using Perturbation Approximation. Journal of Computational Finance 9.
6. Davis M, Schachermayer W. Tompkins R (2001) Pricing, No-arbitrage Bounds and Robust Hedging of Installment Options. Quantitative Finance 1:597–610.
7. Brown B, Lovato J, Russell K (2004) CDFLIB - C++ - library. http://www.csit.fsu.edu/ burkardt/cpp_src/dcdflib/dcdflib.html
8. Geske R (1979) The Valuation of Compound Options. Journal of Financial Economics 7:63–81.
9. Henrard M (2005) Bermudan Swaptions in a Hull-White One-Factor Model: Analytical and Numerical Approaches. Ewp-fin 0505023, Economics Working Paper Archive.
10. El Karoui N, Lepeltier JP, Millet A (1992) A Probabilistic Approach of the Reduite. Probability and Mathematical Statistics 13:97–121.
11. Kimura T, Kikuchi K (2007) Valuing Continuous Installment Options via Laplace Transforms. Working Paper, Hokkaido University.

12. Pietersz R, Pelsser AA (2003) Risk Managing Bermudan Swaptions in the LI-BOR BGM Model. Working Paper 383580, SSRN.
13. R project, http://www.r-project.org.
14. Schilling H (2002) Compound Options.
In http://www.mathfinance.com/FXRiskBook/ Foreign Exchange Risk. Risk Publications. London.
15. Thomassen L, van Wouve M (2002) A Sensitivity Analysis for the N-fold Compound Option. Research Paper, Faculty of Applied Economics, University of Antwerpen.
16. Wystup U (2006) http://fxoptions.mathfinance.com FX Options and Structured Products. Wiley Finance.

# Existence and Lipschitzian Regularity for Relaxed Minimizers

Manuel Guerra[1*] and Andrey Sarychev[2†]

[1] CEOC and ISEG–T.U.Lisbon,
   R. do Quelhas 6, 1200-781 Lisboa, Portugal. mguerra@iseg.utl.pt.
[2] DiMaD, University of Florence,
   via Cesare Lombroso 6/17, 50134–Firenze, Italia. asarychev@unifi.it

**Summary.** In this contribution we follow two main goals: to reconstruct a result announced in [4] about existence of relaxed minimizers for (nonconvex) Lagrange problems of optimal control (Theorem 1); to derive conditions for Lipschitzian regularity of trajectories corresponding to relaxed minimizers (Theorem 3). In passing, elaborating on the approach used in [10], we provide a condition for Lipschitzian regularity of non relaxed minimizers (Theorem 2).

## 1 Introduction

We study a Lagrange problem of optimal control with a minimized functional

$$\mathcal{J}(x,v) = \int_0^T \ell(t,x,v)dt \to \min \tag{1}$$

and a dynamics

$$\dot{x} = f(t,x,v), \qquad v \in \mathbb{R}^r. \tag{2}$$

Here $T \in\, ]0, +\infty[$ is fixed and for simplicity we impose fixed boundary conditions

$$x(0) = x_0, \ x(T) = x_T. \tag{3}$$

We assume $\ell, f$ to be continuous with respect to all variables.

Classical results on existence of minimizers for such problems go back to L.Tonelli (see [4, Ch. 10,11], [6]) and references therein). These results establish existence of a minimizer under certain assumptions, the main of which regard coercivity and convexity of the data. Loosely speaking the coercivity

* Supported by "Fundação para a Ciência e a Tecnologia" FCT, cofinanced by the European Union Fund FEDER/POCI 2010
† Supported by the research grant PRIN No. 2006019927, MIUR, Italy.

assumption obliges $\ell$ to grow faster than $f$ as $\|v\| \rightarrow \infty$. This guarantees weak convergence of the derivatives of the elements of a minimizing sequence. The convexity assumption guarantees lower semicontinuity of the functional $\mathcal{J}(x, v)$ with respect to this weak convergence.

Various coercivity and convexity assumptions, which could be imposed on the data of the problem in order to guarantee existence are discussed in [4, 6].

In what follows we are interested in the nonconvex case. Although in this case there still may exist classical minimizers (see, for example, [3]), it is well understood that one should rather expect *relaxed* minimizers. Recall that relaxed controls ([8]) take their values (at a given instant of time) in the space of probability measures supported on the set of control parameters. It is a rich set of generalized controls, which admit nice approximation by classical controls (e.g., piecewise continuous, or bounded measurable, or square integrable, or integrable controls) in the so-called 'relaxation metric'.

In the case when *controls take its values in a bounded set*, the corresponding existence result for the relaxed controls is well known [8]. The case of unbounded set of controls is more intricate; one of the reasons is that the coercivity assumptions imposed on the functional and the dynamics are not inherited by their relaxed counterparts. Still result by L.Cesari [4, Remark 3, §11.4] claims that relaxed minimizer exists under some of these coercivity assumptions.

We start with a brief reconstruction of the corresponding proof, by tracing the sequence of results dispersed in Chapters 8-11 of [4]. This is the contents of Section 2.

Our main interest though centers on the Lipschitzian property of minimizing *relaxed* trajectories. As it is seen from the existence result of Section 2, one gets a minimizing relaxed trajectory, which is absolutely continuous but not necessarily Lipschitzian.

In the classical (convex and coercive) case there are many examples where the minimizers happen to be non-Lipschitzian. Moreover in the presence of such minimizers a Lavrentiev gap may occur: the infimum of the functional over the space of absolutely continuous functions can be strictly less, than the infimum over the space of Lipschitzian functions. Non-Lipschitzian minimizers may lead to an additional irregularity - sometimes they fail to be extremals, i.e. they do not satisfy standard necessary optimality conditions in either Euler-Lagrange, or Hamilton-Pontryagin form. Therefore these minimizers are hard to calculate by either analytic or numerical methods.

The first condition which provides Lipschitzian property (regularity) of the minimizing trajectory goes back to the work of L.Tonelli at the beginning of the last century. There have been much progress since then and an excellent account of the results up to 1985 can be found in [6]. The subject of these classical studies was the Basic Problem of the Calculus of Variations and, more recently, its version with higher-order derivatives.

The issue of Lipschitzian regularity in control-theoretic formulation has been treated in publication by F.H.Clarke and R.Vinter [7], where the case of *time-invariant linear dynamics* was considered.

Few years ago A.Sarychev and D.Torres ([10]) derived conditions of Lipschitzian regularity for nonrelaxed minimizers of Lagrange problems with *control-affine nonlinear dynamics*. To achieve this, a new approach has been used.

The core of this approach, was introduced by R.V.Gamkrelidze ([8]) in the context of existence theory. It amounts to a transformation of the Lagrange problem into a time-optimal control problem. The set $\mathbb{R}^r$ of control parameters of this latter is subject to (one-point) compactification. If one is able to extend the time-optimal problem into the compactified control set $\bar{\mathbb{R}}^r = \mathbb{S}^r$ ('extend onto infinity') then the existence result can be derived (see [8]) from Filippov's theorem on existence of optimal control in the bounded case. If, moreover, the conditions imposed on the data of the problem, make the Pontryagin Maximum Principle valid for the minimizers of the extended problem, then one can conclude Lipschitzian regularity of minimizing trajectories, which correspond to *normal* Pontryagin extremals. This is a brief account of what has been accomplished in [10]. More progress in this direction has been achieved in [12]. In Section 3 of this contribution we formulate another criterion of Lipschitzian regularity of non relaxed (classical) minimizers for Lagrange variational problem with a nonlinear dynamics (2).

In Section 4 we deal with the main subject of our interest - Lipschitzian regularity of relaxed minimizing trajectories in nonconvex case. We readmit the coercivity assumptions introduced in Section 2, in order to guarantee the existence of a *relaxed* absolutely-continuous minimizer.

Advancing with the approach, sketched above, we show at once that one can not employ one-point compactification of the set of control parameters since, in general the problem does not admit a continuous extension onto such compactification. To overcome this difficulty we invoke a compactification introduced by P.Loeb ([9]), which succeeds the notion of *unified space*, introduced earlier by G.T.Whyburn ([14]).

Roughly speaking, given a family of continuous bounded functions, this construction provides a *minimal* compactification onto which these functions can be extended by continuity in such a way that they separate the points of the remainder. The definition is rather abstract, but in our case, under the imposed coercivity assumptions, we are able to represent the compactification as a fibred space. We also prove that the extended set of control parameters can be chosen $x$-independent.

Formulating the Pontryagin maximum principle and the second Erdmann's condition for the extended problem we derive from the latter a criterion of Lipschitzian regularity of minimizing relaxed trajectories.

The authors are grateful to D. Seabra, who read the manuscript, pointed out several misprints and made suggestions of improvement of the text.

# 2 Existence of relaxed minimizers in unbounded nonconvex case

As we said, an existence result for *relaxed minimizers* of nonconvex optimal control problems (1)-(2) with unbounded control set is referred to [4, Remark 3, §11.4].

The formulation includes one of three coercivity assumptions. In what follows we choose a modification of one of these assumptions, namely:

- For each compact set $X \subset \mathbb{R}^n$ there exists a bounded below scalar function $\phi : [0, +\infty[\mapsto \mathbb{R}$, such that:

$$\phi(\xi)/\xi \to +\infty, \text{ as } \xi \to +\infty, \qquad \ell(t, x, v) \geq \phi\left(\|f(t, x, v)\|\right) \quad (4)$$

  holds for every $(t, x) \in [0, T] \times X$, $v \in \mathbb{R}^r$;

- For each compact set $X \subset \mathbb{R}^n$

$$\ell(t, x, v) \to +\infty, \text{ as } \|v\| \to +\infty, \quad (5)$$

  uniformly with respect to $(t, x) \in [0, T] \times X$.

The proof of existence passes through various intermediate results dispersed in the Sections 1,8-11 of [4]. We will outline the proof.

Consider a (noncovex) optimal control problem (1)-(2)-(3), where the functions $\ell, f$ are continuous on $\mathbb{R} \times \mathbb{R}^n \times \mathbb{R}^r$ and $\ell$ is bounded below. Without lack of generality we may add a positive constant to $\ell$ and assume $\ell \geq 1$.

The relaxation of the problem rests upon convexification of the Lagrangian and of the dynamic equations. Consider (for $(t, x)$ fixed) the map

$$v \mapsto (\ell(t, x, v), f(t, x, v)),$$

of $\mathbb{R}^r$ into $\mathbb{R}^{n+1}$. Any point in the convex hull of $\{(\ell(t, x, v), f(t, x, v), v \in \mathbb{R}^r)\}$ (with fixed $(t, x)$) can be represented as $(L(t, x, p, V), F(t, x, p, V))$, where $V = (v^1, \ldots, v^{n+2}) \in \mathbb{R}^{r(n+2)}$, and $p = (p_1, \ldots, p_{n+2}) \in \Sigma^{n+1}$, the $(n+1)$-dimensional simplex: $p_j \geq 0$, $\sum_{j=1}^{n+2} p_j = 1$,

$$L(t, x, p, V) = \sum_{j=1}^{n+2} p_j \ell(t, x, v^j), \qquad F(t, x, p, V) = \sum_{j=1}^{n+2} p_j f(t, x, v^j).$$

The convexified dynamics is defined by

$$\dot{x} = F(t, x, p, V) = \sum_{j=1}^{n+2} p_j f(t, x, v^j). \quad (6)$$

The relaxed problem consists of minimization of the functional

$$\mathcal{J}^c(p(\cdot), V(\cdot)) = \int_0^T L(t, x(t), p(t), V(t))dt =$$

$$= \int_0^T \sum_{j=1}^{n+2} p_j(t)\ell(t, x(t), v^j(t))dt \to \min, \qquad (7)$$

under the 'dynamic constraint' (6) and the boundary conditions (3). The control $V(\cdot)$ takes its values in $\mathbb{R}^{r(n+2)}$, while $p(\cdot)$ takes its values in $\Sigma^{n+1}$.

It is known that under broad assumptions the infimum of the original problem and of its relaxation coincide ([4]). Obviously the trajectories of the original control system (2) are contained in a richer class of trajectories of the relaxed dynamics (6). Still the trajectories of the relaxed dynamics can be uniformly approximated by trajectories of (2).[3]

Let us introduce the sets

$$Q(t, x) = \left\{ z \,\middle|\, z = F(t, x, p, V), \ p \in \Sigma^{n+1}, \ V \in \mathbb{R}^{r(n+2)} \right\},$$

$$\tilde{Q}(t, x) = \left\{ (z_0, z) \,\middle|\, z_0 \ge L(t, x, p, V), z = F(t, x, p, V), p \in \Sigma^{n+1}, V \in \mathbb{R}^{r(n+2)} \right\}.$$

Evidently $Q(t, x)$ is the projection of $\tilde{Q}(t, x)$ onto $z$-component:

$$Q(t, x) = \{z \,|\, (z_0, z) \in \tilde{Q}(t, x)\}.$$

Following [4] we define the "unparameterized Lagrangian"

$$\mathcal{L}(t, x, z) = \begin{cases} \inf\{z_0 | \ (z_0, z) \in \tilde{Q}(t, x)\}, & \text{if } z \in Q(t, x); \\ +\infty, & \text{otherwise,} \end{cases} \qquad (8)$$

and consider the "unparameterized variational problem"

$$\int_0^T \mathcal{L}(t, x, \dot{x})dt \to \min, \qquad \dot{x} \in Q(t, x). \qquad (9)$$

It is known ([4, §1.14]) that if:

i) the sets $\tilde{Q}(t, x)$ are closed for $x \in \mathbb{R}^n$, and

ii) the infimum in (8) turns to be the minimum,

then for any triple $(x(\cdot), p(\cdot), V(\cdot))$, satisfying the equation (6), the trajectory $x(\cdot)$ satisfies the differential inclusion in (9) and

$$\int_0^T \mathcal{L}(t, x, \dot{x})dt \le \int_0^T L(t, x, p, V)dt.$$

---

[3] In some cases one is not able to maintain the boundary conditions for the approximating trajectories and then the infimum of the relaxed system could be only achieved by a minimizing sequence of classical trajectories whose boundary data are loosened and converge to the data (3) of the relaxed system

Vice versa, by the lemma on measurable selection ([4, Th. 8.2.iii]), for any trajectory $x(\cdot)$ of (9) there exists a measurable pair $(p(\cdot), V(\cdot))$ taking values in $\Sigma^{n+1} \times \mathbb{R}^{r(n+2)}$ such that the triple $(x, p, V)$ satisfies the equation $\dot{x} = F(t, x, p, V)$ and

$$\int_0^T \mathcal{L}(t, x, \dot{x}) dt \geq \int_0^T L(t, x, p, V) dt.$$

Therefore under the above mentioned conditions the parameterized (control-type) and unparameterized problems are equivalent.

We proceed under the growth assumptions (4)-(5) imposed on $\ell$ and $f$. *These assumptions are not valid for the relaxed Lagrangian $L$ and dynamics $F$: after relaxation the nonconvex coercive problem (1)-(2)-(3) becomes noncoercive.* Still one is able to proceed further using the unparameterized formulation (9).

The following property of (weakened) upper semicontinuity is valid for the set-valued map $x \mapsto Q(t, x)$ (see [4, §10.5, Lemma 10.5.iii, Remark 4]).

**Lemma 1.** *Under coercivity assumptions (4)-(5) there holds for each $x_0 \in \mathbb{R}^n$, $t \in [0, T]$:*

$$\tilde{Q}(t, x_0) = \bigcap_{\delta > 0} \text{clos}\tilde{Q}(t, \mathcal{O}_\delta(x_0)). \tag{10}$$

*(Here $\mathcal{O}_\delta(x_0)$ is $\delta$-neighborhood of $x_0$, and $\tilde{Q}(t, \mathcal{O}_\delta(x_0)) = \bigcup_{x \in \mathcal{O}_\delta(x_0)} \tilde{Q}(t, x)$.)*
□

**Corollary 1.** *Under assumptions of the Lemma the sets $\tilde{Q}(t, x_0)$ are closed.*
□

*Remark 1.* There also holds

$$\tilde{Q}(t, x_0) = \bigcap_{\delta > 0} \text{clos conv}\tilde{Q}(t, \mathcal{O}_\delta(x_0)). \ \square$$

The following result relates lower semicontinuity of the unparameterized Lagrangian $\mathcal{L}(t, x, z)$ with property (10).

**Lemma 2.** *([4, §8.5.C]) If the infimum is attained in the definition of $\mathcal{L}(t, x, z)$ (8) whenever $x \in Q(t, x)$ and the map $(x, z) \mapsto \mathcal{L}(t, x, z)$ is lower semicontinuous, then property (10) holds.*

*If property (10) holds then the unparameterized Lagrangian $(x, z) \mapsto \mathcal{L}(t, x, z)$ is lower semicontinuous.* □

Closedness of $\tilde{Q}(t, x_0)$ implies that the infimum in (8) is attained. Under condition (10) the parameterized and unparameterized optimal control problems are equivalent.

The existence of minimizer for the unparameterized problem can be established by virtue of [4, Th. 11.1.i]. One can assume that the trajectories, we consider, are contained in a *bounded* subset of $\mathbb{R}^n$. In fact it can be proven, that fixing the initial point of the trajectories one can establish their equi-boundedness from the assumption (4), as it is done in [4, §11.2].

The concluding result reads as follows.

**Theorem 1.** *Let $\ell, f$ be continuous, $\ell$ bounded below and coercivity assumptions (4)-(5) hold. Then the problem (1)-(2)-(3) possesses a relaxed minimizer.* □

*Remark 2.* We believe that the existence result can be obtained on the basis of a compactification technique, which is developed in Section 4. We will comment on it elsewhere. □

## 3 Lipschitzian regularity of non relaxed minimizers for Lagrange variational problem: brief account

Sarychev and Torres in [10] suggested an approach, which is based on:

i)  transformation of a Lagrange optimal control problem with nonlinear control-affine dynamics into an autonomous time-optimal control problem;

ii)  (one-point) compactification of the set of its control parameters;
iii) application of Pontryagin maximum principle to the problem with the compact set of controls.

The second Erdmann's condition together with growth conditions for the Lagrangian and the dynamics of the Lagrange problem allow to conclude Lipschitzian regularity of its minimizing trajectories. Additional conditions need to be imposed in order to guarantee the possibility of extending the equations of maximum principle up to the infinity point in the set of controls.

Later D.Torres [12] used a different technique of time-reparameterization to derive a criterion of Lipschitzian regularity of minimizers for the problems with nonlinear controlled dynamics (see [12, Theorem 24]).

If one sticks to the techniques invoked in [10], then it is possible to derive another condition for Lipschitzian regularity of classical (nonrelaxed) minimizers of the problem with nonlinear controlled dynamics. This result is not distant from those presented in [12, Theorem 24]). We do not provide a proof, since it comes out as a particular case of the more general result for relaxed minimizers discussed in the next Section.

The following differentiability and growth assumptions are made.

•  the functions $\ell, f$ and their partial derivatives with respect to $t, x$ are continuous with respect to $(t, x, v)$;

- For each compact set $X \subset \mathbb{R}^n$, there exists a scalar function $\phi(\xi)$, $\xi \geq 0$, monotonically increasing in $[0, +\infty[$, bounded below, such that $\lim_{\xi \to +\infty} \frac{\phi(\xi)}{\xi} = +\infty$ and

$$\ell(t, x, v) \geq \phi\left(\|f(t, x, v)\| + \|D_x f(t, x, v)\| + \|D_t f(t, x, v)\|\right) \tag{11}$$

  holds for all $(t, x, v) \in [0, T] \times X \times \mathbb{R}^r$;

- For each compact set $X \subset \mathbb{R}^n$, there are constants $C_1, C_2 < +\infty$ such that the inequality

$$\left(\|D_x \ell(t, x, v)\| + \|D_t \ell(t, x, v)\|\right) \leq C_1 \ell(t, x, v) \tag{12}$$

  holds for every $(t, x) \in [0, T] \times X$ whenever $\|v\| > C_2$.

In the conditions above $D_x, D_t$ denote the differentials with respect to $x$ and $t$, respectively.

The following result holds.

**Theorem 2.** *Let $\ell, f$ and their derivatives $D_t \ell, D_x \ell, D_t f, D_x f$ be continuous. Let growth assumptions (5), (11) and (12) hold. Then any non relaxed minimizer of problem (1)-(2)-(3) satisfies the Pontryagin maximum principle. The corresponding trajectory $\tilde{x}(\cdot)$ is Lipschitzian, unless the corresponding integrable control $\tilde{u}(\cdot)$ is a strictly abnormal extremal control.* $\square$

Being strictly abnormal extremal control for $\tilde{u}(\cdot)$ means satisfying an *abnormal version* of Pontryagin maximum principle and not satisfying any *normal version* of this principle.

Recall what are normal and abnormal versions of maximum principle like. Introduce the (pre)Hamiltonian

$$H(t, x, v, \lambda, \psi) = \lambda \ell(t, x, v) + \psi \cdot f(t, x, v), \tag{13}$$

where $\lambda$ is a nonpositive constant, $\psi \in (\mathbb{R}^n)^*$ is a covector.

We say that a pair $\tilde{x}(\cdot), \tilde{v}(\cdot)$ satisfies *Pontryagin maximum principle* for the problem (1)-(2)-(3), if there exists a nonzero pair $(\tilde{\lambda}, \tilde{\psi}(\cdot))$ such that the quadruple $\tilde{x}(\cdot), \tilde{v}(\cdot), \tilde{\lambda}, \tilde{\psi}(\cdot)$ satisfies the equations (2)-(3) and

$$\dot{\psi} = -\frac{\partial H}{\partial x}, \tag{14}$$

and satisfies the maximality condition

$$H\left(t, \tilde{x}(t), \tilde{v}(t), \tilde{\lambda}, \tilde{\psi}(t)\right) \overset{a.e}{=} \max_{v \in \mathbb{R}^r} H\left(t, \tilde{x}(t), v, \tilde{\lambda}, \tilde{\psi}(t)\right). \tag{15}$$

Note that (14) is Hamiltonian adjoint to the equation (2), which can be written as: $\dot{x} = \frac{\partial H}{\partial \psi}$.

Maximum principle is normal if $\lambda < 0$ and abnormal, if $\lambda = 0$; in the latter case the Lagrangian does not appear in (14)-(15). The respective quadruples $\left(\tilde{x}(\cdot), \tilde{v}(\cdot), \tilde{\lambda}, \tilde{\psi}(\cdot)\right)$ are called, normal or abnormal, extremals, while the respective $\tilde{v}(\cdot)$ are (normal or abnormal) extremal controls. It may happen that to an extremal control there correspond different pairs $\tilde{\lambda}, \tilde{\psi}(\cdot)$. An extremal control is *strictly abnormal* if for all such pairs $\lambda$ vanishes.

Note that if the dynamics (2) is control-affine:

$$\dot{x} = f(t, x, v) = h(t, x) + G(t, x)v(t), \tag{16}$$

with $G$ being a full column rank matrix, then one can formulate a simpler condition for Lipschitzian regularity of non relaxed minimizers.

**Corollary 2.** *Consider a Lagrange problem (1)-(2)-(3) with the control-affine dynamic equation (16). Let the functions $\ell, h, G$ and their derivatives with respect to $t, x$ be continuous. Let assumption (12) hold, and assume that for any compact $X \subset \mathbb{R}^n$ there exists a function $\phi_1(\xi)$ such that:*

$$\phi_1(\xi)/\xi \to +\infty, \text{ as } \xi \to +\infty, \quad and \quad \ell(t, x, v) \geq \phi_1(\|v\|), \text{ as } \|v\| \to +\infty,$$

*uniformly with respect to $(t, x) \in [0, T] \times X$. Then the conclusion of the Theorem 2 holds and the minimizer $\tilde{v}(\cdot)$ is essentially bounded unless it is a strictly abnormal extremal.* $\square$

*Remark 3.* The result established by the Corollary 2 is related to a general condition obtained in [10]. Interested readers can compare this criterion with the conditions (H1) of [10, Theorem 2] and the first one of the conditions listed in Remark 2 in [10]; this latter considered with $\mu = \beta - 2$, $\beta \in (1, 2)$. $\square$

# 4 Lipschitzian regularity of relaxed minimizing trajectory

## 4.1 Main result on Lipschitzian regularity

From now on we consider again the problem (1)-(2)-(3) and assume the functions $f(t, x, v), \ell(t, x, v)$ and their partial derivatives with respect to $t$ and $x$ to be continuous.

We formulate our main result - a criterion of Lipschitzian regularity for the relaxed minimizing trajectory, whose existence has been established in Section 2.

**Theorem 3.** *Consider a Lagrange optimal control problem (1)-(2)-(3). Let the functions $\ell, f$ and their partial differentials with respect to $t, x$ be continuous. Let the growth assumptions (5)-(11)-(12) hold. Then:*

*i)* *there exists a relaxed minimizer of the problem;*

*ii)* *all relaxed minimizers satisfy the Pontryagin maximum principle;*

*iii)* *any minimizing relaxed trajectory is Lipschitzian and the corresponding extremal control is essentially bounded, unless they correspond to a strictly abnormal extremal.* □

Statement *i)* follows from Theorem 1 of Section 2. The rest of the contribution contains the proof of statements *ii)* and *iii)* of Theorem 3. For an explanation of the notions of normality and abnormality see the previous Section. In Subsection 4.6 we explain how normality and Lipschitzian regularity are correlated in the relaxed case.

### 4.2 Reduction of the relaxed optimal control problem to a time-optimal control problem

Let us consider the relaxed Lagrange problem, defined by the functional (7) and the dynamics (6). We introduce the new time variable

$$\tau(t) = \int_0^t L(s, x(s), p(s), V(s))ds. \tag{17}$$

Recall that by our assumptions $\ell \geq 1$ and hence also $L \geq 1$. Therefore $\tau(t)$ is strictly increasing. Obviously $\tau(T)$ corresponds to the value of the relaxed functional $\mathcal{J}^c$ calculated along the relaxed trajectory $(x(\cdot), p(\cdot), V(\cdot))$.

Considering $t$ as a new state variable we derive for it the dynamic equation:

$$\frac{dt}{d\tau} = \frac{1}{L(t, x, p, V)}.$$

The relaxed optimal control problem can be represented as time optimal control problem

$$\tau^* \to \min, \tag{18}$$

with dynamics

$$\frac{dt}{d\tau} = \frac{1}{L(t, x, p(t(\tau)), V(t(\tau)))}, \qquad \frac{dx}{d\tau} = \frac{F(t(\tau), x, p(t(\tau)), V(t(\tau)))}{L(t(\tau), x, p(t(\tau)), V(t(\tau)))}, \tag{19}$$

and boundary conditions

$$t(0) = 0, \qquad t(\tau^*) = T, \qquad x(0) = x_0, \qquad x(\tau^*) = x_T. \tag{20}$$

We denote $(1, F)$ by $\hat{F}$ and $(1, f)$ by $\hat{f}$. Using $z$ to denote the pair $(t, x)$, the dynamics of the time-optimal control problem can be described by the equation

$$\frac{dz}{d\tau} = \frac{\hat{F}(z, p(\tau), V(\tau))}{L(z, p(\tau), V(\tau))}. \tag{21}$$

## 4.3 Loeb's compactification of the set of control parameters

According to the approach sketched at the beginning of the Section 3 our next step will be compactification of our set of control parameters $\Sigma^{n+1} \times \mathbb{R}^{r(n+2)}$.

Although $\Sigma^{n+1}$ is compact, a mere passage to the one point compactification of $\mathbb{R}^{r(n+2)}$ will not do, as long as for example the map

$$(p, V) \mapsto \frac{\hat{F}(z, p, V)}{L(z, p, V)}$$

can not be extended by continuity to such one-point compactification. Indeed if $p \to \hat{p}$, and $\hat{p}$ having some vanishing component $\tilde{p}_j$, and $\|V_j\| \to +\infty$, then the limit (for $z$ fixed) of $p_j \ell(z, V^j)$ and hence of $L(z, p, V)$ and of $\frac{\hat{F}(z,p,V)}{L(z,p,V)}$ may fail to exist.

To construct a proper extension we proceed with a more "sensitive" type of compactification following the construction suggested by P.Loeb in [9].

We can find a compact set $A \subset \mathbb{R}^{n+1}$, such that $\text{int}(A)$ contains (the points of) the optimal trajectory $\tilde{z}(\cdot)$ of (18)-(19)-(20) under consideration.

Consider the functions:

$$\frac{\hat{F}(z, p, V)}{L(z, p, V)}, \quad \frac{\partial}{\partial \hat{x}_k} \left( \frac{\hat{F}(z, p, V)}{L(z, p, V)} \right), \quad (k = 0, \ldots, n), \tag{22}$$

defined on $A \times \Sigma^{n+1} \times \mathbb{R}^{r(n+2)}$.

**Lemma 3.** *The functions (22) are continuous and uniformly bounded provided that the growth assumptions (5)-(11)-(12) hold.* $\square$

*Proof.* As long as $L \geq 1$, continuity is obvious. We prove boundedness. There holds

$$\left\| \frac{\partial}{\partial z_k} \left( \frac{\hat{F}(z, p, V)}{L(z, p, V)} \right) \right\| \leq$$

$$\leq \frac{\left\| \frac{\partial \hat{F}}{\partial x_k}(z, p, V) \right\|}{L(z, p, V)} + \frac{\left\| \hat{F}(z, p, V) \right\|}{L(z, p, V)} \frac{\left\| \frac{\partial L}{\partial z_k}(z, p, V) \right\|}{L(z, p, V)}. \tag{23}$$

For the first addend in the right-hand side of (23) we derive

$$\frac{\left\| \frac{\partial \hat{F}}{\partial z_k}(z, p, V) \right\|}{L(z, p, V)} = \frac{\left\| \sum\limits_{i=1}^{n+2} p_i \frac{\partial \hat{f}}{\partial z_k}(z, v^i) \right\|}{\sum\limits_{i=1}^{n+2} p_i \ell(z, v^i)} \leq \sum\limits_{i=1}^{n+2} \frac{p_i \left\| \frac{\partial \hat{f}}{\partial z_k}(z, v^i) \right\|}{\sum\limits_{j=1}^{n+2} p_j \ell(z, v^j)} \leq$$

$$\leq \circ \sum\limits_{i=1}^{n+2} \frac{\left\| \frac{\partial \hat{f}}{\partial z_k}(z, v^i) \right\|}{\ell(z, v^i)} \leq \sum\limits_{i=1}^{n+2} \frac{\left\| \frac{\partial \hat{f}}{\partial z_k}(z, v^i) \right\|}{\ell(z, v^i)}, \tag{24}$$

where $\odot\sum_{i=1}^{n+2}$ stands for summation with respect to those $i$'s for which $p_i \neq 0$. Analogously to (24)

$$\frac{\left\|\hat{F}(z,p,V)\right\|}{L(z,p,V)} \leq \sum_{i=1}^{n+2} \frac{\left\|\hat{f}(z,v^i)\right\|}{\ell(z,v^i)}, \quad \frac{\left|\frac{\partial L}{\partial z_k}(z,p,V)\right|}{L(z,p,V)} \leq \sum_{i=1}^{n+2} \frac{\left|\frac{\partial \ell}{\partial z_k}(z,v^i)\right|}{\ell(z,v^i)}. \quad (25)$$

By growth assumptions (11) and (12)

$$\frac{\left\|\frac{\partial \hat{f}}{\partial z_k}(z,v^i)\right\|}{\ell(z,v^i)}, \frac{\left\|\hat{f}(z,v^i)\right\|}{\ell(z,v^i)} \longrightarrow 0, \qquad \text{as } \|v^i\| \to \infty, \quad (26)$$

and $\frac{\left|\frac{\partial \ell}{\partial z_k}(z,v^i)\right|}{\ell(z,v^i)}$ is bounded by a constant. All relations are uniform with respect to $z \in [0,T] \times A$.

This provides uniform bounds for $\left\|\frac{\partial}{\partial z_k}\left(\frac{\hat{F}(z,p,V)}{L(z,p,V)}\right)\right\|$ and of $\frac{\hat{F}(z,p,V)}{L(z,p,V)}$. $\square$

Now for each $z \in A$ we will define compactification of the set $\mathcal{U} = \Sigma^{n+1} \times \mathbb{R}^{r(n+2)}$ of control parameters. This compactification $\mathcal{CU}$ is determined by the vector-function

$$\mathcal{E}_z(p,V) : (p,V) \mapsto \frac{\hat{F}(z,p,V)}{L(z,p,V)},$$

and is the minimal compactification, onto which this function is continuously extendable in such a way that it separates points of the remainder $\Delta = \mathcal{CU} \backslash \mathcal{U}$. In fact the compactification $\mathcal{CU}$ should depend on $z$ but we will show below that under our assumptions it can be parameterized in $z$-independent way.

The construction of P.Loeb, which is close to an earlier construction of unified space by G.T.Whyburn ([14]), goes as follows. For each $z \in [0,T] \times A$, the 'vector function' $\mathcal{E}_z(p,V)$ maps $\mathcal{U}$ into a cube in $\mathbb{R}^{n+1}$. The compactification $\mathcal{CU}$ is defined as $\mathcal{CU} = \mathcal{U} \bigcup \Delta$, where the remainder

$$\Delta = \bigcap\{\overline{\mathcal{E}_z(\mathcal{U} \backslash K)}| K \text{ -compact}, K \subset \mathcal{U}\} \subset \mathbb{R}^N. \quad (27)$$

Roughly speaking points of the remainder are accumulation points of $\mathcal{E}_z(p,V)$ as $\|V\| \to +\infty$.

The topology in $\mathcal{CU}$ is defined by open sets in $\mathcal{U}$ and by a system of neighborhoods $N_{U,K}(z)$ of points $z \in \Delta$, where $U$ is a neighborhood from a standard base of $\mathbb{R}^N$ and $K$ is a compact in $\mathcal{U}$:

$$N_{U,K}(z) = \left(U \bigcap \Delta\right) \bigcup \left(\mathcal{E}_z^{-1}(U) \backslash K\right).$$

The extended map $\mathcal{E}_z^e$ is defined on the remainder $\Delta$ as: $\mathcal{E}_z^e(z) = z, \forall z \in \Delta$, while obviously $\mathcal{E}_z^e = \mathcal{E}_z$ on $\mathcal{U}$. Evidently the vector-function $\mathcal{E}_z(p,V)$ can be extended by continuity onto $\mathcal{CU}$.

*Remark 4.* Another seemingly natural kind of compactification, which comes to mind is (maximal) Stone-Cech compactification onto which *any* continuous bounded function can be extended. Unfortunately this compactification lacks many properties which are important for us. In particular it is *not* sequentially compact, there are no sequences of points of $\mathcal{U}$, which converge to a point of the remainder. Besides it can not 'be modeled by a finite-dimensional space' on the contrast to the Loeb's compactification for which it will be done in the next Subsection. $\square$

By construction the compactified set of control parameters may depend on $z$, but we will show in the next subsection, that it can be parameterized in $z$-independent way.

Note that a priori the partial derivatives $\frac{\partial}{\partial \hat{x}_k}\left(\frac{\hat{F}(z,p,V)}{L(z,p,V)}\right)$ can not be extended by continuity onto $\mathcal{CU}$, they are not defined on the remainder $\Delta$. Still by Lemma 3 they are bounded on $\mathcal{U}$.

## 4.4 Parameterization of the compactification

According to (27) in order to construct the remainder one has to study 'limit points of the map $\mathcal{E}_z(p,V)$ at infinity. It suffices to consider sequences of points $(p^j, V^j)$ such that $\|V^j\| \to \infty$. Recall that $p^j = (p^j_1, \ldots, p^j_{n+2}) \in \Sigma^{n+1}$; $V^j = (v^j_1, \ldots, v^j_{n+2})$, where each $v^j_k$ belongs to $\mathbb{R}^r$.

Without lack of generality we may assume $p^j \to \tilde{p} \in \Sigma^{n+1}$ as $j \to \infty$. Subdivide the set $I = \{1, \ldots, n+2\}$, which indexes the coordinates of $p$, into two subsets $I = I^+ \bigcup I^0$ such that

$$i \in I^+ \Leftrightarrow \tilde{p}_i > 0, \qquad i \in I^0 \Leftrightarrow \tilde{p}_i = 0.$$

Assume that the limit $\lim_{j \to \infty} \frac{\hat{F}(z,p^j,V^j)}{L(z,p^j,V^j)}$ exists. To study this limit we represent the function under the limit as

$$\frac{\hat{F}(z,p^j,V^j)}{L(z,p^j,V^j)} = \frac{\sum\limits_{i=1}^{n+2} p^j_i \hat{f}(z,v^j_i)}{\sum\limits_{s=1}^{n+2} p^j_s \ell(z,v^j_s)} = \frac{\sum\limits_{i \in I^+} p^j_i \hat{f}(z,v^j_i)}{\sum\limits_{s=1}^{n+2} p^j_s \ell(z,v^j_s)} + \frac{\sum\limits_{i \in I^0} p^j_i \hat{f}(z,v^j_i)}{\sum\limits_{s=1}^{n+2} p^j_s \ell(z,v^j_s)}. \qquad (28)$$

**Lemma 4.**

$$\lim_{j \to \infty} \frac{\hat{F}(z,p^j,V^j)}{L(z,p^j,V^j)} = \lim_{j \to \infty} \frac{\sum\limits_{i \in I^+} p^j_i \hat{f}(z,v^j_i)}{\sum\limits_{i \in I^+} p^j_i \ell(z,v^j_i) + \sum\limits_{i \in I^0} p^j_i \ell(z,v^j_i)}. \quad \square \qquad (29)$$

*Proof.* Consider the sum

$$\sum_{i \in I^0} \frac{p^j_i \hat{f}(z,v^j_i)}{\sum\limits_{s=1}^{n+2} p^j_s \ell(z,v^j_s)}.$$

For each sufficiently small $\varepsilon > 0$ there exists $N_\varepsilon$ such that $p_i^j \le \varepsilon^2$ holds for every $j > N_\varepsilon, i \in I^0$. The norm of any addend of the previous sum can be estimated from above for $j > N_\varepsilon$ as:

$$\frac{p_i^j \|\hat{f}(z, v_i^j)\|}{\sum\limits_{i=1}^{n+2} p_i^j \ell(z, v_i^j)} \le \begin{cases} \varepsilon, & \text{if } \|\hat{f}(z, v_i^j)\| \le \varepsilon^{-1}; \\ \varepsilon^{-1}/\phi(\varepsilon^{-1}), & \text{if } \|\hat{f}(z, v_i^j)\| > \varepsilon^{-1}, \end{cases} \tag{30}$$

where $\phi$ is the function from the coercivity assumption (11). To derive the first estimate in (30) we took into account the lower bound $\sum\limits_{i=1}^{n+2} p_i^j \ell(z, v_i^j) \ge 1$, while the second estimate in (30) follows from the inequality

$$\frac{p_i^j \hat{f}(z, v_i^j)}{\sum\limits_{s=1}^{n+2} p_i^j \ell(z, v_s^j)} \le \frac{\hat{f}(z, v_i^j)}{\ell(z, v_i^j)} \le \varepsilon^{-1}/\phi(\varepsilon^{-1}),$$

according to the growth assumption (11).

Taking $\varepsilon \to 0$ in (30) we conclude that the second sum in (28) tends to 0.

**Lemma 5.** *If for some $i \in I^+$, the sequence $\|v_i^j\|$ is unbounded, then the limit (29) vanishes.* $\square$

*Proof.* Indeed, if we assume (passing to a subsequence when needed) $\|v_i^j\| \overset{j \to \infty}{\longrightarrow} \infty$, then for each $k \in I^+$:

$$\frac{p_k^j \|\hat{f}(z, v_k^j)\|}{\sum\limits_{i \in I^+} p_i^j \ell(z, v_i^j) + \sum\limits_{i \in I^0} p_i^j \ell(z, v_i^j)} \le \min\left\{ \frac{\|\hat{f}(z, v_k^j)\|}{p_i^j \ell(z, v_i^j)}, \frac{\|\hat{f}(z, v_k^j)\|}{\ell(z, v_k^j)}, \ i \in I^+ \right\},$$

and for sufficiently large $j$ one can continue:

$$\le \min\left\{ \frac{2\|\hat{f}(z, v_k^j)\|}{\tilde{p}_i \ell(z, v_i^j)}, \frac{\|\hat{f}(z, v_k^j)\|}{\ell(z, v_k^j)} \right\}. \tag{31}$$

By coercivity assumption (5) $\forall \varepsilon > 0$ there exists $N_\varepsilon$ such that $\forall j > N_\varepsilon$ : $\tilde{p}_i \ell(z, v_i^j) \ge \varepsilon^{-2}$. Then for $\|\hat{f}(z, v_k^j)\| \le \varepsilon^{-1}$ the first fraction in (31) is $\le 2\varepsilon$, while for $\|\hat{f}(z, v_k^j)\| \ge \varepsilon^{-1}$ the second expression in (31) is $\le \varepsilon^{-1}/\phi(\varepsilon^{-1})$. Hence the minimum of the two expressions tends to 0 as $j \to \infty$. $\square$

Now we restrict our consideration to the case where for all $i \in I^+$ the sequences $\|v_i^j\|$ are bounded. Passing to subsequences, if needed, we may think that $v_i^j$ converge to corresponding $\tilde{v}_i$.

In this case a nonzero limit (29) exists only if there exists $\lim\limits_{j \to \infty} \sum\limits_{i \in I^0} p_i^j \ell(z, v_i^j)$. Recall that $\forall i \in I^0 : \lim\limits_{j \to \infty} p_i^j = 0$.

Using the coercivity condition (5) it is elementary to prove existence of sequences $(p_i^j, v_i^j)$ such that the latter limit equals to any preassigned value $w \in [0, +\infty]$. For $w = +\infty$ the limit (29) turns 0.

Thus if the sequences $\|v_i^j\|$ are bounded then possible values of the limits of $\frac{\hat{F}(z,p,V)}{L(z,p,V)}$ form the set

$$\left\{ \frac{\sum_{i \in I^+} \tilde{p}_i \hat{f}(z, \tilde{v}_i)}{\sum_{i \in I^+} \tilde{p}_i \ell(z, \tilde{v}_i) + w}, \ w \in [0, +\infty] \right\}. \tag{32}$$

Reparameterizing $w$ as $w = \omega/(1 - \omega)$, $\omega \in [0, 1]$ we arrive to the following representation of the set (32)

$$\left\{ \frac{(1 - \omega) \sum_{i \in I^+} \tilde{p}_i \hat{f}(z, \tilde{v}_i)}{(1 - \omega) \sum_{i \in I^+} \tilde{p}_i \ell(z, \tilde{v}_i) + \omega}, \ \omega \in [0, 1] \right\}. \tag{33}$$

Therefore one concludes that $(\tilde{p}, \tilde{V}, \omega)$ parameterizes the compactified set of control parameters.

We will represent this set as a fibred space over $\Sigma^{n+1}$. The topology of fibers will depend on the number of positive components of $p_i$.

For $p \in \text{int } \Sigma^{n+1}$, with all the components being positive, the space $\mathcal{V}_p$ is the one-point compactification of $\mathbb{R}^{r(n+2)}$, i.e. the $r(n+2)$-dimensional sphere $\mathbb{S}^{r(n+2)}$ with distinguished infinite point (north pole) $\iota_p$.

If $p$ possesses $k < n+2$ positive components then the corresponding topological space $\mathcal{V}_p$ is obtained as follows. We take the one point compactification $\mathbb{S}^{kr}$ of $\mathbb{R}^{kr}$ with distinguished north pole $\iota_p^n$. Then we construct the cone $\mathbb{S}^{kr} \times [0, 1]/\mathbb{S}^{kr} \times \{1\}$ over this sphere, which is homeomorphic to the $kr + 1$-dimensional ball $B^{kr+1}$. After it we take the quotient over the subset $\iota_p \times [0, 1]$, i.e. over the points of the radius going from the center of the ball to the north pole $\iota_p$, and take it as the distinguished point $\iota_p$ of the quotient $\mathcal{V}_p$.

In particular, if $p$ is a vertex of the simplex $\Sigma^{n+1}$ with only one nonzero component, then the space $\mathcal{V}_p$ is the quotient of the ball $B^{r+1}$ over the radius going from its center to the (north pole) $\iota_p$.

Finally we glue together the points $(p, \iota_p)$ taking the quotient of the union $\bigcup_{p \in \Sigma^{n+1}} (\mathcal{V}_p, \iota_p)$ over the set $\{(p, \iota_p), \ p \in \Sigma^{n+1}\}$.

The distinguished point is now denoted by $\iota$.

### 4.5 Solutions of the relaxed Lagrangian and of the compactified time-optimal problems

From now on we deal with time-optimal control problem with a compact set $\mathcal{CU}$ of control parameters defined by (33) or (32).

It is easy to establish correspondence between the solutions of the relaxed Lagrangian problem (6)-(7)-(3) and the solutions of the problem (18)-(19)-(20) with compactified set of controls $\mathcal{CU}$.

Consider any trajectory $x(t)$ of the system (6), which: i) satisfies boundary conditions (3), ii) is driven by an admissible relaxed control $(p(t), V(t))$, and iii) provides finite value $\bar{\tau}$ to the functional (1).

Defining strictly monotonous function $\tau(t)$ by (17), we take the inverse function $t(\tau)$ and put $z(\tau) = (t(\tau), x(t(\tau)))$, $u(\tau) = (p(t(\tau)), V(t(\tau)))$. As far as $V(\cdot)$ is finite a.e., the control $u(\cdot)$ takes its values in $\mathcal{U}$ for almost all $\tau$. The trajectory $z(\tau)$ satisfies boundary conditions (20) with transfer time $\bar{\tau}$.

Vice versa let trajectory $z(\tau) = (\underline{t}(\tau), \underline{x}(\tau))$ of the system (19) satisfy (20) for a transfer time $\bar{\tau}$ and let the trajectory be driven by a control $\underline{u}(\cdot)$ with the values in $\mathcal{CU}$, such that $u(\tau) \in \mathcal{U}$ a.e. The function $\underline{t}(\tau)$ is strictly monotonous and invertible. Taking $x(t) = \underline{x}(\tau(t)))$ we obtain trajectory of (6) driven by the relaxed control $(p(t), V(t)) = \underline{u}(\tau(t))$, which is defined for almost all $t \in [0, T]$. The value of the functional (7) equals $\bar{\tau}$.

It remains to prove that any control $u(\cdot)$, which takes values in the remainder $\mathcal{CU} \setminus \mathcal{U}$ on a subset $\mathcal{T}$ of positive measure in $[0, \bar{\tau}]$, can not be optimal. Indeed according to (32) the extension of the differential equation (19) onto $\mathcal{CU}$ can be written as

$$\frac{dz}{d\tau} = \frac{\sum_{i \in I^+} p_i(\tau) \hat{f}(z, v_i(\tau))}{\sum_{i \in I^+} p_i(\tau) \ell(z, v_i(\tau)) + w(\tau)}, \quad w \in (0, +\infty]. \tag{34}$$

Here $(p(\tau), V(\tau), w(\tau))$ provides a parameterization of $u(\tau)$.

It is obvious that if one substitutes $w$ by the zero value on $\mathcal{T}$, this will result in a reparameterization of a trajectory of (34) by a strictly smaller time interval, i.e. $u(\cdot)$ can not be time-optimal.

All this means that a relaxed minimizer of the optimal control problem (1)-(2)-(3), i.e. a minimizer $(\tilde{p}(\cdot), \tilde{V}(\cdot))$ of the problem (7)-(6)-(3), corresponds to a minimizer $(\tilde{p}(\cdot), \tilde{V}(\cdot), \tilde{w}(\cdot))$ of the autonomous time-optimal control problem (18)-(19)-(20) with $\tilde{w}(\cdot) = 0$ almost everywhere.

One can change the values of $\tilde{w}(\cdot)$ on a set of zero measure without affecting the value of the minimal time. Therefore from now on we will assume $\tilde{w}(\cdot)$ to be identically vanishing or, in other words, the time-optimal control of the compactified problem (18)-(19)-(20) to take its values in $\mathcal{U}$.

## 4.6 Pontryagin maximum principle, second Erdmann condition, normality and Lipschitzian regularity

We want to write down the equations of Pontryagin's maximum principle (PMP) for the minimizer of the compactified time-optimal problem.

In order to ensure validity of the PMP for the time-minimizing control $(\tilde{p}(\cdot), \tilde{V}(\cdot), 0)$ we verify that after substitution of this control into the right-hand side of (21) the resulting function is *integrally Lipschitzian* with respect to $z$ (cf. [11], [5, Chapter 5]) in a neighborhood of the minimizing trajectory $\tilde{z}(\cdot)$. In fact the situation is 'more smooth' and 'more classical': after the substitution we end up with the function which is continuously differentiable with respect to $z$ for each fixed $\tau$, and the norm of the respective Jacobians are equibounded for all $\tau \in [0, \tau^*]$. This follows from Lemma 3.

It will be convenient for us now to reestablish old variables $t, x$ in place of $z$. The respective (pre)Hamiltonian of the Pontryagin maximum principle for the time-optimal problem (18)-(19)-(20) is

$$H(t, x, p, V, w, \hat{\lambda}) = \frac{\sum_{i=1}^{n+2} p_i(\hat{\lambda} \cdot \hat{f}(t, x, v^i))}{\sum_{i=1}^{n+2} p_i \ell(t, x, v^i) + w}.$$

Splitting the covector $\hat{\lambda}$ as $\hat{\lambda} = (\lambda_1, \lambda)$ in accordance to the splitting $\hat{f} = (1, f)$, so that $\hat{\lambda} \cdot \hat{f} = \lambda_1 + \lambda \cdot f$, we get the Hamiltonian

$$H(t, x, p, V, w, \lambda_1, \lambda) = \frac{\sum_{i=1}^{n+2} p_i(\lambda_1 + \lambda \cdot f(t, x, v^i))}{\sum_{i=1}^{n+2} p_i \ell(t, x, v^i) + w}. \qquad (35)$$

According to the PMP the absolutely continuous functions $\left(\tilde{\lambda}_1(\tau), \tilde{\lambda}(\tau)\right)$, satisfy the adjoint Hamiltonian equations

$$d\lambda_1/d\tau = -\partial H/\partial t, \qquad d\lambda/d\tau = -\partial H/\partial x, \qquad (36)$$

and the control $(\tilde{p}(\cdot), \tilde{V}(\cdot), 0)$ maximizes the value of the Hamiltonian function $(p, V, w) \mapsto H(\tilde{t}(\tau), \tilde{x}(\tau), p, V, w, \tilde{\lambda}_1(\tau), \tilde{\lambda}(\tau))$, for almost all $\tau$.

Since the time-optimal problem is autonomous - its dynamics does not depend on $\tau$ - the maximized Hamiltonian is known to be constant according to the second Erdmann's condition.

Therefore for the minimizing control $(\tilde{p}(\cdot), \tilde{V}(\cdot), 0)$ and for almost all $\tau$:

$$\frac{\sum_{i=1}^{n+2} \tilde{p}_i(\tau)\left(\tilde{\lambda}_1(\tau) + \tilde{\lambda}(\tau) \cdot f(\tilde{t}(\tau), \tilde{x}(\tau), \tilde{v}^i(\tau))\right)}{\sum_{i=1}^{n+2} \tilde{p}_i(\tau)\ell(\tilde{t}(\tau), \tilde{x}(\tau), \tilde{v}^i(\tau))} = c \geq 0. \qquad (37)$$

The proof of this condition in our case goes along the classical line. A key point is global Lipschitzian continuity of the function $\mathcal{M}(t, x, \lambda_1, \lambda)$ which results from maximization of the Hamiltonian (35) with respect to $(p, V)$. Recall again that due to our growth assumptions the extension of the dynamics (21) onto $\mathcal{CU}$ is Lipschitzian. The proof of the nullity of the derivative $\frac{d}{d\tau}\mathcal{M}\left(\tilde{t}(\tau), \tilde{x}(\tau), \tilde{\lambda}_1(\tau), \tilde{\lambda}(\tau)\right)$ proceeds as in [8],[1].

Coming back to the condition (37) we treat first the case $c > 0$. In this case the Lipschitzian property of the minimizing trajectory of the original problem can be derived from the following

**Lemma 6.** *Let (37) hold with $c > 0$. Then $\exists M_c < +\infty$ depending only on $c$ such that:*

$$\tilde{p}_i(\tau) \neq 0 \Rightarrow \|\tilde{v}^i(\tau)\| \leq M_c \qquad (38)$$

*holds for $i = 1, 2, ..., (n+2)$ and almost every $\tau \in [0, \tau^*]$. Hence the minimizing control $\tilde{v}(\cdot) = \left(\tilde{v}^1(\cdot), ..., \tilde{v}^{n+2}(\cdot)\right)$ is essentially bounded and the minimizing trajectory is Lipschitzian.* $\square$

*Proof.* Fix $\tau_0$ satisfying (37) with $c > 0$, and fix $i \in \{1, 2, ..., (n+2)\}$ such that $\tilde{p}_i(\tau_0) \neq 0$, and the maximality condition of the Pontryagin maximum principle is satisfied at this point. To simplify, we can assume without loss of generality that $i = 1$.

If $\tilde{p}_1(\tau_0) = 1$, then it is clear that $\frac{\tilde{\lambda}_1(\tau_0) + \tilde{\lambda}(\tau_0) \cdot f(\tilde{t}(\tau_0), \tilde{x}(\tau_0), \tilde{v}^1(\tau_0))}{\ell(\tilde{t}(\tau_0), \tilde{x}(\tau_0), \tilde{v}^1(\tau_0))} = c$. To consider the case when $\tilde{p}_1(\tau_0) \in ]0, 1[$, let

$$f_1 = f(\tilde{t}(\tau_0), \tilde{x}(\tau_0), \tilde{v}^1(\tau_0)), \qquad \ell_1 = \ell(\tilde{t}(\tau_0), \tilde{x}(\tau_0), \tilde{v}^1(\tau_0))$$

$$f_2 = \sum_{i=2}^{n+2} \frac{\tilde{p}_i(\tau_0)}{\sum_{j=2}^{n+2} \tilde{p}_j(\tau_0)} f(\tilde{t}(\tau_0), \tilde{x}(\tau_0), \tilde{v}^i(\tau_0)),$$

$$\ell_2 = \sum_{i=2}^{n+2} \frac{\tilde{p}_i(\tau_0)}{\sum_{j=2}^{n+2} \tilde{p}_j(\tau_0)} \ell(\tilde{t}(\tau_0), \tilde{x}(\tau_0), \tilde{v}^i(\tau_0))$$

The Hamiltonian (37) at time $\tau = \tau_0$ is

$$H(\tilde{t}(\tau_0), \tilde{x}(\tau_0), \tilde{p}(\tau_0), \tilde{V}(\tau_0), \tilde{\lambda}_1(\tau_0), \tilde{\lambda}(\tau_0)) =$$
$$= \frac{\tilde{p}_1(\tau_0)(\tilde{\lambda}_1(\tau_0) + \tilde{\lambda}(\tau_0) \cdot f_1) + (1 - \tilde{p}_1(\tau_0))(\tilde{\lambda}_1(\tau_0) + \tilde{\lambda}(\tau_0) \cdot f_2)}{\tilde{p}_1(\tau_0)\ell_1 + (1 - \tilde{p}_1(\tau_0))\ell_2}.$$

Hence the maximum principle implies that

$$\frac{\tilde{p}_1(\tau_0)(\tilde{\lambda}_1(\tau_0) + \tilde{\lambda}(\tau_0) \cdot f_1) + (1 - \tilde{p}_1(\tau_0))(\tilde{\lambda}_1(\tau_0) + \tilde{\lambda}(\tau_0) \cdot f_2)}{\tilde{p}_1(\tau_0)\ell_1 + (1 - \tilde{p}_1(\tau_0))\ell_2} =$$
$$= \max_{p_1 \in [0,1]} \frac{p_1(\tilde{\lambda}_1(\tau_0) + \tilde{\lambda}(\tau_0) \cdot f_1) + (1 - p_1)(\tilde{\lambda}_1(\tau_0) + \tilde{\lambda}(\tau_0) \cdot f_2)}{p_1\ell_1 + (1 - p_1)\ell_2}. \quad (39)$$

Since

$$\frac{\partial}{\partial p_1} \left( \frac{p_1(\tilde{\lambda}_1(\tau_0) + \tilde{\lambda}(\tau_0) \cdot f_1) + (1 - p_1)(\tilde{\lambda}_1(\tau_0) + \tilde{\lambda}(\tau_0) \cdot f_2)}{p_1\ell_1 + (1 - p_1)\ell_2} \right) =$$
$$= \frac{(\tilde{\lambda}_1(\tau_0) + \tilde{\lambda}(\tau_0) \cdot f_1)\ell_2 - (\tilde{\lambda}_1(\tau_0) + \tilde{\lambda}(\tau_0) \cdot f_2)\ell_1}{(p_1\ell_1 + (1 - p_1)\ell_2)^2},$$

condition (39) implies that

$$\frac{\tilde{\lambda}_1(\tau_0) + \tilde{\lambda}(\tau_0) \cdot f_1}{\ell_1} = \frac{\tilde{\lambda}_1(\tau_0) + \tilde{\lambda}(\tau_0) \cdot f_2}{\ell_2} = c.$$

Thus we proved that the equality

$$\frac{\tilde{\lambda}_1(\tau_0) + \tilde{\lambda}(\tau_0) \cdot f(\tilde{t}(\tau_0), \tilde{x}(\tau_0), \tilde{v}^i(\tau_0))}{\ell(\tilde{t}(\tau_0), \tilde{x}(\tau_0), \tilde{v}^i(\tau_0))} = c$$

must hold for every $i$ such that $\tilde{p}_i(\tau_0) \neq 0$. Hence

$$\frac{|\tilde{\lambda}_1(\tau_0)|}{\ell(\tilde{t}(\tau_0), \tilde{x}(\tau_0), \tilde{v}^i(\tau_0))} + \|\tilde{\lambda}(\tau_0)\| \frac{\|f(\tilde{t}(\tau_0), \tilde{x}(\tau_0), \tilde{v}^i(\tau_0))\|}{\ell(\tilde{t}(\tau_0), \tilde{x}(\tau_0), \tilde{v}^i(\tau_0))} \geq c$$

must hold for every $i$ such that $\tilde{p}_i(\tau_0) \neq 0$. Since the trajectory $(\tilde{\lambda}_1, \tilde{\lambda}, \tilde{t}, \tilde{x})$ lies in a compact, the growth assumptions (5),(11) guarantee that $\tilde{v}^i(\tau_0)$ is bounded by a constant $M_c$ which depends only on $c > 0$.

What remains is to clarify what happens when $c$ in (37) vanishes. This gives the relationship between normality and Lipschitzian regularity.

**Lemma 7.** *If $c = 0$ in (37) then $\left(\tilde{p}(\cdot), \tilde{V}(\cdot)\right)$ is an abnormal extremal relaxed control for the original problem, i.e. $\left(\tilde{p}(\cdot), \tilde{V}(\cdot)\right)$ satisfies the abnormal version of Pontryagin maximum principle for this problem.* □

*Proof.* Recall that the denominator of (37) equals $L + w$ and is positive. Therefore, vanishing of $c$ in (37) together with the maximality condition for the Hamiltonian (35) implies that for almost all $\tau$ and all $(p, V)$, we have:

$$0 = \tilde{\lambda}_1(\tau) + \sum_{i=1}^{n+2} \tilde{p}_i(\tau)(\tilde{\lambda}(\tau) \cdot f(\tilde{t}(\tau), \tilde{x}(\tau), \tilde{v}^i(\tau))) \geq \qquad (40)$$

$$\geq \tilde{\lambda}_1(\tau) + \sum_{i=1}^{n+2} p_i(\tilde{\lambda}(\tau) \cdot f(\tilde{t}(\tau), \tilde{x}(\tau), v^i)).$$

Introduce the Hamiltonian

$$h(t, x, p, V, \lambda) = \sum_{i=1}^{n+2} p_i(\lambda \cdot f(t, x, v^i)). \qquad (41)$$

The second equation in (36) takes the form

$$\frac{d\lambda}{d\tau} = -\frac{\partial H}{\partial x} = -\frac{\partial}{\partial x} \frac{\lambda_1 + h}{L + w} = \frac{-\frac{\partial h}{\partial x}}{L + w} + \frac{(\lambda_1 + h)\frac{\partial L}{\partial x}}{(L + w)^2}.$$

Given the fact that $\lambda_1 + h(t, x, p, V, \lambda)$ vanishes along $\left(\tilde{x}(\cdot), \tilde{p}(\cdot), \tilde{V}(\cdot), \tilde{\lambda}_1(\cdot), \lambda(\cdot)\right)$ and also $w$ vanishes, the latter equation takes the form

$$\frac{d\lambda}{\frac{1}{L}d\tau} = -\frac{\partial h}{\partial x}, \quad \text{or, given (17),} \quad \frac{d\lambda}{dt} = -\frac{\partial h}{\partial x}.$$

Besides by (40), (41)

$$h(t(\tau), \tilde{x}(\tau), \tilde{p}(\tau), \tilde{V}(\tau), \tilde{\lambda}(\tau)) \overset{a.e.\tau}{=} -\tilde{\lambda}_1(\tau) \geq h(t(\tau), \tilde{x}(\tau), p, V, \tilde{\lambda}(\tau)).$$

The two latter equations mean that $\left(\tilde{x}(\tau), \tilde{p}(\tau), \tilde{V}(\tau)\right)$ satisfy the abnormal version of Pontryagin maximum principle with the multiplier $\tilde{\lambda}(\cdot)$ and the (abnormal) Hamiltonian (41).

# References

1. Agrachev AA, Sachkov YuL (2004) Control Theory from the Geometric Viewpoint. Springer
2. Alexeev VM, Tikhomirov VM, Fomin SV (1987) Optimal Control. Consultants Bureau, N.Y.
3. Cellina A, Colombo G (1990) On a classical problem of the clalculus of variations without convexity assumptions. Annales Inst. Henri Poincare, Analyse Non Lineaire, 7:97–106.
4. Cesari L (1983) Optimization-theory and applications. Problems with ordinary differential equations. Springer-Verlag, New York
5. Clarke FH (1983) Optimization and Nonsmooth Analysis. Wiley–Interscience Publication
6. Clarke FH (1989) Methods of dynamic and nonsmooth optimization. CBMS-NSF Regional Conference Series in Applied Mathematics, 57. Society for Industrial and Applied Mathematics (SIAM), Philadelphia, PA
7. Clarke FH, Vinter R (1990) Regularity properties of optimal controls. SIAM J. Control Optim., 28(4):980–997.
8. Gamkrelidze RV (1978) Foundations of Control Theory. Plenum Press
9. Loeb P (1967) A Minimal Compactification for Extending Continuous Functions. Proc. Amer. Math. Soc., 18:282–283.
10. Sarychev A, Torres DFM (2000) Lipschitzian regularity, Applied Mathematics and Optimization, 41:237–254.
11. Sussmann HJ (2007) Set separation, approximating multicones, and the Lipschitz maximum principle. J. Differential Equations, 243:446–488.
12. Torres DFM (2003) Lipschitzian Regularity of the Minimizing Trajectories for Nonlinear Optimal Control Problems. Mathematics of Control, Signals and Systems, 16:158–174.
13. Vinter R (2000) Optimal Control. Birkhäuser, Boston
14. Whyburn GT (1953) A Unified Space for Mappings. Trans. Amer. Mathem. Society, 74:344–350.

# Pricing of Defaultable Securities under Stochastic Interest

Nino Kordzakhia[1] and Alexander Novikov[2]

[1] Macquarie University, NSW 2109, Australia. `nino.kordzakhia@mq.edu.au`.
[2] University of Technology, Sydney, NSW 2007, Australia.
`alex.novikov@uts.edu.au`.

**Summary.** We reduce a problem of pricing *continuously monitored* defaultable securities (barrier options, corporate debts) in a stochastic interest rate framework to calculations of boundary crossing probabilities (BCP) for Brownian Motion (BM) with stochastic boundaries. In the case when the interest rate is governed by a linear stochastic equation (Vasicek model) we suggest a numerical algorithm for calculation of BCP based on a piece-wise linear approximation for the stochastic boundaries. We also find an estimation of the rate of convergence of the suggested approximation and illustrate results by numerical examples.

## 1 Introduction

Practitioners often acknowledge an existence of a common problem with pricing schemes of exotic (e.g. barrier, lookback) options for the contracts with long maturities. The prices of these instruments significantly depend on an interest rate term-structure. The case of deterministic interest rates have been studied by Roberts/Shortland (1997), ([16]), Novikov/Frishling/Kordzakhia (1999, 2003), ([12], [13], see also other references in these papers). For deterministic interest rates the pricing problem of barrier options is reduced to calculations of BCP for BM with deterministic boundaries. In Section 2 we present a modification of the algorithm from [12] and [13] to handle a general setup of stochastic interest rates. In Sections 3 and 4 we describe the numerical algorithm in details and provide an estimation for the rate of convergence of the suggested approximation as a function of number of nodes. In Section 5 the results are illustrated by numerical examples.

Further we use the standard notation $S_t$ for a price of an underlying asset and $r_t$ for a default-free short interest rate. We assume that $S_t$ is a diffusion process of the form

$$S_t = S_0 e^{Y_t}$$

with the log-return process $Y_t$ governed by the equation

$$dY_t = \mu(t)dt + \sigma dW_t,$$

where $W_t$ is a standard BM with respect to a 'real-world' measure $P$ and a filtration $\mathcal{F}_t$, $\sigma$ is a constant volatility, $\mu(t)$ is a historical trend.

Let $f_T$ be a payoff of an option at maturity $T$. In this paper we concentrate on two fundamental examples of payoff functions (there are many other examples that may be treated in a similar way).

**Case 1.** The payoff of *Up-and-Out European Barrier Call* option is

$$f_T = (S_T - K)^+ I\{\tau > T\},$$

where $T$ is a maturity time, a first passage (default) time $\tau = \inf\{t : S_t \geq G_+(t)\}$, $G_+(t)$ is a continuous *deterministic* barrier, and $I\{\tau > T\}$ is an indicator function.

**Case 2.** The payoff at maturity $T$ of a *defaultable zero-coupon bond* $S_t$ is

$$f_T = 1 - wI\{\tau \leq T\}, \ 0 < w = const < 1,$$

where $\tau = \inf\{t : Y_t \leq b_t\}$, the default threshold $b_t$ is modeled as follows

$$b_t = b_0 + \lambda \int_0^t (Y_s - v - b_s)ds,$$

with some constants $b_0$, $v$ and $\lambda > 0$. An economic rationale behind the model is well justified by Collin-Dufresne/Goldstein (2001), ([7]) who accounted to the fact that firms tend to decrease the time-dependent debt level $b_t$ when the return on a firm's value, in this case $Y_t$ falls below $b_t + v$, and vice-versa. Put for convenience

$$l_t = Y_t - b_t.$$

Then

$$\tau = \inf\{t : l_t \leq 0\}, \ dl_t = \mu(t)dt + \lambda(v - l_t)dt + \sigma dW_t. \tag{1}$$

The case of a constant threshold $b_t = b_0$ (that is when $\lambda = 0$ in (1)) has been studied by Longstaff/Schwartz (1997), ([8]). In [8] and [7] BCP were evaluated using the first-passage density of a two-factor Gaussian-Markov process. This methodology has been extended by Bernard et al (2005),(2007), ([2], [3]), to pricing of life insurance contracts and barrier options.

It must be noted that the problem of approximating of BCP for BM with one-sided stochastic boundary has been considered by Peskir and Shiryaev (1997) [14], Vondraĉek (2000) [17], Abundo (2003) [1] under asymptotic setting (as $T \to \infty$ in our notation), but their results can not be used under our framework because the parameter $T$ takes, typically, moderate values.

## 2 Pricing formulae

A fair price of an option with a payoff $f_T$ is (under a free-arbitrage assumption)

$$C_f = E^* \left( e^{-\int_0^T r_s ds} f_T \right), \tag{2}$$

where $E^*$ is a symbol of expectation with respect to the equivalent risk-neutral (martingale) measure $P^*$ such that the process

$$e^{-\int_0^t r_s ds} S_t$$

is $(P^*, \mathcal{F}_t)-$martingale. It implies (see e.g. Liptser/Shiryaev (2003), [9]) that

$$Y_t = \sigma W_t^* + \int_0^t r_s ds - \frac{\sigma^2}{2} t, \tag{3}$$

where $W_t^*$ is a standard BM with respect $P^*$.

In general, the fair price (2) can be approximated by Monte Carlo methods via discretization of a time parameter. However, such approximations typically have a significant bias; this aspect is discussed in Section 5 for the case of *discretely monitored* options.

The price of a risk-free zero-coupon bond with maturity $T$ is

$$P(0,T) = E^*(e^{-\int_0^T r_s ds}).$$

Define the *forward measure* $P^F$ as follows:

$$P^F(A) = E^*(I(A)e^{-\int_0^T r_s ds})/P(0,T), \ A \in \mathcal{F}_T.$$

Then

$$C_f = E^* \left( e^{-\int_0^T r_s ds} f_T \right) = P(0,T)E^F(f_T), \tag{4}$$

where $E^F$ is a symbol of expectation with respect to $P^F$.

For affine models (e.g. Vasicek, CIR models) the function $P(0,T)$ can be calculated analytically. Therefore, a pricing problem is reduced to evaluation of $E^F(f_T)$. Further we assume that the risk-free short rate $r_t$ is the Ornstein-Uhlenbeck (or, Vasicek) process governed by a linear stochastic equation

$$dr_t = a_r(\bar{r} - r_t)dt + \sigma_r dW_t^{(r)}, \tag{5}$$

where $a_r > 0$, $W_t^{(r)}$ is another standard BM with respect to the risk-neutral measure $P^*$ and given filtration $F_t$ such that $E^*(W_t^{(r)} W_t^*) = 0$. A general case of correlated $S_t$ and $r_t$ will be discussed elsewhere.

The solution of equation (5) has the following representation

$$r_t = a(t) + \sigma_r \xi_t, \ a(t) = \bar{r} + (r_0 - \bar{r})e^{-a_r t},$$

where

$$d\xi_t = -a_r \xi_t dt + dW_t^{(r)}, \ \xi_0 = 0$$

and, therefore,

$$\xi_t = e^{-a_r t} \int_0^t e^{a_r s} dW_s^{(r)}. \tag{6}$$

By direct calculations

$$P(0,T) = E^*(e^{-\int_0^T r_s ds}) = e^{-A(T)+D^2(T)/2}$$

with

$$A(T) = \int_0^T a(s)ds = \bar{r}T + (r_0 - \bar{r})(1 - e^{-a_r T})/a_r,$$

$$D(T) = \sigma_r^2 Var(\int_0^T \xi_s ds) = \frac{\sigma_r^2}{2a_r^3}\left(2Ta_r - e^{-2Ta_r} - 4a_r^2 + 4a_r^2 e^{-Ta_r} + 1\right).$$

Also, by the Girsanov theorem one could show that under the forward measure $P^F$ the BM $W_t^{(r)}$ has a drift term

$$q(t,T) = -\frac{\sigma_r}{a_r}\int_0^t (1 - e^{-a_r(T-u)})du \tag{7}$$

(see e.g. Brigo/Mercurio (2006), ([6], p. 886)).

**Case 1.** Using (3) we obtain

$$\tau = \inf\{t : S_t > G_+(t)\} = \inf\{t : \sigma W_t^* > \frac{\sigma^2}{2}t + \log(G_+(t)/S_0) - \int_0^t r_s ds\}.$$

The fair price $C_f$ can be written as follows

$$C_f = E^*[e^{-\int_0^T r_s ds}(S_T - K)I\{\tau > T, S_T > K\}].$$

In view of (4) and (7) we have

$$= S_0 E^*\left(e^{\sigma W_T^* - \frac{\sigma^2}{2}T}I\{\tau > T, S_T > K\}\right) - KP(0,T)P^F\{\tau > T, S_T > K\} =$$

$$S_0 q_1 - P(0,T)Kq_2 \tag{8}$$

with

$$q_1 = \tilde{P}\left\{\sigma W_t^* < \frac{\sigma^2}{2}t + \log(G_+(t)/S_0) - \int_0^t r_s ds, \sigma W_T^* \right.$$
$$\left. > \frac{\sigma^2}{2}T + \int_0^T r_s ds + \log(K/S_0)\right\},$$

where the measure $\tilde{P}$ is defined as follows

$$\tilde{P}(A) = E^*(I\{A\}e^{\sigma W_T^* - \frac{\sigma^2}{2}T}) , \ A \in \mathcal{F}_T.$$

Note that the BM $W_t^*$ has a drift $\sigma t$ under the measure $\tilde{P}$.
The second probability in (8) has a similar form:

$$q_2 = P^F \left\{ \sigma W_t^* < \frac{\sigma^2}{2} t + \log(G_+(t)/S_0) - \int_0^t r_s ds, \sigma W_T^* \right.$$

$$\left. > \frac{\sigma^2}{2} T + \int_0^T r_s ds + \log(K/S_0) \right\},$$

(recall that under measure $P^F$ the BM $W_t^r$ has the drift $q(t,T)$ defined in (7)).

**Case 2.** We have

$$C_f = E^* e^{-\int_0^T r_s ds} (1 - w + w I\{\tau > T\}) = P(0,T)[1 - w + P^F\{l_t \geq 0, t \leq T\}],$$

where due to (3) and (1) we have

$$dl_t = r_t dt + \lambda(v - l_t)dt + \sigma dW_t^*, l_0 = -b_0.$$

With a simple algebra one can show that the problem of computation of $P^F\{l_t \geq 0, t \leq T\}$ can be reduced to finding of probabilities

$$P\{\eta_t < g(t) - \sigma_r e^{-\lambda t} \int_0^t e^{\lambda s} \xi_s ds, \ t \leq T\},$$

where $g(t)$ is a deterministic function, $\xi_s$ is defined in (6) and $\eta_t$ is the Ornstein-Uhlenbeck process defined by the equation

$$d\eta_t = -\lambda \eta_t dt + \sigma dW_t, \eta_0 = 0$$

and therefore

$$\eta_t = \sigma e^{-\lambda t} \int_0^t e^{\lambda s} dW_s.$$

Applying the change of time

$$u = \int_0^t e^{2\lambda s} ds$$

we obtain

$$\int_0^{t(u)} e^{\lambda s} dW_s = \tilde{W}_u, \ t(u) = \frac{\log(2\lambda u + 1)}{2\lambda},$$

where $\tilde{W}_u$ is another standard BM. Hence, we can reduce the pricing problem to computation of the probability

$$P\{\sigma \tilde{W}_u < \sqrt{2\lambda u + 1}\, g(t(u)) - \sigma_r \int_0^{t(u)} e^{\lambda s} \xi_s ds, \ u \leq \int_0^T e^{2\lambda s} ds\}.$$

## 3 Approximations to BCP

For **Cases** 1 and 2 discussed above we have reduced the option pricing problem to finding BCP of the form

$$P(b, h) := P\{W_T > b; W_t < h(t), t \le T\},$$

where: $b$ is a random variable, $h(t)$ is a stochastic boundary such that

$$h(t) = g(t) - zR_t, R_t = \int_0^t \xi_s ds,$$

$g(t)$ is a smooth deterministic function, $z = const$ and the process $\xi_t$ is defined in (6) with a standard BM $W_t$.

Further we study in details **Case 1**. In this case the parameter $\lambda = 0$ and the probabilities $q_i$ defined in (8) can be written in the form $q_i = P(b_i, h_i), i = 1, 2$ where

$$\sigma b_i = \sigma_r \int_0^T \xi_s ds + H_i, \ \sigma h_i(t) = g_i(t) - \sigma_r R_t^{(i)},$$

$$g_1(t) = -\frac{\sigma^2}{2} t + \log(G_+(t)/S_0) - \int_0^t a(s) ds,$$

$$H_1 = -\frac{\sigma^2}{2} T + \int_0^T a(s) ds + \log(K/S_0)\}, \ R_t = \int_0^t \xi_s ds,$$

the process $\xi_t$ has the representation (6) with a standard BM $W_t^{(r)}$. Furthermore,

$$g_2(t) = \frac{\sigma^2}{2} t + \log(G_+(t)/S_0) - \int_0^t a(s) ds,$$

$$H_2 = \frac{\sigma^2}{2} T + \int_0^T a(s) ds + \log(K/S_0)\}.$$

We suggest to use approximating probabilities $P(\hat{b}, \hat{h})$ with piece-wise linear stochastic boundaries $\hat{h}(t)$ instead of $h(t) = g(t) - zR_t$ such that for some partition $\{t_i\}, 0 = t_0 < t_1 < ... < t_m = T,$

$$\hat{h}(t_j) = h(t_j) = g(t_j) - zR_{t_j}, j = 0, ..., m. \tag{9}$$

Further we assume that $z = 1$.

Note that the vector $\mathbf{R} = (R_{t_1}, ..., R_{t_m})$ and the process $W_t^*$ are independent due to the assumption on independency of $S_t$ and $r_t$.

The approximating probabilities $P(\hat{b}, \hat{h})$ can be calculated using the following formula

$$P(\hat{b}, \hat{h}) = E[I\{W_T > \hat{b}\} \prod_{j=0}^{m-1} (1 - e^{-\frac{2(\hat{h}(t_j) - W_{t_j})^+ (\hat{h}(t_{j+1}) - W_{t_{j+1}})^+}{t_{j+1} - t_j}})]. \tag{10}$$

This formula was derived by Wang/Pötzelberger (1997), ([18]), for the special case when $b = -\infty$ and boundaries $\hat{h}(t)$ are deterministic piece-wise linear functions. In a more general context, including double barriers, although restricted to deterministic boundaries, a similar result was derived in [12]. To prove formula (10) for stochastic boundaries one could follow the proof in [12] or, alternatively, Proposition 1 formulated below.

Let us consider

$$\eta_j(u) := W_{t_j+u} - W_{t_j} - u\frac{W_{t_j+1} - W_{t_j}}{\Delta t_j}, \ 0 \leq u \leq \Delta t_j = t_{j+1} - t_j, \ j = 0, ..., m-1$$

and

$$\mathbf{W} = \{W_{t_1}, ..., W_{t_m}\}.$$

**Proposition 1.** *For any $u$ and $v$*

$$Cov(\eta_i(u), \eta_j(v)|\mathbf{W}) = 0, i \neq j,$$

$$Cov(\eta_i(u), \eta_i(v)|\mathbf{W}) = \min(u, v) - u\,v/\Delta t_i.$$

The proof of this result is based on standard properties of Gaussian random variables.

As a simple consequence of Proposition 1 one can see that the random processes $\{\eta_j(u), j = 0, ..., m - 1\}$ are jointly independent Brownian bridges conditioned on the vector $\mathbf{W}$. We recall also another well-known fact that for a linear nonrandom function $h(t)$

$$P\{W_t < h(t), t_i \leq t \leq t_{i+1}|(W_{t_i}, W_{t_{i+1}})\} = 1 - e^{-\frac{2(h(t_i)-W_{t_i})^+(h(t_{i+1})-W_{t_{i+1}})^+}{t_{i+1}-t_i}}.$$

Now using conditioning on the vector $(\mathbf{W}, \mathbf{R})$ with $\mathbf{R} = (R_{t_1}, ..., R_{t_m})$ one can derive (10) for the general case under discussion.

Note that to compute probabilities (10) we need to simulate the random variables $W_{t_j}$ and $R_{t_j}, j = 1, ..., m$ or, alternatively, one may attempt to use $2m$ repeated integrations with respect to state variables.

## 4 Accuracy of the approximation

Our estimation of accuracy of the suggested approximation will be based on the following result proved in Borovkov/Novikov (2005), ([5]).

**Notation.** $Lip(\kappa)$ is the class of Lipschitz functions $h(t)$ on $[0, T]$ :

$$|h(t + h) - h(t)| \leq \kappa h, 0 = t < t + h = T,$$

where $\kappa$ is a finite nonrandom constant.

**Proposition 2 (Borovkov-Novikov (2005)).** *Let stochastic boundaries $h(t) \in Lip(\kappa)$. Then for any $\varepsilon > 0$*

$$P(b, h + \varepsilon) - P(b, h) \leq (5\kappa/2 + 1/\sqrt{T})\varepsilon.$$

This result is proved in Lemma 1 of [5]. The latter was formulated for the case of nonstochastic boundaries but one can easily check that the result does hold for any stochastic boundaries $h(t)$ from the class $\in Lip(\kappa)$.

Further we assume: $T = 1$, the process $\xi_t$ is defined in (6) with a standard BM $W_t^{(r)}$,

$$h(t) = g(t) - \int_0^t \xi_s ds,$$

$g(t)$ is a twice continuously differentiable deterministic function, $b$ is a constant.

Let $\hat{g}(t)$ be a piece-wise linear function such that $\hat{g}(t_j) = g(t_j)$, $R_t$ be a piece-wise linear process and the stochastic boundary $\hat{h}$ be defined in (9).

**Theorem 1.** *Let $t_j = \frac{j}{m}$. Then*

$$|P(b, \hat{h}) - P(b, h)| = O(\frac{\log(m)}{m^{3/2}}).$$

For the proof we will need the following

**Proposition 3.** *Let $x_m = C\sqrt{\log(m)}$ with a large enough constant $C$. Then*

$$P\{\max_{t \leq T} |\int_0^t \xi_s ds - R_t| > \frac{x_m}{m^{3/2}}\} = o(\frac{1}{m^{3/2}}).$$

**Proof of Theorem 1.** Since $g(t)$ is a twice continuously differentiable function, for the linear piece-wise function $\hat{g}_t$ such that $\hat{g}(t_j) = g(t_j)$ we have[3]

$$|\hat{g}(t) - g(t)| \leq \frac{C}{m^2}.$$

By virtue of Proposition 3 proved below with $x_m = C\sqrt{\log(m)}$ we have

$$P(b, \hat{h} - \frac{x_m}{m^{3/2}} - \frac{C}{m^2}) - o(\frac{1}{m^{3/2}}) \leq P(b, g) \leq P(b, \hat{h} + \frac{x_m}{m^{3/2}} + \frac{C}{m^2}) + o(\frac{1}{m^{3/2}}).$$

Now applying Proposition 2, for any $\kappa > 0$ we obtain

$$|P(b, g) - P(b, \hat{h})| \leq (5\kappa/2 + 1)(\frac{2x_m}{m^{3/2}} + \frac{C}{m^2}) + P\{\max_{s \leq 1} |\xi_s| > \kappa\} + o(\frac{1}{m^{3/2}}).$$
$$(11)$$

Furthermore, using properties of the Gaussian distribution one can show that for some $\lambda > 0$

$$P(\max_{s \leq 1} |\xi_s| > \kappa) \leq Ce^{-\lambda\kappa^2}/\kappa.$$

Choosing $\kappa = \kappa(m) = \sqrt{\frac{2\log(m)}{\lambda}}$ we obtain

---
[3] $C$ is a generic constant

$$P\{\max_{s\leq 1}|\xi_s| > \kappa(m)\} = O(\frac{1}{m^{3/2}}).$$

Combining the latter with (11) the proof of Theorem 1 is completed.
**Proof of Proposition 3.** The definition of $R_t$ implies that

$$\max_{t\leq 1}|\int_0^t \xi_s ds - R_t| = \max_j \int_{t_{j-1}}^{t_j}|\xi_s - m\int_{t_{j-1}}^{t_j}\xi_u du|ds \leq$$

$$\frac{2}{m}\max_j(\max_{t_{j-1}\leq s\leq t_j}|\xi_s - \xi_{t_{j-1}}|).$$

Applying the Girsanov transformation, for any $x > 0$ we have

$$P\{\max_j(\max_{t_{j-1}\leq s\leq t_j}|\xi_s - \xi_{t_{j-1}}|) > x\} =$$

$$E[I\{\max_j(\max_{t_{j-1}\leq s\leq t_j}|W_s - W_{t_{j-1}}|) > x\}e^{-a_r\int_0^1 W_s dW_s - \frac{a_r^2}{2}\int_0^1 W_s^2 ds}]$$

$$\leq e^{a_r/2}P\{\max_j(\max_{t_{j-1}\leq s\leq t_j}|W_s - W_{t_{j-1}}|) > x\}, (as \int_0^1 W_s dW_s = (W_1^2 - 1)/2).$$

The random variables $\max_{t_{j-1}\leq s\leq t_j}|W_s - W_{t_{j-1}}|$, $j = 1,...,m$, are independent and identically distributed (iid). Then by scaling and homogeneity properties of BM we have

$$\max_{t_{j-1}\leq s\leq t_j}|W_s - W_{t_{j-1}}| \overset{d}{=} \max_{0\leq s\leq 1}|W_s|/\sqrt{m},$$

Note that

$$\max_{s\leq 1}|W_s| \leq \max_{s\leq 1}(W_s) + \max_{s\leq 1}(-W_s)$$

and

$$\max_{0\leq s\leq 1}(-W_s) \overset{d}{=} \max_{0\leq s\leq 1}(W_s) \overset{d}{=} |W_1|.$$

It implies that, for any $x > 0$

$$P\{\max_j(\max_{t_{j-1}\leq s\leq t_j}|W_s - W_{t_{j-1}}|) > x\} \leq 2P\{\max_j \gamma_j > x/2\},$$

where iid r.v.'s $\gamma_j \overset{d}{=} |W_1|$ and, hence,

$$P\{\max_{t\leq 1}|\int_0^t \xi_s ds - R_t| > \frac{x}{m^{3/2}}\} \leq 2e^{a_r/2}P\{\max_j(\gamma_j) > x/2\} =$$

$$= 2e^{a_r/2}(1 - (2\Phi(x/2) - 1)^m) \leq 2e^{a_r/2}(1 - (1 - \frac{2e^{-\frac{x^2}{8}}}{x\sqrt{2\pi}})^m),$$

here we use the well-known inequality $1 - \Phi(x) \leq \frac{e^{-\frac{x^2}{2}}}{x\sqrt{2\pi}}$ for the standard normal distribution $\Phi(x)$.

For $x = x_m \to \infty$

$$\left(1 - \frac{4e^{-\frac{x_m^2}{8}}}{x_m\sqrt{2\pi}}\right)^m = e^{-\frac{4me^{-\frac{x_m^2}{8}}}{x_m\sqrt{2\pi}}}(1+o(1))$$

We select now $x_m = C\sqrt{\log(m)}$ with a large enough constant $C$. Then $\frac{4me^{-\frac{x_m^2}{2}}}{x_m\sqrt{2\pi}} = o(\frac{1}{m^{3/2}})$, thus

$$1 - e^{-\frac{2e^{-\frac{x_m^2}{2}}}{x_m\sqrt{2\pi}}m(1+o(1))} = o\left(\frac{1}{m^{3/2}}\right).$$

This completes the proof of Proposition 3.

# 5 Numerical examples

Here we consider model (5) with the parameters

$$T = 1, a_r = 1, \bar{r} = 0.1, r_0 = 0.15.$$

For the asset price $S_t$ we use the same set of parameters as in [16], namely,

$$S_0 = 10, \ \sigma = 0.1, \ K = 11, G_+(t) = 12.$$

In terms of our notation from Section 2

$$r_t = a(t) + \sigma_r \xi_t, E(r_t) = a(t) = 0.1 + 0.05e^{-t}, \int_0^t a(s)ds = 0.1t + 0.05(1 - e^{-t}),$$

$$d\xi_t = -\xi_t dt + \sigma_r dW_t^{(r)}, \xi_0 = 0,$$

where $W_t^{(r)}$ is a standard BM.

## 5.1 Deterministic interest rate

Roberts/Shortland (1990) have set $\sigma_r = 0$ in ([16]) and by using probabilistic arguments they obtained the following bounds for the fair price of continuously monitored *Up-and-In European call* option

$$0.516758 \leq C_f \leq 0.517968. \tag{12}$$

Denote by $C_f^d(m)$ the fair prices of the discretely monitored options with $m$ monitoring dates and let $\hat{C}_f^d(m)$ be a Monte Carlo estimation for $C_f^d(m)$. With the number of paths $N = 10^7$ we obtained the following results

$$\hat{C}_f^d(1000) = 0.5125,$$

$$\hat{C}_f^d(100) = 0.5036,$$

$$\hat{C}_f^d(50) = 0.4985. \tag{13}$$

The significant difference of 0.8%, even for $m = 1000$, compared to the bounds for the price of continuously monitored option from (12), can be explained by a slow rate of convergence of prices of the discretely monitored option to continuously monitored one. The rate of convergence is known to be of order $O(\frac{1}{\sqrt{m}})$, (it can be derived from results of Nagaev (1970), [11] and Borovkov (1982), [4]).

Denote by $\hat{C}_f(m)$ a Monte Carlo estimation for the fair prices of the *continuously monitored* options obtained through the numerical procedure described in Section 3; here $m$ is a number of node points for a piece-wise linear boundary. With the number of paths $N = 10^7$ we obtained the following results

$$\hat{C}_f(1000) = 0.5167,$$

$$\hat{C}_f(100) = 0.5167,$$

$$\hat{C}_f(50) = 0.5168. \tag{14}$$

A disparity with the results for discretely monitored options is due to a better rate of convergence which, according to Theorem 1, is at least of order $O(\frac{\log(m)}{m^{3/2}})$. As a matter of fact, for smooth deterministic boundaries the rate of convergence is even faster, namely, it is of order $O(\frac{1}{m^2})$ as shown in [15], [5].

In practice, the sequence $C_f^d(m)$, $m = 1, 2, ...$, is also of interest. Based on the known rate of convergence $C_f^d(m)$ to the limit as $m \to \infty$, we suggest to use the following simple approximation $\tilde{C}_f^d(m)$ defined by the following formula

$$\tilde{C}_f^d(m) = \hat{C}_f(50) + \frac{A}{\sqrt{m}}, \tag{15}$$

where $A$ is a constant; further we define the constant $A$ from the equation

$$\hat{C}_f^d(50) = \hat{C}_f(50) + \frac{A}{\sqrt{50}}.$$

The following table contains the simulated prices $\hat{C}_f^d(m)$ and the relative error of approximation (15) in the range $m \in [100, 250]$.

**Table 1. Prices of discretely monitored option, $\sigma_r = 0$,**
$$\hat{C}_f(50) = 0.51680$$

| m | Price $\hat{C}_f^d(m)$ | Relative Error(%) |
|---|---|---|
| 100 | 0.5036 | 0.06 |
| 150 | 0.5058 | 0.09 |
| 200 | 0.5073 | 0.08 |
| 250 | 0.5083 | 0.07 |

## 5.2 Stochastic interest rate

For a comparison, we included results of simulation of prices $\hat{C}_f^d(m)$ and the relative error of approximation (15) when $\sigma_r = 0.2$. Note that the relative errors of approximating formula (15) in the range $[10, 250]$ are consistently less than 0.1%.

**Table 2. Prices of discretely monitored option,**
$$\sigma_r = 0.2, \ \hat{C}_f(50) = 0.62524$$

| m | Price $\hat{C}_f^d(m)$ | Relative Error(%) |
|---|---|---|
| 100 | 0.6148 | 0.05 |
| 150 | 0.6172 | -0.04 |
| 200 | 0.6183 | -0.03 |
| 250 | 0.6191 | -0.03 |

## Acknowledgement

The authors are very grateful to Dr. Thorsten Schmidt for his valuable comments, which improved the exposition of the paper.

# References

1. Abundo M (2003) On the first-passage time of a diffusion process over a one-sided stochastic boundary. Stochastic Anal. Appl. 21(1):1–23. Erratum: ibid, 21(4):953–954.
2. Bernard CL, Le Courtois OA, Quittard-Pinon FM (2005)Market Value of Life Insurance Contracts under Stochastic Interest Rates and Default Risk. Insurance: Mathematics and Economics, 36(3):499-516.
3. Bernard CL, Le Courtois OA, Quittard-Pinon FM (2007) Pricing Derivatives with Barriers in a Stochastic Interest Rate Enviroment. Preprint.
4. Borovkov KA (1982) Rate of Convergence in a Boundary Problem. Theory. Prob. Appl., 27:148–149.
5. Borovkov K, Novikov A (2005) Approximations of boundary crossing probabilities for a Brownian motion. J. Appl. Prob, 1:82-92.
6. Brigo D, Mercurio F (2006) Interest Rate Models - Theory and Practice. Second Ed. Springer, Berlin
7. Collin-Dufrusne P, Goldstein R (2001) Do Credit Spreads Reflect Stationary Leverage Rations? J. Finance 56:1929-1957.
8. Longstaff F, Schwartz E (1995) A simple approach to valuing fixed and floating rate debt. J. Finance 50:789–819.
9. Liptser R, Shiryaev A (2003) Statistics of Random Processes. Springer
10. Merton RC (1974) On the pricing of corporate debt: The risk structure of interest rates. J. Finance 29:449–470.
11. Nagaev SV (1970) The rate of convergence in a certain boundary value problem, part I. Theory. Prob. Appl., 15:179-199; part II, ibid, 15:419–44.

12. Novikov A, Frishling V, Kordzakhia N (1999) Approximations of boundary crossing probabilities for a Brownian motion. J. Appl. Prob. 36(4):1019–1030.
13. Novikov A, Frishling V, Kordzakhia N (2003) Time-dependent barrier options and boundary crossing probabilities. Georgian Math. Journal, 10:325-334.
14. Peskir G, Shiryaev A (1997) On the Brownian first passage time over a one-sided stochastic boundary. TVP, 42(3):591–602 (English translation: Theory Probab. Appl., 42(3):444–453).
15. Pötzelberger K, Wang L (2001) Boundary Crossing Probability for Brownian Motion. J. Appl. Prob., 38(1):152–164.
16. Roberts GO, Shortland CF (1997) Pricing Barrier Options with Time-Dependent Coefficients. Mathematical Finance, 7(1):83–93.
17. Vondraĉek Z. (2000) Asymptotics of first-passage time over a one-sided stochastic boundary. J. Theoret. Probab. 13(1):279–309.
18. Wang L, Pötzelberger K (1997) Boundary Crossing Probability for Brownian Motion and General Boundaries. Journal Applied Probability, 34(1):54–65.

# Spline Cubatures for Expectations of Diffusion Processes and Optimal Stopping in Higher Dimensions (with Computational Finance in View)

Andrew Lyasoff

Mathematical Finance Program, Boston University, USA. alyasoff@bu.edu

**Summary.** We develop certain cubature (quadrature) rules for expectations of diffusion processes in $\mathbb{R}^N$ that are analogous to the well known spline interpolation quadratures for ordinary integrals. By incorporating such rules in appropriate backward induction procedures, we develop new numerical algorithms for solving free-boundary (optimal stopping) problems, or ordinary fixed-boundary problems. The algorithms developed in the paper are directly applicable to pricing contingent claims of both American and European types on multiple underlying assets.

## 1 Preliminaries

The present paper is concerned with certain computational aspects of the optimal stopping of a generic diffusion process $(X_t \in \mathcal{D})_{t \in [0,\infty[}$, $\mathcal{D} \subseteq \mathbb{R}^N$, with a given termination payoff rule $\lambda \colon \mathcal{D} \mapsto \mathbb{R}$ and termination deadline $0 < T < \infty$. The study of such problems, from both analytical and numerical points of view, has a long history and has been crucial for many domains of science, engineering and, of course, control theory – see [12] for one of the most recent expositions on the subject of optimal stopping.

A concrete example of an optimal stopping problem is the exercise of the rights guaranteed by a financial contract known as an *American-style put option.* Such a contract allows its holder to sell fixed number of shares of a particular stock at some pre-specified price $K$ (exercise price) but no later than the – also pre-specified – termination date $T$. At time $t < T$ the holder of such a contract observes the price $X_t$ of the stock for which the option was underwritten and must decide whether to exercise the option, i.e., sell shares at price $K$ or take no action (if the holder of the option does not own shares of the underlying stock he/she may purchase such shares at price $X_t$ and sell them immediately at price $K$ – the right to do so is guaranteed by the contract). The payoff from exercising (and also terminating) this option is $\lambda(X_t)$ and the termination payoff function $\lambda(\cdot)$ is given by

$\lambda(x) = \text{Max}[K - x, 0]$, $x \in \mathbb{R}_{++}$. At the same time financial contracts of this type are tradable: if it has not been exercised at or prior to time $t < T$, the option can be sold to an agent who would like to purchase the right (without any obligation) to sell shares of the underlying stock at price $K$ some time in the future, prior to the termination date $T$. The problem is *at what price should such an option be sold?* Intuitively, it is clear that the market price of the option must depend on the moment in time $t < T$ and also on the observed price $x = X_t$. Consequently, to "price the option" means to construct a family of valuation maps $F(t, \cdot)$, $t \leqslant T$, that are defined on *the entire range of possible prices*, so that when the observed price at time $t$ is $x = X_t$ then at time $t$ the option is priced at the amount $F(t, x)$. It is very important to recognize that the entire family of pricing maps $x \longrightarrow F(t, x)$, $t \leqslant T$, is to be calculated "ahead of time," i.e., before the contract was underwritten. The precise form of the maps $F(t, \cdot)$ must be extracted from information about the dynamics of the price process $(X_t)$ and also from the principles of pricing by arbitrage (under certain assumptions about the nature of the financial market). Fig. 1 below shows the graphs (the solid lines) of 3 different valuation maps for 3 different values of $t \leqslant T$ against the graph (the dashed line) of the termination payoff function $\lambda(\cdot)$, assuming that the price process $(X_t)$ satisfies certain conditions. These graphs are entirely consistent with basic intuition: the option price is always greater than the termination payoff and when the pricing date $t$ approaches the termination date $T$ the associated pricing map at date $t$ converges to the termination payoff. More importantly, the knowledge of the family of pricing maps $F(t, \cdot)$, $t \leqslant T$ yields a precise stopping rule: at time $t < T$ the option is exercised only if the observed stock price $x$ is in the range where $\lambda(x) = F(t, x)$ (the option is more valuable than the immediate termination only when immediate exercise is *not* optimal).

It should be clear from the above example that – as general as possible – methods for optimal stopping are rather important in the realm of finance. In fact, the importance of such methods goes well beyond the domain of simple stock options, of the type that we just described. Indeed, in one form or another, most asset-valuation and investment issues come down to computing the price of some contract that allows its holder to exercise an optimal stopping policy with respect to some pre-specified (generally, multi-dimensional and stochastic) price process and to collect certain random payoff(s) according to certain pre-specified termination rules. What distinguishes the optimal stopping problems encountered in the realm of finance from the optimal stopping problems in most other areas is that, unlike the impression that the above example may create, most financial phenomena involve diffusion processes in spaces of a much higher dimension, well beyond the point where higher dimensionality can be treated as a more or less straight-forward generalization of analogous one-dimensional situations. Several new methods were developed recently, apparently, with high-dimensional finance in view – see [1], [2], [7], [8], [9] and [11], for example. Conceptually, these new methods belong to the general rubric "approximation of expectations of diffusion processes". Even

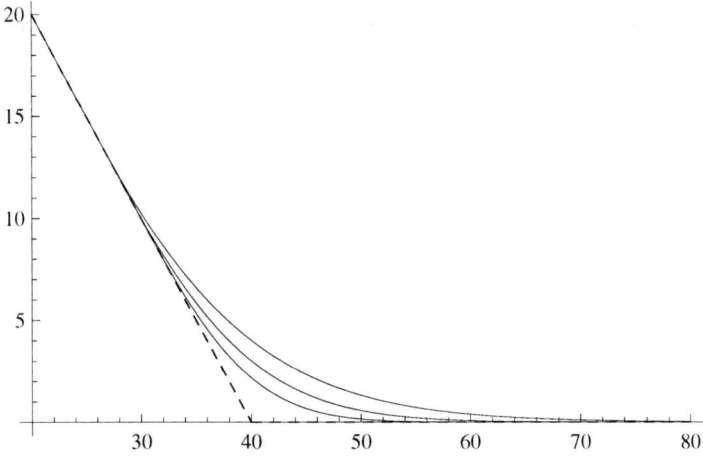

**Fig. 1.** The price of an American style put option with exercise price $K = 40$, shown as a function of the observed stock price at 3 different dates prior to the expiration date $T$, assuming that the stock price $(X_t)$ is governed by $\mathrm{d}X_t = \sigma X_t \mathrm{d}W_t + \mu X_t \mathrm{d}t$, for some fixed constants $\sigma$ and $\mu$ and some standard Brownian motion $(W_t)$. The dashed line is the graph of the termination payoff function $\lambda(\cdot)$. The area where the termination payoff coincides with the valuation map represents the range of stock prices for which exercising the option is optimal (on the respective date).

the classical Monte Carlo methods had to be re-tooled and improved considerably before they can be used in the context of computational finance – see [4], [5] and [14], for example.

The method developed in the present paper has two aspects. The first one may be viewed as a yet another variation on the theme "approximation of expectations of diffusion processes". In this regard, the method is conceptually similar to the methods of Longstaff and Schwartz [10], the quantization algorithm of Bally and Pagès [1] and [2], and the method of Lyons and Victoir [11]. What all these methods have in common is the idea that expectations of certain functionals of diffusion processes can be expressed in terms of "universal" quantities that can be computed, in some sense, once and for all. In the world of computing this idea is not new. Indeed, the most common meaning of the term "solution" is some algebraic expression that involves standard functions like $\exp(x)$, $\log(x)$, $\sin(x)$ and so on, which are nothing but objects that have been computed once and for all. It is to be noted that the methods of Bally and Pagès and Longstaff and Schwartz still rely on Monte Carlo techniques, while the method described in the present paper and the method of Lyons and Victoir can be viewed as fully deterministic methods (in the same sense in which the finite difference method is understood to be a fully deterministic method).

The second aspect of the method developed in the paper is the use of certain approximation techniques for expectations of diffusion processes in the context of optimal stopping; more specifically, for solving certain fixed- and free-boundary problems. In this respect, the method is similar – conceptually – to the method of Longstaff and Schwartz. It is important to recognize that, generally, the optimal stopping problems impose severe restrictions on the type of approximation technique for the expected values that can be used throughout the procedure. There are two principle reasons why such restrictions are encountered. The first one is that in the context of optimal stopping one must compute – dynamically – a whole cluster of expected values, which is practical only if such expected values can be computed reasonably fast. The second one is that – just in principle – one may be able to compute only a finite number of expected values from which the value function has to be restored and further integrated. In other words, merely computing expected values of functions of the underlying diffusion process is not enough, in that one must be able to compute – and do this reasonably fast – expected values of special objects that arise as approximations (constructed from finite lists of tabulated values) of the value function.

In its most general formulation, the solution to the problem for optimal stopping of the diffusion process $(X_t \in \mathcal{D})_{t \in [0,\infty[}$ with termination payoff $\lambda : \mathcal{D} \mapsto \mathbb{R}$ and termination deadline $0 < T < \infty$, is given by the *value function*

$$]-\infty, T] \times \mathcal{D} \ni (t, x) \longrightarrow F(t, x) \in \mathbb{R},$$

which represents the "value" of having the underlying process in state $x \in \mathcal{D}$ at time $t < T$, assuming that the optimal termination policy is exercised, with the understanding that if termination can occur only at date $T$, then there is only one "optimal" policy. Clearly, if stopping can occur at any time prior to the termination deadline $T > 0$, then for any $t < T$ one has $F(t, x) \geqslant \lambda(x)$ and for $t = T$ one has $F(T, x) = \lambda(x)$. In this case stopping is justified at time $t < T$ if and only if

$$X_t \in \{x \in \mathcal{D} \,;\, F(t, x) = \lambda(x)\} \iff F(t, X_t) = \lambda(X_t),$$

i.e., once the value function $F(\cdot, \cdot)$ has been computed, the optimal stopping policy comes down to: terminate the process and collect the termination payoff at the first moment $t$ when the process enters the set $\{x \in \mathcal{D}; F(t, x) = \lambda(x)\}$. Unfortunately, there are very few situations where the value function $(t, x) \longrightarrow F(t, x)$ can be computed explicitly. This is especially true in higher dimensions. As a result, one is often forced to resort to one numerical algorithm or another. It is important to recognize that such algorithms are very different in nature from the general Monte Carlo methods. Indeed, just in principle, the objective of the various Monte Carlo methods is to approximate the distribution law of the random variables $F(t, X_t)$ for some range of values of the time-parameter $t$, without providing any information whatsoever about the valuation map $F(\cdot, \cdot)$ itself. Such a task is considerably less ambitious (and useful) than approximating the actual valuation map $F(\cdot, \cdot)$, from which one can extract information about the greeks and so on. This explains why financial engineers still prefer the finite difference technique whenever this technique is possible to use. Thus, in terms of its principle objective and philosophy, the methodology developed in the present paper is very different in nature from the general Monte Carlo methods and can be viewed as a variation of a sort of the finite difference method.

There are three levels of approximation that one must deal with when computing the valuation map$(t, x) \longrightarrow F(t, x)$. Level-1 approximation is simply discretization of the time-parameter, i.e., given some fixed $n \gg$, one must replace the value function $F(\cdot, \cdot)$ with a function of the form

$$]-\infty, T] \times \mathcal{D} \ni (t, x) \longrightarrow F(T - 2^{-n}\lceil (T - t)2^n \rceil, x),$$

or, which amounts to the same, with the sequence of functions

$$\mathcal{D} \ni x \longrightarrow F_{i,n}(x) := F(T - i\, 2^{-n}, x), \quad i = 0, 1, \dots ,$$

that are calculated recursively for $i = 0, 1, \dots$. Since, in general, the functions $F_{i,n}(\cdot)$ are to be treated as infinite dimensional objects, in any practical situation one must use some compression algorithm that replaces these objects with finite lists of computable symbols. Thus, instead of constructing sequentially the functions $F_{0,n}(\cdot) := \lambda(\cdot)$, $F_{1,n}(\cdot)$, $F_{2,n}(\cdot)$, $\dots$ one must construct (sequentially) the finite lists of symbols $\ell_1$, $\ell_2$, $\dots$ every one of which can be treated as an encryption of some computable function on the domain $\mathcal{D}$. This is the level-2 approximation. Level-3 approximation occurs in the recursive rule for constructing the list $\ell_i$ from the list $\ell_{i-1}$. This calculation inevitably involves conditional expected values of certain functions of the underlying diffusion and therefore – in one form or another – requires information about the associated transition probability density. Since closed form expressions – i.e., algebraic expressions involving only standard functions – for that density may exist only in some very special cases, the density must be "compressed" into a finite list of computable symbols, too.

The key concept on which the present paper rests is to interpret each of the finite lists $\ell_i$ as a representation of some interpolating function $S_i(\cdot)$ defined on the domain $\mathcal{D}$. In general, interpolating functions are constructed from finite lists of tabulated values (for the function and, depending on the order of interpolation, certain derivatives) on some finite mesh in the respective domain. We will think of each $\ell_i$ as being one such list of tabulated values. Suppose that all future values are discounted at some – fixed, from now on – discount rate $r > 0$. Since the interpolating function $S_{i-1}(\cdot)$ is a replacement for the value function $F_{i-1}(\cdot)$, the list $\ell_i$ will be computed by tabulating the function

$$\mathcal{D} \ni x \longrightarrow \mathrm{Max}\big[e^{-r2^{-n}}\mathbb{E}\lfloor S_{i-1}(X_{2^{-n}})|X_0 = x], \lambda(x)\big]$$

and, possibly, some of its derivatives, at some mesh inside the set $\mathcal{D}$ that is chosen accordingly. If stopping can occur only at the termination date $T$, then one must tabulate the following function and, possibly, some of its derivatives

$$\mathcal{D} \ni x \longrightarrow e^{-r2^{-n}} \mathbb{E}[S_{i-1}(X_{2^{-n}}) \mid X_0 = x].$$

The reason why such a procedure happens to be practical is that, whatever the order of interpolation, one can always write

$$S_i(x) = \sum_{j=1}^{k} \ell_i(j) \times e_j(x),$$

where $e_j(x)$ are universal interpolating functions, i.e., interpolating functions that depend only on the mesh; in other words, any interpolating function may be viewed as a linear combination (with weights given by the actual tabulated values) of certain universal interpolating functions (note that, in general, $\ell_i(j)$ may refer to the tabulated value of some mixed derivative). As a result, we can write

$$\mathbb{E}[S_i(X_{2^{-n}}) | X_0 = x] = \sum_{j=1}^{k} \ell_i(j) \times \mathbb{E}[e_j(X_{2^{-n}}) \mid X_0 = x].$$

The key point here is that the functions $x \longrightarrow \mathbb{E}[e_j(X_{2^{-n}}) \mid X_0 = x]$ can be computed once and for all; in fact, one would only need the (finitely many) tabulated values of these functions and, possibly, certain mixed derivatives over the interpolation mesh. Consequently, the list $\ell_i$ can be obtained by acting on the list $\ell_{i-1}$ with some linear operator, i.e., by multiplying $\ell_{i-1}$ with a matrix that is computed once and for all – as long a the process $(X_t)$ and the interpolation mesh remain the same. Of course, due to the Level-3 approximation, if no explicit formula for the distribution of the process $(X_t)$ exists, just in principle, one may be able to compute the functions $x \longrightarrow \mathbb{E}[e_j(X_{2^{-n}}) \mid X_0 = x]$, and therefore the linear operator itself, only approximately.

## 2 Spline cubatures for expectations of diffusion processes

Following [15], we use the term "cubatures" as a reference to integration rules for multiple integrals and the term "quadratures" as a reference to integration rules for single integrals. Let $(X_t \in \mathcal{D})_{t \in [0,\infty[}$, $\mathcal{D} \subseteq \mathbb{R}^N$, be the diffusion process introduced in the previous section. From now on we will suppose that $\mathcal{D}$ is some Cartesian domain in $\mathbb{R}^N$ – such as the orthant $\mathbb{R}_+^N$, or the entire $\mathbb{R}^N$, for example. Consider, next, some Cartesian mesh $(\alpha) \equiv \{\alpha_j \equiv \{\alpha_{1,j_1}, \ldots, \alpha_{N,j_N}\} ; j \in \mathbb{J}\}$ inside the domain $\mathcal{D}$, where $j$ stands for a generic multi-index of the form $\{j_1, \ldots, j_N\}$, $\mathbb{J}$ is the (rectangular) collection of all such multi-indices with $0 \leqslant j_i \leqslant k(i)$, for some fixed list of strictly positive integers $\{k(1), \ldots, k(N)\}$ and we have

$$\alpha_{i,0} < \alpha_{i,1} < \alpha_{i,2} < \ldots < \alpha_{i,k(i)} , 1 \leqslant i \leqslant N.$$

Let $\mathbb{J}_{++} \subset \mathbb{J}$ denote the collection of all multi-indices $\{j_1, \ldots, j_N\}$ with $1 \leqslant j_i \leqslant k(i)$. We can identify the elements of $\mathbb{J}_{++}$ with the individual rectangular regions (or "cells") associated with the partition $\alpha$: for each $\jmath \in \mathbb{J}_{++}$, the vector $\alpha_\jmath \in \mathbb{R}^N$ is the "top-right-up-..." corner of one individual cell in the partition. This individual region (or "cell") can be expressed as

$$R_\jmath \equiv R_{j_1, \ldots, j_N} = \left\{x \in \mathbb{R}^N; \alpha_{1, j_1 - 1} \leqslant x_1 \leqslant \alpha_{1, j_1}, \ldots, \alpha_{N, j_N - 1} \leqslant x_N \leqslant \alpha_{N, j_N}\right\}.$$

With each $\jmath \in \mathbb{J}_{++}$ (i.e., with each cell in the partition) we associate some $N$-linear function $p_\jmath(\cdot)$ and some $N$-cubic function $P_\jmath(\cdot)$, both defined on the unit cube

$$U = \{x \in \mathbb{R}^N; 0 \leqslant x_1 \leqslant 1, \ldots, 0 \leqslant x_N \leqslant 1\},$$

respectively, by

$$p_\jmath(x) = \xi_\jmath \cdot \{1, x_N\} \cdot \ldots \cdot \{1, x_1\},$$

and by

$$P_\jmath(x) = \Xi_\jmath \cdot \{1, x_N, x_N^2, x_N^3\} \cdot \ldots \cdot \{1, x_1, x_1^2, x_1^3\},$$

for some real (Cartesian) tensor $\xi_\jmath$ of rank$(\xi_\jmath) = N$ and dim$(\xi_\jmath) = \{2, \ldots, 2\}$ and some real (Cartesian) tensor $\Xi_\jmath$ of rank$(\Xi_\jmath) = N$ and dim$(\Xi_\jmath) = \{4, \ldots, 4\}$. In the above expressions "$\cdot$" stands for the usual dot-product between tensors, i.e., the contraction of the last index of the tensor on the left of the symbol "$\cdot$" with the first index of the tensor on the right of the symbol "$\cdot$", with the understanding that for rank 1 tensors (i.e., for vectors) the last index is also the first and that the operation is performed left to right. On each individual cell $R_\jmath$, $\jmath \in \mathbb{J}_{++}$, we define the $N$-linear function

$$s_\jmath(x) := p_\jmath\left(\frac{x_1 - \alpha_{1, j_1 - 1}}{\alpha_{1, j_1} - \alpha_{1, j_1 - 1}}, \ldots, \frac{x_N - \alpha_{N, j_N - 1}}{\alpha_{N, j_N} - \alpha_{N, j_N - 1}}\right), x \in R_\jmath,$$

and the $N$-cubic function

$$S_\jmath(x) := P_\jmath\left(\frac{x_1 - \alpha_{1, j_1 - 1}}{\alpha_{1, j_1} - \alpha_{1, j_1 - 1}}, \ldots, \frac{x_N - \alpha_{N, j_N - 1}}{\alpha_{N, j_N} - \alpha_{N, j_N - 1}}\right), x \in R_\jmath,$$

(for simplicity we write $x \equiv \{x_1, \ldots, x_N\} \in \mathbb{R}^N$, $y \equiv \{y_1, \ldots, y_N\} \in \mathbb{R}^N$, and so on). Note that $s_\jmath(\cdot)$ – or, which amounts to the same, the tensor $\xi_\jmath$ – is uniquely determined by the values that $s_\jmath(\cdot)$ takes at the corners of the region $R_\jmath$. In contrast, in order to determine the $N$-cubic function $S_\jmath(\cdot)$, for each corner of the region $R_\jmath$ one must obtain information not only about the value of $S_\jmath(\cdot)$, but also about the values of all mixed partial derivatives (note that there are $2^N - 1$ such derivatives, and that $R_\jmath$ has $2^N$ corners)

$$(\partial_{i_1} \ldots \partial_{i_l} S_\jmath)(x) \equiv \frac{\partial^{i_1 + \ldots + i_l}}{\partial x_{i_1} \ldots \partial x_{i_l}} S_\jmath(x), \quad 1 \leqslant i_1 < \ldots < i_l \leqslant N, \quad 1 \leqslant l \leqslant N.$$

For example, in the case $N = 2$, in order to construct the bi-cubic function $S_\jmath(\cdot)$ one must prescribe the values of the following 3 derivatives at each of the four corners of the rectangle $R_\jmath$:

$$\frac{\partial}{\partial x_1} S_\jmath(x), \ \frac{\partial}{\partial x_2} S_\jmath(x), \ \frac{\partial^2}{\partial x_1 \partial x_2} S_\jmath(x) \,,$$

in addition to the actual value of $S_\jmath(\cdot)$ at the four corners – this gives the total of 16 conditions that uniquely determine all 16 entries in the $4 \times 4$ matrix $\Xi_\jmath$. It is very important to recognize that the entries in the tensor $\xi_\jmath$ are universal linear functions of the values that $s_\jmath(\cdot)$ takes at the corners of $R_\jmath$. Similarly, all entries in the tensor $\Xi_\jmath$ are universal linear functions of the values of $S_\jmath(\cdot)$ and the values of

$$h_\jmath(i_1) \ldots h_\jmath(i_l) \ (\partial_{i_1} \ldots \partial_{i_l} S_\jmath)(\cdot) \,,$$

at the corners of the region $R_\jmath$, where $h_\jmath(i)$ is the length of the projection of $R_\jmath$ on the $i^{\text{th}}$ coordinate axis. In particular, if $K(R_\jmath)$ stands for the collection of all corner points of the region $R_\jmath$, then there are universal constants $c_N^{(1)}$ and $c_N^{(3)}$ for which we can write for any $x \in R_\jmath$ (these are very crude estimates!)

$$|s_\jmath(x)| \leqslant c_N^{(1)} \times \text{Max}\,[|s_\jmath(y)| \ ; \ y \in K(R_\jmath)]$$

and

$$|S_\jmath(x)| \leqslant c_N^{(3)} \times \text{Max}\Big[h_\jmath(i_1) \ldots h_\jmath(i_l) \,|(\partial_{i_1} \ldots \partial_{i_l} S_\jmath)(y)| \ ; $$
$$y \in K\,(R_\jmath)\,, 1 \leqslant i_1 < i_2 < \ldots < i_l \leqslant N, \ 0 \leqslant l \leqslant N\Big] \,.$$

Now consider the region

$$\mathcal{R} = \big\{x \in \mathbb{R}^N; \alpha_{1,0} \leqslant x_1 \leqslant \alpha_{1,k(1)}, \ldots, \alpha_{N,0} \leqslant x_N \leqslant \alpha_{N,k(N)}\big\} \subseteq \mathcal{D} \,,$$

which is simply the union of the regions $R_\jmath$, and let $f \colon \mathcal{D} \mapsto \mathbb{R}$ be any function. We will suppose that $f$ can be differentiated as many times as required by the context. Clearly, the $N$-linear functions $s_\jmath(\cdot)$, $\jmath \in \mathbb{J}_{++}$, can be defined so that the values of $s_\jmath(\cdot)$ at all corners of $R_\jmath$ coincide with the respective values of the function $f(\cdot)$. With this choice, the functions $s_\jmath(\cdot)$ are automatically glued along the edges of the regions $R_\jmath$ in such a way that the aggregate function $x \longrightarrow s(f, \alpha; x)$, i.e., the function defined on $\mathcal{D}$ so that $s(f, \alpha; x) := s_\jmath(x)$ for $x \in R_\jmath$ and $s(f, \alpha; x) = f(x)$ for $x \in \mathcal{D} \backslash \mathcal{R}$, is continuous on $\mathcal{R}$ (but, obviously, not on $\mathcal{D}$). Similarly, one can define the N-cubic functions $S_\jmath(\cdot)$, $\jmath \in \mathbb{J}_{++}$, in such a way that the values of all derivatives $(\partial_{i_1} \ldots \partial_{i_l} S_\jmath)(\cdot)$ at all corners of $R_\jmath$ coincide with the respective mixed derivatives $(\partial_{i_1} \ldots \partial_{i_l} f)(\cdot)$ (note that by "mixed derivatives" we mean derivatives in which the differentiation in each variable occurs at most once, so that all first order derivatives and the function itself are just special cases of "mixed derivatives"). With this choice, the functions $S_\jmath(\cdot)$ are glued along the edges of the regions $R_\jmath$ in such a way that the aggregate function $x \longrightarrow S(f, \alpha; x)$, i.e., the function defined on $\mathcal{D}$ by $S(f, \alpha; x) := S_\jmath(x)$ for $x \in R_\jmath$ and by $S(f, \alpha; x) = f(x)$ for $x \in \mathcal{D} \setminus \mathcal{R}$, is not only continuous, but also has continuous mixed derivatives of all orders everywhere in $\mathcal{R}$ (but obviously not in $\mathcal{D}$), i.e., represents a spline-like surface on $\mathcal{R}$. In fact, if $\chi_{\mathcal{D}\backslash\mathcal{R}}$ stands for the indicator of the set $\mathcal{D}\backslash\mathcal{R}$ then for any $x \in \mathcal{D}$ we can write

$$S(f, \alpha; x) = \sum_{\jmath \in \mathbb{J}} \sum_{l=0}^{N} \sum_{1 \leqslant i_1 < i_2 < \ldots < i_l \leqslant N} (\partial_{i_1} \ldots \partial_{i_l} f)(\alpha_\jmath) \times \mathcal{E}(\alpha_\jmath, i_1, i_2, \ldots, i_l; x)$$

$$+ f(x)\chi_{\mathcal{D}\backslash\mathcal{R}}(x),$$

$$\text{and} \qquad s(f, \alpha; x) = \sum_{\jmath \in \mathbb{J}} f(\alpha_\jmath) \times e(\alpha_\jmath; x) + f(x)\chi_{\mathcal{D}\backslash\mathcal{R}}(x),$$

where $\mathcal{D} \ni x \longrightarrow \mathcal{E}(\alpha_\jmath, i_1, i_2, \ldots, i_l; x)$ and $\mathcal{D} \ni x \longrightarrow e(\alpha_\jmath; x)$ are universal interpolating functions that depend on the mesh $\alpha$ and nothing else and, furthermore, share the following properties (this is the only place in the paper where the symbol $\delta$ is used to denote Dirac's delta function):

$$\left(\partial_{m_1} \ldots \partial_{m_q} \mathcal{E}\right)(\alpha_\jmath, i_1, i_2, \ldots, i_l; \alpha_\imath) = \delta_{\jmath,\imath}\delta_{l,q}\delta_{m_1,i_1} \ldots \delta_{m_l,i_l}, \quad e(\alpha_\jmath; \alpha_\imath) = \delta_{\jmath,\imath}$$

with $\mathcal{E}(\alpha_\jmath, i_1, i_2, \ldots, i_l; x) = 0$, $x \in \mathcal{D}\backslash\mathcal{R}$, and $e(\alpha_\jmath; x) = 0$, $x \in \mathcal{D}\backslash\mathcal{R}$. In fact, the last two identities hold for all $x \in \mathcal{R}_\imath$, for any region $\mathcal{R}_\imath$ for which $\alpha_\jmath \notin \mathcal{R}_\imath$. Consequently, for every multi-index $\jmath \in \mathbb{J}$, both functions $\mathcal{E}(\alpha_\jmath, i_1, i_2, \ldots, i_l; \cdot)$ and $e(\alpha_\jmath; \cdot)$ are supported on the set

$$\mathcal{Q}(\alpha_\jmath) := \bigcup_{\imath \in \mathbb{J}, \alpha_\jmath \in \mathcal{R}_\imath} \mathcal{R}_\imath.$$

Given any multi-index $\jmath \in \mathbb{J}_{++}$ we will write

$$\epsilon_\jmath^{(1)}(f) := \max\left\{|f(x) - s(f, \alpha; x)| \, ; x \in R_\jmath\right\},$$

$$\epsilon_\jmath^{(3)}(f) := \max\left\{|f(x) - S(f, \alpha; x)| \, ; x \in R_\jmath\right\},$$

and set

$$\epsilon^{(1)}(f, \alpha) := \max_{\jmath \in \mathbb{J}_{++}} \epsilon_\jmath^{(1)}(f), \qquad \epsilon^{(3)}(f, \alpha) := \max_{\jmath \in \mathbb{J}_{++}} \epsilon_\jmath^{(3)}(f).$$

As is well known the size of the error terms $\epsilon_\jmath^{(1)}(f)$ and $\epsilon_\jmath^{(3)}(f)$ is controlled by the size of the region $R_\jmath$, defined as

$$|R_\jmath| := \max_{1 \leqslant i \leqslant N} h_\jmath(i),$$

(recall that $h_\jmath(i)$ is the length of the projection of $R_\jmath$ on the $i^{\text{th}}$ coordinate axis) and by some upper bound on certain derivatives of the function $f$ – more precisely, for the error term $\epsilon_\jmath^{(1)}(f)$ we need an upper bound on certain derivatives of $f$ of order $\leqslant 2$, and for the error term $\epsilon_\jmath^{(3)}(f)$ we need an upper bound on certain derivatives of $f$ of order $\leqslant 4$. These are very powerful estimates – especially in the context in which we intend to use them. Just as an example, with $N = 1$ we have

$$\epsilon_\jmath^{(1)}(f) \leqslant \frac{1}{8} |R_\jmath|^2 \max\{|f''(x)| \; ; \; x \in R_\jmath\} \qquad \text{and}$$

$$\epsilon_\jmath^{(3)}(f) \leqslant \frac{1}{384} |R_\jmath|^4 \max\{|f^{(4)}(x)| \; ; \; x \in R_\jmath\}.$$

For the sake of simplicity we will suppose that the diffusion process $(X_t \in \mathcal{D})_{t \in [0,\infty[}$ admits a smooth transition density $\mathbb{R}_+ \times \mathcal{D} \times \mathcal{D} \ni (t, x, y) \longrightarrow \psi_t(x, y)$, so that we can write

$$\mathbb{E}[f(X_t) \mid X_0 = x] = \int_\mathcal{D} f(y)\psi_t(x, y)\mathrm{d}y, \quad x \in \mathcal{D}. \tag{1}$$

Furthermore – again, in order to avoid the discussion of certain technical conditions for the diffusion $(X_t)$ – we will suppose that for a reasonably large class of "nice" functions $f(\cdot)$ (i.e., for all functions that we will be dealing with) we can write

$$\frac{\partial^{k_1 + \cdots + k_l}}{\partial x_{k_1} \ldots \partial x_{k_l}} \mathbb{E}\left[f(X_t)|X_0 = x\right] = \int_\mathcal{D} f(y)\frac{\partial^{k_1 + \cdots + k_l}}{\partial x_{k_1} \ldots \partial x_{k_l}}\psi_t(x, y)\mathrm{d}y, x \in \mathcal{D},$$

where, just as before, we write $x \equiv \{x_1, \ldots, x_N\} \in \mathbb{R}^N$ and $y \equiv \{y_1, \ldots, y_N\} \in \mathbb{R}^N$. Consider now the functions

$$\Theta_t(\alpha_\jmath, i_1, i_2, \ldots, i_l; x) := \int_\mathcal{D} \mathcal{E}(\alpha_\jmath, i_1, i_2, \ldots, i_l; y)\psi_t(x, y)\mathrm{d}y$$

$$\text{and } \theta_t(\alpha_\jmath; x) := \int_\mathcal{D} e(\alpha_\jmath; y)\psi_t(x, y)\mathrm{d}y,$$

for various choices of $\jmath \in \mathbb{J}$. These functions are universal, in the sense that they depend only on the mesh $\alpha$ and the transition density $\psi_t(x, y)$. Furthermore, they provide cubature rules for the integrals in (1), as stated in the following

**Proposition 1.** *For any finite function $f \colon \mathcal{D} \mapsto \mathbb{R}$ one has*

$$\int_\mathcal{D} f(y)\psi_t(x, y)\mathrm{d}y \approx$$

$$\approx \int_\mathcal{D} s(f, \alpha; y)\psi_t(x, y)\mathrm{d}y \equiv \sum_{\jmath \in \mathbb{J}} f(\alpha_\jmath) \times \theta_t(\alpha_\jmath; x) + \int_{\mathcal{D}\setminus\mathcal{R}} f(y)\psi_t(x, y)\mathrm{d}y$$

*with error in the approximation (uniformly for all $x$!) no greater than $\epsilon^{(1)}(f, \alpha)$, and, for any $N$-times differentiable function $f \colon \mathcal{D} \mapsto \mathbb{R}$, one has*

$$\int_{\mathcal{D}} f(y)\psi_t(x,y)\mathrm{d}y \approx \int_{\mathcal{D}} S(f,\alpha;y)\psi_t(x,y)\mathrm{d}y$$

$$= \sum_{j\in\mathbb{J}} \sum_{l=0}^{N} \sum_{1\leqslant i_1 < i_2 < \ldots < i_l \leqslant N} (\partial_{i_1}\ldots\partial_{i_l} f)(\alpha_j) \times \Theta_t(\alpha_j, i_1, i_2, \ldots, i_l; x)$$

$$+ \int_{\mathcal{D}\setminus\mathcal{R}} f(y)\psi_t(x,y)\mathrm{d}y$$

*with approximation error (uniformly for all x!) no greater than* $\epsilon^{(3)}(f,\alpha)$.

Of course, as is common for most cubature (quadrature) rules, we suppose that outside the region $\mathcal{R}$, i.e., outside the region where the actual cubature (quadrature) rule applies, the function is either null, or can be estimated or somehow guessed.

## 3 Termination at fixed time: A numerical recipe for fixed-boundary problems

Let everything be as in the previous section. If the diffusion process $(X_t \in \mathcal{D})_{t\in[0,\infty[}$, $\mathcal{D} \subseteq \mathbb{R}^N$, can be terminated and, consequently, the termination payoff $\lambda(X_t)$ can be collected, only at some fixed date $T > 0$, there is nothing left to "control." In this case, one is only interested in the value function

$$F(t,x) := e^{-r(T-t)}\mathbb{E}[\lambda(X_{T-t}) \mid X_0 = x], \quad t \leqslant T, \quad x \in \mathcal{D},$$

which solves the equation (here $\mathcal{L}_x$ denotes the infinitesimal generator for the Markovian semi-group associated with $(X_t)$ )

$$\partial_t F(t,x) + \mathcal{L}_x F(t,x) = rF(t,x),$$

in the domain $(t,x) \in \,]-\infty, T] \times \mathcal{D}$ with boundary condition $F(T,x) \equiv \lambda(x)$, $x \in \mathcal{D}$. If we now define the operator semi-group $\{A_t \,;\, t \geqslant 0\}$ that acts on functions $f: \mathcal{D} \mapsto \mathbb{R}$ according to the rule

$$A_t f(x) := e^{-rt}\mathbb{E}\left[f(X_t)|X_0 = x\right] = e^{-rt}\int_{\mathbb{R}^N} f(y)\psi_t(x,y)\mathrm{d}y, t \geqslant 0,$$

then the value function can be expressed as $F(t,x) = A_{T-t}\lambda(x) \equiv A_{T-t}F(T,x)$, $t \leqslant T$, $x \in \mathcal{D}$.

Suppose that one is only interested in computing the functions

$$\mathcal{D} \ni x \longrightarrow F_{i,n}(x) := F\left(T - i2^{-n}, x\right) \equiv \left(A_{2^{-n}}\right)^i \lambda(x),$$

for $i = 0, 1, \ldots$ and for some large $n \in \mathbb{Z}_{++}$. Consider the 1-parameter family of operators $\{\mathcal{A}_t \,;\, t \geqslant 0\}$ defined by

$$(\mathcal{A}_t f)(x) := S(A_t f, \alpha; x) \text{ or by } (\mathcal{A}_t f)(x) := s(A_t f, \alpha; x),$$

depending on whether the interpolation used in the definition is $N$-cubic or $N$-linear. Note that if $N$-cubic interpolation is used, then the definition of $(\mathcal{A}_t f)(x)$ involves the tabulated values of all mixed derivatives of the function $x \longrightarrow A_t f(x)$ over the mesh $\alpha$. In contrast, if $(\mathcal{A}_t f)(x)$ is defined by way of $N$-linear interpolation, then the definition involves only the tabulated values of the function $x \longrightarrow A_t f(x)$ over the mesh $\alpha$. In general, the family $\{\mathcal{A}_t \; ; \; t \geqslant 0\}$ will not be a semi-group and what we want to know is how far from a semi-group it really is; more specifically, with $\|\cdot\|_\infty$ defined to be the usual sup-norm for functions on the domain $\mathcal{D}$, we want to estimate the differences

$$\rho_{i,n} := \left\| (\mathcal{A}_{2^{-n}})^i \lambda(\cdot) - (\mathcal{A}_{2^{-n}})^i \lambda(\cdot) \right\|_\infty , i = 0, 1, 2, \dots ,$$

(we suppose that for $i = 0$ one has $(A_{2^{-n}})^i \lambda(\cdot) = (\mathcal{A}_{2^{-n}})^i \lambda(\cdot) = \lambda(\cdot)$, so that $\rho_{0,n} = 0$). In the rest of this section we will consider only the case of $N$-cubic interpolation, as the case of $N$-linear interpolation can be dealt with essentially the same argument and is actually easier. In order to estimate $\rho_{i,n}$ we first expand the above difference as

$$(A_{2^{-n}})^i \lambda - (\mathcal{A}_{2^{-n}})^i \lambda =$$
$$= A_{2^{-n}}((A_{2^{-n}})^{i-1} \lambda - (\mathcal{A}_{2^{-n}})^{i-1} \lambda) + (A_{2^{-n}} - \mathcal{A}_{2^{-n}})(\mathcal{A}_{2^{-n}})^{i-1} \lambda.$$

Since – just by definition – the function $x \longrightarrow (\mathcal{A}_{2^{-n}})(\mathcal{A}_{2^{-n}})^{i-1} \lambda(x) \equiv (\mathcal{A}_{2^{-n}})^i \lambda(x)$ interpolates the function $x \longrightarrow (A_{2^{-n}})(\mathcal{A}_{2^{-n}})^{i-1} \lambda(x)$, from the last relation we can write

$$\rho_{i,n} \leqslant e^{-r(2^{-n})} \rho_{i-1,n} + \epsilon^{(3)}(A_{2^{-n}}(\mathcal{A}_{2^{-n}})^{i-1} \lambda, \alpha). \tag{2}$$

In general, the error term in the $N$-cubic interpolation is controlled by finitely many derivatives of the form (for $N = 1$ this would be just the $4^{\text{th}}$ derivative in the case of spline interpolation and the $2^{\text{nd}}$ derivative in the case of linear interpolation):

$$\mathbb{D}_x := (\partial_{x_i})^4 \partial_{x_{m_1}} \dots \partial_{x_{m_l}} .$$

Now we need to make the following assumption about the diffusion $(X_t)$ and the termination payoff $\lambda(\cdot)$.

**Assumption 1** *For each of the above derivatives $\mathbb{D}_x$ the functions*

$$\mathcal{D} \ni x \longrightarrow \sup_{t<T} \int_{\mathcal{D}} |\mathbb{D}_x \psi_t(x, y)| \, \mathrm{d}y \quad \text{and} \quad \mathcal{D} \ni x \longrightarrow \sup_{t<T} |\mathbb{D}_x A_t \lambda(x)|$$

*are globally bounded.*

As a result of the above assumption, for some finite constant C we can write

$$\left|\mathbb{D}_x A_{2^{-n}} (\mathcal{A}_{2^{-n}})^{i-1} \lambda(x)\right| = e^{-r(2^{-n})} \left|\int_{\mathbb{R}^N} ((\mathcal{A}_{2^{-n}})^{i-1} \lambda(y)) \mathbb{D}_x \psi_{2^{-n}}(x,y) \mathrm{d}y\right|$$

$$\leqslant e^{-r(2^{-n})} \left|\int_{\mathbb{R}^N} ((\mathcal{A}_{2^{-n}})^{i-1} \lambda(y)) \mathbb{D}_x \psi_{2^{-n}}(x,y) \mathrm{d}y\right| + C e^{-r(2^{-n})} \rho_{i-1,n}$$

$$= \left|\mathbb{D}_x (A_{i 2^{-n}}) \lambda(x)\right| + C e^{-r(2^{-n})} \rho_{i-1,n}.$$

Suppose, next, that $\vartheta > 0$ is arbitrarily chosen. Due to the above relation, for any sufficiently fine mesh $\alpha$, one has the following estimate, which is uniform for all $i$ and $n$ (the constant $C'$ below can be made arbitrarily small by refining the mesh and, in general, refining the mesh will not violate the estimate):

$$\left|\epsilon^{(3)} \left(A_{2^{-n}} (\mathcal{A}_{2^{-n}})^{i-1} \lambda, \alpha\right)\right| \leqslant \vartheta + C' e^{-r(2^{-n})} \rho_{i-1,n}.$$

In conjunction with (2), for any sufficiently fine mesh we have (uniformly for all $i$ and $n$)

$$\rho_{i,n} \leqslant e^{-r(2^{-n})} (1 + C') \rho_{i-1,n} + \vartheta.$$

This leads to the following result

**Proposition 2.** *There is a constant $C'$ for which the following claim can be made: for any $n \in \mathbb{Z}_{++}$ with $e^{-r(2^{-n})}(1 + C') < 1$ and any $\vartheta > 0$ one has, uniformly for all $i = 1, 2, \ldots$,*

$$\rho_{i,n} \leqslant \frac{\vartheta}{1 - e^{-r(2^{-n})}(1 + C')},$$

*for any sufficiently dense interpolation mesh $\alpha$.*

Note that, in general, $\rho_{i,n}$ cannot be made small uniformly for all $n$, i.e., a smaller time step requires also a finer mesh.

We now turn to the study of stability issues in the procedure that we just described. By definition, each quantity $S(A_t f, \alpha; x)$ is a linear combination of the universal $N$-cubic spline surfaces $\mathcal{E}(\alpha_j, i_1, i_2, \ldots, i_l; x)$ and contains also the term $A_t f(x) \chi_{\mathcal{D} \setminus \mathcal{R}}(x)$. Suppose next that the coefficients in this linear combination are replaced by values that differ from the actual values by no more than some fixed $\varepsilon > 0$. There are two reasons why such an inaccuracy may occur. The most obvious one is the limitation of the numerical precision in the computing device. The second reason is more interesting. It has to do with the level-3 approximation: instead of computing the function $A_t f(x)$ (and one only needs to know this function and its mixed derivatives on the mesh $\alpha$) one may be able to compute only the function

$$x \longrightarrow e^{-rt} \mathbb{E}\big[f(\tilde{X}_t) \mid \tilde{X}_0 = x\big], \tag{3}$$

with $(\tilde{X}_t)$ being some suitable approximation of the underlying diffusion. For example, for small $t$ one may be able to replace

$$X_t = x + \int_0^t \sigma(X_s)\mathrm{d}W_s + \int_0^t \mu(X_s)\mathrm{d}s \quad ((W_t) \equiv \text{standard BM}) ,$$

$$\text{with } \tilde{X}_t = x + \sigma(x)W_t + \mu(x)t ,$$

which reduces the computation of the expression in (3) to the (possibly numerical) evaluation of a Gaussian integral. Now we need to make the following

**Assumption 2** *For some $t_0 < T$ the interpolation region $\mathcal{R}$ can be chosen so that*

$$|A_t f(x) - \lambda(x)| \leqslant \varepsilon, \quad \text{for all } t \in [t_0, T] \text{ and for all } x \in \mathcal{D} \backslash \mathcal{R} .$$

Such an assumption is certainly reasonable and is more or less unavoidable in any approximation procedure; for example, with finite time left to expiry, the value of a put option is very close to 0 for prices of the underlying that are very large. In any case, we are going to denote this approximate spline-surface by $S_\varepsilon(A_t f, \alpha; x)$ and set

$$\left(\mathcal{A}_t^\varepsilon f\right)(x) := S_\varepsilon\left(A_t f, \alpha; x\right), t \leqslant T, x \in \mathcal{D} ,$$

with the understanding that – just by definition – $S_\varepsilon(A_t f, \alpha; x) = \lambda(x)$ for $x \in \mathcal{D} \backslash \mathcal{R}$. Plainly, $(\mathcal{A}_t^\varepsilon f)(x)$ is obtained by perturbing the data from which $(A_t f)(x)$ is constructed by no more than $\varepsilon$. The next step is to examine the difference

$$(A_{2^{-n}})^i f - (\mathcal{A}_{2^{-n}}^\varepsilon)^i \lambda = A_{2^{-n}}\left((A_{2^{-n}})^{i-1}\lambda - (\mathcal{A}_{2^{-n}}^\varepsilon)^{i-1}\lambda\right)$$
$$+ (A_{2^{-n}} - \mathcal{A}_{2^{-n}})(\mathcal{A}_{2^{-n}}^\varepsilon)^{i-1}\lambda + (A_{2^{-n}} - \mathcal{A}_{2^{-n}}^\varepsilon)(\mathcal{A}_{2^{-n}}^\varepsilon)^{i-1}\lambda \quad (4)$$

and we set

$$\rho_{i,n}^\varepsilon := \left\|(A_{2^{-n}})^i\lambda(\cdot) - (\mathcal{A}_{2^{-n}}^\varepsilon)^i\lambda(\cdot)\right\|_\infty, \quad 0 \leqslant i \leqslant (T - t_0)\, 2^n .$$

The first term in the right side of (4) cannot exceed $e^{-r(2^{-n})}\rho_{i-1,n}^\varepsilon$. Consider next the third term. On the domain $\mathcal{D}\backslash\mathcal{R}$, i.e., outside the interpolation region the function $\mathcal{A}_{2^{-n}}\left(\mathcal{A}_{2^{-n}}^\varepsilon\right)^{i-1}\lambda$ is the same as

$$A_{2^{-n}}\left(\mathcal{A}_{2^{-n}}^\varepsilon\right)^{i-1}\lambda = \lambda + A_{2^{-n}}\left((\mathcal{A}_{2^{-n}}^\varepsilon)^{i-1}\lambda - (A_{2^{-n}})^{i-1}\lambda\right) + \left((A_{2^{-n}})^i\lambda - \lambda\right),$$

while $\mathcal{A}_{2^{-n}}^\varepsilon\left(\mathcal{A}_{2^{-n}}^\varepsilon\right)^{i-1}\lambda = \lambda$ by definition. Consequently, by a very rough estimate, on the domain $\mathcal{D} \backslash \mathcal{R}$ the absolute value of the third term in (4) cannot exceed

$$e^{-r(2^{-n})}\rho_{i-1,n}^\varepsilon + \varepsilon .$$

On the domain $\mathcal{R}$ the third term in (4) is simply the difference between two interpolating functions for which the respective interpolated values differ by no more than $\varepsilon$. Consequently, on the domain $\mathcal{R}$ one must have

$$\left|(A_{2^{-n}} - A_{2^{-n}}^{\varepsilon})(A_{2^{-n}}^{\varepsilon})^{i-1}\lambda(x)\right| \leqslant C\varepsilon$$

where the constant $C$ is universal. As for the second term in the right side of (4), it is simply the difference between the function $A_{2^{-n}}(A_{2^{-n}}^{\varepsilon})^{i-1}\lambda$ and its spline-surface interpolation on the mesh $\alpha$. By definition this difference is 0 in the domain $\mathcal{D}\setminus\mathcal{R}$, and on each region $R_j$ it is controlled by powers of the quantity $|R_j|$ times quantities of the form

$$e^{-r2^{-n}}\sup_{x\in R_j}\left|\int_{\mathcal{D}}(A_{2^{-n}}^{\varepsilon})^{i-1}\lambda(y)\mathbb{D}_x\psi_{2^{-n}}(x,y)\mathrm{d}y\right|$$

$$= e^{-r2^{-n}}\sup_{x\in R_j}\left|\int_{\mathcal{D}}(A_{2^{-n}})^{i-1}\lambda(y)\mathbb{D}_x\psi_t(x,y)\mathrm{d}y\right|$$

$$+ \rho_{i-1}e^{-r2^{-n}}\sup_{x\in R_j}\int_{\mathcal{D}}|\mathbb{D}_x\psi_{2^{-n}}(x,y)|\,\mathrm{d}y.$$

Consequently, for any sufficiently dense mesh $\alpha$ one has

$$\rho_{i,n}^{\varepsilon} \leqslant e^{-r(2^{-n})}\rho_{i-1,n}^{\varepsilon} + e^{-r(2^{-n})}\left(1+\rho_{i-1,n}^{\varepsilon}\right)\vartheta + (1+C)\varepsilon + e^{-r(2^{-n})}\rho_{i-1,n}^{\varepsilon}$$

$$= (2+\vartheta)e^{-r(2^{-n})}\rho_{i-1,n}^{\varepsilon} + e^{-r(2^{-n})}\vartheta + (1+C)\varepsilon,$$

for any $0 \leqslant i \leqslant (T-t_0)2^n$. The above consideration leads to the following

**Proposition 3.** *Given any (fixed) $n \in \mathbb{Z}_{++}$ with $e^{-r2^{-n}} < 1/3$ and any (arbitrarily small) $\vartheta \in\ ]0,1[$ one can claim that, uniformly for $0 \leqslant i \leqslant (T-t_0)2^n$, one has*

$$\rho_{i,n}^{\varepsilon} \leqslant \frac{e^{-r(2^{-n})}\vartheta + (1+C)\varepsilon}{1 - (2+\vartheta)e^{-r(2^{-n})}} \leqslant \frac{e^{-r(2^{-n})}\vartheta + (1+C)\varepsilon}{1 - 3e^{-r(2^{-n})}}, \quad 1 \leqslant i \leqslant (T-t_0)\,2^n,$$

*for any sufficiently dense interpolation mesh $\alpha$.*

It is important to recognize that, while $\vartheta$ can be chosen to be arbitrarily small after $n$ has been fixed, the same cannot be said about the precision level $\varepsilon$, which may be exogenous; to wit, there is no point in taking a smaller time step unless one can increase the precision in the calculation. In any case, assuming that the above estimate is acceptable, the following numerical recipe for approximating the sequence of functions

$$F_{i,n}(x) := F\left(T - i2^{-n}, x\right) = (A_{2^{-n}})^i\lambda(x), x \in \mathcal{D}, 0 \leqslant i \leqslant (T-t_0)\,2^n$$

can be prescribed: construct the sequence of spline-surfaces,

$$(A_{2^{-n}}^{\varepsilon})^i\lambda(x), \quad 1 \leqslant i \leqslant (T-t_0)\,2^n,$$

or, which amounts to the same, construct the sequence of finite lists of data

$$\left\{ (\partial_{i_1} \ldots \partial_{i_l} (\mathcal{A}_{2^{-n}}^{\varepsilon})^i \lambda)(\alpha_j) ; j \in \mathbb{J}, \ 1 \leqslant i_1 < i_2 < \ldots < i_l \leqslant N, \ 0 \leqslant l \leqslant N \right\},$$

recursively for $i = 2, 3, \ldots, \lceil (T - t_0)2^n \rceil$, by applying a special (computed once and for all!) linear operation to the $(i-1)^{\text{st}}$ list in order to obtain the $i^{\text{th}}$ list (the very first list is obtained by integrating – perhaps numerically – the termination payoff function directly). It is particularly easy to illustrate this recipe in the case of $N$-linear interpolation, i.e., when the interpolation does not require any information about the derivatives (which is the price to pay for having an approximation for the value function in terms of functions that are continuous but are not $\mathcal{C}^1$-smooth). In this case one has to update the lists of data

$$\left\{ (\mathcal{A}_{2^{-n}}^{\varepsilon})^i \lambda(\alpha_j) ; j \in \mathbb{J} \right\},$$

successively, for $i = 2, 3, \ldots$, according to the rule

$$(\mathcal{A}_{2^{-n}}^{\varepsilon})^i \lambda(\alpha_j) = \sum_{i \in \mathbb{J}} (\mathcal{A}_{2^{-n}}^{\varepsilon})^{i-1} \lambda(\alpha_i) \times \tilde{\theta}_{2^{-n}}(\alpha_i; \alpha_j) + \tilde{\zeta}_{2^{-n}}(\alpha_j), \qquad (5)$$

where, just as before, we write

$$\theta_{2^{-n}}(\alpha_i; \alpha_j) := \int_{\mathcal{R}} e(\alpha_i; y) \psi_{2^{-n}}(\alpha_j, y) dy \qquad \text{and}$$

$$\zeta_{2^{-n}}(\alpha_j) = \int_{\mathcal{D} \setminus \mathcal{R}} \lambda(y) \psi_{2^{-n}}(\alpha_j, y) dy$$

and assume that $\tilde{\theta}_{2^{-n}}(\alpha_i; \alpha_j)$ and $\tilde{\zeta}_{2^{-n}}(\alpha_j)$ are some approximate values for these integrals. The important point, of course, is that all quantities $\tilde{\theta}_{2^{-n}}(\alpha_i; \alpha_j)$ and $\tilde{\zeta}_{2^{-n}}(\alpha_j)$, which define the affine transformation in (5), can be computed once and for all. This recipe is discussed further in the next section. Several concrete examples of this procedure, with actual computer code, can be found in [9].

## 4 Termination at arbitrary time: A numerical recipe for optimal stopping and free-boundary problems

Let everything be as in the previous two sections, except that we now suppose that the diffusion process $(X_t \in \mathcal{D})_{t \in [0,\infty[}$, $\mathcal{D} \subseteq \mathbb{R}^N$, can be terminated, and, consequently, the termination payoff $\lambda(X_t)$ can be collected, at any time $t \leqslant T$, but no later than the termination date $T$. Now we define the family of non-linear operators

$$H_t f(x) := \text{Max}\left[ (A_t f)(x), \lambda(x) \right], t \geqslant 0,$$

and the associated value function

$$F(t,x) := \overline{\lim}_{n \nearrow \infty} (H_{2^{-n}})^{\lceil 2^n (T-t) \rceil} \lambda(x), t \leqslant T, x \in \mathcal{D}. \qquad (6)$$

Not only are the operators $H_t$ non-linear and the semi-group property $H_s \circ H_t = H_{s+t}$ fails, but, as is immediate from their definition, these operators actually destroy smoothness. Certain types of regularity must be preserved, however, and we now make the following

**Assumption 3** *The function $\lambda(\cdot)$ is piece-wise smooth and is Lipschitz continuous on the entire domain $\mathcal{D}$ with some finite global Lipschitz constant $c$. Furthermore, this property is unharmed by the operators $H_t$, i.e., all functions of the form $H_{t_1}(\dots (H_{t_l} \lambda))$ are also piece-wise smooth and Lipschitz continuous on $\mathcal{D}$ with the same (global) Lipschitz constant $c$, for any choice of $t_1, \dots, t_l \in \mathbb{R}_{++}$.*

It was essential in the last section to suppose that, roughly speaking, the operation $A_t$ does not increase the size of the derivatives; more specifically, we needed an uniform (with respect to $t$) global bound on those derivatives that control the error term in the interpolation. Although the operators $H_t$ cannot have such property (as they destroy smoothness) we must still require that the operators $H_t$ "do not increase the size of the derivatives" in some weaker sense. In order to formulate such a condition, define the mollified operators $\overline{H}_t^\delta$ by

$$\overline{H}_t^\delta f(x) := \frac{1}{(2\pi\delta)^{N/2}} \int_{\mathcal{D}} H_t f(y) e^{-\frac{1}{2\delta}\|y-x\|^2} dy, \quad \delta > 0,$$

and make the following

**Assumption 4** *Given any fixed $\delta > 0$, all functions of the form $\overline{H}_{t_0}^\delta (H_{t_1}(\dots (H_{t_l} \lambda)))$, for all possible choices of $t_1, \dots, t_l \in \mathbb{R}_{++}$, have bounded derivatives $\mathbb{D}_x$ on the entire domain $\mathcal{D}$, the bounds being independent from the choice of $t_0, t_1, \dots, t_l \in \mathbb{R}_{++}$.*

We now state without a proof the following

**Proposition 4.** *Suppose that the termination payoff $\lambda(\cdot)$ has the property*

$$\sup\left\{ A_t \lambda(x) - \lambda(x); x \in \mathcal{D}, A_t \lambda(x) > \lambda(x) \right\} = o(t).$$

*Then the value function $F(t,x)$ (see 6) is finite everywhere in $]-\infty, T] \times \mathcal{D}$ and satisfies the equation*

$$\partial_t F(t,x) + \mathcal{L}_x F(t,x) = r F(t,x) \qquad (7)$$

*in the interior of the continuation set*

$$\mathcal{S} := \{(t,x); t < T, x \in \mathcal{D}, F(t,x) > \lambda(x)\} \subseteq \mathcal{D}.$$

*In addition, $F(t,x)$ satisfies what is known as "the smooth fit condition": on the free (and unknown) boundary $\partial \mathcal{S}$ the function $F(t,x)$ coincides with the payoff $\lambda(x)$ and all $1^{st}$ derivatives $\partial_{x_i} F(t,x)$ coincide with the respective first order derivatives $\partial_{x_i} \lambda(x)$.*

We refer to [12], Ch. IV, §8, for a detailed discussion of the smooth fit and the continuous fit conditions. The classical approach to free boundary problems is to recover – perhaps only numerically – the value function $F(t,x)$ from equation (7) and from the smooth fit condition on the free (and unknown) boundary $\partial S$. The approach that we are going to develop is completely different. Setting,

$$(\mathcal{H}_t f)(x) := S(\text{Max}\,[A_t f, \lambda]\,, \alpha; x) \equiv S(H_t f, \alpha; x), \quad t \geqslant 0\,.$$

we again want to obtain a uniform bound on the sequence

$$\rho_{i,n} := \left\| (\mathcal{H}_{2^{-n}})^i \lambda(\cdot) - (\mathcal{H}_{2^{-n}})^i \lambda(\cdot) \right\|_\infty, \quad i = 0, 1, 2, \ldots\,.$$

Such a bound is now harder to obtain because of the nature of the operators $H_t$. In fact, strictly speaking, the spline-surfaces $S(H_t f, \alpha; x)$ cannot be defined, as their construction requires information about all mixed derivatives of $H_t F$ on the mesh. The only exception is the case $N = 1$. In that case information about the first derivative is needed only at the end-points of the interpolation interval and the spline is uniquely determined by the values of the function on the mesh and by the requirement for the second derivative to be continuous. There are some well known methods that allow one to define spline-surfaces in higher dimension by guessing – as opposed explicitly prescribing – all derivatives needed in the construction. In order to avoid complications of this sort, we will work only with $N$-linear interpolating functions – as stated in [13], p.123, $N$-linear interpolating functions are "close enough for government work." Thus, we now change the definition of the operators $\mathcal{H}_t$ to

$$(\mathcal{H}_t f)(x) := s(\text{Max}\,[A_t f, \lambda]\,, \alpha; x) \equiv s(H_t f, \alpha; x\,), \quad t \geqslant 0\,,$$

and consider the differences

$$(\overline{H}^\delta_{2^{-n}})^i \lambda - (H_{2^{-n}})^i \lambda = \left( \overline{H}^\delta_{2^{-n}} (\overline{H}^\delta_{2^{-n}})^{i-1} \lambda - \overline{H}^\delta_{2^{-n}} (H_{2^{-n}})^{i-1} \lambda \right)$$
$$+ \left( \overline{H}^\delta_{2^{-n}} - H_{2^{-n}} \right)(H_{2^{-n}})^{i-1} \lambda. \tag{8}$$

As a consequence of the following simple relation

$$\sup_{y \in \mathcal{D}} \frac{e^{-rt}}{(2\pi\delta)^{N/2}} \left| \int_{\mathcal{D}} \left( \text{Max}\,[(A_t f)(y), \lambda(y)] - \text{Max}\,[(A_t g)(y), \lambda(y)] \right) e^{-\frac{1}{2\delta}\|y-x\|^2} dy \right|$$

$$\leqslant \sup_{y \in \mathcal{D}} \frac{e^{-rt}}{(2\pi\delta)^{N/2}} \int_{\mathcal{D}} |(A_t f)(y) - (A_t g)(y)| \, e^{-\frac{1}{2\delta}\|y-x\|^2} dy$$

$$\leqslant e^{-rt} \|f - g\|_\infty,$$

if we now set

$$d_{i,n} := \left\| (\overline{H}^\delta_{2^{-n}})^i \lambda(\cdot) - (H_{2^{-n}})^i \lambda(\cdot) \right\|_\infty, \quad i = 1, 2, \ldots\,,$$

then the first term in the right side of (8) can be estimated by

$$\left\|\overline{H}^{\delta}_{2^{-n}}(\overline{H}^{\delta}_{2^{-n}})^{i-1}\lambda(\cdot) - \overline{H}^{\delta}_{2^{-n}}(H_{2^{-n}})^{i-1}\lambda(\cdot)\right\|_{\infty} \leqslant e^{-r2^{-n}} d_{i-1,n}.$$

In order to estimate the second term in (8), notice first that if $f(\cdot)$ is Lipschitz continuous with constant $c$ then

$$\frac{1}{(2\pi\delta)^{N/2}} \int_{\mathcal{D}} |f(y) - f(x)| e^{-\frac{\|y-x\|^2}{2\delta}} dy \leqslant \frac{c}{(2\pi\delta)^{N/2}} \int_{\mathbb{R}^N} \|y-x\| e^{-\frac{1}{2\delta}\|y-x\|^2} dy$$
$$= c \times C_N \times \delta.$$

Since the second term in (8) is nothing but the difference between a Lipschitz continuous function and its mollification, a straight-forward application of the last relation gives

$$\left\|\left(\overline{H}^{\delta}_{2^{-n}} - H_{2^{-n}}\right)(H_{2^{-n}})^{i-1}\lambda(\cdot)\right\|_{\infty} \leqslant c \times C_N \times \delta,$$

as a result of which we can write

$$d_{i,n} \leqslant e^{-r2^{-n}} d_{i-1,n} + c \times C_N \times \sqrt{\delta} \implies d_{i,n} \leqslant \left(e^{-r2^{-n}}\right)^{i-1} d_{1,n} + \frac{c \times C_N \times \delta}{1 - e^{-r2^{-n}}}.$$

Thus, for any fixed $n \in \mathbb{Z}_{++}$ we can choose the mollification parameter $\delta$ so that the sequence

$$(\overline{H}^{\delta}_{2^{-n}})^{i}\lambda(\cdot), \quad i = 1, 2, \dots,$$

is arbitrarily uniformly (for $i$) close to the sequence $(H_{2^{-n}})^{i}\lambda(\cdot)$, $i = 1, 2, \dots$.

Suppose now that $n \in \mathbb{Z}_{++}$ has already been chosen, so that, in conjunction with (6), the sequence of functions $(H_{2^{-n}})^{i}\lambda(\cdot)$, $i = 1, 2, \dots$, gives an acceptable discrete time approximation for the value function $F(t, x)$. Now we want to estimate the differences

$$(H_{2^{-n}})^{i}\lambda - (\mathcal{H}_{2^{-n}})^{i}\lambda = \left(H_{2^{-n}}(H_{2^{-n}})^{i-1}\lambda - H_{2^{-n}}(\mathcal{H}_{2^{-n}})^{i-1}\lambda\right)$$
$$+ (H_{2^{-n}} - \mathcal{H}_{2^{-n}})(\mathcal{H}_{2^{-n}})^{i-1}\lambda \qquad (9)$$

for $i = 1, 2, \dots$ and set

$$\rho_{i,n} := \left\|(H_{2^{-n}})^{i}\lambda(\cdot) - (\mathcal{H}_{2^{-n}})^{i}\lambda(\cdot)\right\|_{\infty}, i = 0, 1, 2, \dots$$

($\rho_{i,0} = 0$ by definition). Since, given any $y \in \mathcal{D}$, we have

$$|\text{Max}\left[(A_t f)(y), \lambda(y)\right] - \text{Max}\left[(A_t g)(y), \lambda(y)\right]| \leqslant |(A_t f)(y) - (A_t g)(y)|$$
$$\leqslant e^{-rt}\|f - g\|_{\infty},$$

it is not hard to see that for the first term in the right side of (9) we have

$$\left\|H_{2^{-n}}(H_{2^{-n}})^{i-1}\lambda(\cdot) - H_{2^{-n}}(\mathcal{H}_{2^{-n}})^{i-1}\lambda(\cdot)\right\|_{\infty} \leqslant e^{-r2^{-n}}\rho_{i-1,n}.$$

The second term we can express as $H_{2^{-n}} f(\cdot) - s(H_{2^{-n}} f, \alpha; \cdot)$, with $f = (\mathcal{H}_{2^{-n}})^{i-1} \lambda$. Setting $g = (H_{2^{-n}})^{i-1} \lambda$, we get

$$\|H_{2^{-n}} f(\cdot) - s(H_{2^{-n}} f, \alpha; \cdot)\|_\infty \leqslant \|H_{2^{-n}} g(\cdot) - s(H_{2^{-n}} g, \alpha; \cdot)\|_\infty$$
$$+ \|H_{2^{-n}} f(\cdot) - H_{2^{-n}} g(\cdot)\|_\infty + \|s(H_{2^{-n}} f - H_{2^{-n}} g, \alpha; \cdot)\|_\infty .$$

The last two terms in the right side of the above inequality are dominated by

$$C e^{-r 2^{-n}} \rho_{i-1,n} ,$$

with some universal constant $C$, while the first term can be estimated by

$$\|H_{2^{-n}} g(\cdot) - s(H_{2^{-n}} g, \alpha; \cdot)\|_\infty \leqslant \left\| H_{2^{-n}} g(\cdot) - \overline{H}^\delta_{2^{-n}} g(\cdot) \right\|_\infty$$
$$+ \left\| \overline{H}^\delta_{2^{-n}} g(\cdot) - s\left(\overline{H}^\delta_{2^{-n}} g, \alpha; \cdot\right) \right\|_\infty + \left\| s(\overline{H}^\delta_{2^{-n}} g - H_{2^{-n}} g, \alpha; \cdot) \right\|_\infty .$$

If necessary, we can now increase $n$ so that $e^{-r 2^{-n}}(1+C) < 1$ and keep $n$ fixed from now on. Let $\vartheta > 0$ be arbitrarily chosen. As the function $H_{2^{-n}} g(\cdot)$ is Lipschitz continuous with a Lipschitz constant c, we can choose the mollification parameter $\delta$ so that (independently from the choice of the mesh) we have

$$\left\| H_{2^{-n}} g(\cdot) - \overline{H}^\delta_{2^{-n}} g(\cdot) \right\|_\infty + \left\| s(\overline{H}^\delta_{2^{-n}} g - H_{2^{-n}} g, \alpha; \cdot) \right\|_\infty \leqslant \frac{1}{2} \vartheta .$$

Now we keep $\delta$ fixed and remind that all second derivatives of the functions

$$\overline{H}^\delta_{2^{-n}} g(\cdot) \equiv \overline{H}^\delta_{2^{-n}} (H_{2^{-n}})^{i-1} \lambda(\cdot)$$

that control the error in the $N$-linear interpolation are bounded by some constant that depends on $\delta$ but not on $i$. Consequently, we can now choose the interpolation mesh $\alpha$ in such a way that

$$\left\| \overline{H}^\delta_{2^{-n}} g(\cdot) - s(\overline{H}^\delta_{2^{-n}} g, \alpha; \cdot) \right\|_\infty \leqslant \frac{1}{2} \vartheta .$$

Finally, from (9) we have

$$\rho_{i,n} \leqslant (1+C) e^{-r 2^{-n}} \rho_{i-1,n} + \vartheta ,$$

and arrive at the following counterpart of Proposition 2:

**Proposition 5.** *There is a constant $C$ for which the following claim can be made: for any sufficiently large $n \in \mathbb{Z}_{++}$ with $e^{-r(2^{-n})}(1+C) < 1$ and any $\vartheta > 0$ one has, uniformly for all $i = 1, 2, \ldots,$*

$$\rho_{i,n} \leqslant \frac{\vartheta}{1 - e^{-r(2^{-n})}(1+C)} ,$$

*for any sufficiently dense interpolation mesh $\alpha$.*

In this setting, too, the study of stability issues in the approximation procedure is of paramount importance. We need to show that one can approximate reasonably well the (theoretically computable) sequence $(\mathcal{H}_{2-n})^i \lambda(\cdot)$, $i = 1, 2, \ldots$, with the (actually computed) sequence $(\mathcal{H}_{2-n}^\varepsilon)^i \lambda(\cdot)$, $i = 1, 2, \ldots$, defined from

$$(\mathcal{H}_t^\varepsilon f)(x) := s_\varepsilon(\text{Max}\,[A_t f, \lambda]\,, \alpha; x) \equiv s_\varepsilon(H_t f, \alpha; x), t \geqslant 0\,,$$

where, just as before, $s_\varepsilon$ stands for the associated $N$-linear interpolating function defined from values that may differ from the prescribed ones by no more than $\varepsilon$. In this regard, just as we did in the previous section, we have to assume that the interpolation region $\mathcal{R}$ is chosen so that uniformly for all $x \in \mathcal{D} \setminus \mathcal{R}$ and for all $i$ with $T - i2^{-n} \in [t_0, T]$ the quantities $(H_{2-n})^i \lambda(x)$ differ from the termination payoff $\lambda(x)$, by no more than $\varepsilon$ and, consequently, will suppose that $s_\varepsilon(H_t f, \alpha; x) = \lambda(x)$ for $x \in \mathcal{D} \setminus \mathcal{R}$ (this is how the quantities $s_\varepsilon(H_t f, \alpha; x)$ are defined for $x \in \mathcal{D} \setminus \mathcal{R}$). In conjunction with the approximation developed in the present section, the study of all stability issues is completely analogous to the one presented in the previous section and will be omitted.

Adapting the numerical recipe from the previous section to the case of optimal stopping and free-boundary problems is now completely straight-forward: we must update the finite lists of data

$$\left\{ (\mathcal{A}_{2-n}^\varepsilon)^i \lambda(\alpha_\jmath) \; ; \jmath \in \mathbb{J} \right\},$$

successively, for $i = 2, 3, \ldots$, according to the rule

$$(\mathcal{A}_{2-n}^\varepsilon)^i \lambda(\alpha_\jmath) =$$
$$= \text{Max}\left[\sum_{\imath \in \mathbb{J}} (\mathcal{H}_{2-n}^\varepsilon)^{i-1} \lambda(\alpha_\imath) \times \tilde{\theta}_{2-n}(\alpha_\imath; \alpha_\jmath) + \tilde{\zeta}_{2-n}(\alpha_\jmath), \lambda(\alpha_\jmath)\right], \quad (10)$$

where, just as before, we write

$$\theta_{2-n}(\alpha_\imath; \alpha_\jmath) := \int_{\mathcal{R}} e(\alpha_\imath; y) \psi_{2-n}(\alpha_\jmath, y) \mathrm{d}y, \quad \zeta_{2-n}(\alpha_\jmath) = \int_{\mathcal{D} \setminus \mathcal{R}} \lambda(y) \psi_{2-n}(\alpha_\jmath, y) \mathrm{d}y$$

and suppose that $\tilde{\theta}_{2-n}(\alpha_\imath \alpha_\jmath)$ and $\tilde{\zeta}_{2-n}(\alpha_\jmath)$ are some approximate values for these integrals.

As it turns out, for some special choices of the diffusion process $(X_t)$ it becomes possible to obtain explicit formulas for the universal weights $\theta_{2-n}(\alpha_\imath; \alpha_\jmath)$, $\imath, \jmath \in \mathbb{J}$. Fortunately, one of the most widely used processes in financial modeling, namely a process in which all $N$ coordinates follow independent geometric Brownian motions, happens to be in this category. In fact, as we are about to illustrate, the calculation of the universal weights $\theta_{2-n}(\alpha_\imath; \alpha_\jmath)$ for a process of this type is completely straight-forward. To see this, notice first that for $x \in \mathcal{Q}(\alpha_\imath)$ the universal interpolating function $e(\alpha_\imath; x)$, $\imath \equiv \{i_1, \ldots, i_N\}$, can be written as

$$e(\alpha_i; x) \equiv e(\alpha_i; x_1, \ldots, x_N)$$

$$= \sum_{v \in \mathcal{V}_0,\ \imath + \vec{1} - 2v \in \mathbb{J}} \prod_{m=1}^{N} \frac{x_m - \alpha_{m,i_m+1-2v_m}}{\alpha_{m,i_m} - \alpha_{m,i_m+1-2v_m}} I_m^{\imath,v}(x_m),$$

where $\mathcal{V}_0$ is the collection of all vertices in the unit cube in $\mathbb{R}^N$, i.e., $\mathcal{V}_0$ is the set of all $2^N$ vectors $v \equiv \{v_1, \ldots, v_N\} \in \mathbb{R}^N$ with $v_m \in \{0,1\}$, $1 \leqslant m \leqslant N$, $\vec{1} := \{1, \ldots, 1\} \in \mathbb{R}^N$ and $I_m^{\imath,v}(\cdot)$ denotes the indicator of the interval with end-points $\alpha_{m,i_m}$ and $\alpha_{m,i_m+1-2v_m}$. Consequently, for the process

$$X_t = X_0 \cdot \left\{ e^{\sigma_1 W_t^{(1)} + (r - \frac{1}{2}\sigma_1^2)t}, \ldots, e^{\sigma_N W_t^{(N)} + (r - \frac{1}{2}\sigma_N^2)t} \right\},$$

defined in terms of the independent Brownian motions $(W_t^{(m)})$, $1 \leqslant m \leqslant N$, and for any two multi-indices $\imath \equiv \{i_1, \ldots, i_N\}$ and $\jmath \equiv \{j_1, \ldots, j_N\}$, we can write

$$\theta_{2^{-n}}(\alpha_\imath; \alpha_\jmath) =$$

$$= \sum_{v \in \mathcal{V}_0,\ \imath + \vec{1} - 2v \in \mathbb{J}} \prod_{m=1}^{N} \mathbb{E}\left[ \frac{\mathcal{X}_{m,n}(\Gamma) - \alpha_{m,i_m+1-2v_m}}{\alpha_{m,i_m} - \alpha_{m,i_m+1-2v_m}} \times I_m^{\imath,v}(\mathcal{X}_{m,n}(\Gamma)) \right]$$

where $\Gamma$ is some standard normal $\mathcal{N}(0,1)$-random variable and

$$\mathcal{X}_{m,n}(\Gamma) = \alpha_{m,j_m} e^{\sqrt{2^{-n}}\sigma_m \Gamma + 2^{-n}(r - \frac{1}{2}\sigma_m^2)}.$$

After a somewhat tedious – but otherwise completely straight-forward – calculation we arrive at the following explicit formula

$$\theta_{2^{-n}}(\alpha_\imath; \alpha_\jmath) = \sum_{v \in \mathcal{V}_0,\ \imath + (\vec{1} - 2v) \in \mathbb{J}} \prod_{m=1}^{N} \frac{1}{2(\alpha_{m,i_m} - \alpha_{m,i_m+1-2v_m})}$$

$$\times \left( \alpha_{m,j_m} e^{r\,2^{-n}} \left( \mathrm{erf}\left[ \frac{B_m^{\imath,v}}{\sqrt{2}} - \frac{\sigma_m}{\sqrt{2^{n+1}}} \right] - \mathrm{erf}\left[ \frac{A_m^{\imath,v}}{\sqrt{2}} - \frac{\sigma}{\sqrt{2^{n+1}}} \right] \right) \right.$$

$$\left. - \alpha_{m,i_m+1-2v_m} \left( \mathrm{erf}\left[ \frac{B_m^{\imath,v}}{\sqrt{2}} \right] - \mathrm{erf}\left[ \frac{A_m^{\imath,v}}{\sqrt{2}} \right] \right) \right),$$

where

$$A_m^{\imath,v} := \frac{1}{\sigma_m \sqrt{2^{-n}}} \left( \log\left( \frac{\alpha_{m,i_m+1-2v_m}}{\alpha_{m,j_m}} \right) - \left( r - \frac{1}{2}\sigma_m^2 \right) 2^{-n} \right),$$

and $\quad B_m^{\imath,v} := \frac{1}{\sigma_m \sqrt{2^{-n}}} \left( \log\left( \frac{\alpha_{m,i_m}}{\alpha_{m,j_m}} \right) - \left( r - \frac{1}{2}\sigma_m^2 \right) 2^{-n} \right).$

Thus – just for this special choice of the process $(X_t)$ – the actual implementation of the numerical recipe described above comes down to evaluating some standard functions and, of course, involves the manipulation of (possibly very large) lists of data. Note that this method involves neither numerical integration nor Monte-Carlo simulation of any kind – just evaluation of standard functions and manipulation of lists of data.

We conclude this section with a simple graphical illustration of the method in dimension $N = 2$. In this example the process $(X_t)$ is taken to be a two-dimensional diffusion whose coordinates are exponents of two independent Brownian motions. The termination payoff function is $\lambda(x_1, x_2) = \text{Max}[K - \text{Min}[x_1, x_2], 0]$ – this is the payoff from an American-style put option on the choice of any one of two given underlying stocks. This is the simplest example of a stock option on multiple assets.

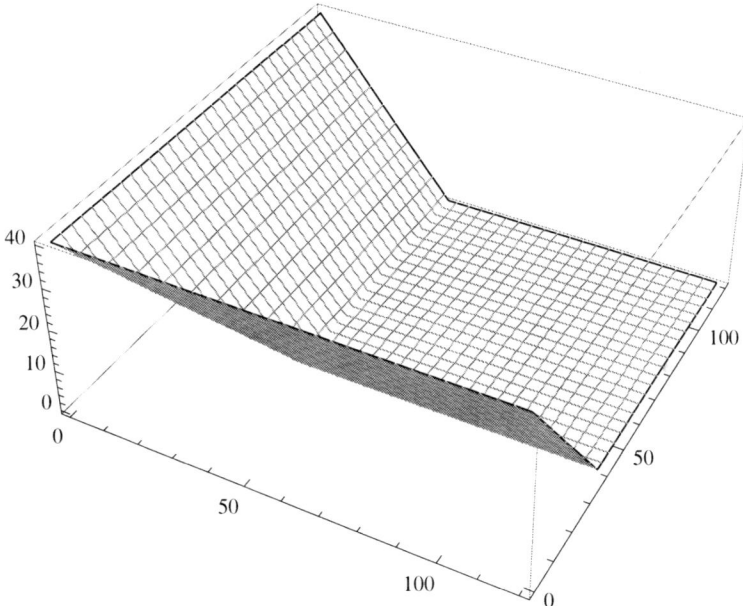

**Fig. 2.** The termination payoff from an American-style put option with exercise price $K = 40$ on the choice of any one of two underlying stocks.

Fig. 3 below shows the graph of the associated valuation map 1 year before the termination date.

With regard to applications to finance, we stress that the technique that we have developed here is much more than a tool for simulating the sample-paths of the option price. Of course, once the valuation maps $F(t, \cdot)$, $t \leqslant T$, have been computed, one can certainly simulate the sample paths $t \longrightarrow F(t, X_t)$ by using completely straight-forward conventional methods. However, the val-

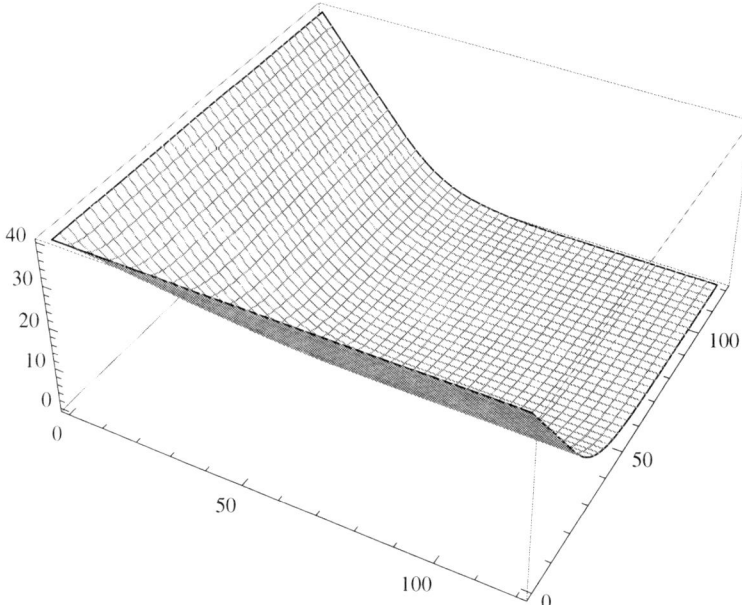

**Fig. 3.** The actual price of an American-style put option on the choice of any one of two underlying stocks with one year left to expiry.

uation maps $F(t, \cdot)$, $t \leqslant T$, contain valuable information about the sensitivity of the option price to changes in the underlying prices and such information is difficult to obtain by way of simulation. In fact, in many practical situations this information is more valuable than the actual option price. Since we construct the valuation maps as interpolating functions, the calculation of the derivatives (the so called "delta" of the option) is completely straightforward. As an example, the plot on Fig. 4 below shows the graph of the derivative $\partial_{x_1} F(T - 1, x_1, x_2)$, associated with the above valuation map (this is the marginal delta of the option relative to the first asset).

Other similar examples (with actual computer code) can be found in [9].

## 5 Concluding remarks

It can be argued that the most natural – and certainly the most common – tool for compressing a general surface into a finite list of data is by way of interpolation (this is what makes digital photography possible, for example). In this regard, the method described in the present paper is quite natural and is hardly original. Indeed, the finite element method – to give just one example – rests on the same main idea, except that in the context of optimal stopping one must construct interpolation surfaces dynamically. However , the respective dynamic procedure involves integration and the most obvious

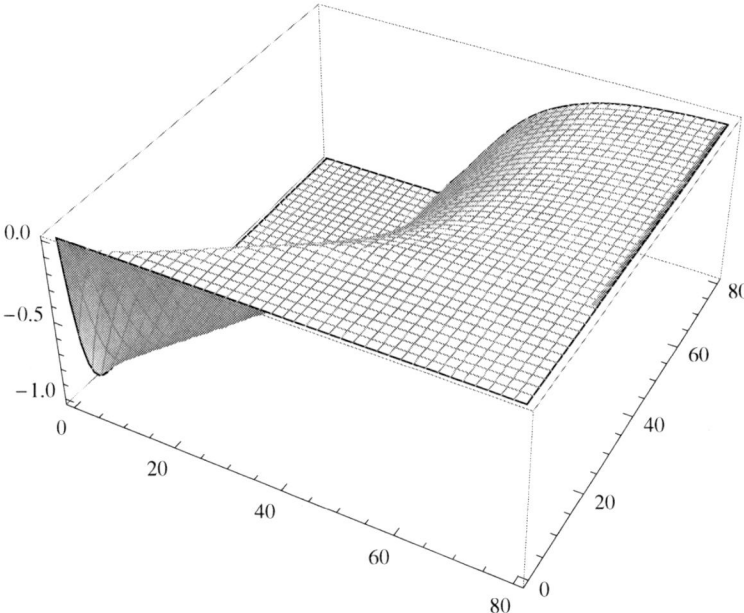

**Fig. 4.** The sensitivity to changes in the price of the first stock for an American-style put option on the choice of any one of two underlying stocks.

way to integrate an interpolating function is to use the associated quadrature rule directly, which comes down to a simple linear operation over the interpolated data. Such methods are particularly suitable for most optimal stopping problems that are commonly encountered in the realm of finance. This is because, in general, the payoffs from most financial contracts can be expressed as reasonably regular functions – nothing even close to a generic digital image, say. To compress such objects by using zero-order interpolation – as in the binomial model, for example – is somewhat similar to recording an image of a circle pixel-by-pixel, as opposed to recording just 3 floating-point numbers. Furthermore, in many situations an upper bound on certain higher order derivatives of the value function in various parts of the state space is not hard to guess – for example, the price of a put option should have very low curvature away from expiry. Since these derivatives and the size of the mesh are the only factors that control the error in the interpolation, one can place many fewer interpolation points in those parts of the state space where the derivatives that control the error are relatively small. To give a concrete example, if one needs to approximate with precision $\varepsilon$ the price of an American put option on a single asset by using cubic splines, then one can afford to take interpolation intervals of width $\sqrt[4]{384\,\varepsilon}/\sqrt[4]{M_4}$, where $M_4$ is some upper bound on the 4$^{\text{th}}$ derivative, which happens to be quite small, especially away from the strike-price in the state space and away from the expiration date in the time domain. In fact, since there is nothing in the procedure that requires the use of one and the same mesh throughout all iterations, there is no reason to not take advantage of the fact that, in most situations, the $2^d$ derivative, which controls the error in the $N$-linear interpolation, and the $4^{\text{th}}$ derivative, which controls the error in the N-cubic interpolation, decrease quite rapidly throughout the iteration process and this means that one can get away with fewer interpolation points later in the iteration process.

Finally, it may be useful to note that the practical implementation of the method described in the paper on several parallel processors is an essentially trivial task. Indeed, the calculation of the expression in (10) for any particular multi-index $j$ does not have to wait for the calculation associated with some other multi-index $j'$ to finish. Similarly, the weights $\theta_{2-n}(\alpha_i; \alpha_j)$ can be computed simultaneously for different choices of the multi-indices $i$ and $j$, by merely sending the tasks to several different processors. In addition, the summation in (10), which essentially comes down to computing the dot-product between two (possibly very large) lists of data, can be executed on parallel processors, too – in fact, the execution of dot-products on parallel processors is a standard function in some widely available computing systems.

# References

1. Bally V, Pagès G (2003) A Quantization Algorithm for Solving Discrete Time Multidimensional Optimal Stopping Problems. Bernoulli 6(9):1–47
2. Bally V, Pagès G (2003) Error Analysis of the Quantization Algorithm for Obstacle Problems. Stochastic Processes and their Applications 106(1):1–40
3. Dayanik S, Karatzas I (2003) On the Optimal Stopping Problem for Onedimensional Diffusions. Stochastic Processes & Applications 107:173–212
4. Fournie E, Lasry JM, Lebuchoux J, Lions P-L, Touzi N (1999) Some Applications of Malliavin Calculus to Monte Carlo Methods in Finance. Finance and Stochastics 3:391–412
5. Glasserman P (2004) Monte Carlo Methods in Financial Engineering. Springer-Verlag
6. El Karoui N, Karatzas I (1995) On the Optimal Stopping Problem Associated with an American Put-Option. IMA Lecture Notes in Mathematics & Applicatioins 65:35–46
7. Kusuoka S (2001) Approximation of Expectation of Diffusion Process and Mathematical Finance. In: Advanced Studies in Pure Mathematics 31:147–165. Taniguchi Conference on Mathematics Nara'98
8. Kusuoka S (2004) Approximation of Expectation of Diffusion Processes Based on Lie Algebra and Malliavin Calculus. In: Advances in Mathematical Economics 6:69–83. Springer-Verlag
9. Lyasoff A (2007) Dynamic Integration of Interpolating Functions and Some Concrete Optimal Stopping Problems. The Mathematica Journal 10(4) (to appear)
10. Longstaff FA, Schwartz RS (2001) Valuing American Options By Simulation: A simple Least-Square Approach. Review of Financial Studies 14:113–147

11. Lyons T, Victoir N (2004) Cubature on Wiener Space. Proceedings of the Royal Society Lond. ser. A 460:169–198
12. Peskir G, Shiryaev A (2006) Optimal Stopping and Free-Boundary Problems. Lectures in Mathematics – ETH Zürich, Birkhäuser
13. Press WH, Teukolsky SA, Vetterling WT, Flannery B (1992) Numerical Recipes in C: the Art of Scientific Computing, 2nd ed. Cambridge University Press
14. Rogers LCG (2002) Monte Carlo Valuation of American Options. Mathematical Finance 12(3):271–286
15. Stroud AH (1971) Approximate Calculation of Multiple Integrals. Prentice-Hall

# An Approximate Solution for Optimal Portfolio in Incomplete Markets

Francesco Menoncin

Dipartimento di Scienze Economiche, Università di Brescia,
Via S. Faustino, 74/B – 25122 – Brescia, Italia. `menoncin@eco.unibs.it`

**Summary.** In an incomplete financial market where an investor maximizes the expected constant relative risk aversion utility of his terminal wealth, we present an approximate solution for the optimal portfolio. We take into account a set of assets and a set of state variables, all of them described by general diffusion processes. Finally, we supply an easy test for checking the goodness of the approximate result.

## 1 Introduction

In this paper we study the problem of an investor who maximizes the expected utility of his terminal wealth. The utility function is supposed to belong to the CRRA (Constant Relative Risk Aversion) family. Furthermore, the values of the financial assets are supposed to depend on a set of stochastic state variables.

In this paper, we follow the traditional stochastic dynamic programming approach [20, 21] leading to the Hamilton-Jacobi-Bellman (HJB) equation (for a complete derivation of the HJB equation, see [22, 1]). For the alternative "martingale approach" the reader is referred to [7, 8, 18].

In the literature about optimal asset allocation two main fields of research can be found. On the one hand some authors check the existence (and uniqueness) of a viscosity solution for the Hamilton-Jacobi-Bellman equation in a very general framework (see for instance [10, 3, 4]). On the other hand, some authors find a closed form solution to the optimization problem but in a very restrictive framework. In particular, we refer to the works of Kim and Omberg[16], Wachter[24], Chacko and Viceira[6], Deelstra et al.[11], Boulier et al.[2], Zariphopoulou[26] and Menoncin[19]. The two last papers use a solution approach based on the Feynman-Kač theorem,[1] in an incomplete market and in a complete market with a background risk, respectively.

---

[1] For a review of the financial applications of the Feynman-Kač theorem the reader is referred to [12, 1, 22].

Unfortunately, the former literature is not suitable for an immediate application since it does not provide any closed form solution to the optimal portfolio. Instead, the latter can be easily applied but lies on the assumption that the asset values and the state variables behave in a very particular way.

Our work is aimed at finding a "third way" to the investment problem by supplying, on a very general framework, an approximate closed form solution for the optimal portfolio.

Wherever a closed form solution is obtained the market structure is as follows: (i) there exists only one state variable (the riskless interest rate or the risk premium) following a Vasiček process[23] or a Cox et al. process[9]; (ii) there exists only one risky asset; and (iii) a bond may exist. Some works consider a complete financial market [24, 11, 2, 19] while others deal with an incomplete market [16, 6, 26]. Furthermore, all these articles consider a CRRA utility function, with the exception of Kim and Omberg[16] who deal with a HARA (Hyperbolic Absolute Risk Aversion) utility function and Menoncin[19] who considers a CARA (Constant Absolute Risk Aversion) utility function.

In this work we take into account a quite general framework where the prices of a set of assets are driven by a set of state variables, all of them following general diffusion processes. We do not need the hypothesis of completeness for the financial market, and our approximate solution will keep valid even in an incomplete market driven by some risk sources which cannot be perfectly hedged.

We show that the optimal portfolio is a function of some matrices whose elements are given by a combination of preference parameters, drift and diffusion terms of both asset prices and state variables. We approximate the value of these matrices through a Taylor expansion. After this approximation the value function solving the HJB equation turns out to be log-linear in the state variables. Thus, the optimal portfolio becomes very easy to compute and we also present an easy way for checking the goodness of this approximate solution.

In the literature there exists another example where an approximate solution to the optimal portfolio is computed. We refer to Kogan and Uppal[17] who solve the HJB equation by approximating it around a given value of the risk aversion index. Their work takes into account a CRRA utility function and it is valid for a value of the Arrow-Pratt risk aversion index close to zero. On the contrary, in our work, we allow for a more general pattern of consumer preferences. In fact, we compute the Taylor approximation around given values of the state variables.

Through this work we consider agents trading continuously in a frictionless, arbitrage-free, and incomplete market with a finite time horizon.

The paper is structured as follows. Section 2 details the general economic framework, presents the stochastic differential equations describing the behaviour of asset prices and state variables and shows the dynamic behaviour of the investor's real. In Section 3, both the optimal portfolio in an implicit form and the HJB equation are computed. Finally, we show an exact solution in the case of a complete market. Section 4 presents our main result, i.e. an approximate solution for the optimal portfolio. This section ends by presenting an easy way for computing the goodness of the approximation. Section 5 compares our approximate solution with two closed form solutions existing in the literature for an incomplete market. Section 6 concludes.

## 2 The market structure

Let the financial market be described by the following stochastic differential equations:

$$
\begin{cases}
\underset{s\times 1}{dX} = \underset{s\times 1}{f\,(t,X)}dt + \underset{s\times k}{g\,(t,X)'}\underset{k\times 1}{dW}, & X\,(t_0) = X_0, \\[2mm]
\underset{n\times n}{I_S^{-1}}\,\underset{n\times 1}{dS} = \underset{n\times 1}{\mu\,(t,X)}dt + \underset{n\times k}{\Sigma\,(t,X)'}\underset{k\times 1}{dW}, & S\,(t_0) = S_0, \\[2mm]
dG = Gr\,(t,X)\,dt, & G\,(t_0) = G_0,
\end{cases}
\tag{1}
$$

where $X$ is a vector containing all the state variables affecting the asset prices which are listed in vector $S$. $I_S$ is a diagonal matrix containing the asset prices $(S)$. For a review of all variables which can affect the asset prices the reader is referred to Campbell[5] who offers a survey of the most important contributions in this field. $G$ is the value of a riskless asset paying the instantaneously riskless (spot) interest rate $r$. Hereafter, the prime denotes transposition.

All the functions $f\,(t,X)$, $g\,(t,X)$, $\mu\,(t,X)$, $\Sigma\,(t,X)$, and $r\,(t,X)$ are supposed to be $\mathcal{F}_t$−measurable. The $\sigma$−algebra $\mathcal{F}$ is defined on a set $\Theta$ wherethrough the complete probability space $(\Theta, \mathcal{F}, \mathbb{P})$ is defined. Here, $\mathbb{P}$ can be considered as the "historical" probability measure.

Two sufficient conditions for a solution of (1) to exist, are that the above mentioned functions are Lipschitz continuous and square integrable. Duffie and Kan[13] define a different condition on the diffusion terms which may be worth mentioning.

**Definition 1.** *A function $\sigma : \mathbb{R}_+ \to \mathbb{R}$ satisfied the Yamada condition if bounded and measurable, and if there exists a function $\rho : \mathbb{R}_+ \to \mathbb{R}_+$, strictly increasing, continuous, with $\rho\,(0) = 0$, $\int_{0+}^{1} \rho\,(u)^{-2}\,du = +\infty$, and $|\sigma\,(u) - \sigma\,(v)| \le \rho\,(|u-v|)$ for all $u$ and $v$.*

**Lemma 1.** *Suppose that $g\,(X)$ satisfies the Yamada condition, and $f\,(X)$ is Lipschitz, then there exists a unique (strong) solution to*

$$
dX\,(t) = f\,(X)\,dt + g\,(X)\,dW,
$$
$$
X\,(t_0) > 0.
$$

The stochastic equations in System (1) are driven by a set of risks listed in $dW \in \mathbb{R}^k$ which is a vector of standard independent Brownian motions (with zero mean and variance $dt$).[2]

---

[2] Independence can be imposed without any loss of generality since the case of dependent Brownian motions can be obtained via a Cholesky decomposition.

The set of risk sources is the same for both the state variables and the asset prices. This hypothesis is not restrictive because setting the elements of matrices $g$ and $\Sigma$ allows us to take into account many different frameworks. For instance, if we set $dW = \begin{bmatrix} dW_1 & dW_2 \end{bmatrix}$, $g' = \begin{bmatrix} g_1 & 0 \end{bmatrix}$, and $\Sigma' = \begin{bmatrix} 0 & \sigma_2 \end{bmatrix}$ then $X$ and $S$ are instantaneously uncorrelated even if they formally have the same risk sources.

We recall the main result concerning completeness and arbitrage in this kind of market (for the proof of the following theorem see [22].

**Theorem 1.** *A market* $\{S(t, X)\}_{t \in [t_0, H]}$ *is arbitrage free (complete) if and only if there exists a (unique) $k-$dimensional vector $\xi(t, X)$ such that*

$$\Sigma(t, X)' \xi(t, X) = \mu(t, X) - r(t, X),$$

*and such that*

$$\mathbb{E} \left[ e^{\frac{1}{2} \int_{t_0}^{H} \| \xi(t, X) \|^2 dt} \right] < \infty.$$

If on the market there are less assets than risk sources ($n < k$), then the market cannot be complete even if it is arbitrage free. In this work, we assume that $n \leq k$ and the rank of matrix $\Sigma$ is maximum (i.e. it equals $n$). Thus, the results we obtain in this work are valid for a financial market which is incomplete as well as for a complete market ($n = k$ and no redundant assets).

## 2.1 The investor's wealth

Let $\theta(t) \in \mathbb{R}^{n \times 1}$ and $\theta_G(t) \in \mathbb{R}$ be the number of risky assets and riskless asset held, respectively. Then the investor's wealth can be written as

$$R = \theta(t)' S + \theta_G(t) G. \tag{2}$$

After differentiating the budget constraint (2) and taking into account the self-financing condition[3] we obtain

$$dR = \theta(t)' dS + \theta_G(t) dG,$$

and, after substituting the differentials from System (1), we have

$$dR = (\theta' I_S \mu + \theta_G Gr) dt + \theta' I_S \Sigma' dW. \tag{3}$$

Once the value of $\theta_G$ is taken from (2) and substituted into (3) we finally obtain

$$dR = (Rr + w'M) dt + w' \Sigma' dW, \tag{4}$$

where

$$w \equiv I_S \theta, \quad M \equiv \mu - r\mathbf{1},$$

and $\mathbf{1}$ is a vector of 1s.

---

[3] The self-financing condition can be written as

$$d\theta' (S + dS) + d\theta_G G = 0.$$

## 3 The optimal portfolio

Given the market structure (1) and the wealth differential equation (4), the optimization problem for an investor maximizing the expected CRRA utility of his terminal wealth over his (deterministic) time horizon $H$, can be written as

$$\begin{cases} \underset{w}{\max}\mathbb{E}_{t_0}\left[\frac{1}{1-\delta}R\left(H\right)^{1-\delta}\right] \\ d\begin{bmatrix} z \\ R \end{bmatrix} = \begin{bmatrix} \mu_z \\ Rr + w'M \end{bmatrix}dt + \begin{bmatrix} \Omega' \\ w'\Sigma' \end{bmatrix}dW, \\ z\left(t_0\right) = z_0, \qquad R\left(t_0\right) = R_0, \qquad \forall t_0 \leq t \leq H, \end{cases} \quad (5)$$

where

$$\underset{(s+n)\times 1}{z} \equiv \begin{bmatrix} X \\ S \end{bmatrix}, \qquad \underset{(s+n)\times 1}{\mu_z} \equiv \begin{bmatrix} f \\ I_S\mu \end{bmatrix}, \qquad \underset{k\times(s+n)}{\Omega'} \equiv \begin{bmatrix} g' \\ I_S\Sigma' \end{bmatrix}.$$

The vector $z$ contains all the state variables but the investor's wealth. The parameter $\delta$ measures the investor's (constant) relative risk aversion and, then, it must be strictly positive. Furthermore, we assume $\delta \geq 1$ in order to have a well defined associated partial differential equation for the value function solving Problem (5).[4]

The form we have chosen for the objective function (a CRRA utility) and the assumptions that must hold on the state variable differential equations, guarantee that there exists an optimal portfolio solving Problem (5) (see [15, Theorem 3.7.3] and [25, Theorem 3.5.2]).

The Hamiltonian of Problem (5) is

$$\mathcal{H} = \mu_z'J_z + J_R\left(Rr + w'M\right) + \frac{1}{2}tr\left(\Omega'\Omega J_{zz}\right) + w'\Sigma'\Omega J_{zR} + \frac{1}{2}J_{RR}w'\Sigma'\Sigma w, \quad (6)$$

where the subscripts on $J$ indicate partial derivatives, and $J\left(R, z, t\right)$ is the value function solving the Hamilton-Jacobi-Bellman partial differential equation (see Section 3.1) and verifying

$$J\left(R, z, t\right) = \underset{w}{\sup}\mathbb{E}_t\left[\frac{1}{1-\delta}R\left(H\right)^{1-\delta}\right].$$

The first order condition on $\mathcal{H}$ is[5]

---

[4] We will explain the reason for this choice better in the next section.

[5] The second order condition holds if the Hessian matrix of $\mathcal{H}$

$$\frac{\partial \mathcal{H}}{\partial w'\partial w} = J_{RR}\Sigma'\Sigma$$

is negative definite. Because $\Sigma'\Sigma$ is a quadratic form it is always positive definite and so the second order conditions are satisfied if and only if $J_{RR} < 0$, that is if the value function is concave in $R$. This is actually the case if the utility function is strictly concave (as in our framework).

$$\frac{\partial \mathcal{H}}{\partial w} = J_R M + \Sigma' \Omega J_{zR} + J_{RR} \Sigma' \Sigma w = 0,$$

from which we obtain the optimal portfolio

$$w^* = \underbrace{-\frac{J_R}{J_{RR}} \left(\Sigma' \Sigma\right)^{-1} M}_{w^*_{(1)}} \underbrace{-\frac{1}{J_{RR}} \left(\Sigma' \Sigma\right)^{-1} \Sigma' \Omega J_{zR}}_{w^*_{(2)}}. \tag{7}$$

In order to have a unique solution to the optimization problem, the matrix $\Sigma' \Sigma \in \mathbb{R}^{n \times n}$ must be invertible. This condition is satisfied if $\Sigma' \in \mathbb{R}^{n \times k}$ has rank equal to $n$ and $n \le k$, as we have already assumed in the previous section.

We just outline that $w^*_{(1)}$ increases if the net returns on assets $(M)$ increase and decreases if the risk aversion $(-J_{RR}/J_R)$ or the asset variance $(\Sigma' \Sigma)$ increase. ¿From this point of view, we can argue that this optimal portfolio component has just a speculative role.

The second optimal portfolio component $w^*_{(2)}$ is the only one which explicitly depends on the diffusion of the state variables $(\Omega)$. We will investigate the precise role of this component after computing the functional form of the value function.

We recall that Kogan and Uppal[17] call $w^*_{(1)}$ the "myopic" component and $w^*_{(2)}$ the "hedging" component of the optimal portfolio. In fact, in the next section we will see that $w^*_{(2)}$ is the only part of $w^*$ depending on the financial time horizon $(H)$. From this point of view $w^*_{(1)}$ can be properly called "myopic". Instead, the hedging nature of $w^*_{(2)}$ depends on its characteristics to contain the volatility matrix of state variables. Accordingly, we can say that the second portfolio component $w^*_{(2)}$ can hedge the optimal portfolio against the risk of changes in the state variables.

## 3.1 The value function

In order to study the exact form of the portfolio components we have called $w^*_{(1)}$ and $w^*_{(2)}$ (see Equation (7)), we need to compute the value function $J(R, z, t)$. By substituting the optimal value of $w$ into the Hamiltonian (6) we have

$$\mathcal{H}^* = \mu'_z J_z + J_R R r - \frac{1}{2} \frac{J_R^2}{J_{RR}} M' \left(\Sigma' \Sigma\right)^{-1} M - \frac{J_R}{J_{RR}} M' \left(\Sigma' \Sigma\right)^{-1} \Sigma' \Omega J_{zR}$$
$$+ \frac{1}{2} tr \left(\Omega' \Omega J_{zz}\right) - \frac{1}{2} \frac{1}{J_{RR}} J'_{zR} \Omega' \Sigma \left(\Sigma' \Sigma\right)^{-1} \Sigma' \Omega J_{zR},$$

from which we can formulate the PDE whose solution is the value function. This is the so-called Hamilton-Jacobi-Bellman equation (hereafter HJB) and it can be written as follows:

$$\begin{cases} J_t + \mathcal{H}^* = 0, \\ J(H, R, z) = \frac{1}{1-\delta} R(H)^{1-\delta}. \end{cases} \tag{8}$$

One of the most common ways to solve this kind of PDE is to try a separability condition (through a so-called guess function). In the literature [20, 21], a separability by product is generally found. Here, we suppose that the value function $J(z, R, t)$ inherits the form of the utility function according to the following form:

$$J(z, R, t) = \frac{1}{1-\delta} h(z,t)^{\delta} R(t)^{1-\delta}, \tag{9}$$

where $h(z,t)$ is a function that must be computed. After substituting this functional form into the HJB equation (8) and dividing by $J$ we obtain

$$0 = h_t + \left( \mu_z' - \frac{\delta-1}{\delta} M' (\Sigma'\Sigma)^{-1} \Sigma'\Omega \right) h_z + \frac{1}{2} tr \left( \Omega'\Omega h_{zz} \right) \tag{10}$$

$$- \frac{\delta-1}{\delta} \left( r + \frac{1}{2}\frac{1}{\delta} M' (\Sigma'\Sigma)^{-1} M \right) h$$

$$+ \frac{1}{2}\frac{\delta-1}{h} h_z'\Omega' \left( I - \Sigma (\Sigma'\Sigma)^{-1} \Sigma' \right) \Omega h_z,$$

and the boundary condition becomes $h(z, H) = 1$. Equation (10) can be written as

$$\begin{cases} h_t + a(z,t)' h_z + b(z,t) h + \frac{1}{2} tr(\Omega'\Omega h_{zz}) + \frac{1}{2}\frac{1}{h} h_z' C(z,t) h_z = 0, \\ \qquad\qquad\qquad\qquad\qquad\qquad\qquad\qquad\qquad\qquad h(z, H) = 1, \end{cases} \tag{11}$$

where

$$a(z,t)' \equiv \mu_z' - \frac{\delta-1}{\delta} M' (\Sigma'\Sigma)^{-1} \Sigma'\Omega, \tag{12}$$

$$b(z,t) \equiv -\frac{\delta-1}{\delta} \left( r + \frac{1}{2}\frac{1}{\delta} M' (\Sigma'\Sigma)^{-1} M \right), \tag{13}$$

$$C(z,t) \equiv (\delta-1) \Omega' \left( I - \Sigma (\Sigma'\Sigma)^{-1} \Sigma' \right) \Omega. \tag{14}$$

In the usual economic setting, the coefficient $b(z,t)$ takes the role of a discount rate and it must be non positive. Here, it is evident that the coefficient $b(z,t)$ is non positive if and only if $\delta \geq 1$.

By using the separability condition, we can write the optimal portfolio as in the following proposition.

**Proposition 1.** *The optimal portfolio solving Problem (5) is given by*

$$w^* = \underbrace{\frac{R}{\delta} (\Sigma'\Sigma)^{-1} M}_{w_{(1)}^*} + \underbrace{\frac{R}{h(z,t)} (\Sigma'\Sigma)^{-1} \Sigma'\Omega \frac{\partial h(z,t)}{\partial z}}_{w_{(2)}^*}, \tag{15}$$

*where $h(z,t)$ solves the HJB equation (11).*

If we want to compute the optimal portfolio in a closed form, we accordingly have to compute the function $h(z,t)$ solving Equation (11). We are going to present both a closed form solution for a complete market and an approximate solution for an incomplete market.

We just underline that the optimal portfolio (15) is a linear transformation of wealth. This means that the percentage of wealth invested in each asset does not depend on the wealth level itself.

Menoncin[19] takes into account a framework where there exists also a set of background risks. He shows that if a CARA utility function is taken into account in an incomplete financial market (where the matrix $\Sigma^{-1}$ does not exist) then the HJB equation (11) can be solved thanks to the Feynman-Kač theorem.

## 3.2 An exact solution for complete market

When the financial market is complete then the volatility matrix $\Sigma$ is invertible and so we have

$$I - \Sigma\left(\Sigma'\Sigma\right)^{-1}\Sigma' = \mathbf{0}.$$

Accordingly, Equation (11) becomes

$$\begin{cases} h_t + a\left(z,t\right)'h_z + b\left(z,t\right)h + \frac{1}{2}tr\left(\Omega'\Omega h_{zz}\right) = 0, \\ \qquad\qquad\qquad\qquad\qquad\qquad h\left(z,H\right) = 1, \end{cases}$$

and its solution can be represented through the Feynman-Kač theorem as stated in the following proposition. Here, we stress that a Feynman-Kač type representation holds even for non linear PDE's as demonstrated in [25] (see, in particular, Chapter 7, Section 4, and Theorem 7.4.6).

**Proposition 2.** *When the financial market is complete (i.e. $\exists!\Sigma^{-1}$) the function $h\left(z,t\right)$ solving Equation (11) can be written as*

$$h\left(z,t\right) = \mathbb{E}_t\left[\int_t^H e^{\int_t^s b(Z(u),u)du}ds\right]$$

*where $b\left(Z,u\right)$ is as in (13) and $Z\left(u\right)$ solves the stochastic differential equation*

$$dZ\left(u\right) = a\left(Z\left(u\right),u\right)du + \Omega\left(Z\left(u\right),u\right)'dW,$$
$$Z\left(t\right) = z.$$

# 4 A general approximate solution for an incomplete market

In order to find an approximate solution to the HJB equation (11) we propose to develop in Taylor series the matrices $a\left(z,t\right)$, $b\left(z,t\right)$, $\Omega\left(z,t\right)'\Omega\left(z,t\right)$, and

$C(z,t)$ as defined in (12)-(14). In particular, our proposal in based on the literature computing the optimal portfolio in a closed form and in a particular market framework. In this literature (see for instance [6, 11, 2]) all the above-mentioned matrices are linear in $z$. In [16] the scalar $b(z,t)$ is a second order polynomial in $z$ but, on the other way, $\Omega(z,t)'\Omega(z,t)$ and $C(z,t)$ are both constant.

Accordingly, we propose the following simplification, based on the expansion in Taylor series around the value $z_0$:

$$a(z,t) \approx a(z_0,t) + \left.\frac{\partial a(z,t)}{\partial z}\right|_{z=z_0}(z-z_0) \equiv a_0(t) + A_1(t)'z,$$

$$b(z,t) \approx b(z_0,t) + \left.\frac{\partial b(z,t)}{\partial z'}\right|_{z=z_0}(z-z_0) \equiv b_0(t) + b_1(t)'z,$$

$$\Omega(z,t)'\Omega(z,t) \approx \Omega(z_0,t)'\Omega(z_0,t) \equiv \Omega_0(t),$$

$$C(z,t) \approx C(z_0,t) \equiv C_0(t).$$

The choice of approximating the matrices $\Omega(z,t)'\Omega(z,t)$ and $C(z,t)$ to order zero in $z$ allows us to avoid the problem of solving a matrix Riccati equation. Since we let all the matrices depend on time, then we would not be able to solve this Riccati matrix equation without knowing a particular solution to it, and this is not the case.

It is worth noting that Boulier et al. (2001) take into account a financial market where the matrices $a(z,t)$, $b(z,t)$, $\Omega(z,t)'\Omega(z,t)$, and $C(z,t)$ have exactly the form we use here as an approximation (i.e. the two first functions are linear in $z$ while the last two do not depend on $z$).

After substituting the Taylor approximations in (11), our guess function is[6]

$$h(z,t) = e^{F_0(t)+F_1(t)'z}, \tag{16}$$

where $F_0(t) \in \mathbb{R}$ and $F_1(t) \in \mathbb{R}^{s+n}$ solve

$$0 = \frac{\partial F_0(t)}{\partial t} + z'\frac{\partial F_1(t)}{\partial t} + \left(a_0(t)' + z'A_1(t)\right)F_1(t) + b_0(t) + b_1(t)'z$$
$$+ \frac{1}{2}F_1(t)'\left(\Omega_0(t) + C_0(t)\right)F_1(t).$$

This polynomial equation in $z$ can be split into two equations: one containing all the constant terms with respect to $z$ and one containing the terms in $z$. Thus, we can write down the following system:

$$\begin{cases} \frac{\partial F_0(t)}{\partial t} + a_0(t)'F_1(t) + b_0(t) + \frac{1}{2}F_1(t)'\left(\Omega_0(t) + C_0(t)\right)F_1(t) = 0, \\ \frac{\partial F_1(t)}{\partial t} + A_1(t)F_1(t) + b_1(t) = 0, \end{cases} \tag{17}$$

---

[6] This kind of guess function is quite usual in the optimization frameworks where the state variables follow affine processes.

and the original boundary condition on $h(z,t)$ is now defined on the two new functions $F_0(t)$ and $F_1(t)$ as follows:

$$\begin{cases} F_0(H) = 0, \\ F_1(H) = \mathbf{0}. \end{cases} \tag{18}$$

Accordingly, $F_1(t)$ is the (unique) solution of the second equation in (17), i.e.

$$F_1(t) = \int_t^H e^{\int_t^s A_1(\tau)d\tau} b_1(s)\,ds, \tag{19}$$

where the exponential matrix is computed, as usual, as[7]

$$e^{\int_t^s A_1(\tau)d\tau} = \sum_{n=0}^{\infty} \frac{1}{n!} \left( \int_t^s A_1(\tau)\,d\tau \right)^n.$$

Once the value of $F_1(t)$ is obtained, the value of $F_0(t)$ can be easily computed by solving the first equation in (17).

Since the solutions $F_0(t)$ and $F_1(t)$ to system (17) are unique, then the function $h(z,t)$ is unique too. Furthermore, since $h(z,t)$ is a classical solution to the HJB approximate equation, then it is also a viscosity solution. Yong and Zhou demonstrate such a result together with a uniqueness theorem [25, Sections 5 and 6 of Chapter 4]. The differentiability of the function $h(z,t)$ allows us to write the optimal portfolio hedging component as

$$w_{(2)}^* = R\left(\Sigma'\Sigma\right)^{-1}\Sigma'\Omega\frac{1}{h(z,t)}\frac{\partial h(z,t)}{\partial z} = R\left(\Sigma'\Sigma\right)^{-1}\Sigma'\Omega F_1(t),$$

and its approximate value can be formulated as in Proposition 3.

**Proposition 3.** *The second component $(w_{(2)}^*)$ of the optimal portfolio solving Problem (5) can be approximated as follows:*

$$w_{(2)}^* \approx R\left(\Sigma'\Sigma\right)^{-1}\Sigma'\Omega \int_t^H e^{\int_t^s A_1(\tau)d\tau} b_1(s)\,ds, \tag{20}$$

*where*

---

[7] Thus, for instance, if

$$A_1(t) = \begin{bmatrix} 1 & 2 \\ 2 & 1 \end{bmatrix},$$

then

$$\int_t^s A_1(\tau)\,d\tau = \begin{bmatrix} 1 & 2 \\ 2 & 1 \end{bmatrix}(s-t),$$

and

$$e^{\int_t^s A_1(\tau)d\tau} = e^{(s-t)}\begin{bmatrix} \cosh(2(s-t)) & \sinh(2(s-t)) \\ \sinh(2(s-t)) & \cosh(2(s-t)) \end{bmatrix}.$$

$$A_1(t) \equiv \frac{\partial}{\partial z} \left( \mu_z - \frac{\delta-1}{\delta} \Omega' \Sigma \left( \Sigma' \Sigma \right)^{-1} M \right) \Bigg|_{z=z_0},$$

$$b_1(t) \equiv -\frac{\delta-1}{\delta} \frac{\partial}{\partial z} \left( r + \frac{1}{2}\frac{1}{\delta} M' \left( \Sigma' \Sigma \right)^{-1} M \right) \Bigg|_{z=z_0}.$$

The approximate solution for $w^*_{(2)}$ given in Equation (20) is very simple to apply and to implement with a mathematical software. The weak point of this approximate solution lies on the difference $z - z_0$. When this difference increases the approximation becomes more and more inaccurate. Nevertheless, the integrals in Equation (20) can be easily computed numerically, and so the strategy of recomputing the optimal portfolio when $z$ becomes farther off $z_0$ does not seem to be too expensive.

We are now going to present how to estimate the approximation error implied in Proposition 3.

### 4.1 The maximum error

In the previous section we have presented the approximate optimal portfolio solving Problem 5. Here, we recall that the Taylor approximation of the matrices $a(z,t)$, $b(z,t)$, $\Omega(z,t)' \Omega(z,t)$, and $C(z,t)$ implies a maximum error $(\varepsilon)$ which is respectively:[8]

$$\varepsilon_{a_i}(t,z) \equiv \max_{\lambda \in [0,1]} \left\{ \frac{1}{2}(z-z_0)' \left( \frac{\partial^2 a_i(z,t)}{\partial z' \partial z} \Bigg|_{z=z_0+\lambda(z-z_0)} \right) (z-z_0) \right\}, \quad (21)$$

$$\varepsilon_b(t,z) \equiv \max_{\lambda \in [0,1]} \left\{ \frac{1}{2}(z-z_0)' \left( \frac{\partial^2 b(z,t)}{\partial z' \partial z} \Bigg|_{z=z_0+\lambda(z-z_0)} \right) (z-z_0) \right\}, \quad (22)$$

$$\varepsilon_\Omega(t,z) \equiv \max_{\lambda \in [0,1]} \left\{ \frac{\partial}{\partial z} \left( \Omega'(z,t) \Omega(z,t) \right) \Bigg|_{z=z_0+\lambda(z-z_0)} (z-z_0) \right\}, \quad (23)$$

$$\varepsilon_C(t,z) \equiv \max_{\lambda \in [0,1]} \left\{ \frac{\partial C(z,t)}{\partial z} \Bigg|_{z=z_0+\lambda(z-z_0)} (z-z_0) \right\}, \quad (24)$$

where $a_i$ is the $i^{th}$ element ($i \in [1,...,s+n+1]$) of vector $a$. Now, after setting

$$\varepsilon_a(z,t) = \{\varepsilon_{a_i}(z,t)\}_{i=1,...,s+n+1}, \quad (25)$$

we can substitute the error values into the HJB equation and conclude what follows.

**Proposition 4.** *Let function $h(z,t)$ be as in (16) where $F_0(t)$ and $F_1(t)$ solve the system (17) with boundary conditions (18), and the maximum error terms $\varepsilon_a(z,t)$, $\varepsilon_b(z,t)$, and $\varepsilon_C(z,t)$ as in (21)-(25), then*

---

[8] We consider here the so-called Lagrange's error term.

$$\lim_{z \to z_0} \left( \varepsilon_a \left( z,t \right)' h_z + \varepsilon_b \left( z,t \right) h + \frac{1}{2} tr \left( \varepsilon_\Omega \left( z,t \right) h_{zz} \right) + \frac{1}{2} \frac{1}{h} h_z' \varepsilon_C \left( t,z \right) h_z \right) = 0.$$

(26)

*Proof.* Let us substitute the values $a \left( z,t \right)$, $b \left( z,t \right)$, $\Omega \left( z,t \right)' \Omega \left( z,t \right)$, and $C \left( z,t \right)$, by their approximations augmented by the maximum error. Then, the HJB equation (11) can be written as

$$0 = h_t + \left( a_0 \left( t \right) + A_1 \left( t \right)' z + \varepsilon_a \left( z,t \right) \right)' h_z + \left( b_0 \left( t \right) + b_1 \left( t \right)' z + \varepsilon_b \left( z,t \right) \right) h$$

$$+ \frac{1}{2} tr \left( \Omega_0 \left( t \right) h_{zz} + \varepsilon_\Omega \left( z,t \right) h_{zz} \right) + \frac{1}{2} \frac{1}{h} h_z' \left( C_0 \left( t \right) + \varepsilon_C \left( t,z \right) \right) h_z,$$

or

$$0 = h_t + \left( a_0 \left( t \right) + A_1 \left( t \right)' z \right)' h_z + \left( b_0 \left( t \right) + b_1 \left( t \right)' z \right) h + \frac{1}{2} tr \left( \Omega_0 \left( t \right) h_{zz} \right)$$

$$+ \frac{1}{2} \frac{1}{h} h_z' C_0 \left( t \right) h_z + \varepsilon_a \left( z,t \right)' h_z + \varepsilon_b \left( z,t \right) h + \frac{1}{2} tr \left( \varepsilon_\Omega \left( z,t \right) h_{zz} \right)$$

$$+ \frac{1}{2} \frac{1}{h} h_z' \varepsilon_C \left( t,z \right) h_z.$$

Since the function $h \left( z,t \right)$ is as in (16) and solves (17), then it makes the first five terms of this equation equal to zero. The remaining terms just contain the error terms. As it is well known, these terms tend to zero while $z$ tends to $z_0$ according to the Taylor expansion with the so-called Peano's remainder.

Thus, after choosing an initial level $z_0$ for approximating the matrices $a \left( z,t \right)$, $b \left( z,t \right)$, $\Omega \left( z,t \right)' \Omega \left( z,t \right)$, and $C \left( z,t \right)$, the goodness of these approximations can be easily checked by computing the absolute value of (26).

From what we have presented in Proposition 4, the convergence of the matrices $a \left( z,t \right)$, $b \left( z,t \right)$, $\Omega \left( z,t \right)' \Omega \left( z,t \right)$, and $C \left( z,t \right)$, implies the convergence of the function $h$. Because of the form (9) of the value function, then we also know that the convergence of $h$ implies the convergence of the value function itself.

## 5 Two incomplete markets displaying an exact solution

### 5.1 Stochastic volatility

The market structure studied in [6] can be summarized as

$$\begin{cases} dX = k \left( \theta - X \right) dt + \sigma \sqrt{X} dW_X, & X \left( t_0 \right) = X_0, \\ \frac{dS}{S} = \mu dt + \frac{\sigma_{SX}}{\sigma \sqrt{X}} dW_X + \frac{1}{\sqrt{X}} \sqrt{1 - \frac{\sigma_{SX}^2}{\sigma^2}} dW_S, & S \left( t_0 \right) = S_0, \\ dG = Gr dt, & G \left( t_0 \right) = G_0, \end{cases}$$

where $k$, $\theta$, $\sigma$, $\mu$, and $r$ are positive constant. $dW_X$ and $dW_S$ are two independent Wiener processes and $\sigma_{SX} dt$ is the covariance between $dX$ and $\frac{dS}{S}$ (here, we do not impose any sign on $\sigma_{SX}$)[9]. The correlation index between $dX$ and $\frac{dS}{S}$ is given by $\frac{\sigma_{SX}}{\sigma}$. The only acceptable values for $\sigma_{SX}$ are those verifying

$$-\sigma \leq \sigma_{SX} \leq \sigma,$$

for which the correlation index belongs to the set $[-1, 1]$.

Since the variance of $\frac{dS}{S}$ is $\frac{1}{X}dt$, then this is a stochastic volatility model where the (inverse of the) stock volatility follows a diffusion process.

This is also an incomplete market since we have two risk sources (driven by $dW_X$ and $dW_S$) but just one tradeable risky asset (we recall that volatility cannot be traded and, accordingly, cannot be hedged).

This model can be traced back to ours by setting

$$z \equiv X, \quad \mu_z \equiv k(\theta - X), \quad M \equiv (\mu - r),$$

$$\Omega \equiv \begin{bmatrix} \sigma\sqrt{X} \\ 0 \end{bmatrix}, \quad \Sigma \equiv \begin{bmatrix} \frac{\sigma_{SX}}{\sigma\sqrt{X}} \\ \frac{1}{\sqrt{X}}\sqrt{1 - \frac{\sigma_{SX}^2}{\sigma^2}} \end{bmatrix}.$$

The functions $a(z,t)$ and $b(z,t)$ as in (12) and (13) are, respectively,

$$a(X,t) = k\theta - \left(k + \frac{\delta - 1}{\delta}(\mu - r)\sigma_{SX}\right)X,$$

$$b(X,t) = -\frac{\delta - 1}{\delta}r - \frac{\delta - 1}{\delta}\frac{1}{2}\frac{1}{\delta}(\mu - r)^2 X,$$

and so we have

$$A_1 = -\left(k + \frac{\delta - 1}{\delta}(\mu - r)\sigma_{SX}\right),$$

$$b_1 = -\frac{\delta - 1}{\delta}\frac{1}{2}\frac{1}{\delta}(\mu - r)^2.$$

Since $A_1$ and $b_1$ do not depend on time, and the time horizon in [6] tends towards infinity (i.e. $H \to \infty$), then Equation (20) simplifies to[10]

$$\frac{w^*_{(2)}}{R} \approx -\left(\Sigma'\Sigma\right)^{-1}\Sigma'\Omega\frac{b_1}{A_1}$$

$$= -X\sigma_{SX}\frac{1}{2}\frac{\delta - 1}{\delta}\frac{1}{\delta}\frac{(\mu - r)^2}{k + \frac{\delta - 1}{\delta}(\mu - r)\sigma_{SX}}.$$

---

[9] In the original paper by Chacko and Viceira[6] the two Wiener processes are correlated. Here, we have accordingly changed the volatility coefficient in order to keep the same statistics property as Chacko and Viceira's and to adapt their model to our framework.

[10] In order to have a convergent integral, it must be true that

$$k + \frac{\delta - 1}{\delta}(\mu - r)\sigma_{SX} > 0.$$

The exact solution shown in [6] is

$$\frac{w_{(2)}^{**}}{R} = X \sigma_{SX} \frac{1}{\delta} y,$$

where $y$ is the negative root[11] of

$$\frac{1}{2} \left( \frac{\delta - 1}{\delta} \sigma_{SX}^2 - \sigma^2 \right) y^2 + \left( k + \frac{\delta - 1}{\delta} (\mu - r) \sigma_{SX} \right) y + \frac{1}{2} \frac{\delta - 1}{\delta} (\mu - r)^2 = 0.$$

We can define the difference

$$\phi \equiv \frac{w_{(2)}^{**} - w_{(2)}^*}{R} = X \frac{\sigma_{SX}}{\delta} \left( y + \frac{1}{2} \frac{\frac{\delta-1}{\delta} (\mu - r)^2}{k + \frac{\delta-1}{\delta} (\mu - r) \sigma_{SX}} \right),$$

which is of course 0 when the approximation perfectly matches the true solution. A trivial case when $\phi = 0$ is that with $\sigma_{SX} = 0$, i.e. when the stochastic volatility process is not correlated with the risky asset price.

Now, we study the value of $\phi$ with respect to $\sigma_{SX}$ by neglecting the value of $X$ and given all the other parameters. In particular, we assume that

$$\mu - r = 0.07, \quad \sigma = 0.2, \quad \delta = 3, \quad k = 0.2.$$

The behaviour of the function $\phi$ is drawn in Figure 1 where we see that the maximum error is of the order $10^{-7}$.

## 5.2 Stochastic market price of risk

The market structure studied in [16] can be summarized as

$$\begin{cases} dX = k (\theta - X) \, dt + \sigma_X dW_X, & X (t_0) = X_0, \\ \frac{dS}{S} = (r + \sigma_S X) \, dt + \sigma_S \rho dW_X + \sigma_S \sqrt{1 - \rho^2} dW_S, & S (t_0) = S_0, \\ dG = Grdt, & G (t_0) = G_0, \end{cases}$$

where $k$, $\theta$, $\sigma_X$, $\sigma_S$, and $r$ are positive constant. $dW_X$ and $dW_S$ are two independent Wiener processes and $\rho$ is the correlation coefficient between $dX$ and $\frac{dS}{S}$. Here, $X$ measures the (stochastic) market price of risk.

This model can be traced back to ours by setting

$$z \equiv X, \quad \mu_z \equiv k (\theta - X), \quad M \equiv \sigma_S X,$$

$$\Omega \equiv \begin{bmatrix} \sigma_X \\ 0 \end{bmatrix}, \quad \Sigma \equiv \begin{bmatrix} \sigma_S \rho \\ \sigma_S \sqrt{1 - \rho^2} \end{bmatrix}.$$

The functions $a (z, t)$ and $b (z, t)$ as in (12) and (13) respectively are

---

[11] This holds for $\delta > 1$. If $\delta < 1$, instead, we must take the positive root.

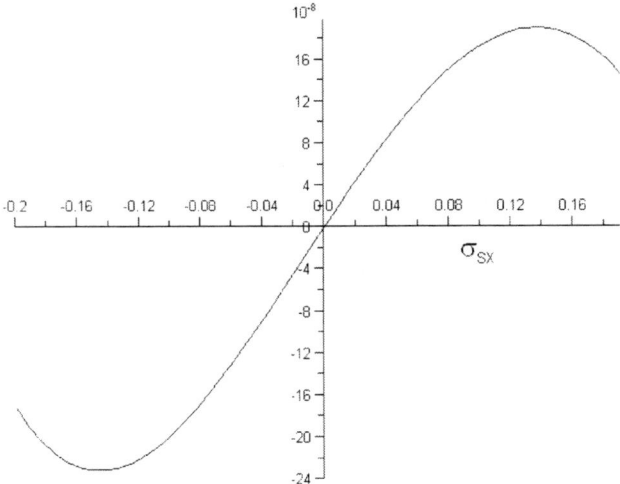

**Fig. 1.** Value of the difference between the actual and the approximate solution as a function of $\sigma_{SX}$

$$a\left(X,t\right) = k\theta - \left(k + \frac{\delta - 1}{\delta}\rho\sigma_X\right)X,$$

$$b\left(X,t\right) = -\frac{\delta - 1}{\delta}r - \frac{1}{2}\frac{1}{\delta}\frac{\delta - 1}{\delta}X^2,$$

and so we have

$$A_1 = -\left(k + \frac{\delta - 1}{\delta}\rho\sigma_X\right),$$

$$b_1 = -\frac{1}{\delta}\frac{\delta - 1}{\delta}X_0,$$

where $X_0$ is the value of $X$ around which we have approximated.

Since $A_1$ and $b_1$ do not depend on time the Equation (20) simplifies to

$$\frac{w^*_{(2)}}{R} \approx \left(\Sigma'\Sigma\right)^{-1}\Sigma'\Omega b_1 \frac{e^{A_1\tau} - 1}{A_1}$$

$$= -\frac{1}{\delta}\frac{\rho\sigma_X}{\sigma_S}\frac{\delta - 1}{\delta}X_0 \frac{1 - e^{-\left(k + \frac{\delta-1}{\delta}\rho\sigma_X\right)\tau}}{k + \frac{\delta-1}{\delta}\rho\sigma_X},$$

where $\tau \equiv H - t$.

The exact solution shown in [16] is

$$\frac{w^{**}_{(2)}}{R} = \frac{\rho\sigma_X}{\sigma_S}\frac{1}{\delta}\left(B\left(\tau\right) + C\left(\tau\right)X\right),$$

where

$$B\left(\tau\right) \equiv \frac{4\frac{1-\delta}{\delta}k\theta\left(1 - e^{-\frac{\eta}{2}\tau}\right)^{2}}{2\eta^{2} - \eta\left(\eta + \sigma_{X}^{2}\left(1 + \frac{1-\delta}{\delta}\rho^{2}\right)\right)\left(1 - e^{-\eta\tau}\right)},$$

$$C\left(\tau\right) \equiv \frac{2\frac{1-\delta}{\delta}\left(1 - e^{-\eta\tau}\right)}{2\eta - \left(\eta + \sigma_{X}^{2}\left(1 + \frac{1-\delta}{\delta}\rho^{2}\right)\right)\left(1 - e^{-\eta\tau}\right)},$$

$$\eta \equiv 2k\sqrt{1 - \frac{1-\delta}{\delta}\left(\frac{\sigma_{X}^{2}}{k^{2}} + 2\rho\frac{\sigma_{X}}{k}\right)}.$$

Since the optimal strategy we have found is in the feed-back form, then we can assume $X_0 = X$ at each time. In fact, we can adjust our approximation at each instant.

The index $\phi$, in this case, has the form

$$\phi \equiv \frac{w_{(2)}^{**} - w_{(2)}^{*}}{R} = \frac{1}{\delta}\frac{\rho\sigma_{X}}{\sigma_{S}}\left(B\left(\tau\right) + C\left(\tau\right)X + X\frac{1 - e^{-\left(k + \frac{\delta - 1}{\delta}\rho\sigma_{X}\right)\tau}}{k + \frac{\delta - 1}{\delta}\rho\sigma_{X}}\right).$$

In Figure 2 we show the behaviour of $\phi$ with respect to $\rho$ and $X$ given the following values:

$$X = 0.35, \quad \sigma_S = 0.2, \quad \delta = 3, \quad k = 0.2, \quad \sigma_X = 0.1, \quad \theta = 0.35.$$

As in the previous example, the closer the correlation to 0, the better the approximation. Furthermore, in this case, there exists a time horizon $\tau$ for which the approximation is best. We underline that our approximation does not work well for any time horizon. In particular, while the time horizon becomes longer, our approximation becomes worse.

# 6 Conclusion

In this paper we have studied the problem of an investor maximizing the expected CRRA utility of his terminal wealth (for a deterministic time horizon). All the variables taken into account are supposed to follow general diffusion processes. In particular, we consider a set of financial assets and a set of state variables in a incomplete financial market.

We have computed the Hamilton-Jacobi-Bellman (HJB) equation solving the investor's dynamic programming problem and we have proposed an approximate solution to it. In particular, we have approximated in Taylor series some matrices whose values are given by combinations of preference parameters and drift and diffusion terms for both asset prices and state variables. After substituting the approximations into the HJB equation, the value function solving it has turned out to be a log-linear function of the state variables.

Finally, we have presented both an easy test for checking the goodness of the approximate optimal portfolio and the comparison of our approximate

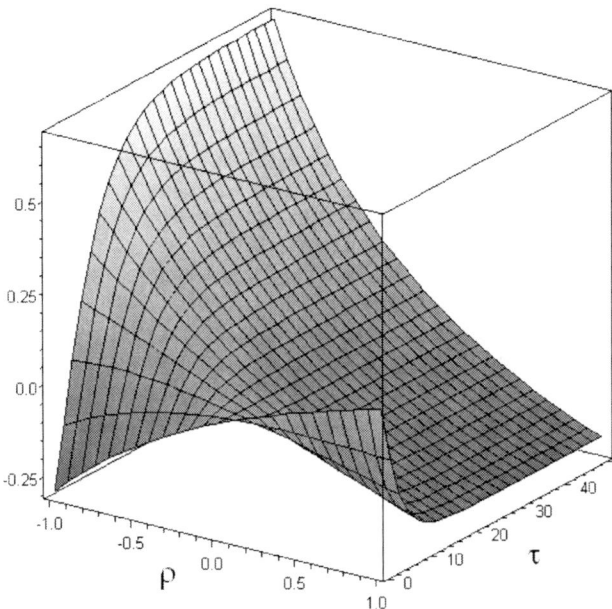

**Fig. 2.** Value of the difference between the actual and the approximate solution as a function of $\tau$ and $\rho$

solution with two exact solutions presented in the literature for two particular cases of incomplete markets.

With respect to the literature, our model presents a higher degree of generality in terms of financial market structure it deals with. In particular, while the literature is mainly concerned with the problem of the existence of a solution without providing an actual form for it, our model supplies an approximation which can be useful for computing the actual optimal portfolio for the proposed problem.

# References

1. Björk T (1998) Arbitrage Theory in Continuous Time. Oxford University Press, New York
2. Boulier JF, Huang SJ, Taillard G (2001) Optimal Management Under Stochastic Interest. Insurance: Mathematics and Economics 28:173-189.
3. Buckdahn R, Ma J (2001a) Stochastic viscosity solutions for nonlinear stochastic partial differential equations - Part I. Stochastic Processes and their Applications 93:181-204.
4. Buckdahn R, Ma J (2001b) Stochastic viscosity solutions for nonlinear stochastic partial differential equations - Part II. Stochastic Processes and their Applications 93:205-228.

5. Campbell JY (2000) Asset Pricing at the Millennium. The Journal of Finance 55:1515-1567.
6. Chacko G, Viceira LM (2005) Dynamic consumption and portfolio choice with stochastic volatility in incomplete markets. The Review of Financial Study 18:1369 1402.
7. Cox JC, Huang CF (1989) Optimal consumption and portfolio policies when asset prices follow a diffusion process. Journal of Economic Theory 49:33-83.
8. Cox JC, Huang CF (1991) A variational problem arising in financial economics. Journal of Mathematical Economics 20:465-487.
9. Cox JC, Ingersoll JE Jr., Ross SA (1985) A Theory of the Term Structure of Interest Rates. Econometrica 53:385-407.
10. Crandall MG, Ishii H, Lions P (1992) User's guide to viscosity solutions of second order partial differential equations. Bulletin of the American Mathematical Society 27:1-67.
11. Deelstra G, Grasselli M, Koehl PF (2000) Optimal investment strategies in a CIR framework. Journal of Applied Probability 37:1-12.
12. Duffie D (1996) Dynamic Asset Pricing Theory. Second edition, Princeton University Press.
13. Duffie D, Kan R (1996) A Yield-Factor Model of Interest Rates. Mathematical Finance 6:379-406.
14. Freiling G (2002) A survey of nonsymmetric Riccati equations. Linear Algebra and its Applications 351-352:243-270.
15. Karatzas I, Shreve SE (1998) Methods of Mathematical Finance. Springer
16. Kim TS, Omberg E (1996) Dynamic Nonmyopic Portfolio Behavior. The Review of Financial Studies 9:141-161.
17. Kogan L, Uppal R (1999) Risk Aversion and Optimal Portfolio Policies in Partial and General Equilibrium Economies. Working Paper, Wharton.
18. Lioui A, Poncet P (2001) On Optimal Portfolio Choice under Stochastic Interest Rates. Journal of Economic Dynamics and Control 25:1841-1865.
19. Menoncin F (2002) Optimal Portfolio and Background Risk: An Exact and an Approximated Solution. Insurance: Mathematics and Economics 31:249-265.
20. Merton RC (1969) Lifetime Portfolio Selection under Uncertainty: the Continuous-Time Case. Review of Economics and Statistics 51:247-257.
21. Merton RC (1971) Optimum Consumption and Portfolio Rules in a Continuous-Time Model. Journal of Economic Theory 3:373-413.
22. Øksendal B (2000) Stochastic Differential Equations - An Introduction with Applications - Fifth edition. Springer, Berlin
23. Vasiček O (1977) An Equilibrium characterization of the Term Structure. Journal of Financial Economics 5:177-188.
24. Wachter JA (1998) Portfolio and Consumption Decisions Under Mean-Reverting Returns: An Exact Solution for Complete Markets. Working Paper, Harvard University
25. Yong J, Zhou XY (1999) Stochastic Controls - Hamiltonian Systems and HJB Equations. Springer
26. Zariphopoulou T (2001) A solution approach to valuation with unhedgeable risks. Finance and Stochastics 5:61-82.

# Carleman Linearization of Linearly Observable Polynomial Systems

Dorota Mozyrska and Zbigniew Bartosiewicz

Białystok Technical University, Faculty of Computer Science,
Wiejska 45A, Białystok, Poland. `admoz@w.tkb.pl`, `bartos@pb.bialystok.pl`

**Summary.** Carleman linearization is used to transform a polynomial control system with output, defined on $n$-dimensional space, into a linear or bilinear system evolving in the space of infinite sequences. Such a system is described by infinite matrices with special properties. Linear observability of the original system is studied. It means that all coordinate functions can be expressed as linear combinations of functions from the observation space. It is shown that this property is equivalent to a rank condition involving matrices that appear in the Carleman linearization. This condition is equivalent to observability of the first $n$ coordinates of the linearized system.

## 1 Introduction

Carleman linearization is a procedure that allows to embed a finite dimensional system of differential equations, with analytic or polynomial data, into a system of linear differential equations on an infinite dimensional space. Thus we trade polynomials (or analytic functions) that describe the system for the infinite matrices of the Carleman linearization. The reader may consult [12], which gives a general introduction to the subject.

There were several attempts to apply Carleman linearization in control theory. Let us mention [19], where this technique was used for linear systems with polynomial output. We consider a more general situation, where also the system's dynamics is polynomial. Our goal is to relate observability of the original system and its Carleman linearization. We study linear observability of a polynomial system on $\mathbb{R}^n$, which means that all the coordinate functions can be expressed as linear combinations of functions from the observation space of the system. Two cases are studied. The simpler one concerns a system without control, which leads to a linear system with output, also without control. We show that the original system is linearly observable if and only if the first $n$ coordinates of its Carleman linearized system are observable, and express this property by a rank condition involving the matrices of the linearized system. The other case, where the original system contains control, leads to a bilinear infinite dimensional system. A similar rank condition for linear observability is presented.

Checking observability of infinite dimensional linear systems is not easy as we have to deal with infinite matrices. Though the Kalman condition of observability holds for this class of systems, it must be expressed in a different way and there is no finite algorithm to check this. Here we have a simpler task, as only finitely many coordinates are to be observed. Several results on observability of systems described by infinite matrices were given in our papers [3, 4] and duality between observability and controllability was studied in [17]. In [3] we considered discrete-time systems, which are easier to study since existence and uniqueness of solutions is always guaranteed. The continuous-time case, examined in [4], is much harder as even for linear infinite systems of differential equations, solutions of initial value problems may not exist or be nonunique. Concerning this subject more can be found in [10, 13, 14]. In the next section we provide the reader with basic definitions and facts. The Banach space case, studied in [6, 7], is more regular. Observability of nonlinear infinite dimensional systems was studied in [15, 16]. Such systems may appear as infinite extensions of finite dimensional nonlinear control systems (see e.g. [11]). One can be also interested in an embedding or an immersion of a finite-dimensional nonlinear system into a linear *finite-dimensional* system. Fliess and Kupka [9] constructed such immersion under the assumption that the observation space of the system without control is finite-dimensional. For systems with control this led to state-affine or bilinear systems. If we replace this assumption by the condition that the observation algebra is finitely generated, the nonlinear system can be immersed into a polynomial (with respect to state) finite-dimensional system (see [2]).

## 2 Preliminaries

Let $\mathbb{R}^\infty$ denote the linear space of all real sequences denoted by infinite columns. Let $\Pi_n : \mathbb{R}^\infty \to \mathbb{R}^n$ denotes the projection on the first $n$ coordinates, that is if $z = (z_1, \ldots, z_n, \ldots)^T \in \mathbb{R}^\infty$ then $\Pi_n(z) = (z_1, \ldots, z_n)^T$. We say that a function $f : \mathbb{R}^\infty \to \mathbb{R}$ is *finitely presented* on $\mathbb{R}^\infty$ if there is $n \in \mathbb{N}$ and a function $\tilde{f} : \mathbb{R}^n \to \mathbb{R}$ such that $f = \tilde{f} \circ \Pi_n$. If we consider $\mathbb{R}^\infty$ as a topological space we use the product (Tikhonov) topology. A basis of this topology consists of the sets $U = \prod_{i \in \mathbb{N}} U_i$, where $U_i$ is the open subset of $\mathbb{R}$ and $U_i = \mathbb{R}$ for all $i$, but finite number values of $i$. It is the weakest topology for which projections $\Pi_n$ are continuous.

From [1] we have the following:

**Proposition 1.** *Let $L(\mathbb{R}^\infty, \mathbb{R})$ be the space of linear and continuous functions on $\mathbb{R}^\infty$. If $f \in L(\mathbb{R}^\infty, \mathbb{R})$ then $f$ is finitely presented and there are $n_f \in \mathbb{N}$ and $c_1, \ldots, c_{n_f} \in \mathbb{R}$ such that for all $z \in \mathbb{R}^\infty$, $f(z) = \sum_{i=1}^{n_f} c_i z_i$.*

We deal with differential systems described by infinite matrices which can be interpreted as functions from $\mathbb{R}^\infty \times \mathbb{R}^\infty$ to $\mathbb{R}$. We say that a matrix $A = (a_{ij})_{i,j \in \mathbb{N}}$ is *row-finite* if for each $i \in \mathbb{N}$ there is $n(i) \in \mathbb{N}$ such that for $j > n(i)$, $a_{ij} = 0$. The matrix is *upper-diagonal* if $a_{ij} = 0$ for $j < i$. The set of row-finite matrices forms an algebra over $\mathbb{R}$ with a unit $E = (\delta_{ij})_{i,j \in \mathbb{N}}$. Hence if $A$ is a row-finite matrix, then for each $k \in \mathbb{N}$, $A^k$ is well defined and it is a row-finite matrix as well.

Let $A$ be an infinite row-finite matrix. Then the system of differential equations $\dot{z}(t) = \frac{dz}{dt} = Az(t)$ is called a row-finite system. If together with this system we consider the initial condition $z(0) = z^0 \in \mathbb{R}^\infty$ then the discussion of existence and uniqueness of solutions of the initial value problem can be found, e.g., in [4, 7, 15, 16]. In particular the concept of formal solutions is there presented.

**Proposition 2.** *Let* $\frac{dz}{dt} = Az, z(0) = z^0 \in \mathbb{R}^\infty$, *be the initial value problem with $A$ being a row-finite matrix. Then it has the unique formal solution* $\Gamma_{z^0,A} := \sum_{k=0}^{\infty} \frac{t^k}{k!} A^k z^0$.

We are concerned with the system with output:

$$(\Sigma): \begin{array}{rl} \dot{z}(t) = & Az(t) \\ y(t) = & Cz(t), \end{array} \tag{1}$$

where $z : [0, \infty) \to \mathbb{R}^\infty, y : [0, \infty) \to \mathbb{R}^p$, and $A \in \mathbb{R}^\infty \times \mathbb{R}^\infty$ and $C \in \mathbb{R}^p \times \mathbb{R}^\infty$ are row-finite. Let $z^0 \in \mathbb{R}^\infty$. Given a formal solution $\Gamma_{z^0,A}$ of the dynamical part of the system and corresponding to the initial condition $z^0$ we define the formal output: $\mathcal{Y}_{z^0} = C\Gamma_{z^0,A}$.

**Definition 1.** *We say that $z^1, z^2 \in \mathbb{R}^\infty$ are indistinguishable (with respect to $(\Sigma)$) if $\mathcal{Y}_{z^1} = \mathcal{Y}_{z^2}$. Otherwise $z^1, z^2$ are distinguishable. We say that the system $(\Sigma)$ is observable if any two distinct points are distinguishable.*

**Proposition 3.** ([4])
*The points $z^1, z^2 \in \mathbb{R}^\infty$ are indistinguishable iff for all $k \in \mathbb{N} \cup \{0\}$: $CA^k z^1 = CA^k z^2$.*

**Corollary 1.** ([4])
*System $(\Sigma)$ is observable if and only if $\forall \, n \in \mathbb{N} \; \exists k \in \mathbb{N} \cup \{0\}$:*

$$\text{rank} \begin{pmatrix} C \\ \vdots \\ CA^k \end{pmatrix} = \text{rank} \begin{pmatrix} C \\ \vdots \\ CA^k \\ E_n^T \end{pmatrix}, \tag{2}$$

*where $E_n^T$ denotes the infinite row with 1 at the $n$-th position.*

Let $D := \begin{pmatrix} C \\ CA \\ \vdots \end{pmatrix}$. Since the rows of $D$ correspond to derivatives of the output, one can characterize observability as possibility to compute every state variable as a linear combination of finitely many outputs and their derivatives.

As using formal solutions and formal outputs brought us to the characterization of observability based on matrices of the systems, we extend now this concept by defining some kind of observability of infinite bilinear systems.

**Definition 2.** *Let us consider an infinite bilinear system*

$$\begin{aligned} \dot{z} &= (A + uB)z, \\ y &= Cz, \end{aligned} \tag{3}$$

*where* $z : [0, \infty) \to \mathbb{R}^\infty, y : [0, \infty) \to \mathbb{R}^p$, $A, B \in \mathbb{R}^\infty \times \mathbb{R}^\infty$ *and* $C \in \mathbb{R}^p \times \mathbb{R}^\infty$ *are row–finite matrices. Let*

$$\Gamma(C, A, B)_k := col[C, CA, CB, CA^2, CAB, CBA, CB^2, \dots, CB^k].$$

*System (3) is said to be* formally observable *if* $\forall\ n \in \mathbb{N}\ \exists k \in \mathbb{N} \cup \{0\}$ *such that*

$$\mathrm{rank}\Gamma(C, A, B)_k = \mathrm{rank}\begin{pmatrix} \Gamma(C, A, B)_k \\ E_n^T \end{pmatrix}. \tag{4}$$

*Remark 1.* Condition (4) becomes the same as (2) for $B = 0$.

We will consider the situation when condition (4) is satisfied only for some number of variables.

**Definition 3.** *We say that system (3) is* observable with respect to the variable $z_i$ *if for* $n = i$ *there is* $k$ *such that (4) holds.*

*Remark 2.* System (3) is formally observable iff it is observable with respect to each variable.

## 3 Carleman linearization

By $\mathcal{M}(m, n)$ we denote the set of matrices of dimensions $m \times n$ with real elements.

Let the function $x : \mathbb{R} \supset J \to \mathbb{R}^n, x \in C^1(J)$, be a solution of the first order system of ordinary differential equations:

$$\Sigma : \frac{dx}{dt} = F(x), \tag{5}$$

where $F = (f_1, \dots, f_n)^T$ is a vector field whose components are polynomials without constant term, i.e. $F(0) = 0$. Then we can write $F(x) = \sum_{k=1}^m F_k(x)$, where each $F_k$ is a vector of homogeneous polynomials of degree $k$ and $m \in \mathbb{N}$, $m \geq \max_i \deg f_i$.

For every integer $k \geq 1$, let $H_k$ denote the space of homogeneous polynomials of degree $k$ in $n$ variables $x_1, x_2, \ldots, x_n$. In $H_k$ we choose the canonical bases $\{x^q = x_1^{q_1} \ldots x_n^{q_n}\}$ with $q = (q_1, \ldots, q_n)$, where $q_i \in \mathbb{N}$ and $|q| = \sum_{i=1}^n q_i = k$. We use the lexicographic order in the set of monomials $x^q$ induced by the order $x_1 < x_2 < \ldots < x_n$. We use the following notation:

$$
\begin{aligned}
e_1^k = x_1^k, e_2^k = x_1^{k-1}x_2, \ldots, e_{d_k}^k = x_n^k, \\
\zeta_k = \left(e_1^k, \ldots, e_{d_k}^k\right)^T,
\end{aligned}
\tag{6}
$$

where $d_k = \binom{n+k-1}{k} = \dim H_k$. Hence if $\varphi \in H_k$ then $\varphi(x) = \sum_{|q|=k} \beta_q x^q = \sum_{i=1}^{d_k} \alpha_i e_i^k$. Let $H^\infty$ be the space of all polynomials (without constant term) in variables $x_1, \ldots, x_n$. Then $H^\infty$ may be represented by the direct sum of the family $\{H_k\}_{k \in \mathbb{N}}$ of the spaces of homogeneous polynomials, i.e. $H^\infty := \bigoplus_{k \in \mathbb{N}} H_k$. Let us mention that the direct sum of an infinite family of modules is defined to be the set of all functions $w$ with domain $\mathbb{N}$ such that $w(k) \in H_k$ for all $k \in \mathbb{N}$ and $w(k) = 0$ for all but finitely many indices $k$.

Let $P$ be a polynomial of degree $r$ with $P(0) = 0$. So $P \in H^\infty$ and there are homogeneous polynomials $\varphi_k \in H_k, k = 1, \ldots, m$, such that $P(x) = \sum_{k=1}^r \varphi_k(x)$. Using the above notation we can write that

$$
P(x) = \sum_{k=1}^r \sum_{i=1}^{d_k} p_{ki} e_i^k = \sum_{k=1}^r (p_{k1}, \ldots, p_{kd_k}) \zeta_k.
\tag{7}
$$

Then system (5) can be written in the form

$$
\frac{dx}{dt} = A_{11}\zeta_1 + \cdots + A_{1m}\zeta_m,
\tag{8}
$$

where $F_k(x) = A_{1k}\zeta_k$ and matrices $A_{1k} \in \mathcal{M}(n, d_k)$, $i = 1, \ldots, m$. Let us observe that $F_1(x) = A_{11}\zeta_1$, where the matrix $A_{11} \in M(n, n)$, forms the linear part of system (5). Additionally we can obtain matrices $A_{1k}$ by the formula $A_{1k} = \left(\frac{1}{q_1! \cdots q_n!} \frac{\partial^k f_i}{\partial x_1^{q_1} \cdots \partial x_n^{q_n}}(0)\right)$, where $q_i \in \mathbb{N}$ and $\sum_{i=1}^n q_i = k$.

Let $\zeta = (\zeta_1^T, \zeta_2^T, \ldots)^T$ be the infinite vector of elements of the basis of $H^\infty$. Then by (7)

$$
P(x) = (p_{11}, \ldots, p_{1d_1}, \ldots, p_{r1}, \ldots, p_{rd_r}, 0, \ldots)\zeta.
\tag{9}
$$

The Lie derivative in the direction of the vector field $F$ of system (5) defines the linear map $D_\Sigma : H^\infty \to H^\infty$ by $(D_\Sigma P)(x) = \nabla P(x) \cdot F(x)$. Let $P(x)$ be in the form (7). Then we have $D_\Sigma P(x) = \sum_{k=1}^m \sum_{i=1}^{d_k} p_{ki} D_\Sigma e_i^k = \sum_{k=1}^m \sum_{i=1}^{d_k} p_{ki} \nabla e_i^k \cdot F(x)$.

As in particular $D_\Sigma x_i = f_i(x)$ then $\begin{pmatrix} D_\Sigma x_1 \\ \vdots \\ D_\Sigma x_n \end{pmatrix} = \sum\limits_{j=1}^{m} A_{1j}\zeta_j$ and there are uniquely determined matrices $A_{kj} \in \mathcal{M}(d_k, d_j)$ such that

$$\begin{pmatrix} D_\Sigma e_1^k \\ \vdots \\ D_\Sigma e_{d_k}^k \end{pmatrix} = \sum_{j=k}^{m+k-1} A_{kj}\zeta_j. \tag{10}$$

Applying (10) to $D_\Sigma P(x)$ we obtain

$$D_\Sigma P(x) = \sum_{k=1}^{r}(p_{k1}, \ldots, p_{kd_k}) \sum_{j=k}^{m+k-1} A_{kj}\zeta_j. \tag{11}$$

This yields

$$D_\Sigma P(x) = (p_{11}, \ldots, p_{1d_1}, \ldots, p_{rd_r}, 0\ldots)M_F\zeta, \tag{12}$$

where

$$M_F = \begin{pmatrix} A_{11} & A_{12} & \ldots & A_{1m} & 0 & 0 & \ldots \\ 0 & A_{22} & \ldots & A_{2m} & A_{2,m+1} & 0 & \ldots \\ & \ddots & \ddots & \ddots & & \ddots & \ddots \end{pmatrix}. \tag{13}$$

In the following example we show how matrices $A_{kj}$ could be determined for the case $n = 2$.

*Example 1.* We have that $A_{1k} = \left( \frac{1}{q_1! q_2!} \frac{\partial^k f_i}{\partial x_1^{q_1} \partial x_2^{q_2}}(0) \right)$, where $q_i \in \mathbb{N}$ and $q_1 + q_2 = k$. For calculating matrices $A_{2k}, k = 1, \ldots, m+1$ we can use the following. Let us observe that

$$\dot{\zeta}_2 = \begin{pmatrix} D_\Sigma x_1^2 \\ D_\Sigma x_1 x_2 \\ D_\Sigma x_2^2 \end{pmatrix} = \begin{pmatrix} 2x_1 & 0 \\ x_2 & x_1 \\ 0 & 2x_2 \end{pmatrix}\dot{\zeta}_1 = \begin{pmatrix} 2x_1 & 0 \\ x_2 & x_1 \\ 0 & 2x_2 \end{pmatrix}\sum_{k=1}^{m} A_{1k}\zeta_k.$$

The last equality can be written as

$$\dot{\zeta}_2 = \begin{pmatrix} 2 & 0 \\ 0 & 1 \\ 0 & 0 \end{pmatrix}\sum_{k=1}^{m} A_{1k}x_1\zeta_k + \begin{pmatrix} 0 & 0 \\ 1 & 0 \\ 0 & 2 \end{pmatrix}\sum_{k=1}^{m} A_{1k}x_2\zeta_k = \sum_{k=2}^{m+1} A_{2k}\zeta_k,$$

where:

$$A_{22} = \begin{pmatrix} 2 & 0 \\ 0 & 1 \\ 0 & 0 \end{pmatrix} A_{11}\left( I_3\ \mathbf{0} \right) + \begin{pmatrix} 0 & 0 \\ 1 & 0 \\ 0 & 2 \end{pmatrix} A_{11}\left( \mathbf{0}\ I_3 \right),$$

where $I_3$ is the identity matrix of the degree 3. And for $k = 2, \ldots, m$ :

$$A_{2,k+1} = \begin{pmatrix} 2 & 0 \\ 0 & 1 \\ 0 & 0 \end{pmatrix} A_{1k}\left( I_{d_k}\ \mathbf{0} \right) + \begin{pmatrix} 0 & 0 \\ 1 & 0 \\ 0 & 2 \end{pmatrix} A_{1m}\left( \mathbf{0}\ I_{d_k} \right)$$

with $I_{d_k}$ being identity matrices also. Next matrices can be produced in a similar way.

**Definition 4.** *Let $M_F$ denotes matrix given by (13). The system of equations*

$$\frac{dz}{dt} = M_F z \tag{14}$$

*for an infinite sequence of functions $z = (z_1, z_2, \ldots)^T$ with $z_i \in C^1(J), i \in \mathbb{N}$ and with $J$ an open interval, is called the associated infinite linear system to the finite nonlinear system (5). The setting up the associated infinite linear system to a given finite system is called the Carleman linearization procedure (Carleman embedding).*

**Definition 5.** *Let $s = \sum_{i=1}^{k} \dim H_i$. By the truncation of the order $s \geq 1$ of system (14) we mean the following finite dimensional system:*

$$\frac{d}{dt} \begin{pmatrix} z_1 \\ \vdots \\ z_s \end{pmatrix} = \begin{pmatrix} A_{11} & A_{12} & \ldots & A_{1k} \\ 0 & A_{22} & \ldots & A_{2k} \\ & & \ddots & \vdots \\ & & & A_{kk} \end{pmatrix} \begin{pmatrix} z_1 \\ \vdots \\ z_s \end{pmatrix}. \tag{15}$$

*Remark 3.* [12] Each solution $x$ of system (5) gives a solution of (14). And conversely each solution $z$ of (14) gives $x = (z_1, \ldots, z_n)$ as a solution of the system (8).

*Remark 4.* If $\frac{dx}{dt} = x + a_2 x^2 + \cdots + a_m x^m$, $a_2, \ldots, a_m \in \mathbb{R}$, then the matrix $M_F$ in the system (14) is triangular, with $m - 1$ filled diagonal lines above the main diagonal.

*Example 2.* Let us consider one-dimensional system: $\frac{dx}{dt} = -x + x^2$ with the initial condition: $x(0) = c \in \mathbb{R}$. The solution of the initial value problem is the following: $x(t) = \frac{c}{c+(1-c)\exp(t)}$. Using the Carleman technique we take $z_1 := x, z_2 := x^2, \ldots, z_n := x^n, \ldots$. Then $\frac{dz_n}{dt} = \frac{dx^n}{dt} = nx^{n-1}\frac{dx}{dt}$ and $\frac{dz_n}{dt} = -nz_n + nz_{n+1}$. Hence $\frac{dz}{dt} = M_F z$, where the matrix

$$M_F = \begin{pmatrix} -1 & 1 & 0 & 0 & 0 & 0 \ldots \\ 0 & -2 & 2 & 0 & 0 & 0 \ldots \\ 0 & 0 & -3 & 3 & 0 & 0 \ldots \\ \vdots & \vdots & & \ddots & \ddots & \ddots \end{pmatrix}$$

is infinite-dimensional, triangular and row-finite. Additionally $z_n(0) = c^n$ and $x(t) = z_1(t) = \frac{c}{c+(1-c)\exp(t)}$. We can look at truncated versions of the Carleman linearization, e.g. the system

$$\frac{d}{dt} \begin{pmatrix} z_1 \\ z_2 \\ z_3 \end{pmatrix} = \begin{pmatrix} -1 & 1 & 0 \\ 0 & -2 & 2 \\ 0 & 0 & -3 \end{pmatrix} \begin{pmatrix} z_1 \\ z_2 \\ z_3 \end{pmatrix}$$

is an approximation of $\dot{x} = -x + x^2$.

Now the vector field of system (14) defines the map

$$D_{\tilde{\Sigma}} : L(\mathbb{R}^\infty, \mathbb{R}) \to L(\mathbb{R}^\infty, \mathbb{R})$$

by

$$D_{\tilde{\Sigma}} f(z) = C M_F z, \tag{16}$$

where $f(z) = Cz$ and $C = (c_1, c_2, \ldots, c_{n_f}, 0, \ldots)$.

Let $\zeta = (\zeta_1^T, \zeta_2^T, \ldots)^T$ be a basis in $H^\infty$. Let us consider $P \in H^\infty$ in the form (9). Then we define the map $\alpha : H^\infty \to L(\mathbb{R}^\infty, \mathbb{R})$ in the following way

$$\alpha(P)(z) = \sum_{s=1}^{l} c_s z_s = Cz, \tag{17}$$

where $l = \sum_{j=1}^{d_m} d_j$ and $c_s = p_{ki}$ for $s = i + \sum_{j=1}^{k-1} d_j$.

**Proposition 4.** *The map $\alpha$ defined by (17) is a linear bijective mapping from $H^\infty$ to $L(\mathbb{R}^\infty, \mathbb{R})$.*

**Proposition 5.** *Let $\alpha$ be the map defined by (17). Then*

$$D_{\tilde{\Sigma}} \circ \alpha = \alpha \circ D_\Sigma. \tag{18}$$

*Proof.* Let $P \in H^\infty$ be a polynomial of degree $r$ in $n$-variables, in the form (9). Then by the definition (17) of the map $\alpha$ and the definition (16) of $D_{\tilde{\Sigma}}$ we get $(D_{\tilde{\Sigma}} \circ \alpha)(P)(z) = (p_{11}, \ldots, p_{rd_r}, 0, \ldots) M_F z$. Hence by (12) we get that (18) is true.

Let $D_\Sigma^0 P := P$ and $D_\Sigma^k P := D_\Sigma(D_\Sigma^{k-1} P)$. Then by induction we conclude that

**Corollary 2.** $\alpha(D_\Sigma^k P)(z) = C M_F^k z.$

## 4 Linear observability of polynomial dynamical system

Let $(\Sigma)$ be a polynomial system with output:

$$\dot{x} = F(x) \tag{19}$$
$$y = h(x), \tag{20}$$

where $x \in \mathbb{R}^n, y \in \mathbb{R}^p$ and $F = (f_1, f_2, \ldots, f_n)^T, h = (h_1, \ldots, h_p)^T$ are vectors of polynomials without constant terms. Let $F(x) = A_{11}\zeta_1 + \cdots + A_{1m}\zeta_m = \sum_{k=1}^{m} F_k(x)$, where $F_k(x) = A_{1k}\zeta_k$ and $h_j(x) = C_j\zeta$, where $C_j = (\beta_{11}^j, \ldots, \beta_{rd_r}^j, 0, \ldots)$.

**Definition 6.** By $\mathcal{O}(\Sigma)$ we denote the smallest linear subspace of $H^\infty$ containing $h_j, j = 1, \ldots, p$, the components of $h$, and invariant under the map $D_\Sigma$, the action of the vector field $F$ of the system $(\Sigma)$. The system $(\Sigma)$ is said to be linearly observable if for each $i = 1, \ldots, n$: $x_i \in \mathcal{O}(\Sigma)$.

*Example 3.* Let $(\Sigma)$ be the following :

$$\begin{pmatrix} \dot{x}_1 \\ \dot{x}_2 \end{pmatrix} = \begin{pmatrix} -x_2 - x_1 x_2 \\ x_1 + x_1 x_2 \end{pmatrix}$$
$$y = x_1 + x_2,$$

where the dynamics is the same as in the Lotka-Volterra model. Then $\mathcal{O}(\Sigma)(x) = \mathrm{span}\{(D_\Sigma^k y)(x), k = 0, 1, \ldots\} = \mathrm{span}\{x_1 + x_2, x_1 - x_2, \ldots\}$. Hence $x_1 = \frac{1}{2}(y + D_\Sigma y)$ and $x_2 = \frac{1}{2}(y - D_\Sigma y)$ and $(\Sigma)$ is linearly observable.

By $(\tilde{\Sigma})$ we denote the associated with $(\Sigma)$ (by Definition 4) infinite-dimensional system:

$$\dot{z} = M_F z \tag{21}$$
$$y = Cz, \tag{22}$$

where $z \in \mathbb{R}^\infty, y \in \mathbb{R}^p$ and the matrix $C = \begin{pmatrix} C_1 \\ \vdots \\ C_p \end{pmatrix}$.

**Corollary 3.**

1. For each $k = 0, 1, \ldots$ the following holds: $\alpha(D_\Sigma^k h_j)(z) = C_j M_F^k z$, where for $j = 1, \ldots, p$ and $h_j(x) = C_j \zeta$ are components of the output of $(\Sigma)$.
2. $\alpha(\mathcal{O}(\Sigma)) = \mathrm{span}\{C_j M_F^k z, j = 1, \ldots, p$ and $k = 0, 1, \ldots\}$.

**Proposition 6.** Let $E_i = (0, \ldots, 0, 1, 0, \ldots)^T$ be the vector from $\mathbb{R}^\infty$ with $1$ at the $i$-th position. The finite-dimensional polynomial system $(\Sigma)$ is linearly observable iff there is $k \in \mathbb{N} \cup \{0\}$ such that

$$\mathrm{rank} \begin{pmatrix} C \\ CM_F \\ \vdots \\ CM_F^k \\ E_1^T \\ \vdots \\ E_n^T \end{pmatrix} = \mathrm{rank} \begin{pmatrix} C \\ CM_F \\ \vdots \\ CM_F^k \end{pmatrix}. \tag{23}$$

*Proof.* Let $i = 1, \ldots, n$. As, by Proposition 4, $\alpha$ is a bijective mapping from $H^\infty$ to $L(\mathbb{R}^\infty, \mathbb{R})$, we have that $x_i \in \mathcal{O}(\Sigma) \Leftrightarrow \alpha(x_i) = E_i z \in \alpha(\mathcal{O}(\Sigma))$. Moreover $E_i^T z \in \alpha(\mathcal{O}(\Sigma)) = \mathrm{span}\{C_j M_F^k z, j = 1, \ldots, p, k \geq 0\}$ iff $E_i^T \in \mathrm{span}\{C_j M_F^k, j = 1, \ldots, p, k = 0, 1, \ldots\}$. It holds for all $i = 1, \ldots, n$ iff the condition (23) is satisfied.

*Remark 5.* Using Definition 3 we can formulate Proposition 6 as follows: system $(\Sigma)$ is linearly observable iff its Carleman linearization is observable with respect to variables $z_1, \ldots, z_n$.

**Corollary 4.** *If the system $(\Sigma)$ is linearly observable then there is $k \geq \deg h$ such that the truncation of the order $k$ of associated infinite linear system is observable on $\mathbb{R}^k$.*

# 5 Carleman bilinearization for polynomial control system

Let $(\Lambda)$ denote the finite dimensional polynomial system with one-dimensional input $u$:

$$\dot{x} = F(x) + G(x)u \tag{24}$$

$$y = h(x), \tag{25}$$

where $F, G$ are vectors of polynomials without constant term and $u \in \mathcal{U}$, where $\mathcal{U}$ denotes the set of piece wise constant functions $u : [0, T_u] \to \mathbb{R}$ and $T_u$ depends on $u$, $T_u \geq 0$. Using the same description as for systems without control, equation (24) of $(\Lambda)$ can be written in the following form: $\dot{x} = \sum_{j=1}^{m} (A_{1j} + uB_{1j}) \zeta_j$, where $m = \max_{j=1}^{n} \deg(f_j, g_j)$ and $F = (f_1, \ldots, f_n)^T, G = (g_1, \ldots, g_n)^T$. Let $D_\Lambda^u$ denote the derivation in the direction of the vector field $F + Gu$, for fixed $u \in \mathbb{R}$. Moreover let $D_\Lambda^u = D_0 + D_u$, where by $D_0$ we mean the derivation in the direction of the vector field $F$ while the derivation $D_u$ is defined by the vector field $uG$. Let $\mathcal{D}_\Lambda = \{D_\Lambda^u = D_0 + D_u : u \in \mathbb{R}\}$. Then the observability of this system depends on properties of the space $\mathcal{O}(\Lambda)$, which is the smallest subspace of $H^\infty$ that contains all functions $h_j, j = 1, \ldots, p$, and is invariant under the action of the maps from $\mathcal{D}_\Lambda$. Hence, according to Definition 6, $(\Lambda)$ is linearly observable if for each $i = 1, \ldots, n : x_i \in \mathcal{O}(\Lambda)$.

Similarly as in the case of system $\Sigma$, we consider the action of the derivation $D_\Lambda^u$ on a polynomial $P \in H^\infty$ given by (7). By (10) and (12):

$$D_\Lambda^u P(x) = \sum_{k=1}^{r} (p_{k1}, \ldots, p_{kd_k}) \sum_{j=k}^{m+k-1} (A_{kj} + uB_{kj}) \zeta_j. \text{ Let}$$

$$M_G = \begin{pmatrix} B_{11} & B_{12} & \ldots & B_{1m} & 0 & 0 & \ldots \\ 0 & B_{22} & \ldots & B_{2m} & B_{2,m+1} & 0 & \ldots \\ & \ddots & \ddots & \ddots & & \ddots & \ddots \end{pmatrix}. \tag{26}$$

By $(\tilde{\Lambda})$ we denote the infinite system associated (by Carleman embedding) to the system $(\Lambda)$.

*Remark 6.* $(\tilde{\Lambda})$ has the form:

$$\frac{dz}{dt} = (M_F + uM_G)z \tag{27}$$

$$y = Cz, \tag{28}$$

where $z \in \mathbb{R}^\infty, y \in \mathbb{R}^p$. $(\tilde{A})$ is a bilinear infinite-dimensional system.

*Example 4.* Let $(A)$ be as follows: $\begin{cases} \dot{x} = -x + x^2 + xu \\ y = x^2 \end{cases}$.

Then let for $z_i = x^i, i \in \mathbb{N}$ we have $\frac{dz_i}{dt} = -iz_i + iz_{i+1} + iz_i u$.
Hence

$$(\tilde{A}) : \frac{d}{dt} \begin{pmatrix} z_1 \\ z_2 \\ z_3 \\ \vdots \end{pmatrix} =$$

$$= \begin{pmatrix} -1 & 1 & 0 & 0 & 0 & 0 \ldots \\ 0 & -2 & 2 & 0 & 0 & 0 \ldots \\ 0 & 0 & -3 & 3 & 0 & 0 \ldots \\ \vdots & \vdots & & \ddots & \ddots & \ddots \end{pmatrix} \begin{pmatrix} z_1 \\ z_2 \\ z_3 \\ \vdots \end{pmatrix} + u \begin{pmatrix} 1 & 0 & 0 & 0 & 0 \ldots \\ 0 & 2 & 0 & 0 & 0 \ldots \\ 0 & 0 & 3 & 0 & 0 \ldots \\ \vdots & \vdots & & \ddots & \ddots & \ddots \end{pmatrix} \begin{pmatrix} z_1 \\ z_2 \\ z_3 \\ \vdots \end{pmatrix},$$

$$y = (0, 1, 0, \ldots)z.$$

**Proposition 7.** *Let*

$$\Gamma(C, M_F, M_G)_k =$$
$$= col[C, CM_F, CM_G, CM_F^2, CM_F M_G, CM_G M_F, CM_G^2, \ldots, CM_G^k].$$

*System $(A)$ is linearly observable iff there is $k \in \mathbb{N} \cup \{0\}$ :*

$$\text{rank}\Gamma(C, M_F, M_G)_k = \text{rank} \begin{pmatrix} \Gamma(C, M_F, M_G)_k \\ E_1^T \\ \vdots \\ E_n^T \end{pmatrix}. \tag{29}$$

*Proof.* Let us observe that similarly as in the proof of Proposition 6 we conclude the thesis by the fact that the map $\alpha$ is bijective and $\alpha(D_0^l D_u^k h_j)(z) = C_j M_F^l M_G^k z$.

*Example 5.* Let $(A) : \dot{x}_1 = -x_2 u, \dot{x}_2 = x_1 u, y = x_1^2 + x_2^2$. $(A)$ is not observable. Let $z_1 = x_1, z_2 = x_2, z_3 = x_1^2, z_4 = x_1 x_2, z_5 = x_2^2$. Then from the truncation of the order $s = 5$:

$$\frac{d}{dt} \begin{pmatrix} z_1 \\ z_2 \\ z_3 \\ z_4 \\ z_5 \end{pmatrix} = \begin{pmatrix} 0 & -1 & 0 & 0 & 0 \\ 1 & 0 & 0 & 0 & 0 \\ 0 & 0 & 0 & -2 & 0 \\ 0 & 0 & 1 & 0 & -1 \\ 0 & 0 & 0 & 2 & 0 \end{pmatrix} \tilde{z}u,$$

$y = \begin{pmatrix} 0 & 0 & 1 & 0 & 1 \end{pmatrix} \tilde{z}$, $\tilde{z} = (z_1, z_2, z_3, z_4, z_5)^T$ of the corresponding system $(\tilde{\Lambda})$ we can establish that $CM_F = CM_G = 0$. Hence the equation (29) is not satisfied.

*Example 6.* Let $(\Lambda) : \dot{x} = x^2 + xu, y = -x + x^2$. Then $(\tilde{\Lambda})$:

$$
\dot{z} = \begin{pmatrix} 0 & 1 & 0 & 0 & \cdots \\ 0 & 0 & 2 & 0 & \cdots \\ \vdots & \vdots & \ddots & \ddots & \ddots \end{pmatrix} z + \begin{pmatrix} 1 & 0 & \cdots \\ 0 & 2 & \cdots \\ & & \ddots & \ddots \end{pmatrix} zu, \qquad y = (-1, 1, 0, \ldots)z.
$$

Hence it is enough to take $k = 1$ to have the equality (29) true.

## Acknowledgment

This work was supported by the Bialystok Technical University grants:
Dorota Mozyrska: W/WI/7/07
Zbigniew Bartosiewicz: W/WI/1/07.

# References

1. Banach S (1932) Théorie des opérations linéaires, Warsaw
2. Bartosiewicz Z (1986) Realizations of polynomial systems, In: Fliess M, Hazewinkel M, Reidel D (eds) Algebraic and Geometric Methods in Nonlinear Control Theory, Dordrecht
3. Bartosiewicz Z, Mozyrska D (2001) Observability of infinite–dimensional finitely presented discrete–time linear systems. Zeszyty Naukowe Politechniki Białostockiej, Matematyka-Fizyka-Chemia 20:5–14.
4. Bartosiewicz Z, Mozyrska D (2005) Observability of row–finite countable systems of linear differential equations. In: Proceedings of 16th IFAC Congress, 4–8 July 2005, Prague
5. Chen G, Dora JD (1999) Rational normal form for dynamical systems by Carleman linearization. In Proceedings of the 1999 International Symposium on Symbolic and Algebraic Computation 165–172, Vancouver, British Columbia, Canada
6. Curtain RF, Pritchard AJ (1978) Infinite Dimensional Linear Systems Theory, Springer-Verlag, Berlin
7. Deimling R (1977) Ordinary Differential Equations in Banach Spaces, Lecture Notes in Mathematics, vol. 596, Springer-Verlag
8. Elliott DL (1999) Bilinear Systems, Encyclopedia of Electrical Engineering, edited by John Webster; J. Wiley and Sons
9. Fliess M, Kupka I (1983) A finiteness criterion for nonlinear input-output differential systems. SIAM J. Control Optim. 21(5):721–728.
10. Herzog G (1998) On Lipschitz conditions for ordinary differential equations in Fréchet spaces, Czech. Math. J. 48: 95–103.
11. Jakubczyk B (1992) Remarks on equivalence and linearization of nonlinear systems. In: Proc. Nonlinear Control Systems Design Symposium, Bordeaux, France

12. Kowalski K, Steeb WH (1991) Nonlinear dynamical systems and Carleman linearization. World Scientific Publishing Co. Pte. Ltd., Singapore
13. Lemmert R (1986) On ordinary differential equations in locally convex spaces. Nonlinear Anal. 10:1385–1390.
14. Moszyński K, Pokrzywa A (1974) Sur les systémes infinis d'équations différentielles ordinaires dans certain espaces de Fréchet. Dissert. Math. 115.
15. Mozyrska D, Bartosiewicz Z (2000) Local observability of systems on $\mathbb{R}^\infty$. In: Proceedings of MTNS'2000, Perpignan, France
16. Mozyrska D (2000) Local observability of infinitely-dimensional finitely presented dynamical systems with output (in Polish), Ph.D. thesis, Technical University of Warsaw, Poland
17. Mozyrska D, Bartosiewicz Z (2006) Dualities for linear control differential systems with infinite matrices. Control & Cybernetics, vol. 35
18. Sen P (1981) On the choice of input for observability in bilinear systems. IEEE Transactions on Automatic Control. vol.AC-26 no. 2.
19. Zhou Y, Martin C (2003) Carleman linearization of linear systems with polynomial output. Report 2002/2003.

# Observability of Nonlinear Control Systems on Time Scales - Sufficient Conditions

Ewa Pawłuszewicz*

Białystok Technical University, Faculty of Computer Sciences,
Wiejska 45A, Białystok, Poland. epaw@pb.edu.pl

**Summary.** In the paper the problem of observability of nonlinear control systems defined on time scales is studied. For this purpose it is introduced a family of operators which in the continuous-time case coincides with Lie derivatives associated to the given system. Then it is shown that set of functions generated by this operator distinguishes states that are distinguishable. The proved sufficient condition for observability is classical, but it works not only for continuous-time case but also for the other models of time.

**Keywords:** time scale, indistinguishability relation, observability sets, observability, rank condition on time scales.

## 1 Introduction

*"A major task of mathematics today is to harmonize the continuous and discrete, to include them in one comprehensive mathematics and eliminate obscurity from both."*, E.T. Bell, 1937

It is known that in engineering practice there are situations in which one cannot measure the state function but some of its parts, i.e. output function. This means that all information about the state should in principle be recoverable from knowledge of output and control. The key tool to study such issue is the notion of observability introduced by Kalman for linear control systems [2] and extended to nonlinear systems by Herman and Krener [7]. In the linear case necessary and sufficient conditions for observability as well in the continuous- as for discrete-time systems look very similar. In [5] it is shown that the standard Kalman conditions are still valid for linear time-invariant systems defined on any time scale. In [6] it is shown that these conditions are also valid for linear time-variant systems on time scales. Some interesting results concerning unification of linear continuous- and discrete-time systems

---

* Supported by Białystok Technical University grant No W/WI/18/07

were obtained by Goodwin at al. [10], but restricted to the time sets $\mathbb{R}$, $\mathbb{Z}$ and $h\mathbb{Z}$.

A time scale is a model of time. It is an arbitrary closed subset of the real line. Besides the standard cases of whole line (continuous-time case) and the set of integers (discrete-time case), there are many examples of time models, that may be partly continuous and partly discrete. Calculus on time scales, originated in 1988 by S. Hilger [12], seems to be a perfect language for unifying the continuous- and the discrete-time cases. One of the main concepts in the time scale theory is the delta derivative, which is a generalization of the classical (time) derivative for continuous time and the finite forward difference for the discrete time. Similarly the integral of real functions defined on time scale is an extension of the Riemann integral in the continuous-time and the finite sum in the discrete-time. Consequently, differential equations as well as difference equations are naturally accommodated into this theory.

For nonlinear systems the main tool for studying observability problem is given by the indistinguishability relation and by a family of functions that distinguish states that are distinguishable. Well known Herman-Krener condition, roughly speaking, says that if the dimension of the space spanned by Lie derivatives of outputs with respect to the vector fields of the control system coincides with the state space dimension, then this system is observable.

For studying observability of nonlinear control systems on time scales we introduce operators acting on real functions on $\mathbb{R}^n$ which in the continuous-time case coincide with Lie derivatives associated to the given control system. Using this operator we construct observability sets of functions that distinguish states that are distinguishable. In the last step we show that classical Herman-Krener condition for observability of control system can be extended to systems defined on time scales.

The paper is organized as follows: in Section 2 there is given a short introduction to differential calculus on time scales. In Section 3 we define and studied the indistinguishability relation for nonlinear control systems defined on time scales. In Section 4 a sufficient condition for observability of nonlinear systems on time scales is proved.

## 2 Calculus on time scales

We give here a short introduction to differential calculus on time scales. This is a generalization on one hand, of the standard differential calculus, and of the other hand the calculus of finite differences. This will allow to solve differential equations on time scales. More on this subject can be found in [1, 4]. It can be also stressed that differential calculus on time scales can be implemented in MATHEMATICA [11].

A *time scale* $\mathbb{T}$ is an arbitrary nonempty closed subset of the set $\mathbb{R}$ of real numbers. The standard cases comprise $\mathbb{T} = \mathbb{R}$, $\mathbb{T} = \mathbb{Z}$, $\mathbb{T} = h\mathbb{Z}$ for $h > 0$. We assume that $\mathbb{T}$ is a topological space with the relative topology induced from $\mathbb{R}$.

For $t \in \mathbb{T}$ we define

- the *forward jump operator* $\sigma : \mathbb{T} \to \mathbb{T}$ by $\sigma(t) := \inf\{s \in \mathbb{T} : s > t\}$
- the *backward jump operator* $\rho : \mathbb{T} \to \mathbb{T}$ by $\rho(t) := \sup\{s \in \mathbb{T} : s < t\}$
- the *graininess function* $\mu : \mathbb{T} \to [0, \infty)$ by $\mu(t) := \sigma(t) - t$

Using these operators we can classified points in real line. Namely

- if $\sigma(t) > t$, then $t$ is called *right-scattered*
- if $\rho(t) < t$ is called *left-scattered*
- if $t < \sup \mathbb{T}$ and $\sigma(t) = t$ then $t$ is called *right-dense*
- if $t > \inf \mathbb{T}$ and $\rho(t) = t$, then $t$ is *left-dense*.

We define also the set $\mathbb{T}^k$ as:

$$\mathbb{T}^k := \begin{cases} \mathbb{T} \setminus (\rho(\sup \mathbb{T}), \sup \mathbb{T}] & \text{if } \sup \mathbb{T} < \infty \\ \mathbb{T} & \text{if } \sup \mathbb{T} = \infty \end{cases}$$

i.e. $\mathbb{T}^k$ does not contains maximal points.

**Definition 1.** Let $f : \mathbb{T} \to \mathbb{R}$ and $t \in \mathbb{T}^k$. The *delta derivative* of $f$ at $t$, denoted by $f^\triangle(t)$ (or by $\frac{\triangle}{\triangle t} f$), is the real number (provided it exists) with the property that given any $\varepsilon$ there is a neighborhood $U = (t - \delta, t + \delta) \cap \mathbb{T}$ (for some $\delta > 0$) such that

$$|(f(\sigma(t)) - f(s)) - f^\triangle(t)(\sigma(t) - s)| \le \varepsilon |\sigma(t) - s|$$

for all $s \in U$. We say that $f$ is *delta differentiable* on $\mathbb{T}^k$ provided $f^\triangle(t)$ exists for all $t \in \mathbb{T}^k$.

**Proposition 1.** *[1] Let $f : \mathbb{T} \to \mathbb{R}$, $g : \mathbb{T} \to \mathbb{R}$ be delta differentiable functions at $t \in \mathbb{T}^k$. Let $t \in \mathbb{T}$. Then*

1. *if $t \in \mathbb{T}^k$ then $f$ has at most one $\triangle$-derivative at $t$*
2. *if $f^\triangle$ exists, then $f(\sigma(t)) = f(t) + \mu(t) f^\triangle(t)$*
3. *for any constants $a, b$ holds $(af(t) + bg(t))^\triangle = af^\triangle(t) + bg^\triangle(t)$*
4. *$((f(t)g(t))^\triangle = f^\triangle(t)g(\sigma(t)) + f(t)g^\triangle(t) = f(\sigma(t))g^\triangle + f^\triangle(t)g(t)$*
5. *if $g(t)g(\sigma(t)) \ne 0$, then $\left(\frac{f}{g}\right)^\triangle(t) = \frac{f^\triangle(t)g(t) - f(t)g^\triangle(t)}{g(t)g(\sigma(t))}$*

*Remark 1.* In general the function $\sigma$ need not be differentiable ($\sigma$ need not be continuous).

*Example 1.*

- If $\mathbb{T} = \mathbb{R}$, then for any $t \in \mathbb{R}$ we have $\sigma(t) = t = \rho(t)$ and the graininess function $\mu(t) \equiv 0$. A function $f : \mathbb{R} \to \mathbb{R}$ is delta differentiable at $t \in \mathbb{R}$ if and only if $f^\triangle(t) = \lim_{s \to t} \frac{f(t) - f(s)}{t - s} = f'(t)$ i.e. if and only if $f$ is differentiable in the ordinary sense at $t$.

- If $\mathbb{T} = \mathbb{Z}$, then for every $t \in \mathbb{Z}$ we have $\sigma(t) = t + 1$, $\rho(t) = t - 1$ and the graininess function $\mu(t) \equiv 1$. A function $f : \mathbb{Z} \to \mathbb{R}$ is always delta differentiable at every $t \in \mathbb{Z}$ with $f^\triangle(t) = \frac{f(\sigma(t)) - f(t)}{\mu(t)} = f(t+1) - f(t) = \triangle f(t)$ where $\triangle$ is the usual forward difference operator defined by the last equation above.

- Let $q > 1$. We define the time scale $\mathbb{T} = \overline{q^{\mathbb{Z}}} := \{q^k : k \in \mathbb{Z}\} \cup \{0\}$. Then $\sigma(t) = qt$, $\rho(t) = \frac{t}{q}$ and $\mu(t) = (q - 1)t$ for all $t \in \mathbb{T}$. Any function $f : \overline{q^{\mathbb{Z}}} \to \mathbb{R}$ is differentiable and $f^\triangle(t) = \frac{f(qt) - f(t)}{(q-1)t}$ for all $t \in \mathbb{T} \setminus \{0\}$.

A function $f : \mathbb{T} \to \mathbb{R}$ is called *regulated* provided its right-sided limits exist (finite) at all right-dense points at $\mathbb{T}$ and its left-sided limits exist (finite) at all left-dense points in $\mathbb{T}$. A function $f : \mathbb{T} \to \mathbb{R}$ is called *rd-continuous* provided it is continuous at right-dense points in $\mathbb{T}$ and its left-sided limits exist (finite) at left-dense points in $\mathbb{T}$. A function $f$ is *piecewise rd-continuous* if it is regulated and if it is rd-continuous at all, except possibly many, right-dense points $t \in \mathbb{T}$ (see [9]). It can be shown that

$$f \text{ is continuous} \Rightarrow f \text{ is rd-continuous} \Rightarrow f \text{ is regulated}$$

A continuous function $f : \mathbb{T} \to \mathbb{R}$ is called *pre-differentiable* with (the region of differentiation) $D$, provided $D \subset \mathbb{T}^k$, $\mathbb{T}^k \setminus D$ is countable and contains no right-scattered elements of $\mathbb{T}$. It can be proved [1](Th.1.70) that if $f$ is regulated then there exists a function $F$ that is pre-differentiable with region of differentiation $D$ such that $F^\triangle(t) = f(t)$ for all $t \in D$. Any such function is called pre-antiderivative of $f$. Then *indefinite integral* of $f$ is defined by $\int f(t)\triangle t := F(t) + C$ where $C$ is an arbitrary constant. *Cauchy integral* is $\int_r^s f(t)\triangle t = F(s) - F(r)$ for all $r, s \in \mathbb{T}^k$. A function $F : \mathbb{T} \to \mathbb{R}$ is called an *antiderivative* of $f : \mathbb{T} \to \mathbb{R}$ provided $F^\triangle(t) = f(t)$ holds for all $t \in \mathbb{T}^k$. It can be shown that every rd-continuous function has an antiderivative.

*Example 2.*

- If $\mathbb{T} = \mathbb{R}$, then $\int_a^b f(\tau)\triangle\tau = \int_a^b f(\tau)d\tau$ where the integral on the right is the usual Riemann integral.

- If $\mathbb{T} = h\mathbb{Z}$, $h > 0$, then $\int_a^b f(\tau)\triangle\tau = \sum_{t=\frac{a}{h}}^{\frac{b}{h}-1} f(th)h$ for $a < b$.

**Corollary 1.** *For any rd-continuous function $f$ we have*

$$\frac{\triangle}{\triangle t} \int_{t_0}^t f(\tau)\triangle\tau = f(t)$$

*Proof.* Because of $\int_{t_0}^t f(\tau)\triangle\tau = F(t) - F(t_0)$, then from the properties of antiderivative follows that $\frac{\triangle}{\triangle t}(F(t) - F(t_0)) = f(t)$.

# 3 Indistinguishability relation

If $t_0, t_1 \in \mathbb{T}$ and $t \leq t_0$, then $[t_0, t_1]$ denotes the intersection of the ordinary closed interval with $\mathbb{T}$. Similar notation is used for open, half-open or infinite intervals.

Let $\Omega$ be an arbitrary set. It will called the *set of control values*. Let $\omega \in \Omega$ and $t_0, t_1 \in \mathbb{T}$, $t_0 < t_1$. A rd-piecewise constant function $u : [t_0, t_1] \to \Omega$ defined by $u(t) = \omega$ is called a *piecewise constant control* and denoted by $[\omega, t_0, t_1]$. The set of all controls will be denoted by $U$.

Let us consider a nonlinear control system with the output defined on time scale $\mathbb{T}$, denoted by $\Sigma$:

$$x^{\triangle}(t) = f(x(t), u(t))$$
$$y(t) = h(x(t)) \tag{1}$$

where $f : \mathbb{R}^n \times \Omega \to \mathbb{R}^n$, $\Omega \subseteq \mathbb{R}$, $h : \mathbb{R}^n \to \mathbb{R}^p$. We assume that $x(t_0) = x_0$ for a fixed $t_0 \in \mathbb{T}$. The dynamics of the system $\Sigma$, given by relation $x^{\triangle} = f(x, u)$, may be represented by the set $\mathcal{D}_{\Sigma} = \{f_{\omega} : \omega \in \Omega\}$ where $f_{\omega} := f(\cdot, \omega)$.

Let us choose an initial point $x_0 \in \mathbb{R}^n$ and a control $u$. The *trajectory* of $\Sigma$ from $x_0$ corresponding to the control $u$ is a function $x = x(\cdot, x_0, u) : [t_0, t_1] \to \mathbb{R}^n$ defined as follows: if $u = [\omega, t_0, t_1]$, then $x$ is the unique solution to the initial value problem:

$$x^{\triangle}(t) = f_{\omega}(x(t)), \qquad x(t_0) = x_0 \tag{2}$$

provided it is defined.

*Remark 2.* Initial value problem (2) has a local unique forward solution on time scale $\mathbb{T}$. It is the particular case of the theorem 8.16 in [1]. For more details see also [8]

A control $u$ is called *admissible* for $x_0 \in \mathbb{R}^n$ if there exists a trajectory of $\Sigma$ from $x_0$ corresponding to a control $u$. If it exists, such a trajectory is unique. The set of all controls admissible for $x_0$ will be denoted by $U_{\Sigma, x_0}$.

Let us consider the system $\Sigma$ given by (1). We say that points $x_1, x_2 \in \mathbb{R}^n$, $x_1 \neq x_2$, are *indistinguishable by the control* $u \in \mathcal{U}_{\Sigma}$ if

$$h(x(t, x_1, u)) = h(x(t, x_2, u)) \tag{3}$$

for any $t \in \mathbb{T}$, $t \geq t_0$, if both sides of (3) are defined. If two points $x_1, x_2$ are not indistinguishable, then they are *distinguishable* by the control $u \in \mathcal{U}_{\Sigma}$.

*Remark 3.* From the definition of indistinguishability it follows that if states $x_1, x_2$ are indistinguishable by the control $u$, then for every $t_0 \in \mathbb{T}$ also $x_1 + \mu(t_0)f_u(x_1)$ and $x_2 + \mu(t_0)f_u(x_2)$ are indistinguishable by this control.

Let $\varphi : \mathbb{R}^n \to \mathbb{R}$, $f : \mathbb{R}^n \to \mathbb{R}^n$ be analytic real valued functions. Then, for the fixed $t_0 \in \mathbb{T}$, the operator $\Gamma_f^{t_0}$ is defined as follows:

$$(\Gamma_f^{t_0}\varphi)(x) := \int_0^1 \varphi'(x + h\mu(t_0)f(x))dh \cdot f(x), \tag{4}$$

where $\varphi'$ is the gradient of the function $\varphi$. Thus $\Gamma_f^{t_0}\varphi$ is again a real analytic function on $\mathbb{R}^n$. Moreover, $\Gamma_f^{t_0}$ is a family of operators parameterized by time $t_0$. In general, when operator $\Gamma_f^{t_0}\varphi$ does not depend on $t_0$, we will denote it by $\Gamma_f\varphi$.

From definition follows that $\Gamma_f^{t_0}d\varphi = d\Gamma_f^{t_0}\varphi$. It can be noticed also that if $\mu(t_0) \neq 0$ then

$$(\Gamma_f^{t_0}\varphi)(x) = \frac{1}{\mu(t_0)}\int_0^1 \frac{d}{dh}(\varphi(x + h\mu(t_0)f(x)))dh =$$

$$= \frac{1}{\mu(t_0)}(\varphi(x + \mu(t_0)f(x)) - \varphi(x))$$

For $\mu(t_0) = 0$ we have

$$(\Gamma_f^{t_0}\varphi)(x) = \varphi'(x)f(x) = L_f\varphi$$

where $L_f\varphi$ denotes the Lie derivative of the function $\varphi$ with respect to $f$.

*Example 3.*

- If $\mathbb{T} = \mathbb{R}$, then $\Gamma_f = L_f$.
- If $\mathbb{T} = \mathbb{Z}$, then $(\Gamma_f\varphi)(x) = \varphi(x + f(x)) - \varphi(x)$
- If $\mathbb{T} = q^{\mathbb{N}}$, $q > 1$, then

$$(\Gamma_f^{t_0}\varphi)(x) = \frac{\varphi(x + t_0(q - 1)f(x)) - \varphi(x)}{(q - 1)t_0}$$

*Remark 4.* Let $\varphi(x) = x^i$ be the $i$-th coordinate function on $\mathbb{R}^n$. Then $(x^i)' = e_i$ - the vector of the standard basis of $\mathbb{R}^n$ with 1 at the $i$-th position. For any $t_0 \in \mathbb{T}$ we have $(\Gamma_f^{t_0}x^i)(x) = \varphi(x) \circ f(x) = f_i(x)$.

For $h = (h_1, \ldots, h_p)$ and $f_\omega \in \mathcal{D}_\Sigma$ let us put

$$\mathcal{O}_0(x) = \{h_1(x), \ldots, h_p(x)\}$$

$$\mathcal{O}_1(x) = \{h_1(x), \ldots, h_p(x), \Gamma_{f_u}^{t_0}h_1(x), \ldots, \Gamma_{f_u}^{t_0}h_p(x)\}$$

$$\vdots \tag{5}$$

$$\mathcal{O}_j(x) = \mathcal{O}_{j-1}(x) \cup \{\Gamma_{f_u}^{t_0}g(x) : g(x) \in \mathcal{O}_{j-1}(x)\}$$

where $t_0 \in \mathbb{T}$ is fixed and $u \in \mathcal{U}_\Sigma$. The pair $\Sigma_{u,j} = (\mathcal{D}_\Sigma, \mathcal{O}_j)$, $j = 0, 1, \ldots$, will be called the *family of the observability sets* of the system $\Sigma$ corresponding to the control $u \in U_\Sigma$.

*Example 4.* Let us consider a linear system defined on time scale $\mathbb{T}$

$$x^{\Delta}(t) = Ax(t) + Bu(t)$$
$$y(t) = Cx(t)$$

where $x \in \mathbb{R}^n$, $u \in \mathbb{R}^m$, $y \in \mathbb{R}^p$, $A, B, C$ are constant matrices. Let us observe that

- If $\mathbb{T} = \mathbb{R}$ or $\mathbb{T} = \mathbb{Z}$ then we have $\Gamma_{Ax}^{t_0} Cx = CAx$,. Then $\mathcal{O}_0(x) = \{c_1, \ldots, c_p\}$ and $\mathcal{O}_j(x) = \mathcal{O}_{j-1}(x) \cup \{c_1 A^j x, \ldots, c_p A^j x\}$ where $c_1, \ldots, c_p$ are columns of the matrix $C$. Hence, it coincides with classical observation space of the linear control system.
- If $\mathbb{T} = \overline{q^{\mathbb{Z}}}$, $q > 1$, then

$$\Gamma_{Ax}^{t_0} Cx = \frac{C(x + \mu(t_0)Ax) - Cx}{(q-1)t_0} = CAx$$

so $\mathcal{O}_j(x) = \mathcal{O}_{j-1}(x) \cup \{c_1 A^j x, \ldots, c_p A^j x\}$, $j = 0, 1, \ldots$
- If $\mathbb{T} = \bigcup\limits_{k=0}^{\infty} [2k, 2k+1]$, then

$$\mu(t) = \begin{cases} 0 & \text{for } t \in \bigcup\limits_{k=0}^{\infty} [2k, 2k+1) \\ 1 & \text{for } t \in \bigcup\limits_{k=0}^{\infty} \{2k+1\} \end{cases}$$

Hence $\Gamma_{Ax}^{t_0} Cx = CAx$ and again $\mathcal{O}_j(x) = \mathcal{O}_{j-1}(x) \cup \{c_1 A^j x, \ldots, c_p A^j x\}$, $j = 0, 1, \ldots$

*Example 5.* Let us consider nonlinear system defined on time scale $\mathbb{T}$

$$x^{\Delta}(t) = u(t)$$
$$y_1(t) = \sin x(t) \tag{6}$$
$$y_2(t) = \cos x(t)$$

with $x \in \mathbb{R}$ and $u \in \mathbb{R}$. We have $f_u = u$.

- If $\mathbb{T} = \mathbb{R}$ then

$$\Gamma_u^{t_0} \sin x = u \cos x, \qquad \Gamma_u^{t_0} \cos x = -u \sin x$$

Hence $\mathcal{O}_j(x) = \{(-u)^j \sin(\frac{\pi}{2}j + x), (-u)^j \cos(\frac{\pi}{2}j + x)\}$ for $j = 0, 1, \ldots$.
- If $\mathbb{T} = \mathbb{Z}$, then

$$\Gamma_u^{t_0} \sin x = a \cos(x + \frac{u}{2}), \qquad \Gamma_u^{t_0} \cos x = -a \sin(x + \frac{u}{2})$$

where $a = 2 \sin \frac{u}{2}$. Hence,

$$\mathcal{O}_0(x) = \{\sin x, \cos x\}$$

$$\mathcal{O}_1(x) = \{\sin x, \cos x, -a\sin(x + \frac{u}{2}), a\cos(x + \frac{u}{2})\}$$

$$\vdots$$

$$\mathcal{O}_j(x) = \{(-1)^j a^j \sin(x + \frac{u}{2}), (-1)^{j+1} a^j \cos(x + \frac{u}{2})\}$$

for $j = 1, 2, \ldots$.

- If $\mathbb{T} = q^{\mathbb{Z}}$, $q > 1$, then

$$\underbrace{\Gamma_u^{t_0} \circ \ldots \circ \Gamma_u^{t_0}}_{j - times} \sin x = \frac{(-2)^j a^j}{(q-1)^j t_0^j} \sin(\frac{\pi}{2}j + x + \frac{(q-1)j t_0 u}{2})$$

and

$$\underbrace{\Gamma_u^{t_0} \circ \ldots \circ \Gamma_u^{t_0}}_{j - times} \cos x = \frac{(-2)^j a^j}{(q-1)^j t_0^j} \cos(\frac{\pi}{2}j + x + \frac{(q-1)j t_0}{2})$$

where $a = \sin \frac{q-1 t_0}{2} u$. So,

$$\mathcal{O}_0(x) = \{\sin x, \cos x\}$$

$$\vdots$$

$$\mathcal{O}_j(x) = \{ \frac{(-2)^j a^j}{(q-1)^j t_0^j} \sin(\frac{\pi}{2}j + x + \frac{(q-1)j t_0 u}{2});$$

$$\frac{(-2)^j a^j}{(q-1)^j t_0^j} \cos(\frac{\pi}{2}j + x + \frac{(q-1)j t_0}{2})\}, \quad j = 0, 1, \ldots$$

Hence, for the system defined on any time scale $\mathbb{T}$, the family $\mathcal{O}_j$, $j = 0, 1, \ldots$, can be bigger then for classical continuous time systems.

**Lemma 1.** *If states $x_1, x_2$ are indistinguishable by the control $u \in \mathcal{U}_\Sigma$, then for all $t \in \mathbb{T}$ and any $\nu \in \mathcal{O}_j$, $j = 0, 1, \ldots$ the following equation holds*

$$\nu(x(t, x_1, u)) = \nu(x(t, x_2, u)) \qquad (7)$$

*Proof.* We shall use induction with respect to $j$. For $j = 0$ the statement is obvious. Let as assume that $g(x(t, x_1, u)) = g(x(t, x_2, u))$ for $g \in \mathcal{O}_{j-1}$. Then, for $\mu(t_0) \neq 0$

$$\Gamma_{f_u}^{t_0} g(x(t, x_1, u)) = \frac{g(x(t, x_1, u) + \mu(t_0)f_u(x(t, x_1, u))) - g(x(t, x_1, u))}{\mu(t_0)} =$$

$$= \frac{g(x(t, x_2 + \mu(t_0)f_u(x_2), u))) - g(x(t, x_2, u)))}{\mu(t_0)}.$$

Because of indistinguishability of $x(t, x_1, u) + \mu(t_0) f_u(x(t, x_1, u))$ and $x(t, x_2, u) + \mu(t_0) f_u(x(t, x_2, u))$ we have

$$g(x(t, x_1, u)) + \mu(t_0) f_u(x(t, x_1, u)) = g(x(t, x_2, u)) + \mu(t_0) f_u(x(t, x_2, u))$$

Defining

$$\nu(x(t, x_1, u)) = g(x(t, x_1, u) + \mu(t_0) f_u(x(t, x_1, u)))$$

and

$$\nu(x(t, x_2, u)) = g(x(t, x_2, u) + \mu(t_0) f_u(x(t, x_2, u)))$$

we have our result.

For $\mu(t_0) = 0$ operator $\Gamma_{f_u}^{t_0}$ coincides with Lie derivative, so the statement (7) is obvious.

**Corollary 2.** *If states $x_1, x_2$ are indistinguishable by the control $u \in \mathcal{U}_\Sigma$, then they are indistinguishable by the family $\Sigma_{u,j}$, $j = 0, 1, \ldots$.*

*Proof.* Proof follows from lemma 1.

*Example 6.* It can be noticed that for the system $\Sigma$ defined in the example 5 family $\Sigma_{u,j}$, $j = 0, 1, \ldots$, does not distinguish states that are distinguishable on the given time scales.

## 4 Observability

The system $\Sigma$ defined by (1) is *observable* if any two distinct points are not distinguishable by the control $u \in \mathcal{U}_\Sigma$. System $\Sigma$ is *locally* observable at point $x = x(t, x_0, u)$ if there exists a neighborhood $V$ of $x$ such that any point $x_1 = x_1(t, x_0, u) \in V$, $x_1 \neq x$, is distinguishable from $x$ by the control $u \in \mathcal{U}_\Sigma$. System $\Sigma$ is *locally observable*, if it is locally observable at each point of the state space $\mathbb{R}^n$.

For $j = 0, 1 \ldots, p - 1$ let us define

$$d\mathcal{O}_j(x) := \left\{ \left( \frac{\partial g(x)}{\partial x_1}, \ldots, \frac{\partial g(x)}{\partial x_n} \right)^T : g \in \mathcal{O}_j, x \in \mathbb{R}^n \right\}$$

where $(\ldots)^T$ denotes a transposition of a vector $(\ldots)$. By $\dim d\mathcal{O}_j$ we will denote the maximal number of linearly independent vectors in $d\mathcal{O}_j(x)$.

**Theorem 1.** *Let us assume that for the system $\Sigma$ given by (1) holds that*

$$\dim d\mathcal{O}_{n-1}(x_0) = n, \qquad x_0 = x(t_0). \tag{8}$$

*for fixed $t_0 \in \mathbb{T}$. Then for any $t \geq t_0$, $t \in \mathbb{T}$, the system $\Sigma$ is locally observable at point $x_0$ with respect to the control $u \in \mathcal{U}_\Sigma$.*

*Proof.* The classical proof works well for all times scales. Let us assume that $\dim dO_{n-1}(x) = n$ and $\Sigma$ is not observable at $x_0$. Then there exists a neighborhood of point $x_0$ such that the state $x_1$, $x_1 \neq x_0$, from this neighborhood such that $h(x(t, x_0, u)) = h(x(t, x_1, u))$ for $t \geq t_0$, $t, t_0 \in \mathbb{T}$. So, $g(x(t, x_0, u)) = y(x(t, x_1, u))$ for any $g \in O_{n-1}$.

Let us consider a map $\xi : \mathbb{R}^n \to \mathbb{R}^n$, $\xi(x) = (\tilde{g}_1, \ldots, \tilde{g}_n)^T(x)$ for $\tilde{g}_i \in O_i$, $i = 0, 1, \ldots, n-1$. Assumption $\dim dO_{n-1}(x) = n$ implies that in each neighborhood of point $x_0$ matrix $[\frac{\partial \xi}{\partial x}](x_0)$ is not singular. In particular there exists $x_1$ close to $x_0$ such that $\det[\frac{\partial \xi}{\partial x}](x_1) \neq 0$. Then there exists $i$, $i < n$, such that $\tilde{g}_i(x(t, x_0, u)) \neq \tilde{g}_i(x(t, x_1, u))$ for $\tilde{g}_i \in O_i$. So

$$\tilde{g}_i(x(t, x_0, u)) = \Gamma_{f_u}^t \circ \ldots \circ \Gamma_{f_u}^{t_0} h(x(t_0, x_0, u)) \neq$$
$$\neq \Gamma_{f_u}^{t_0} \circ \ldots \circ \Gamma_{f_u}^{t_0} h(x(t_0, x_1, u)) = \tilde{g}_i(x(t, x_1, u))$$

Hence contradiction.

Condition (8) is called the *rank condition*.

*Remark 5.* Theorem 1, for the fixed initial time $t_0$, unifies Herman-Krener [7] condition given in the literature for continuous-time and discrete-time cases and extends it to other more general cases on time scales.

## 5 Remarks

In previous sections, for simplicity, we defined indistinguishability relation with respect to the fixed control $u \in \mathcal{U}_\Sigma$. This definition can be extended for any set of controls $u_1, \ldots, u_k$ from the set of admissible controls $\mathcal{U}_\Sigma$ of the considered system. For this purpose the sets $O_j$, $j = 0, 1, \ldots$, should be defined as follows

$$O_0(x) = \{h_1(x), \ldots, h_p(x)\}$$
$$O_1(x) = \{h_1(x), \ldots, h_p(x), \Gamma_{f_{u_1}}^{t_0} h_1(x), \ldots, \Gamma_{f_{u_1}}^{t_0} h_p(x)\}$$
$$O_2(x) = O_1(x) \cup \{\Gamma_{f_{u_2}}^{t_0} g_1(x) : g_1(x) \in O_1(x)\}$$
$$\vdots$$
$$O_j(x) = O_{j-1}(x) \cup \{\Gamma_{f_{u_{j-1}}}^{t_0} g_{j-1}(x) : g_{j-1}(x) \in O_{j-1}(x)\}$$

For such defined sets $O_j$, $j = 0, 1, \ldots$, lemma 1 and theorem 1 are still true.

## 6 Conclusions and future works

We generalize the notation of observability for nonlinear control systems defined on time scales. It means that sufficient condition for the observability can be applied to the bigger class of control systems not only for the continuous-time one. To this aim we introduce the family of operators $\Gamma_f^{t_0}$ parameterized by fixed $t_0$, $t_0 \in \mathbb{T}$. This family can be viewed, roughly speaking, as generalization of Lie derivative for any model of time. Using this operators we construct the sets $\mathcal{O}_j$, $j = 0, 1, \ldots$, that distinguish the distinguishable states.

Necessary conditions for nonlinear control systems on time scales require some studies on polynomials and on delta differential equations with analytic right hand side.

# References

1. Bohner M, Peterson A (2001) Dynamic Equations on Time Scales. Birkhauser, Boston Basel Berlin
2. Kalman RE, Falb PL, Arbib MA (1969) Topics in Mathematical System Theory. McGraw-Hill, New York
3. Sontag E. (1990) Mathematical Control Theory. Springer-Verlag, Berlin Heidelberg New York
4. Agarwal RP, Bohner M (1999) Basic calculus on time scales and some its applications. Results Math 35(1-2):3–22.
5. Bartosiewicz Z, Pawłuszewicz E (2006) Realizations of linear control systems on time scales. Control and Cybernetics 35:769–786.
6. Fausett LV, Murty KN (2004) Controllability, Observability and Realizability Criteria on Time Scale Dynamical Systems. Nonlinear Studies 11:627–638.
7. Herman R., Krener AJ (1977) Nonlinear controllability and observability. IEEE Trans. Aut. Contr. 728–740.
8. Hilger S. (1990) Analysis on measure chains - a unified appproach to continuous and discrete calculus. Results Math. 18:18–56.
9. Hilscher R., Zeidan V (2004) Calculus of variations on time scales: weak local piecewise $C_{rd}^1$ solutions with variable endpoints. J. Math. Anal. Appl. 289: 143–166.
10. Goodwin GC, Middleton RH, Poor HV (1992) High-Speed Digital Signal Processing and Control. Proceedings of the IEEE 80:240–259.
11. Yantir A (2003) Derivative and integration on time scale with Mathematica. Proceedings of the 5th International Mathematica Symposium: Challenging the Boundaries of Symbolic Computation 325–331.
12. Hilger S (1988) Ein Maßkettenkalkülmit Anwendung auf Zentrumsmannigfaltigkeiten. Ph.D. Thesis, Universität Würzburg

# Sufficient Optimality Conditions for a Bang-bang Trajectory in a Bolza Problem

Laura Poggiolini and Marco Spadini

Dipartimento di Matematica Applicata "G. Sansone", Università di Firenze,
Via S. Marta, 3, I–50139 Firenze, Italia.
laura.poggiolini@math.unifi.it, marco.spadini@math.unifi.it

**Summary.** This paper gives sufficient conditions for a class of bang-bang extremals with multiple switches to be locally optimal in the strong topology. The conditions are the natural generalizations of the ones considered in [4, 11] and [12]. We require both the *strict bang-bang Legendre condition*, a non degeneracy condition at multiple switching times, and the second order conditions for the finite dimensional problems obtained by moving the switching times of the reference trajectory.

## 1 Introduction

We consider a Bolza problem on a fixed time interval $[0, T]$, where the control functions are bounded and enter linearly in the dynamics. Namely:

$$\text{minimize } C(\xi, u) := \beta(\xi(T)) + \int_0^T \left( f_0^0(\xi(t)) + \sum_{i=1}^m u_i f_i^0(\xi(t)) \right) \mathrm{d}t \quad (1a)$$

$$\text{subject to } \dot{\xi}(t) = f_0(\xi(t)) + \sum_{i=1}^m u_i f_i(\xi(t)) \quad (1b)$$

$$\xi(0) = \widehat{x}_0; \quad u = (u_1, \ldots, u_m) \in L^\infty([0, T], [-1, 1]^m). \quad (1c)$$

The state space is a $n$–dimensional manifold $M$, $\widehat{x}_0$ is a given point, the vector fields $f_0, f_1, \ldots, f_m$ and the functions $f_0^0, f_1^0, \ldots, f_m^0$, $\beta$ are $C^\infty$.

Optimal control problems in Economics with the above structure have been considered in [6] and references therein.

The authors aim at giving second order sufficient conditions for a *reference bang-bang couple* $(\widehat{\xi}, \widehat{u})$ *to be a local optimizer in the strong topology*; the strong topology being the one induced by $C([0, T], M)$ on the set of the admissible trajectories. Therefore optimality is with respect to neighboring trajectories, independently of the values of the associated controls.

Recall that a control $\widehat{u}$ (a trajectory $\widehat{\xi}$) is bang-bang if there is a finite number of switching times $0 < \widehat{t}_1 < \cdots < \widehat{t}_r < T$ such that each control function $\widehat{u}_i$ is constantly either $-1$ or $1$ on each interval $(\hat{t}_k, \hat{t}_{k+1})$. A switching time $\hat{t}_k$ is called *simple* if only one control function changes value at $\hat{t}_k$, while it is called *multiple* if at least two control functions change value.

Second order conditions for the optimality of a bang-bang extremal with simple switches only are given in [4, 8, 11, 12], and references therein, while in [13] the author gives sufficient conditions, in the case of the minimum time problem, for $L^1$–local optimality of a bang bang extremal having both simple and multiple switches with the extra assumption that the Lie brackets of the switching vector fields is annihilated by the adjoint covector.

Here we consider the problem of local strong optimality in the case of a Bolza problem, when at most one double switch occurs, but there are finitely many simple ones. More precisely we extend the conditions in [4, 11, 12] requiring the sufficient second order conditions for the finite dimensional sub–problems obtained by allowing the switching times to move. We remark that, while in the case of simple switches the only variables are the switching times, each time when a double switch occurs we have to consider the two possible combinations of the switching controls. In order to complete the proof, the investigation of the invertibility of some Lipschitz continuous, piecewise $C^1$ operators has been done via topological methods described in the Appendix. To apply such methods it is necessary to assume a "non–degeneracy" condition at the double switching time.

## 2 The result

The result is based on some regularity assumption on the vector fields associated to the problem and on a second order condition for a finite dimensional sub–problem.

### 2.1 Notation and regularity

Assume we are given an admissible reference couple $(\widehat{\xi}, \widehat{u})$ satisfying Pontryagin maximum principle (PMP) with adjoint covector $\widehat{\lambda}$. Remark that, since no constraint is given on the final point of admissible trajectories, then $(\widehat{\xi}, \widehat{u})$ must satisfy PMP in normal form. We assume the reference control is regular bang–bang with a finite number of switching times $\hat{t}_1, \ldots, \hat{t}_K$ such that only two kinds of switchings appear:

- *simple switching time*: only one of the control functions $\widehat{u}_1, \ldots, \widehat{u}_m$ switches at time $\hat{t}_i$;
- *double switching time*: two of the control functions $\widehat{u}_1, \ldots, \widehat{u}_m$ switch at time $\hat{t}_i$.

We assume that there is just one double switching time, which we denote by $\widehat{\tau}$. Without loss of generality we may assume that the controls switching at time

$\widehat{\tau}$ are $\widehat{u}_1$ and $\widehat{u}_2$. In the interval $(0, \widehat{\tau})$, $J_0$ simple switches occur (if no simple switch occurs in $(0, \widehat{\tau})$, then $J_0 = 0$), while $J_1$ simple switches occur in the interval $(\widehat{\tau}, T)$ (if no simple switch occurs in $(\widehat{\tau}, T)$, then $J_1 = 0$). We denote the simple switching times by $\widehat{\theta}_{ij}$, $j = 1, \ldots, J_i$, $i = 0, 1$ with a self–evident meaning of the double index. In order to simplify the notation, we also define $\widehat{\theta}_{00} := 0$, $\widehat{\theta}_{0,J_0+1} := \widehat{\theta}_{10} := \widehat{\tau}$, $\widehat{\theta}_{1,J_1+1} := T$, i.e. we have

$$\widehat{\theta}_{00} := 0 < \widehat{\theta}_{01} < \ldots < \widehat{\theta}_{0J_0} < \widehat{\tau} := \widehat{\theta}_{0,J_0+1} := \widehat{\theta}_{10} <$$
$$< \widehat{\theta}_{11} < \ldots < \widehat{\theta}_{1\,J_1} < T := \widehat{\theta}_{1,J_1+1}.$$

For any $m$–uple $u = (u_1, \ldots, u_m) \in \mathbb{R}^m$ let us denote

$$h_u \colon \ell \in T^*M \mapsto \left\langle \ell, f_0(\pi\ell) + \sum_{i=1}^m u_i f_i(\pi\ell) \right\rangle - \left( f_0^0(\pi\ell) + \sum_{i=1}^m u_i f_i^0(\pi\ell) \right) \in \mathbb{R}$$

and let $\widehat{f}_t$, $\widehat{f}_t^0$ and $\widehat{H}_t$ be the reference vector field, the reference running cost and the reference Hamiltonian function, respectively, i.e.

$$\widehat{f}_t(x) := f_0(x) + \sum_{i=1}^m \widehat{u}_i(t) f_i(x) \qquad \widehat{f}_t^0(x) := f_0^0(x) + \sum_{i=1}^m \widehat{u}_i(t) f_i^0(x)$$
$$\widehat{H}_t(\ell) := \left\langle \ell, \widehat{f}_t(\pi\ell) \right\rangle - \widehat{f}_t^0(\pi\ell) = h_{\widehat{u}(t)}(\ell).$$

Throughout the paper, for any Hamiltonian function $K \colon T^*M \to \mathbb{R}$, $\overrightarrow{K}$ will denote the associated Hamiltonian vector field. Also, let $\widehat{x}_d := \widehat{\xi}(\widehat{\tau})$ and $\widehat{x}_f := \widehat{\xi}(T)$. In our situation PMP reads as follows:
*There exists an absolutely continuous function $\widehat{\lambda} \colon [0, T] \to T^*M$ such that*

$$\pi\widehat{\lambda}(t) = \widehat{\xi}(t) \; \forall t \in [0, T] \qquad \widehat{\lambda}(T) = -\mathrm{d}\beta(\widehat{x}_f)$$
$$\dot{\widehat{\lambda}}(t) = \overrightarrow{\widehat{H}}_t(\widehat{\lambda}(t)) \quad a.e. \; t \in [0, T],$$
$$\widehat{H}_t(\widehat{\lambda}(t)) = \max\left\{ h_u(\widehat{\lambda}(t)) \colon u \in [-1, 1]^m \right\} \quad \forall t \in [0, T]. \tag{2}$$

Maximality condition (2) implies $\widehat{u}_i(t)\big(\big\langle \widehat{\lambda}(t), f_i(\widehat{\xi}(t)) \big\rangle - f_i^0(\widehat{\xi}(t))\big) \geq 0$ for any $t \in [0, T]$. We assume the following regularity condition holds:

**Regularity.** *If $t$ is not a switching time for the control $\widehat{u}_i$, then*

$$\widehat{u}_i(t)\big(\big\langle \widehat{\lambda}(t), f_i(\widehat{\xi}(t)) \big\rangle - f_i^0(\widehat{\xi}(t))\big) > 0. \tag{3}$$

Notice that (3) implies that $\operatorname{argmax} h_u(\widehat{\lambda}(t)) = \widehat{u}(t)$ for any $t$ that is not a switching time. Let

$$k_{ij} = \widehat{f}_t|_{(\widehat{\theta}_{ij}, \widehat{\theta}_{i,j+1})}, \quad k_{ij}^0 = \widehat{f}_t^0|_{(\widehat{\theta}_{ij}, \widehat{\theta}_{i,j+1})} \qquad i = 0, 1 \quad j = 0, \ldots, J_i$$

be the restrictions of $\widehat{f}_t$, and $\widehat{f}_t^0$ to each of the time intervals where the reference control $\widehat{u}$ is constant. Let $K_{ij}(\ell) := \langle \ell, k_{ij}(\pi\ell)\rangle - k_{ij}^0(\pi\ell)$ be the associated Hamiltonian function. Then, from maximality condition (2) we get

$$\frac{\mathrm{d}}{\mathrm{d}t}\left(K_{10} - K_{0J_0}\right) \circ \widehat{\lambda}_{|\widehat{\tau}} \geq 0 \quad \text{and} \quad \frac{\mathrm{d}}{\mathrm{d}t}\left(K_{ij} - K_{i,j-1}\right) \circ \widehat{\lambda}_{|\widehat{\theta}_{ij}} \geq 0$$

for any $i = 0, 1, \quad j = 1, \dots, J_i$. We assume that the strong inequality holds at each simple switching time $\widehat{\theta}_{ij}$:

**Strong bang–bang Legendre condition for simple switching times.**

$$\frac{\mathrm{d}}{\mathrm{d}t}\left(K_{ij} - K_{i,j-1}\right) \circ \widehat{\lambda}_{|\widehat{\theta}_{ij}} > 0 \qquad i = 0, 1, \quad j = 1, \dots, J_i. \tag{4}$$

We make a stronger assumption at the double switching time $\widehat{\tau}$. Denoting by $\Delta_\nu := \widehat{u}_\nu(\widehat{\tau} + 0) - \widehat{u}_\nu(\widehat{\tau} - 0)$, $\nu = 1, 2$, the jumps at $\widehat{\tau}$ of the two switching components, we have

$$k_{10} = k_{0J_0} + \Delta_1 f_1 + \Delta_2 f_2, \qquad k_{10}^0 = k_{0J_0}^0 + \Delta_1 f_1^0 + \Delta_2 f_2^0.$$

Define the new vector fields and functions

$$k_\nu := k_{0J_0} + \Delta_\nu f_\nu, \qquad k_\nu^0 := k_{0J_0}^0 + \Delta_\nu f_\nu^0, \qquad \nu = 1, 2,$$

with associated hamiltonian functions $K_\nu(\ell) := \langle \ell, k_\nu(\pi\ell)\rangle - k_\nu^0(\pi\ell)$. We assume that all the following one–side derivatives are strictly positive:

**Strong bang–bang Legendre condition for double switching times.**

$$\frac{\mathrm{d}}{\mathrm{d}t}\left(K_\nu - K_{0J_0}\right) \circ \widehat{\lambda}_{|\widehat{\tau}-0} > 0, \quad \frac{\mathrm{d}}{\mathrm{d}t}\left(K_{10} - K_\nu\right) \circ \widehat{\lambda}_{|\widehat{\tau}+0} > 0, \quad \nu = 1, 2. \tag{5}$$

Equivalently, conditions (4) and (5) can be expressed in terms of the canonical symplectic structure $\sigma(\cdot, \cdot)$ on $T^*M$:

$$\sigma\left(\overrightarrow{K}_{i,j-1}, \overrightarrow{K}_{ij}\right)(\widehat{\lambda}(\widehat{\theta}_{ij})) > 0 \qquad i = 0, 1, \quad j = 1, \dots, J_i, \tag{6}$$

$$\sigma\left(\overrightarrow{K}_{0J_0}, \overrightarrow{K}_\nu\right)(\widehat{\lambda}(\widehat{\tau})) > 0, \quad \sigma\left(\overrightarrow{K}_\nu, \overrightarrow{K}_{10}\right)(\widehat{\lambda}(\widehat{\tau})) > 0 \quad \nu = 1, 2. \tag{7}$$

We also assume the following condition holds at the double switching time:

**Non degeneracy.**

$$\frac{\Delta_1 f_1(\widehat{x}_d)}{\sigma\left(\overrightarrow{K}_{0J_0}, \overrightarrow{K}_1\right)(\widehat{\lambda}(\widehat{\tau}))} \neq \frac{\Delta_2 f_2(\widehat{x}_d)}{\sigma\left(\overrightarrow{K}_{0J_0}, \overrightarrow{K}_2\right)(\widehat{\lambda}(\widehat{\tau}))}. \tag{8}$$

## 2.2 The finite dimensional sub–problem

By allowing the switching times of the reference control function to move we can define a finite dimensional sub–problem of the given one. In doing so we must distinguish between the simple switching times and the double switching time. Moving a simple switching time $\widehat{\theta}_{ij}$ to time $\theta_{ij} := \widehat{\theta}_{ij} + \delta_{ij}$ amounts to using the values $\widehat{u}|_{(\widehat{\theta}_{i,j-1},\widehat{\theta}_{i,j})}$ and $\widehat{u}|_{(\widehat{\theta}_{i,j},\widehat{\theta}_{i,j+1})}$ of the control function in the time intervals $(\widehat{\theta}_{i,j-1},\theta_{ij})$ and $(\theta_{ij},\widehat{\theta}_{i,j+1})$, respectively. On the other hand, when we move the double switching time $\widehat{\tau}$ we change the switching time of two different components of the reference control function and we must allow for each of them to change its switching time independently of the other. This means that between the values of $\widehat{u}|_{(\widehat{\theta}_{0,J_0},\widehat{\tau})}$ and $\widehat{u}|_{(\widehat{\tau},\widehat{\theta}_{01})}$ we introduce a value of the control function which is not assumed by the reference one at least in a neighborhood of $\widehat{\tau}$, and which may assume two different values according to which component switches first between the two available ones. Let $\tau_\nu := \widehat{\tau} + \varepsilon_\nu$, $\nu = 1, 2$. We move the switching time of $\widehat{u}_1$ from $\widehat{\tau}$ to $\tau_1 := \widehat{\tau} + \varepsilon_1$, and the switching time of $\widehat{u}_2$ from $\widehat{\tau}$ to $\tau_2 := \widehat{\tau} + \varepsilon_2$.

Defining $\theta_{ij} := \widehat{\theta}_{ij} + \delta_{ij}$, $j = 1, \ldots, J_i$, $i = 0, 1$; $\theta_{0,J_0+1} := \min\{\tau_\nu, \nu = 1, 2\}$, $\theta_{10} := \max\{\tau_\nu, \nu = 1, 2\}$, $\theta_{00} := 0$ and $\theta_{1,J_1+1} := T$, we have two finite–dimensional sub–problems $P_\nu$, $\nu = 1, 2$ given by

$$\text{minimize } \beta(\xi(T)) + \sum_{i=0}^{1}\sum_{j=0}^{J_i} \int_{\theta_{ij}}^{\theta_{i,j+1}} k_{ij}^0(\xi(t))\mathrm{d}t + \int_{\theta_{0,J_0+1}}^{\theta_{10}} k_\nu^0(\xi(t))\mathrm{d}t \qquad (P_\nu\text{a})$$

$$\text{subject to } \dot{\xi}(t) = \begin{cases} k_{0j}(\xi(t)) & t \in (\theta_{0j},\theta_{0,j+1}) \quad j = 0,\ldots,J_0, \\ k_\nu(\xi(t)) & t \in (\theta_{0,J_0+1},\theta_{10}), \\ k_{1j}(\xi(t)) & t \in (\theta_{0j},\theta_{0,j+1}) \quad j = 0,\ldots,J_1 \end{cases} \qquad (P_\nu\text{b})$$

$$\text{and } \xi(0) = \widehat{x}_0. \qquad (P_\nu\text{c})$$

where $k_\nu = k_1$, $k_\nu^0 = k_1^0$ if $\theta_{0,J_0+1} = \tau_1$, and $k_\nu = k_2$, $k_\nu^0 = k_2^0$ if $\theta_{0,J_0+1} = \tau_2$.

We shall denote the solution, evaluated at time $t$, of $(P_\nu\text{b})$ emanating from a point $x \in \mathbb{R}^n$ at time 0 as $S_t(x, \delta, \varepsilon)$. We remark that $S_t(x, 0, 0)$ is the flow associated to the reference control. We shall denote it by $\widehat{S}_t(x)$.

Notice that $P_1$ is defined only for $\varepsilon_1 \leq \varepsilon_2$, while $P_2$ is defined only for $\varepsilon_2 \leq \varepsilon_1$, and the reference control is the one we obtain when every $\delta_{ij}$ and $\varepsilon_k$ is zero, i.e. in a point on the boundary of the domain of $P_\nu$. From PMP we get that the first variation of both these problems at $\delta_{ij} = 0$, $\varepsilon_1 = \varepsilon_2 = 0$ is null, hence we can consider the second variation for the constrained problems $P_1$ and $P_2$. We shall ask for their second order variations to be positive and prove the following theorem:

**Theorem 1.** *Let* $(\widehat{\xi}, \widehat{u})$ *be a bang–bang regular extremal* (3) *for problem* (1) *with associated covector* $\widehat{\lambda}$. *Assume all the switching times of* $(\widehat{\xi}, \widehat{u})$ *but one are simple, while the only non–simple switching time is double.*

*Assume the Legendre conditions* (6) *and* (7) *hold. Also, assume the non degeneracy condition* (8) *holds at the double switching time. Assume also that each second variation* $J''_\nu$ *is positive definite on the kernel of the first variation of problem* $P_\nu$. *Then* $(\widehat{\xi}, \widehat{u})$ *is a strict strong local optimizer for problem* (1).

## 3 Proof of the result

The proof will be carried out by means of Hamiltonian methods. Namely we shall define a time–dependent maximized Hamiltonian $H$ in $T^*M$ with flow $\mathcal{H}\colon [0,T] \times T^*M \to T^*M$ and consider the restriction of $\mathcal{H}$ to a suitable Lagrangian manifold $\Lambda_0$ containing $\widehat{\ell}_0 := \widehat{\lambda}(0)$. We shall prove that $\psi := \mathrm{id} \times \pi \circ \mathcal{H}\colon (t,\ell) \in [0,T] \times \Lambda_0 \mapsto (t, \pi\mathcal{H}_t(\ell)) \in [0,T] \times M$ is locally invertible around $[0,T] \times \{\widehat{\ell}_0\}$ and we will take advantage of the exactness of $\omega := \mathcal{H}^*\, (pdq - Hdt)$ on $\{(t, \mathcal{H}_t(\ell))\, , \ \ell \in \Lambda_0\}$ (see Section 3.4) to reduce our problem to a local optimization problem for a function $F$ defined in a neighborhood of $\widehat{x}_T$. Finally we shall conclude the proof of Theorem 1 showing that such problem has a local minimum in $\widehat{x}_T$. In proving both the invertibility of $\psi$ and the minimality of $\widehat{x}_T$ for $F$ we shall exploit the positivity of the second variations $J''_\nu$. See [1, 2, 3] for a general introduction to Hamiltonian methods.

### 3.1 The maximized flow

We are now going to define the maximized Hamiltonian and the flow of its associated Hamiltonian vector field. Such flow will turn out to be Lipschitz continuous and piecewise–$C^1$. Define

$$\theta_{00}(\ell) := 0 \qquad \phi_{00}(\ell) := \ell$$

for $j = 1, \ldots, J_0$

$$\theta_{0j}(\ell) := \begin{cases} \theta_{0j}(\widehat{\ell}_0) = \widehat{\theta}_{0j} \\ (K_{0j} - K_{0,j-1}) \circ \exp\theta_{0j}(\ell)\overrightarrow{K}_{0,j-1}\,(\phi_{0,j-1}(\ell)) = 0 \end{cases} \tag{10a}$$

$$\phi_{0j}(\ell) := \exp\big(-\theta_{0j}(\ell)\overrightarrow{K}_{0j}\big) \circ \exp\theta_{0j}(\ell)\overrightarrow{K}_{0,j-1}\,(\phi_{0,j-1}(\ell)) \tag{10b}$$

for $\nu = 1, 2$

$$\tau_\nu(\ell) := \begin{cases} \tau_\nu(\widehat{\ell}_0) = \widehat{\tau} \\ (K_\nu - K_{0J_0}) \circ \exp\tau_\nu(\ell)\overrightarrow{K}_{0J_0}(\phi_{0J_0}(\ell)) = 0 \end{cases} \tag{10c}$$

$$\theta_{0,J_0+1}(\ell) := \min\{\tau_1(\ell), \tau_2(\ell)\} \tag{10d}$$

$$K'(\ell) := \begin{cases} K_1(\ell) & \text{if } \theta_{0,J_0+1}(\ell) = \tau_1(\ell) \\ K_2(\ell) & \text{if } \theta_{0,J_0+1}(\ell) = \tau_2(\ell) \end{cases}$$

$$\phi'(\ell) := \exp\big(-\theta_{0,J_0+1}(\ell)\overrightarrow{K}'\big) \circ \exp\theta_{0,J_0+1}(\ell)\overrightarrow{K}_{0J_0}\,(\phi_{0J_0}(\ell)) \tag{10e}$$

$$\theta_{10}(\ell) := \begin{cases} \theta_{10}(\widehat{\ell}_0) = \widehat{\theta}_{10} = \widehat{\tau} \\ (K_{10} - K') \circ \exp \theta_{10}(\ell) \overrightarrow{K}' (\phi'(\ell)) = 0 \end{cases} \tag{10f}$$

$$\phi_{10} := \exp\left(-\theta_{10}(\ell) \overrightarrow{K}_{10}\right) \exp \theta_{10}(\ell) \overrightarrow{K}' (\phi'(\ell)) \tag{10g}$$

for $j = 1, \ldots, J_1$

$$\theta_{1j}(\ell) := \begin{cases} \theta_{1j}(\widehat{\ell}_0) = \widehat{\theta}_{1j} \\ (K_{1j} - K_{1,j-1}) \circ \exp \theta_{1j}(\ell) \overrightarrow{K}_{1,j-1} (\phi_{1,j-1}(\ell)) = 0 \end{cases} \tag{10h}$$

$$\phi_{1j}(\ell) := \exp\left(-\theta_{1j}(\ell) \overrightarrow{K}_{1j}\right) \exp \theta_{1j}(\ell) \overrightarrow{K}_{1,j-1} (\phi_{i,j-1}(\ell)) \tag{10i}$$

$$\theta_{1,J_1+1}(\ell) = T. \tag{10j}$$

To prove that such flow is well defined, we need to show that the switching times $\theta_{ij}(\ell)$, $\tau(\ell)$ are themselves well defined and that they are ordered as follows $\theta_{0,j-1}(\ell) < \theta_{0j}(\ell) \ldots < \theta_{0J_0}(\ell) < \theta_{0,J_0+1}(\ell) \leq \theta_{10}(\ell) < \theta_{11}(\ell) < \ldots$

The proof that the switching times $\theta_{0j}$ are well defined can be carried out as in [4]. Here we show that $\theta_{0,J_0+1}$ and $\theta_{10}$ are also well defined. Let

$$\Psi_\nu(t, \ell) = (K_\nu - K_{0J_0}) \circ \exp t \overrightarrow{K}_{0J_0} \circ \phi_{0J_0}(\ell)$$

then $\left.\dfrac{\partial \Psi_\nu}{\partial t}\right|_{(\widehat{\tau}, \widehat{\ell}_0)} = \sigma\left(\overrightarrow{K}_{0J_0}, \overrightarrow{K}_\nu\right)(\widehat{\lambda}(\widehat{\tau}))$ which is positive by (7). Now, let

$$\Phi_{10}(t, \ell) = (K_{10} - K') \circ \exp t \overrightarrow{K}' \circ \phi_i(\ell)$$

then $\left.\dfrac{\partial \Phi_{10}}{\partial t}\right|_{(\widehat{\tau}, \widehat{\ell}_0)} = \sigma\left(\overrightarrow{K}', \overrightarrow{K}_{10}\right)(\widehat{\lambda}(\widehat{\tau}))$ which is positive by (7).

Since, by assumption $\widehat{\theta}_{i,j-1} < \widehat{\theta}_{ij}$ and $\widehat{\theta}_{0J_0} < \widehat{\tau}$, then, by continuity, $\theta_{i,j-1}(\ell) < \theta_{ij}(\ell)$ and $\theta_{0J_0}(\ell) < \theta_{0,J_0+1}(\ell)$ for any $\ell$ in a sufficiently small neighborhood of $\widehat{\ell}_0$. Therefore, it suffices to show that $\theta_{0,J_0+1}(\ell) \leq \theta_{10}(\ell)$. Notice that if $\tau_1(\ell) = \tau_2(\ell)$, then $\theta_{10}(\ell) = \theta_{0,J_0+1}(\ell)$, so there is nothing to prove and the choice of $K'(\ell)$ either as $K_1(\ell)$ or as $K_2(\ell)$ gives no contribution to the flow of such $\ell$'s, since for these $\ell$'s the interval $(\theta_{0,J_0+1}(\ell), \theta_{10}(\ell))$ is empty.

Let us assume $\theta_{0,J_0+1}(\ell) = \tau_1(\ell) < \tau_2(\ell)$; at time $\theta_{0,J_0+1}(\ell)$ we have

$$0 = (K_1 - K_{0J_0}) \circ \exp \theta_{0,J_0+1}(\ell) \overrightarrow{K}_{0J_0} \circ \phi_{0J_0}(\ell), \tag{11}$$

$$0 > (K_2 - K_{0J_0}) \circ \exp \theta_{0,J_0+1}(\ell) \overrightarrow{K}_{0J_0} \circ \phi_{0J_0}(\ell). \tag{12}$$

Since $K_2 - K_{0J_0} = K_{10} - K_1$, equation (12) can be written as

$$0 > (K_{10} - K_1) \circ \exp 0 \overrightarrow{K}_1 \circ \exp \theta_{0,J_0+1}(\ell) \overrightarrow{K}_{0J_0} \circ \phi_{0J_0}(\ell),$$

i.e. $\theta_{10}(\ell) - \tau_1(\ell) > 0$. Analogous proof holds if $\theta_{0,J_0+1}(\ell) = \tau_2(\ell) < \tau_1(\ell)$. The proof for the $\theta_{1j}$'s can again be done as in [4].

The maximized flow is thus defined as follows:

$$\mathcal{H}\colon (t,\ell)\in[0,T]\times T^*M\mapsto \mathcal{H}_t(\ell)\in T^*M$$

$$\mathcal{H}_t(\ell):=\begin{cases}\exp t\overrightarrow{K}_{0j}(\phi_{0j}(\ell)) & t\in(\theta_{0j}(\ell),\theta_{0,j+1}(\ell)], \quad j=0,\dots,J_0\\ \exp t\overrightarrow{K}'(\phi'(\ell)) & t\in(\theta_{0,J_0+1}(\ell),\theta_{10}(\ell)]\\ \exp t\overrightarrow{K}_{1j}(\phi_{1j}(\ell)) & t\in(\theta_{1j}(\ell),\theta_{1,j+1}(\ell)], \quad j=0,\dots,J_1.\end{cases} \tag{13}$$

### 3.2 The second variations

In order to write the second variations of the finite dimensional sub–problems $P_\nu$ we write them in Mayer form introducing an auxiliary variable $x^0$, as in [11]. The new state space is $\mathbb{R}\times M$ whose elements we denote by $\widetilde{x}:=(x^0,x)$. Let

$$\widetilde{k}_{ij}:=\begin{pmatrix}k_{ij}^0\\ k_{ij}\end{pmatrix}\quad i=0,1,\quad j=0,\dots,J_i,\qquad \widetilde{k}_\nu:=\begin{pmatrix}k_\nu^0\\ k_\nu\end{pmatrix}\qquad \nu=1,2.$$

Then problem $P_\nu$ is equivalent to

$$\text{minimize}\ \ \beta(\xi(T))+\xi^0(T)\ \ \text{subject to} \tag{14a}$$

$$\dot{\widetilde{\xi}}(t)=\begin{pmatrix}\dot\xi^0(t)\\ \dot\xi(t)\end{pmatrix}=\begin{cases}\widetilde{k}_{0j}(\xi(t)) & t\in(\theta_{0j},\theta_{0,j+1})\quad j=0,\dots,J_0,\\ \widetilde{k}_\nu(\xi(t)) & t\in(\tau,\theta_{10}),\\ \widetilde{k}_{1j}(\xi(t)) & t\in(\theta_{0j},\theta_{0,j+1})\quad j=0,\dots,J_1\end{cases} \tag{14b}$$

$$\widetilde{\xi}(0)=(0,\widehat{x}_0). \tag{14c}$$

We denote the solutions of (14b) evaluated at time $t$, emanating from a point $\widetilde{x}=(x^0,x)$ at time 0, as $\widetilde{S}_t(\widetilde{x},\delta,\varepsilon)=\left(S_t^0(x^0,x,\delta,\varepsilon),S_t(x,\delta,\varepsilon)\right)$ and by $\widehat{\widetilde{S}}_t(\widetilde{x})=\left(\widehat{S}_t^0(x^0,x),\widehat{S}_t(x)\right)=\left(S_t^0(x^0,x,0,0),S_t(x,0,0)\right)$ we denote the flow associated to the reference control. Define

$$a_{00}:=\delta_{01};\qquad a_{ij}:=\delta_{ij+1}-\delta_{ij}\qquad i=0,1\quad j=1,\dots,J_i-1;$$

$$a_{0J_0}:=\varepsilon_1-\delta_{0J_0};\qquad b:=\varepsilon_2-\varepsilon_1;\qquad a_{10}:=\delta_{11}-\varepsilon_2;\qquad a_{1J_1}:=-\delta_{1J_1}.$$

Then $b+\sum_{i=0}^{1}\sum_{j=0}^{J_i}a_{ij}=0$. Let

$$g_{ij}(x)=\left(D\widehat{S}_{\widehat\theta_{ij}}\right)^{-1}k_{ij}\circ\widehat{S}_{\widehat\theta_{ij}}(x),\qquad g_{ij}^0(x)=k_{ij}^0\circ\widehat{S}_{\widehat\theta_{ij}}(x)-g_{ij}\cdot S_{\widehat\theta_{ij}}^0(x),$$

$$h_\nu(x)=\left(D\widehat{S}_{\widehat\tau}\right)^{-1}k_\nu\circ\widehat{S}_{\widehat\tau}(x),\qquad h_\nu^0(x)=f_\nu^0\circ\widehat{S}_{\widehat\tau}(x)-h_\nu\cdot\widehat{S}_{\widehat\tau}^0(x)$$

and put $\widehat{\beta}(x):=\beta\circ\widehat{S}_T(x)$, $\widehat{B}_0(x):=\int_0^T\widehat{f}_t(\widehat{S}_t(x))\mathrm{d}t$, $\alpha:=-\widehat{\beta}$ and $\widehat{\gamma}:=\alpha+\widehat{\beta}+\widehat{B}_0$. Also define $\Lambda_0:=\{\mathrm{d}\alpha(x),\ x\in M\}$. Let $\widetilde{\zeta}_t(\widetilde{x},\delta,\varepsilon):=\left(\widehat{\widetilde{S}}_t\right)^{-1}\circ\widetilde{S}_t(x,\delta,\varepsilon)$. We consider the second–order variations of

$$J_\nu(x, a, b) := \alpha(x) + \widehat{\beta}(\zeta_T^\nu(x, a, b)) + \widehat{S_T^0}\big(\widehat{\zeta}_T(0, x, a, b)\big)$$

at the reference triplet $(x, a, b) = (\widehat{x}_0, 0, 0)$. By assumption, for each $\nu = 1, 2$, $J_\nu''$ is positive definite on

$$\mathcal{N}_0^\nu := \left\{ (\delta x, a, b) \in T_{\widehat{x}_0} M \times \mathbb{R}^{J_0 + J_1} \times \mathbb{R} \colon \delta x = 0, \quad b + \sum_{i=0}^{1} \sum_{j=0}^{J_i} a_{ij} = 0 \right\}.$$

Possibly redefining $\alpha$ by adding a suitable second–order penalty at $\widehat{x}_0$, we may assume that each second variation $J_\nu''$ is positive definite on

$$\mathcal{N}^\nu := \left\{ (\delta x, a, b) \in T_{\widehat{x}_0} M \times \mathbb{R}^{J_0 + J_1} \times \mathbb{R} \colon b + \sum_{i=0}^{1} \sum_{j=0}^{J_i} a_{ij} = 0 \right\}.$$

Let $G_{ij}$, $H_\nu$ be the Hamiltonian functions associated to $(g_{ij}, g_{ij}^0)$ and $(h_\nu, h_\nu^0)$ respectively, and introduce the anti–symplectic isomorphism $i$ as in [4],

$$i \colon (\delta p, \delta x) \in T_{\widehat{x}_0}^* M \times T_{\widehat{x}_0} M \mapsto -\delta p + \mathrm{d}\left(-\widehat{\beta} - \widehat{B}_0\right)_* \delta x \in T\left(T^* M\right). \quad (15)$$

Defining $\overrightarrow{G}_{ij}'' = i^{-1}\left(\overrightarrow{G}_{ij}(\widehat{\ell}_0)\right)$, $\overrightarrow{H}_\nu'' = i^{-1}\left(\overrightarrow{H}_\nu(\widehat{\ell}_0)\right)$, we have that $\overrightarrow{G}_{ij}''$ and $\overrightarrow{H}_\nu''$ are the Hamiltonian vector fields associated to the following linear Hamiltonian functions defined in $T_{\widehat{x}_0}^* M \times T_{\widehat{x}_0} M$

$$G_{ij}''(\omega, \delta x) = \langle \omega, g_{ij}(\widehat{x}_0) \rangle + \delta x \cdot \left(g_{ij} \cdot \left(\widehat{\beta} + \widehat{B}_0 - \widehat{S}_{\theta_{ij}}^0\right) + g_{ij}^0 \circ \widehat{S}_{\theta_{ij}}\right)(\widehat{x}_0), \quad (16)$$

$$H_\nu''(\omega, \delta x) = \langle \omega, h_\nu(\widehat{x}_0) \rangle + \delta x \cdot \left(h_\nu \cdot \left(\widehat{\beta} + \widehat{B}_0 - \widehat{S}_{\widehat{\tau}}^0\right) + h_\nu^0 \circ \widehat{S}_{\widehat{\tau}}\right)(\widehat{x}_0). \quad (17)$$

Moreover $L_0'' := i^{-1} T_{\widehat{\ell}_0} \Lambda_0 = \left\{ \delta \ell = \left(-D^2 \widehat{\gamma}(\widehat{x}_0)(\delta x, \cdot), \delta x\right) : \delta x \in T_{\widehat{x}_0} M \right\}$ and the bilinear form $J_\nu''$ associated to the second variation can be written in a rather compact form: for any $\delta e := (\delta x, a, b) \in \mathcal{N}^\nu$ let

$$\omega_0 := -D^2 \widehat{\gamma}(\widehat{x}_0)(\delta x, \cdot), \quad \delta \ell := (\omega_0, \delta x) = i^{-1} \left(\mathrm{d}\alpha_* \delta x\right),$$

$$(\omega_\nu, \delta x_\nu) := \delta \ell + \sum_{i=0}^{1} \sum_{j=0}^{J_i} a_{ij} \overrightarrow{G}_{ij}'' + b \overrightarrow{H}_\nu'' \quad \text{and} \quad \delta \ell_\nu := (\omega_\nu, \delta x_\nu).$$

Then $J_\nu''$ can be written as

$$J_\nu''\big((\delta x, a, b), (\delta y, c, d)\big) = -\left\langle \omega_\nu, \delta y + \sum_{s=0}^{J_0} c_{0s} g_{0s} + d\, h_\nu + \sum_{s=0}^{J_1} c_{1s} g_{1s} \right\rangle$$

$$+ \sum_{j=0}^{J_0} c_{0j} G_{0j}'' \left(\delta \ell + \sum_{s=0}^{j-1} a_{0s} \overrightarrow{G}_{0s}''\right) + d\, H_\nu'' \left(\delta \ell + \sum_{s=0}^{J_0} a_{0s} \overrightarrow{G}_{0s}''\right) \quad (18)$$

$$+ \sum_{j=0}^{J_1} c_{1j} G_{1j}'' \left(\delta \ell + \sum_{s=0}^{J_0} a_{0s} \overrightarrow{G}_{0s}'' + b \overrightarrow{H}_\nu'' + \sum_{s=0}^{j-1} a_{1s} \overrightarrow{G}_{1s}''\right).$$

We shall study the positivity of $J''_\nu$ as follows: consider the increasing sequence of sub–spaces of

$$V^\nu := \left\{ (\delta x, a, b) \in \mathcal{N}^\nu : \delta x + \sum_{i=0}^{1} \sum_{j=0}^{J_i} a_{ij} g_{ij}(\widehat{x}_0) + b\, h_\nu(\widehat{x}_0) = 0 \right\}.$$

defined as

$$V^\nu_{0j} := \{ (\delta x, a, b) \in V^\nu : a_{0s} = 0 \quad \forall s = j+1, \ldots, J_0, \; a_{1s} = 0$$
$$\forall s = 0, \ldots, J_1, \quad b = 0 \},$$
$$V^\nu_{1j} := \{ (\delta x, a, b) \in V^\nu : a_{1s} = 0 \quad \forall s = j+1, \ldots, J_1 \}.$$

Then $V^1_{0j} = V^2_{0j}$ for any $j = 0, \ldots, J_0$, so we denote these sets as $V_{0j}$. Moreover

$$\dim \left( V_{0j} \cap V^{\perp J''_\nu}_{0,j-1} \right) = \dim \left( V^\nu_{1k} \cap V^{\perp J''_\nu}_{1,k-1} \right) = 1, \quad \dim \left( V^\nu_{10} \cap V^{\perp J''_\nu}_{0J_0} \right) = 2$$

for any $j = 2, \ldots, J_0$ and any $k = 0, \ldots, J_1$.

Using the first order approximations of the quantities $\theta_{ij}(\ell)$, $\phi_{ij}(\ell)$, defined in equations (10) and proceeding as in [4] we can prove the following lemmata

**Lemma 1.** $\delta e = (\delta x, a, b) \in V_{0j} \cap V^{\perp J''_\nu}_{0,j-1}$ *if and only if* $\delta e \in V_{0j}$ *and*

$$G''_{0s}\Big(\delta \ell + \sum_{\mu=0}^{s-1} a_{0\mu} \vec{G}''_{0\mu}\Big) = G''_{0,j-1}\Big(\delta \ell + \sum_{s=0}^{j-2} a_{0s} \vec{G}''_{0s}\Big), \quad \forall s = 0, \ldots, j-2 \quad (19)$$

*i.e.* $a_{0s} = \mathrm{d}\,(\theta_{0,s+1} - \theta_{0s})\,(\mathrm{d}\alpha_* \delta x) \quad \forall s = 0, \ldots, j-2.$

*In this case* $J''[\delta e]^2 = a_{0j}\, \sigma\Big(\delta \ell + \sum_{s=0}^{j-1} a_{0s} \vec{G}''_{0s}, \vec{G}''_{0j} - \vec{G}''_{0,j-1}\Big).$

**Lemma 2.** $\delta e = (\delta x, a, b) \in V^\nu_{10} \cap V^{\perp J''_\nu}_{0J_0}$ *if and only if* $\delta e \in V^\nu_{10}$ *and*

$$G''_{0s}\Big(\delta \ell + \sum_{\mu=0}^{s-1} a_{0\mu} \vec{G}''_{0\mu}\Big) = G''_{0,J_0}\Big(\delta \ell + \sum_{s=0}^{J_0-1} a_{0s} \vec{G}''_{0s}\Big), \quad \forall s = 0, \ldots, J_0 - 1 \quad (20)$$

*i.e.* $a_{0s} = \mathrm{d}\,(\theta_{0,s+1} - \theta_{0s})\,(\mathrm{d}\alpha_* \delta x) \quad \forall s = 0, \ldots, J_0 - 1.$

*In this case*

$$J''[\delta e]^2 = b\, \sigma\Big(\delta \ell + \sum_{s=0}^{J_0} a_{0s} \vec{G}''_{0s}, \vec{H}''_\nu - \vec{G}''_{0,J_0}\Big)$$

$$+ a_{10}\, \sigma\Big(\delta \ell + \sum_{s=0}^{J_0} a_{0s} \vec{G}''_{0s} + b\vec{H}''_\nu, \vec{G}''_{10} - \vec{H}''_\nu\Big).$$

**Lemma 3.** $\delta e = (\delta x, a, b) \in V_{1j}^{\nu} \cap V_{1,j-1}^{\perp J_{\nu}''}$ if and only if $\delta e \in V_{1j}^{\nu}$ and

$$G_{0s}''(\delta\ell + \sum_{i=0}^{s-1} a_{0i}\vec{G}_{0i}'') = H_{\nu}''(\delta\ell + \sum_{i=0}^{J_0} a_{0i}\vec{G}_{0i}'')$$

$$= G_{1k}''(\delta\ell + \sum_{i=0}^{J_0} a_{0i}\vec{G}_{0i}'' + b\vec{H}_{\nu}'' + \sum_{i=0}^{k-1} a_{1i}\vec{G}_{1i}'')$$

$$\forall\, s = 0, \ldots, J_0 \quad \forall\, k = 0, \ldots, j-2$$

i.e. if and only if $\delta e \in V_{1j}^{\nu}$ and

$$a_{0s} = \mathrm{d}\,(\theta_{0,s+1} - \theta_{0s})\,(\mathrm{d}\alpha_*\delta x) \quad \forall s = 0, \ldots, J_0$$
$$b = \mathrm{d}\,(\theta_{10} - \theta_{0,J_0+1})\,(\mathrm{d}\alpha_*\delta x)$$
$$a_{1s} = \mathrm{d}\,(\theta_{1,s+1} - \theta_{1s})\,(\mathrm{d}\alpha_*\delta x) \quad \forall s = 0, \ldots, j-2.$$

In this case

$$J''[\delta e]^2 = a_{1j}\,\sigma\left(\delta\ell + \sum_{s=0}^{J_0} a_{0s}\vec{G}_{0s}'' + b\vec{H}_{\nu}'' + \sum_{i=0}^{j-1} a_{1i}\vec{G}_{1i}'', \vec{G}_{1j}'' - \vec{G}_{1,j-1}''\right).$$

**Lemma 4.** $\delta e = (\delta x, a, b) \in \mathcal{N}^{\nu} \cap V_{1J_1}^{\perp J_{\nu}''}$ if and only if $\delta e \in \mathcal{N}^{\nu}$ and

$$G_{0s}''(\delta\ell + \sum_{i=0}^{s-1} a_{0i}\vec{G}_{0i}'') = H_{\nu}''(\delta\ell + \sum_{i=0}^{J_0} a_{0i}\vec{G}_{0i}'')$$

$$= G_{1k}''(\delta\ell + \sum_{i=0}^{J_0} a_{0i}\vec{G}_{0i}'' + b\vec{H}_{\nu}'' + \sum_{i=0}^{k-1} a_{1i}\vec{G}_{1i}'')$$

$$\forall\, s = 0, \ldots, J_0 \quad \forall\, k = 0, \ldots, J_1$$

i.e. if and only if $\delta e \in \mathcal{N}^{\nu}$ and

$$a_{0s} = \mathrm{d}\,(\theta_{0,s+1} - \theta_{0s})\,(\mathrm{d}\alpha_*\delta x) \quad \forall s = 0, \ldots, J_0$$
$$b = \mathrm{d}\,(\theta_{10} - \theta_{0,J_0+1})\,(\mathrm{d}\alpha_*\delta x)$$
$$a_{1s} = \mathrm{d}\,(\theta_{1,s+1} - \theta_{1s})\,(\mathrm{d}\alpha_*\delta x) \quad \forall s = 0, \ldots, J_1-1.$$

In this case

$$J''[\delta e]^2 = -\left\langle \omega_{\nu}, \delta x + \sum_{i=0}^{1}\sum_{s=0}^{J_i} a_{is}g_{is} + b\,h_{\nu} \right\rangle$$

$$= \sigma\left(\left(0, \delta x + \sum_{i=0}^{1}\sum_{s=0}^{J_i} a_{is}g_{is} + bh_{\nu}\right), -D^2\hat{\gamma}(\delta x, \cdot) + \sum_{i=0}^{1}\sum_{s=0}^{J_i} a_{is}\vec{G}_{is}'' + b\vec{H}_{\nu}''\right).$$

## 3.3 The invertibility of the flow

Lemma 1 allows us to prove the following property (whose proof can be found in [4]) for the linearization of the maximized flow:

**Lemma 5.** *Let* $j \in \{1, \ldots, J_0\}$ *and* $\delta x_1, \delta x_2 \in T_{\hat{x}_0} M$ *such that* $d\theta_{0j}(\delta x_2) < 0 < d\theta_{0j}(\delta x_1)$. *Then* $\left(\pi \circ \mathcal{H}_{\widehat{\theta}_{0j}}\right)_* d\alpha_* \delta x_1 \neq \left(\pi \circ \mathcal{H}_{\widehat{\theta}_{0j}}\right)_* d\alpha_* \delta x_2$.

Lemma 5 implies that the application

$$\psi \colon (t, \ell) \in [0, T] \times \Lambda_0 \mapsto (t, \pi \circ \mathcal{H}_t(\ell)) \in [0, T] \times M \qquad (21)$$

is locally invertible around $[0, \hat{\tau} - \varepsilon] \times \{\widehat{\ell_0}\}$. In fact, $\psi$ is locally one–to–one if and only if $\pi \circ \mathcal{H}_t$ is locally one–to–one in $\widehat{\ell_0}$ for any $t$. On the other hand $\pi \circ \mathcal{H}_t$ is locally one–to–one for any $t < \hat{\tau}$ if and only if it is one–to–one at any $\widehat{\theta}_{0j}$. This property is granted by Lemma 5.

We now want to show that such procedure can be carried out also on $[\hat{\tau} - \varepsilon, T] \times \{\widehat{\ell_0}\}$, so that $\psi$ will turn out to be locally invertible from a neighborhood $[0, T] \times \mathcal{O} \subset [0, T] \times \Lambda_0$ of $[0, T] \times \{\widehat{\ell_0}\}$ onto a neighborhood $\mathcal{U} \subset [0, T] \times M$ of the graph $\widehat{\Xi}$ of $\widehat{\xi}$.

The first step will be proving the invertibility of $\pi \circ \mathcal{H}_{\hat{\tau}}$ at $\widehat{\ell_0}$. In a neighborhood of $\widehat{\ell_0}$, $\pi \circ \mathcal{H}_{\hat{\tau}}$ has the following piecewise representation

| M1 | $\min\{\tau_1(\ell),\, \tau_2(\ell)\} \geq \hat{\tau}$ | $\pi \exp \hat{\tau} \overrightarrow{K}_{0J_0} \circ \phi_{0J_0}(\ell)$ |
|----|----|----|
| M2 | $\min\{\tau_1(\ell), \tau_2(\ell)\} =$ $\tau_1(\ell) \leq \hat{\tau} \leq \theta_{10}(\ell)$ | $\pi \exp \hat{\tau} \overrightarrow{K}_1 \circ \exp(-\tau_1(\ell)\overrightarrow{K}_1) \circ \exp \tau_1(\ell) \overrightarrow{K}_{0J_0} \circ$ $\phi_{0J_0}(\ell)$ |
| M3 | $\min\{\tau_1(\ell), \tau_2(\ell)\} =$ $\tau_2(\ell) \leq \hat{\tau} \leq \theta_{10}(\ell)$ | $\pi \exp \hat{\tau} \overrightarrow{K}_2 \circ \exp(-\tau_2(\ell)\overrightarrow{K}_2) \circ \exp \tau_2(\ell) \overrightarrow{K}_{0J_0} \circ$ $\phi_{0J_0}(\ell)$ |
| M4 | $\min\{\tau_1(\ell), \tau_2(\ell)\} =$ $\tau_1(\ell) \leq \theta_{10}(\ell) \leq \hat{\tau}$ | $\pi \exp (\hat{\tau} - \theta_{10}(\ell)) \overrightarrow{K}_{10} \circ \exp \theta_{10}(\ell) \overrightarrow{K}_1 \circ$ $\exp(-\tau_1(\ell)\overrightarrow{K}_1) \circ \exp \tau_1(\ell) \overrightarrow{K}_{0J_0} \circ \phi_{0J_0}(\ell)$ |
| M5 | $\min\{\tau_1(\ell), \tau_2(\ell)\} =$ $\tau_2(\ell) \leq \theta_{10}(\ell) \leq \hat{\tau}$ | $\pi \exp (\hat{\tau} - \theta_{10}(\ell)) \overrightarrow{K}_{10} \circ \exp \theta_{10}(\ell) \overrightarrow{K}_2 \circ$ $\exp(-\tau_2(\ell)\overrightarrow{K}_2) \circ \exp \tau_2(\ell) \overrightarrow{K}_{0J_0} \circ \phi_{0J_0}(\ell)$ |

The invertibility of $\pi \circ \mathcal{H}_{\hat{\tau}}$ will be proved by the means of Theorem 3 in the Appendix. Notice that the non degeneracy condition (8) implies that the second order penalty on $\alpha$ can be chosen so that $d\tau_1(d\alpha_*(\cdot)) \neq d\tau_2(d\alpha_*(\cdot))$. In order to apply Theorem 3 we write the piecewise linearized map $(\pi \circ \mathcal{H}_{\hat{\tau}})_*$. $(\pi \circ \mathcal{H}_{\hat{\tau}})_* \delta \ell$ is given by

**M1′** if $\min\{d\tau_1(\delta \ell),\, d\tau_2(\delta \ell)\} \geq 0$

$$\exp(\hat{\tau} \overrightarrow{K}_{0J_0})_* \phi_{0J_0} \delta \ell \qquad (22a)$$

**M2′** if $d\tau_1(\delta \ell) \leq 0 \leq d\theta_{10}(\ell)$, $d\tau_1(\delta \ell) \leq d\tau_2(\delta \ell)$

$$-d\tau_1(\delta \ell)\left[\exp(\hat{\tau} \overrightarrow{K}_1)_* \overrightarrow{K}_1 - \overrightarrow{K}_{0J_0}\right] + \exp(\hat{\tau} \overrightarrow{K}_{0J_0})_* \phi_{0J_0 *} \delta \ell \qquad (22b)$$

**M3′** if $d\tau_2(\delta\ell) \leq 0 \leq d\theta_{10}(\delta\ell)$, $d\tau_2(\delta\ell) \leq d\tau_1(\delta\ell)$

$$-d\tau_2(\delta\ell)\left[\exp(\hat{\tau}\vec{K}_2)_*\vec{K}_2 - \vec{K}_{0J_0}\right] + \exp(\hat{\tau}\vec{K}_{0J_0})_*\phi_{0J_0*}\delta\ell \quad (22c)$$

**M4′** if $d\tau_1(\delta\ell) \leq d\theta_{10}(\delta\ell) \leq 0$, $d\tau_1(\delta\ell) \leq d\tau_2(\delta\ell)$

$$d\theta_{10}(\delta\ell)\left(\vec{K}_{10} + d\theta_{10}(\delta\ell)\exp(\hat{\tau}\vec{K}_{10})_*\vec{K}_1\right)$$
$$+ \exp(\hat{\tau}\vec{K}_1)_*\left(-d\tau_1(\delta\ell)\vec{K}_1 + \exp(-\hat{\tau}_1\vec{K}_1)_*d\tau_1(\delta\ell)\vec{K}_{0J_0}\right.$$
$$\left. + \exp(-\hat{\tau}\vec{K}_1)_*\exp(\hat{\tau}\vec{K}_{0J_0})\phi_{0J_0*}\delta\ell\right) \quad (22d)$$

**M5′** if $d\tau_2(\delta\ell) \leq d\theta_{10}(\delta\ell) \leq 0$, $d\tau_2(\delta\ell) \leq d\tau_1(\delta\ell)$

$$d\theta_{10}(\delta\ell)\left(\vec{K}_{10} + d\theta_{10}(\delta\ell)\exp(\hat{\tau}\vec{K}_{10})_*\vec{K}_2\right)$$
$$+ \exp(\hat{\tau}\vec{K}_2)_*\left(-d\tau_2(\delta\ell)\vec{K}_2 + \exp(-\hat{\tau}_2\vec{K}_2)_*d\tau_2(\delta\ell)\vec{K}_{0J_0}\right.$$
$$\left. + \exp(-\hat{\tau}\vec{K}_2)_*\exp(\hat{\tau}\vec{K}_{0J_0})\phi_{0J_0*}\delta\ell\right) \quad (22e)$$

According to Theorem 3, in order to prove the invertibility of our map it is sufficient to prove that both the map and its linearization are continuous in a neighborhood of $\hat{\ell}_0$ and of $0$ respectively, that they maintain the orientation and that there exists a point $\delta\overline{x}$ whose preimage is a singleton not belonging to the above boundaries.

Notice that the continuity of $\pi \circ \mathcal{H}_{\hat{\tau}}$ follows from the very definition of the maximized flow. Discontinuities of $(\pi \circ \mathcal{H}_{\hat{\tau}})_*$ may occur only at the boundaries described above. A direct computation in formulas (22) shows that this is not the case. Let us now prove the last assertion.

Throughout the rest of the section, all the Hamiltonian vector fields $\vec{G}_{ij}$ and $\vec{H}_\nu$ are computed in $\hat{\ell}_0$. For "symmetry" reasons it is convenient to look for $\delta\overline{x}$ among those which belong to the image of the set $\{\delta\ell \in T_{\hat{\ell}_0}\Lambda_0 : 0 < d\tau_1(\delta\ell) = d\tau_2(\delta\ell)\}$. Observe that this implies that $d\theta_{10}(\delta\ell) = d\tau_1(\delta\ell) = d\tau_2(\delta\ell)$: Introducing the quantity $\eta_{\hat{\tau}}(\delta\ell) := \mathcal{H}_{\hat{\tau}*}^{-1}\exp(\hat{\tau}\vec{K}_{0J_0})_*\phi_{0J_0*}(\delta\ell)$, the assertion $d\tau_1(\delta\ell) = d\tau_2(\delta\ell)$ can be written as

$$\frac{\sigma\left(\eta_{\hat{\tau}}(\delta\ell), \vec{H}_1 - \vec{G}_{0J_0}\right)}{\sigma\left(\vec{G}_{0J_0}, \vec{H}_1\right)} = \frac{\sigma\left(\eta_{\hat{\tau}}(\delta\ell), \vec{H}_2 - \vec{G}_{0J_0}\right)}{\sigma(\vec{G}_{0J_0}, \vec{H}_2)}$$

thus, $d\theta_{10}(\delta\ell)\sigma\left(\vec{H}_2, \vec{G}_{10}\right)$ is given by

$$-\sigma\left(\eta_{\hat{\tau}}(\delta\ell), \vec{G}_{10} - \vec{H}_2\right) + d\tau_2(\delta\ell)\sigma\left(\vec{H}_2 - \vec{G}_{0J_0}, \vec{G}_{10} - \vec{H}_2\right)$$

and, using that $d\tau_1(\delta\ell) = d\tau_2(\delta\ell)$ and $\overrightarrow{G}_{10} - \overrightarrow{H}_2 = \overrightarrow{H}_1 - \overrightarrow{G}_{0J_0}$, this is equal to $d\tau_1(\delta\ell)\sigma\left(\overrightarrow{H}_2, \overrightarrow{G}_{10}\right)$.

Thus, we consider $\delta\overline{x} = \pi_* \exp(\widehat{\tau}\overrightarrow{K}_{0J_0})_* \phi_{0J_0*}\delta\ell_1$ with $0 < d\tau_1(\delta\ell_1) = d\tau_2(\delta\ell_1)$. Clearly $\delta\overline{x}$ has at most one preimage per each of the above sectors. Let us prove that actually its preimage is the singleton $\{\delta\ell_1\}$.

Assume by contradiction that there is $\delta\ell_2$ in sector M2$'$ such that

$$\pi_* \exp(\widehat{\tau}\overrightarrow{K}_{0J_0})_* \phi_{0J_0*}(\delta\ell_1) = \pi\left(\exp \widehat{\tau}\overrightarrow{K}_1\right)_*\left(-d\tau_1(\delta\ell_2)\overrightarrow{K}_1\right)$$
$$+ \pi_* d\tau_1(\delta\ell_2)\overrightarrow{K}_{0J_0} + \pi_* \exp(\widehat{\tau}\overrightarrow{K}_{0J_0})_* \phi_{0J_0*}(\delta\ell_2)$$

Taking the pull-back we get

$$\delta x_1 - \delta x_2 + \sum_{s=1}^{J_0-1} d\left(\theta_{0,s+1} - \theta_{0s}\right)\left(d\alpha_*(\delta x_1 - \delta x_2)\right)g_{0s}$$
$$- \left(d\theta_{0J_0}\left(d\alpha_*(\delta x_1 - \delta x_2)\right) + d\tau_1(d\alpha_*\delta x_2)\right)g_{0J_0} + d\tau_1(d\alpha_*\delta x_2)h_1 = 0.$$

Consider $\delta e := (\delta x_1 - \delta x_2, a, b)$, where, for $j = 0, \ldots, J_1$, $a_{1j} = 0$ and, for $s = 0, \ldots, J_0$,

$$a_{0s} = \begin{cases} d\left(\theta_{0,s+1} - \theta_{0s}\right)\left(d\alpha_*(\delta x_1 - \delta x_2)\right) & s = 0, \ldots, J_0 - 1, \\ -d\theta_{0J_0}\left(d\alpha_*(\delta x_1 - \delta x_2)\right) + d\tau_1(d\alpha_*\delta x_2) & s = J_0, \end{cases}$$

and $b = d\tau_1(d\alpha_*\delta x_2) < 0$. Thus $\delta e \in V_{10}^1 \cap V_{0J_0}^{\perp J_1''}$, therefore Lemma 2 applies:

$$\sigma\left(\delta\ell + \sum_{s=0}^{J_0} a_{0s}\overrightarrow{G}_{0s}'', \overrightarrow{H}_1'' - \overrightarrow{G}_{0J_0}''\right) < 0$$

where $\delta\ell = \left(-D^2\widehat{\gamma}(x_0)(\delta x_1 - \delta x_2, \cdot), \delta x_1 - \delta x_2\right)$. Thus, applying $i$,

$$\sigma\left(d\alpha_*(\delta x_1 - \delta x_2) + \sum_{s=0}^{J_0} a_{0s}\overrightarrow{G}_{0s}, \overrightarrow{H}_1 - \overrightarrow{G}_{0J_0}\right) > 0$$

or, linearizing the formula for $\tau_1(\delta\ell)$ in (10),

$$\sigma\left(\eta_{\widehat{\tau}}(d\alpha_*(\delta x_1 - \delta x_2)), \overrightarrow{H}_1 - \overrightarrow{G}_{0J_0}\right) - d\tau_1(d\alpha_*\delta x_2)\sigma\left(\overrightarrow{G}_{0J_0}, \overrightarrow{H}_1\right) > 0$$

which implies $-d\tau_1\left(d\alpha_*(\delta x_1 - \delta x_2)\right) - d\tau_1(d\alpha_*\delta x_2) > 0$ or $d\tau_1(d\alpha_*\delta x_1) < 0$ a contradiction.

Let us now assume by contradiction that there is $\delta\ell_4$ in sector M4$'$ whose image under the linearized map coincides with $\delta\overline{x}$.

Thus, proceeding in the same way as between sectors M1$'$ and M2$'$, we get

$$\delta x_1 - \delta x_4 + \sum_{s=1}^{J_0-1} \mathrm{d}\left(\theta_{0,s+1} - \theta_{0s}\right)\left(\mathrm{d}\alpha_*(\delta x_1 - \delta x_4)\right)g_{0s} - \left(\mathrm{d}\theta_{0J_0}\left(\mathrm{d}\alpha_*(\delta x_1 - \delta x_4)\right)\right.$$

$$\left. + \mathrm{d}\tau_1\left(\mathrm{d}\alpha_*\delta x_4\right)\right)g_{0J_0} - \mathrm{d}\left(\theta_{10} - \tau_1\right)\left(\mathrm{d}\alpha_*\delta x_4\right)h_1 + \mathrm{d}\theta_{10}(\mathrm{d}\alpha_*\delta x_4)g_{10} = 0.$$

Consider $\delta e := (\delta x_1 - \delta x_4, a, b)$, where, for $j = 1, \ldots, J_1$, $a_{1j} = 0$, $a_{10} = \mathrm{d}\theta_{10}(\delta x_4) < 0$ and, for $s = 0, \ldots, J_0$,

$$a_{0s} = \begin{cases} \mathrm{d}\left(\theta_{0,s+1} - \theta_{0s}\right)\left(\mathrm{d}\alpha_*(\delta x_1 - \delta x_4)\right) & s = 0, \ldots, J_0 - 1 \\ -\mathrm{d}\theta_{0J_0}\left(\mathrm{d}\alpha_*(\delta x_1 - \delta x_4)\right) + \mathrm{d}\tau_1(\mathrm{d}\alpha_*\delta x_4) & s = J_0, \end{cases}$$

and $b = \mathrm{d}\left(\theta_{10} - \tau_1\right)(\mathrm{d}\alpha_*\delta x_4) < 0$. Thus $\delta e \in V_{10}^1 \cap V_{0J_0}^{\perp J_1''}$, and Lemma 3 applies

$$b\,\sigma\left(\mathrm{d}\alpha_*(\delta x_1 - \delta x_4) + \sum_{s=0}^{J_0} a_{0s}\vec{G}_{0s}, \vec{H}_1 - \vec{G}_{0J_0}\right)$$

$$+ a_{10}\,\sigma\left(\mathrm{d}\alpha_*(\delta x_1 - \delta x_4) + \sum_{s=0}^{J_0} a_{0s}\vec{G}_{0s} + b\vec{H}_1, \vec{G}_{10} - \vec{H}_1\right) < 0$$

The coefficient of $b$ is equal to

$$\sigma\left(\eta_{\hat{\tau}}(\mathrm{d}\alpha_*(\delta x_1 - \delta x_4)) - \mathrm{d}\tau_1(\delta x_4)\vec{G}_{0J_0}, \vec{H}_1 - \vec{G}_{0J_0}\right)$$

$$= -\mathrm{d}\tau_1\left(\mathrm{d}\alpha_*(\delta x_1 - \delta x_4)\right)\sigma\left(\vec{G}_{0J_0}, \vec{H}_1\right) - \mathrm{d}\tau_1(\mathrm{d}\alpha_*\delta x_4)\sigma\left(\vec{G}_{0J_0}, \vec{H}_1\right)$$

$$= -\mathrm{d}\tau_1(\mathrm{d}\alpha_*\delta x_1)\sigma\left(\vec{G}_{0J_0}, \vec{H}_1, \right) < 0$$

On the other hand, taking the first order approximations in (10), one can show that the coefficient of $a_{10}$ is:

$$\left(-\mathrm{d}\tau_1(\mathrm{d}\alpha_*(\delta x_1)) + \mathrm{d}\theta_{10}(\mathrm{d}\alpha_*(\delta x_4))\left(-\mathrm{d}\tau_1(\mathrm{d}\alpha_*(\delta x_1))\right)\right)\sigma\left(\vec{G}_{0J_0}, \vec{H}_2\right)$$

$$- \mathrm{d}\left(\theta_{10} - \tau_1\right)(\mathrm{d}\alpha_*(\delta x_4))\left(-\mathrm{d}\tau_1(\mathrm{d}\alpha_*(\delta x_1))\right)\sigma\left(\vec{G}_{0J_0}, \vec{H}_1\right) < 0$$

which is impossible.

The orientation preserving condition can be proved by the means of Lemma 6: consider any pair of adjacent cones $M_i'$ and $M_j'$. They are separated by a hyperplane. A similar argument to the one used above shows that any pair of points lying on opposite sides of the separating hyperplane have different images under the maps used in $M_i'$ and $M_j'$, extended to the corresponding half space.

This proves the invertibility of $\pi \circ \mathcal{H}_{\hat{\tau}}$, hence $\psi$ is one–to–one in a neighborhood of $[0, \widehat{\theta}_{10} - \varepsilon] \times \{\widehat{\ell}_0\}$.

We only sketch the idea of the proof of the invertibility of $\pi \circ \mathcal{H}_{\widehat{\theta}_{1j}}$, $j = 1, \ldots, J_1$. Given $j$, there are four regions $N_{1j}, \ldots, N_{4j}$ in $\Lambda_0$, characterized by the following properties

$$(N_{1j})\ \theta_{1j}(\ell) > \widehat{\theta}_{1j} \text{ and } \theta_{0,J_0+1}(\ell) = \tau_1(\ell),$$
$$(N_{2j})\ \theta_{1j}(\ell) > \widehat{\theta}_{1j} \text{ and } \theta_{0,J_0+1}(\ell) = \tau_2(\ell),$$
$$(N_{3j})\ \theta_{1j}(\ell) < \widehat{\theta}_{1j} \text{ and } \theta_{0,J_0+1}(\ell) = \tau_1(\ell),$$
$$(N_{4j})\ \theta_{1j}(\ell) < \widehat{\theta}_{1j} \text{ and } \theta_{0,J_0+1}(\ell) = \tau_2(\ell);$$

as for $\pi \circ \mathcal{H}_{\widehat{\tau}}$, $\pi \circ \mathcal{H}_{\widehat{\theta}_{1j}}$ turns out to be a Lipschitz continuous, piecewise $C^1$ application. Its invertibility can be proved applying again Theorem 3. We will consider first the case $j = 1$ and the following linearization of $\pi \circ \mathcal{H}_{\widehat{\theta}_{11}}$. Here, for the sake of brevity we have already passed to the pullback

$N'_{1j}$     where $d\theta_{11}(\delta\ell) \geq 0$ and $d\tau_1(\delta\ell) \leq d\tau_2(\delta\ell)$,

$$-d\theta_{10}(\delta\ell)g_{10} + d(\theta_{10} - \tau_1)(\delta\ell)h_1 + d\tau_1(\delta\ell)g_{0J_0} + \eta_{\widehat{\tau}}(\delta\ell)$$

$N'_{2j}$     where $d\theta_{11}(\delta\ell) \geq 0$ and $d\tau_2(\delta\ell) \leq d\tau_1(\delta\ell)$,

$$-d\theta_{10}(\delta\ell)g_{10} + d(\theta_{10} - \tau_2)(\delta\ell)h_2 + d\tau_2(\delta\ell)g_{0J_0} + \eta_{\widehat{\tau}}(\delta\ell)$$

$N'_{3j}$     where $d\theta_{11}(\delta\ell) \leq 0$ and $d\tau_1(\delta\ell) \leq d\tau_2(\delta\ell)$,

$$-d\theta_{11}(\delta\ell)g_{11} + d(\theta_{11} - \theta_{10})(\delta\ell)g_{10} + d(\theta_{10} - \tau_1)(\delta\ell)h_1 + d\tau_1(\delta\ell)g_{0J_0} + \eta_{\widehat{\tau}}(\delta\ell)$$

$N'_{4j}$     where $d\theta_{11}(\delta\ell) \leq 0$ and $d\tau_2(\delta\ell) \leq d\tau_2(\delta\ell)$,

$$-d\theta_{11}(\delta\ell)g_{11} + d(\theta_{11} - \theta_{10})(\delta\ell)g_{10} + d(\theta_{10} - \tau_2)(\delta\ell)h_2 + d\tau_2(\delta\ell)g_{0J_0} + \eta_{\widehat{\tau}}(\delta\ell)$$

As above, according to Theorem 3, we only have to prove that both the map and its linearization are continuous in a neighborhood of $\widehat{\ell}_0$ and of 0 respectively, that the linearized pieces are orientation preserving and that there exists a point $\delta\overline{x}$ whose preimage is a singleton. The only nontrivial part is the last statement which can be proved by picking $\delta\overline{x} \in N'_{11} \cap N'_{12}$.

## 3.4 Reduction to a finite–dimensional problem

In this section, in order to shorten the notation, for any $(t, \ell) \in [0, T] \times \Lambda_0$, let us define $\psi_t(\ell) := \pi \circ \mathcal{H}_t(\ell)$. Also we recall that the maximized Hamiltonian function is a lift: $H_t(\ell) = \langle \ell, f(t, \pi\ell) \rangle - f^0(t, \pi\ell)$

In the product space $[0, T] \times M$ consider the path obtained with the concatenation of the graph of a generic trajectory, $\Xi := \{(t, \xi(t)) : t \in [0, T]\}$ (ran backward) contained in $\mathcal{U}$ and the graph of the reference trajectory $\widehat{\Xi} := \{(t, \widehat{\xi}(t)) : t \in [0, T]\}$. We can obtain a close circuit with a path $\gamma$ from

$(T, \widehat{x}_T)$ to $(T, \xi(T))$ whose image is contained in $\{T\} \times M$.
Consider the following sets in $[0, T] \times T^*M$:

$$\mathcal{O}_{0j} = \{(t, \ell) \colon \ell \in \mathcal{O}, \quad t \in [\theta_{0,j-1}(\ell), \theta_{0j}(\ell)]\} \qquad j = 1, \ldots, J_0$$

and, for $\nu = 1, 2$ define

$$\mathcal{O}^\nu_{0,J_0+1} = \{(t, \ell) \colon \ell \in \mathcal{O}, \; \theta_{0,J_0+1}(\ell) = \tau_\nu(\ell), \; t \in [\theta_{0J_0}(\ell), \theta_{0,J_0+1}(\ell)]\}$$
$$\mathcal{O}^\nu_{10} = \{(t, \ell) \colon \ell \in \mathcal{O}, \; \theta_{0,J_0+1}(\ell) = \tau_\nu(\ell), \; t \in [\theta_{0,J_0+1}(\ell), \theta_{10}(\ell)]\}$$
$$\mathcal{O}^\nu_{ij} = \{(t, \ell) \colon \ell \in \mathcal{O}, \; \theta_{0,J_0+1}(\ell) = \tau_\nu(\ell),$$
$$t \in [\theta_{1,j-1}(\ell), \theta_{1j}(\ell)]\} \qquad j = 1, \ldots, J_1 + 1.$$

The one–form $\omega := \mathcal{H}^*(pdq - H_t dt)$ is closed on each of these sets, it is continuous on $[0, T] \times \mathcal{O}$ hence it is exact on $[0, T] \times \mathcal{O}$ (without loss of generality we may assume $\mathcal{O}$ to be simply connected) and we have

$$0 = \oint \omega = \int_{\psi^{-1}(\widehat{\Xi})} \omega + \int_{\psi^{-1}(\gamma)} \omega - \int_{\psi^{-1}(\Xi)} \omega.$$

From the maximality properties of $H$ we get

$$\int_{\psi^{-1}(\widehat{\Xi})} \omega = \int_0^T \widehat{f}^0_t(\widehat{\xi}(t)) dt \qquad \int_{\psi^{-1}(\Xi)} \omega \le \int_0^T f^0(\xi(t), u(t)) dt; \qquad (23)$$

so that

$$\int_0^T f^0(\xi(t), u(t)) dt - \int_0^T \widehat{f}^0_t(\widehat{\xi}(t)) dt \ge \int_{\psi^{-1}(\gamma)} \omega.$$

If we now evaluate the difference of the costs associated to the generic pair $(\xi, u)$ and to the reference pair $(\widehat{\xi}, \widehat{u})$ we have

$$C(\xi, u) - C(\widehat{\xi}, \widehat{u}) \ge \beta(\xi(T)) - \beta(\widehat{x}_T) + \int_{\psi^{-1}(\gamma)} \omega. \qquad (24)$$

Evaluating this last integral we get

$$\int_{\psi^{-1}(\gamma)} \omega = \alpha(\pi\psi_T^{-1}(\xi(T))) - \alpha(\pi\psi_T^{-1}(\widehat{x}_T))$$
$$+ \int_0^T f^0(\psi(r, \psi_T^{-1}\xi(T))) dr - \int_0^T f^0(\psi(r, \psi_T^{-1}\widehat{x}_T)) dr.$$

Defining $F \colon y \in M \mapsto \alpha(\pi \circ \psi_T^{-1}(y)) + \beta(y) + \int_0^T f^0(\psi(r, \psi_T^{-1}(y))) dr$ equation (24) simplifies to $C(\xi, u) - C(\widehat{\xi}, \widehat{u}) \ge F(\xi(T)) - F(\widehat{x}_T)$ i.e. we have reduced optimal control problem (1) to a finite–dimensional one. Thus in order to prove that $(\widehat{\xi}, \widehat{u})$ is a minimum it now suffices to prove that $F$ has a local minimum in $\widehat{x}_T$.

**Theorem 2.** *F has a strict local minimum in $\widehat{x}_T$.*

*Proof.* It suffices to prove that

$$dF(\widehat{x}_T) = 0 \qquad D^2 F(\widehat{x}_T) > 0. \tag{25}$$

The first equality in (25) is an immediate consequence of the definition of $\alpha$ and PMP. Let us prove that also the second one holds.
Since $d\left(\alpha \circ \pi \circ \psi_T^{-1} + \int_0^T f(r, \psi_r \circ \psi_T^{-1}) dr\right) = \mathcal{H}_T \circ \psi_T^{-1}$, we also have

$$dF = \mathcal{H}_T \circ \psi_T^{-1} + d\beta \tag{26}$$

$$D^2 F(\widehat{x}_T)[\delta x_T]^2 = \left((\mathcal{H}_T \circ \psi_T^{-1})_* + D^2\beta\right)(\widehat{x}_T)[\delta x_T]^2$$
$$= \sigma\left((\mathcal{H}_T \circ \psi_T^{-1})_* \delta x_T, d(-\beta)_* \delta x_T\right). \tag{27}$$

From Lemma 4 we get

$$0 < \sigma\left(\left(0, \delta x + \sum_{i=0}^1 \sum_{j=0}^{J_i} a_{ij} g_{ij} + bh_\nu\right), -D^2 \widehat{\gamma}(\delta x, \cdot) + \sum_{i=0}^1 \sum_{j=0}^{J_i} a_{ij} \vec{G}_{ij}'' + b\vec{H}_\nu''\right). \tag{28}$$

Applying $\widehat{\mathcal{H}}_{T*} \circ i^{-1}$ we get $0 < \sigma\left(\mathcal{H}_{T*} d\alpha_* \delta x, d(-\beta)_* (\psi_{T*} d\alpha_* \delta x)\right)$ which is exactly (27) with $\delta x := \pi_* \psi_{T*}^{-1} \delta x_T$. Since $\pi_* \psi_{T*}^{-1}$ is one–to–one, such a choice is always possible.

To conclude the proof of Theorem 1 we have to prove that $\widehat{\xi}$ is a strict minimizer. Assume $C(\xi, u) = C(\widehat{\xi}, \widehat{u})$. Since $\widehat{x}_T$ is a strict minimizer for $F$, then $\xi(T) = \widehat{x}_T$ and equality must hold in (23):

$$\left\langle \mathcal{H}_s(\psi_s^{-1}(\xi(s))), \dot{\xi}(s) \right\rangle - f^0(\xi(s), u(s)) = H_s(\mathcal{H}_s(\psi_s^{-1}(\xi(s)))).$$

By regularity assumption, $u(s) = \widehat{u}(s)$ for any $s$ at least in a left neighborhood of $T$, hence $\xi(s) = \widehat{\xi}(s)$ and $\psi_s^{-1}(\xi(s)) = \widehat{\ell}_0$ for any $s$ in such neighborhood. $u$ takes the value $\widehat{u}_{|(\widehat{\theta}_{1,J_1}, T)}$ until $\mathcal{H}_s \psi_s^{-1}(\xi(s)) = \mathcal{H}_s(\widehat{\ell}_0) = \widehat{\lambda}(s)$ hits the hypersurface $K_{1,J_1} = K_{1,J_1-1}$, which happens at time $s = \widehat{\theta}_{1,J_1}$. At such time, again by regularity assumption, $u$ must switch to $\widehat{u}_{(\widehat{\theta}_{1,J_1-1}, \widehat{\theta}_{1,J_1})}$, so that $\xi(s) = \widehat{\xi}(s)$ also for $s$ in a left neighborhood of $\widehat{\theta}_{1,J_1}$. Proceeding backward in time, with an induction argument we finally get $(\xi(s), u(s)) = (\widehat{\xi}(s), \widehat{u}(s))$ for any $s \in [0, T]$.

# 4 Appendix: Invertibility of piecewise $C^1$ maps

The straightforward proof of the following fact is left to the reader.

**Lemma 6.** *Let $A$ and $B$ be linear automorphisms of $\mathbb{R}^n$. Assume that for some $v \in \mathbb{R}^n$, $A$ and $B$ coincide on the space $\{x \in \mathbb{R}^n : \langle x, v \rangle = 0\}$. Then, the map $\mathcal{L}_{AB}$ defined by $x \mapsto Ax$ if $\langle x, v \rangle \geq 0$, and by $x \mapsto Bx$ if $\langle x, v \rangle \leq 0$, is a homeomorphism if and only if $\det(A) \cdot \det(B) > 0$.*

Let $G : \mathbb{R}^n \to \mathbb{R}^n$ be a continuous, piecewise linear map at $0$, in the sense that $G$ is continuous and there exists a decomposition $S_1, \ldots, S_k$ of $\mathbb{R}^n$ in closed polyhedral cones (intersection of half spaces, hence convex) with common vertex in the origin and such that $\partial S_i \cap \partial S_j = S_i \cap S_j$, $i \neq j$, and linear maps $L_1, \ldots, L_k$ with

$$G(x) = L_i x \qquad x \in S_i, \qquad i = 1, \ldots, k,$$

with $L_i x = L_j x$ for any $x \in S_i \cap S_j$.

It is easily shown that $G$ is proper, and therefore $\deg(G, \mathbb{R}^n, p)$ is well-defined for any $p \in \mathbb{R}^n$ (the construction of [9] is still valid if the assumption on the compactness of the manifolds is replaced with the assumption that $G$ is proper). Moreover $\deg(G, \mathbb{R}^n, p)$ is constant with respect to $p$. So we shall denote it by $\deg(G)$.

We shall also assume that $\det L_i > 0$ for any $i = 1, \ldots, k$.

**Lemma 7.** *If $G$ is as above, then $\deg(G) > 0$. In particular, if there exists $q \neq 0$ such that its preimage belongs to at most two of the convex polyhedral cones $S_i$ and $G^{-1}(q)$ is a singleton, then $\deg(G) = 1$.*

*Proof.* Let us assume in addition that $q \notin \cup_{i=1}^k G(\partial S_i)$. Observe that the set $\cup_{i=1}^k G(\partial S_i)$ is nowhere dense hence $A_1 := G(S_1) \setminus \cup_{i=1}^k G(\partial S_i)$ is non–empty. Take $x \in A_1$ and observe that if $y \in G^{-1}(x)$ then $y \notin \cup_{i=1}^k \partial S_i$. Thus

$$\deg(G) = \sum_{y \in G^{-1}(x)} \operatorname{sign} \det dG(y) = \#G^{-1}(x). \tag{29}$$

Since $G^{-1}(x) \neq \emptyset$ the first part of the assertion is proved. The second part of the assertion follows taking $x = q$ in (29).

Let us now remove the additional assumption. Let $\{p\} = G^{-1}(q)$ be such that $p \in \partial S_i \cap \partial S_j$ for some $i \neq j$. Thus one can find a neighborhood $V$ of $p$, with $V \subset \operatorname{int}(S_i \cup S_j \setminus \{0\})$. By the excision property of the topological degree $\deg(G) = \deg(G, V, p)$. Let $\mathcal{L}_{L_i L_j}$ be a map as in Lemma 6. Observe that, the assumption on the signs of the determinants of $L_i$ and $L_j$ imply that $\mathcal{L}_{L_i L_j}$ is orientation preserving. Also notice that $\mathcal{L}_{L_i L_j}|_{\partial V} = G|_{\partial V}$. The multiplicativity, excision and boundary dependence properties of the degree yield $1 = \deg(\mathcal{L}_{L_i L_j}) = \deg(\mathcal{L}_{L_i L_j}, V, p) = \deg(G, V, p)$. Thus, $\deg(G) = 1$, as claimed.

Let $\sigma_1, \ldots, \sigma_r$ be a family of $C^1$–regular pairwise transversal hypersurfaces in $\mathbb{R}^n$ with $\cap_{i=1}^r \sigma_i = \{x_0\}$ and let $U \subset \mathbb{R}^n$ be an open and bounded neighborhood of $x_0$. Clearly, if $U$ is sufficiently small, $U \setminus \cup_{i=1}^r \sigma_i$ is partitioned into a finite number of open sets $U_1, \ldots, U_k$.

Let $f : \overline{U} \to \mathbb{R}^n$ be a continuous map such that there exist Fréchet differentiable functions $f_1, \ldots, f_k$ in $\overline{U}$ with the property that

$$f(x) = f_i(x) \qquad x \in \overline{U}_i, \qquad i = 1, \ldots, k, \tag{30}$$

with $f_i(x) = f_j(x)$ for any $x \in \overline{U}_i \cap \overline{U}_j$. Notice that such a function is $PC^1(\overline{U})$, hence locally Lipschitz continuous (see [7]).

Let $S_1, \ldots, S_k$ be the tangent cones at $x_0$ to the sets $U_1, \ldots, U_k$, (by the transversality assumption on the hyper–surfaces $\sigma_i$ each $S_i$ is a convex polyhedral cone with non empty interior) and assume $df_i(x_0)x = df_j(x_0)x$ for any $x \in S_i \cap S_j$. Define

$$F(x) = df_i(x_0)x \qquad x \in S_i. \tag{31}$$

so that $F$ is a continuous piecewise linear map (compare [7]).

One can see that $f$ is Bouligand differentiable and that its B-derivative is the map $F$ (compare [7, 10]). Let $y_0 := f(x_0)$. There exists a continuous function $\varepsilon$, with $\varepsilon(0) = 0$, such that $f(x) = y_0 + F(x - x_0) + |x - x_0|\varepsilon(x - x_0)$.

**Lemma 8.** *Let $f$ and $F$ be as in (30)–(31), then there exists $\rho > 0$ such that $\deg(f, B(x_0, \rho), y_0) = \deg(F, B(0, \rho), 0)$. In particular, if $\det df_i(x_0) > 0$, then $F$ is proper and $\deg(f, B(x_0, \rho), y_0) = \deg(F)$.*

*Proof.* Consider the homotopy $H(x, \lambda) = F(x - x_0) + \lambda |x - x_0| \varepsilon(x - x_0)$, $\lambda \in [0, 1]$ and observe that $m := \inf\{|F(v)| : |v| = 1\} > 0$, $F$ being invertible. Thus,

$$|H(x, \lambda)| \geq (m - |\varepsilon(x - x_0)|) |x - x_0|.$$

This shows that in a conveniently small ball centered at $x_0$, homotopy $H$ is admissible. The assertion follows from the homotopy invariance property of the degree.

Let $f$ and $F$ be as in (30)–(31) and assume $\det df_i(x_0) > 0$. Assume also that there exists $p \in \mathbb{R}^n \setminus \cup_{i=1}^k F(\partial S_i)$ such that $F^{-1}(p)$ is a singleton. From Lemmas 7–8, it follows that $\deg(f, B(x_0, \rho), y_0) = 1$ for sufficiently small $\rho > 0$. By Theorem 4 in [10], we immediately obtain

**Theorem 3.** *Let $f$ and $F$ be as in (30)–(31) and assume $\det df_i(x_0) > 0$. Assume also that there exists $p \in \mathbb{R}^n \setminus \cup_{i=1}^k F(\partial S_i)$ such that $F^{-1}(p)$ is a singleton. Then $f$ is a Lipschitzian homeomorphism in a sufficiently small neighborhood of $x_0$.*

# References

1. Agrachev AA, Gamkrelidze RV (1990) Symplectic geometry for optimal control. In: Sussmann HJ (ed) Nonlinear Controllability and Optimal Control, Pure and Applied Mathematics, vol 133. Marcel Dekker
2. Agrachev AA, Gamkrelidze RV (1997) Symplectic methods for optimization and control. In: Jacubczyk B, Respondek W (eds) Geometry of Feedback and Optimal Control, Pure and Applied Mathematics, 1–58, Marcel Dekker, New York

3. Agrachev AA, Sachkov, YuL (2004) Control Theory from the Geometric Viewpoint Springer-Verlag
4. Agrachev AA, Stefani G, Zezza P(2002) Strong optimality for a bang-bang trajectory. SIAM J. Control Optimization, 41(4):991–1014.
5. Arnold VI (1980) Mathematical Methods in Classical Mechanics. Springer, New York
6. Koslik B, Breitner MH (1997) In optimal control problem in economics with four linear control. J. Optim. Theory and App., 94(3):619–634.
7. Kuntz L, Scholtes S (1994) Structural analysis of nonsmooth mappings, inverse functions and metric projections. Journal of Mathematical Analysis and Applications, 188:346–386.
8. Maurer H, Osmolovskii NP (2003) Second order optimality conditions for bang-bang control problems. Control and Cybernetics, 3(32):555–584.
9. Milnor J (1965) Topology from the Differentiable Viewpoint. The University Press of Virginia
10. Jong-Shi Pang J-S, Ralph D (1996) Piecewise smoothness, local invertibility, and parametric analysis of normal maps. Mathematics of Operations Research, 21(2):401–426.
11. Poggiolini L (2006) On local state optimality of bang-bang extremals in a free horizon bolza problem. Rendiconti del Seminario Matematico dell'Universitá e del Politecnico di Torino.
12. Poggiolini L, Stefani G (2004) State-local optimality of a bang-bang trajectory: a Hamiltonian approach. Systems & Control Letters, 53:269–279.
13. Sarychev AV (1997) First and second order sufficient optimality conditions for bang-bang controls. SIAM J. Control Optimization, 1(35):315–340.

# Modelling Energy Markets with Extreme Spikes

Thorsten Schmidt

Department of Mathematics, University of Leipzig,
D-04081 Leipzig, Germany `thorsten.schmidt@math.uni-leipzig.de`

**Summary.** This paper suggests a new approach to model spot prices of electricity. It uses a shot-noise model to capture extreme spikes typically arising in electricity markets. Moreover, the model easily accounts for seasonality and mean reversion. We compute futures prices in closed form and show that the resulting shapes capture a large variety of typically observed term structures. For statistical purposes we show how to use the EM-algorithm. An estimation on spot price data from the European Energy Exchange illustrate the applicability of the model.

## 1 Motivation

It is well-known that as many other commodities electricity prices exhibit strong seasonalities. Besides this, due to the difficulty of storing electricity and inelastic demand, electricity spot prices show extremely strong spikes. The spot price data shown in Figure 1 clearly confirms this. In this paper, we propose a model which naturally captures this spiking behaviour. The model uses a type of shot-noise which is particularly suited for electricity spikes. It is furthermore simple enough to allow for closed-form solutions of futures and other power derivatives.

It is important to mention that electricity markets are young and small markets. For example, in Germany it is possible to trade electricity since 2000 and currently there are about 150 market participants trading at the European Energy Exchange, Leipzig[1]. Electricity prices have a number of features which are necessary to capture by a good model.

First, the necessity for using a model incorporating jumps is underlined in [7] or [20]. There are two approaches, which are closely related to the model presented here. In [10] a model is proposed, where the jump component jumps up until a exogenously level is reached and thereafter jumps down. The approach of [4] is a special case of ours. The authors use a jump-diffusion

---

[1] 158 participants from 19 countries, cited from the webpage www.eex.de on January 2007.

to capture the spikes and the mean reversion. For an overview of existing literature on electricity models we refer to these papers. An approach using Lévy processes may be found in [2]. In contrast to the Lévy approach, the shot-noise modelling allows for an easier estimation: an efficient tool for estimating shot-noise models is the EM-algorithm. We derive the necessary densities and apply the model to electricity prices in Section 4.

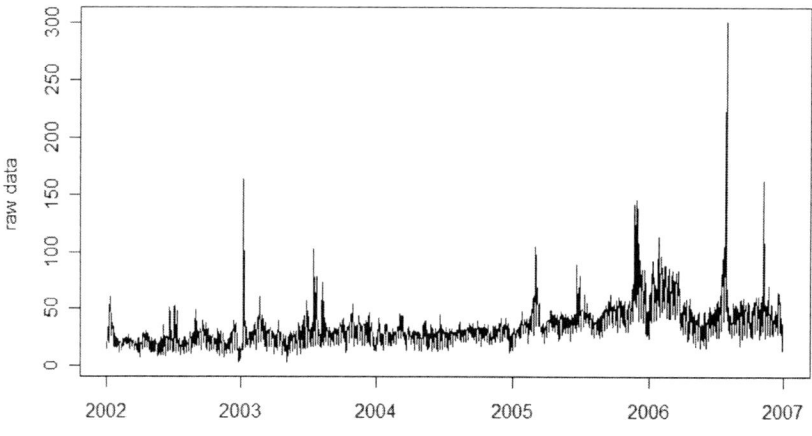

**Fig. 1.** The spot prices of energy (base load) quoted from the European Energy Exchange (www.eex.de).

The proposed model generalizes [4] and offers more flexibility in capturing the statistical properties of the spot price as well as in calibrating to the futures curve. On the other side, the approach to modelling spikes seems more natural as in [10], and in contrast to this model, we are able to compute prices of derivatives in closed form.

It seems important to note the specific characteristics of futures traded on electricity markets in contrast to futures, for example, from interest rate markets[2]. Electricity futures offer delivery of electricity over a certain period, typically a month, a quarter or a year. In a certain way this is a practicable approach to insure against extreme price fluctuations, because the payoff smoothes singular effects like spikes. On the other side, futures with a yearly

---

[2] See, for example, the European Energy Exchange (EEX) Contract Specifications, downloadable from www.eex.de.

delivery period also loose the dependence on the seasonalities. We take this into account and derive prices of futures on electricity markets.

## 2 Setup

Consider a filtered probability space $(\Omega, \mathcal{F}, (\mathcal{F}_t)_{t \geq 0}, \mathbb{P})$ which admits a Brownian motion $(W_t)_{t \geq 0}$, a Poisson process $(N_t)_{t \geq 0}$ and iid rvs $Y_i, i = 1, 2, \ldots$, all independent of each other. We generalize simple shot-noise approaches as eg in [1] in a way suitable for electricity spot prices. A close analysis of electricity prices reveals that the arising spikes either have an up-jump and then a strong decline or a sharp rise followed by a strong decline. The following function $h$ will be able to capture this behaviour. For more general types of shot-noise processes we refer to [16].

For $a, b > 0$ define $h : (\mathbb{R}^+)^2 \times \mathbb{R} :\mapsto \mathbb{R}^+$ by[3]

$$h(t, \gamma, Y) := Y \cdot \begin{cases} \exp(a(t - \gamma)) & \text{if } 0 \leq t < \gamma, \\ \exp(-b(t - \gamma)) & \text{if } t \geq \gamma. \end{cases}$$

$Y$ is the jump height and typically will be positive, while not necessarily. For $\gamma = 0$ this resembles simple shot-noise as a special case. If $\gamma > 0$, then $h$ jumps at zero to $Y \exp(-a\gamma)$, then rises to $Y$ at $\gamma$ and thereafter it declines exponentially. For the shot-noise component we propose

$$J_t := \sum_{\tau_i \leq t} h(t - \tau_i, \gamma_i, Y_i). \tag{1}$$

*Example 1.* A simple example would be to assume that $\gamma_i \in \{0, \tilde{\gamma}\}$ with $p_\gamma := \mathbb{P}(\gamma_1 = 0)$. In this case one has classical shot-noise with probability $p_\gamma$ and the "steep rise followed by sharp decline" case with probability $1 - p_\gamma$.

The diffusive part is responsible for mean-reversion and seasonalities. As the focus of the paper is mainly on the jump part, we stay quite simple in the assumptions on the diffusion. Assume that $D$ is the strong solution of

$$dD_t = \kappa(\theta(t) - D_t)dt + \sigma dB_t, \tag{2}$$

where $B$ is a standard Brownian motion. Under the above specification we say that

$$S = D + J$$

follows a *Vasicek/shot-noise process* with parameters $(a, b, f_\gamma, f_Y, \lambda, \kappa, \theta(\cdot), \sigma)$.

The process given in (2) is the well-known dynamics proposed by [19] for interest rate models. This process has a stochastic mean-reversion to the level $\theta(t)$. A different form of mean-reversion is obtained, if $D_t = \theta(t) + \tilde{D}_t$ is chosen,

---

[3] We set $\mathbb{R}^+ := \{x \in \mathbb{R} : x \geq 0\}$.

where $\tilde{D}_t$ is mean-reverting to the level 0. This was done in [4]. Contrary, in the stochastic mean-reversion, as chosen here, the mean reversion speed depends on the distance of $D$ to the mean reversion level. Thus, if $|\theta(t) - D_t|$ is large, the process is pulled strongly back towards ($\theta$), while if this difference is low, the mean-reversion is not so strong.

It is straightforward to extend the given setup to more general dynamics of $D$. For example, using the formulas obtained in [9], one immediately obtains closed-form solutions for electricity futures using generalized quadratic models for the diffusive part. For example, the well-known CIR-Modell (see [5]) moreover guarantees positivity of $D$. In contrast to interest rate-models, polynomials of order higher than two can also be considered[4].

## 2.1 Changing measure

On one side, statistical estimation, as we consider in Section 4, is always done under the real-world measure $\mathbb{P}$ while on the other side pricing of derivatives takes place under the risk-neutral measure $\mathbb{Q}$. There is a vast of literature on specific choices of the risk-neutral measure. However, in this paper we consider a rather pragmatic approach which serves the need of applicability on one side and retains a reasonable amount of flexibility on the other side: we assume that the chosen model retains its structure while changing from $\mathbb{P}$ to $\mathbb{Q}$ although it of course will have different parameter values under $\mathbb{Q}$.

The Girsanov theorem[5] gives all possible changes of measure. For our purposes, we restrict to a sufficiently flexible measure change. Define $L_t := \mathrm{e}(\frac{d\mathbb{Q}}{d\mathbb{P}}|\mathcal{F}_t), t \geq 0$ and assume that $L$ is given by

$$L_t = \prod_{\tau_i \leq t} \left( \frac{\tilde{\lambda}\tilde{f}_Y(Y_i)}{\lambda f_Y(Y_i)} \right) \exp\left( -\int_0^t a(s)dW_s + \int_0^t (b - \frac{1}{2}a^2(s))ds \right), \quad (3)$$

where $a(s) = (\theta(t) - \tilde{\theta}(t))\sigma\kappa^{-1}$ for a deterministic function $\tilde{\theta}(t)$ and $b = \int(\tilde{\lambda}\tilde{f}_Y(z) - \lambda f_Y(z))dz$. The following result precisely states the obtained model under $\mathbb{Q}$.

**Proposition 1.** *Assume that $S$ is a Vasicek/shot-noise process with parameters $(a, b, f_\gamma, f_Y, \lambda, \kappa, \theta(\cdot), \sigma)$ under $\mathbb{P}$ and the measure change $d\mathbb{Q}/d\mathbb{P}$ is given by the likelihood process in (3). Then $S$ is a Vasicek/shot-noise process under $\mathbb{Q}$ with parameters $(a, b, f_\gamma, \tilde{f}_Y, \tilde{\lambda}, \kappa, \tilde{\theta}(\cdot), \sigma)$.*

Intuitively spoken, this means that under $\mathbb{Q}$, $(W_t + \int_0^t a(s)ds)_{t \geq 0}$ is a standard Brownian motion, $N$ is a Poisson process with intensity $\tilde{\lambda}$. The distribution of $Y$ may be changed in a quite general fashion, provided they are still equivalent. For practical purposes it might be reasonable to choose a parametric family and assume that the parameters change from $\mathbb{P}$ to $\mathbb{Q}$ while the

---

[4] The degree problem in interest rate models was observed in [8].
[5] Compare [14] for a suitably general version.

$Y$ stays in the parametric family. We assume the distribution of $\gamma$ does not change to retain the shot-noise type. However, it is straightforward to also incorporate a change of the distribution of $\gamma$.

*Proof.* The claim follows directly from the Girsanov theorem. First, note that $a$, $b$, $\sigma$ and $\kappa$ do not change under equivalent measure changes. Second, $(W_t + \int_0^t a(s)ds)_{t \geq 0}$ is a $\mathbb{Q}$-Brownian motion and hence

$$dD_t = \kappa(\theta(t) - \sigma\frac{\theta(t) + \tilde{\theta}(t)}{\sigma} - D_t)dt + \sigma(dW_t + a(t)dt)$$
$$= \kappa(\tilde{\theta}(t) - D_t)dt + \sigma(dW_t + a(t)dt).$$

Furthermore, the jump component of $L$ immediately reveals that $Y_i$, $i \geq 1$ are again i.i.d. under $\mathbb{Q}$ with densities $\tilde{f}_Y$; moreover $N$ is a Poisson process with intensity $\tilde{\lambda}$ (see, for example, [3], Section VIII.3, Theorem T10). This yields the claim.    □

It is important to note that there is no kind of no-arbitrage restriction on $\mathbb{Q}$ as the spot price, which is modeled here, is not a traded asset; see also the next section for further details. Any other asset what we consider later on will be an expectation of its discounted payoffs under $\mathbb{Q}$ and hence by definition be in line with no-arbitrage.

Thanks to Proposition 1, we can consider from now on Vasicek/shot-noise processes under $\mathbb{P}$ as well as under $\mathbb{Q}$. Note that, still, estimated parameters are under $\mathbb{P}$ while parameters for pricing as well as calibrated parameters are under $\mathbb{Q}$ throughout and typically do not coincide. A comparison analysis of calibrated prices with estimated parameters could clarify on the market prices or risk chosen by the market and would make a link possible.

## 2.2 Pricing of electricity futures

To price electricity futures we mainly follow Teichmann (2005), hence we assume that futures are traded for time-to-maturity of at least a small value, say $\epsilon$. As fluctuations in electricity markets are quite large in comparison to interest rate markets, it is reasonable to assume zero interest rates. Then the futures price of a contingent claim $\mathcal{X}$ is $e^{\mathbb{Q}}(\mathcal{X}|\mathcal{F}_t)$, where the expectation is taken under an equivalent martingale measure $\mathbb{Q}$.

The futures actually traded in electricity markets are not futures on a single spot rate. Instead, they offer electricity for a certain period of length $L$. More precisely, the future offers delivery of electricity in the period $[T, T+\Delta]$, with the value

$$\sum_{T_i \in [T, T+\Delta]} S_{T_i}$$

where $T_i \in [T, T+\Delta]$ refers to the respective trading days in the period under consideration. We assume that the mesh of the trading days is equidistant, i.e. $T_i - T_{i-1} := \delta$ for all $i$. In the following, we approximate the sum by an integral, $\sum_{T_i \in [T, T+\Delta]} S_{T_i} \approx \frac{1}{\delta} \int_T^{T+\Delta} S_u \, du$. This is not necessary and is just used to simplify the formulas. It is an easy exercise to compute the explicit formulas for $\sum_{T_i \in [T, T+\Delta]} S_{T_i}$ instead of the integral.

Using the approximation we consider the following futures price:

$$F(t, T, \Delta) = \frac{1}{\delta} \, e^{\mathbb{Q}} \Big( \int_T^{T+\Delta} S_u \, du | \mathcal{F}_t \Big).$$

We take this formula as a starting point and compute futures prices under the proposed shot-noise model. First, notice that as $S$ is a sum of a diffusive and a shot-noise part, for pricing the futures it is sufficient to price the diffusive and the shot-noise part separately. As already mentioned, it is therefore straightforward to incorporate more general dynamics for $D$. Later on, in Example 2 case (3.) we also show how to consider an exponential model for $D$. In particular in the german market, spot prices show higher volatilities for higher prices, which can be captured well by an exponential model. This is not the case in the model considered in [2].

From now on, assume that $S$ is a Vasicek/shot-noise model with parameters $(a, b, f_\gamma, f_Y, \lambda, \kappa, \theta(\cdot), \sigma)$ under $\mathbb{Q}$. First, we give an auxiliary lemma. It basically shows how to compute certain expectations of shot-noise processes on different levels of generality. For an $U[0,1]$-distributed rv, independent of $\gamma_1$ and $Y_1$, define

$$\bar{S}(t) := e^{\mathbb{Q}} \Big( h\big( t(1 - U_1), \gamma_1, 1 \big) \Big).$$

Furthermore, we set $\bar{Y} := e(Y_1)$. Throughout we assume $\bar{Y}, \bar{S}(t) < \infty$ for all $t \geq 0$.

**Lemma 1.** *Consider $t, \Delta > 0$ and a function $h : [0, \infty)^2 \times \mathbb{R} \mapsto \mathbb{R}$. For the shot-noise process $J$, defined in (1) we have that*

$$e^{\mathbb{Q}}(J_t) = \lambda t \bar{Y} \bar{S}(t), \quad e^{\mathbb{Q}} \Big( \int_t^{t+\Delta} J_u \, du \Big) = \lambda \bar{Y} \int_t^{t+\Delta} u \bar{S}(u) \, du.$$

This small lemma illustrates the typical procedure for computing expectations of shot-noise processes. First, one conditions on the number of jumps in the desired interval. Second, under this condition the jump-times are distributed as order statistics of i.i.d. uniformly distributed random variables $U_i$. Third, using the i.i.d. property of the other ingredients, on can interchange the order of the $U_i$ and finally ends up with a nice formula.

*Proof.* We have that

$$e^{\mathbb{Q}}(J_t) = \sum_{k \geq 0} e^{-\lambda t} \frac{(\lambda t)^k}{k!} e^{\mathbb{Q}} \Big( \sum_{i=1}^k h\big( t - t U_{i:k}, \gamma_i, Y_i \big) \Big),$$

where $U_1, U_2, \dots$ are i.i.d. $U[0,1]$. As the random variables $\gamma_i$ and $Y_i$ are also i.i.d. one can interchange the order of the second sum and obtains

$$e^{\mathbb{Q}}(J_t) = \sum_{k \geq 0} e^{-\lambda t} \frac{(\lambda t)^k}{k!} e^{\mathbb{Q}} \left( \sum_{i=1}^{k} Y_i\, h\Big(t - tU_i, \gamma_i, 1\Big) \right).$$

The expectation equals $k\bar{Y}\bar{S}(t)$ and the first result follows. The second assertion follows by interchanging expectation and the integral.   □

Denote the Laplace-Transform of $\gamma_1$ by $\varphi_\gamma(c) := e^{\mathbb{Q}}(\exp(-c\gamma_1))$ and assume $\phi_\gamma(c) < \infty$ at least for $c \in \{a, -b\}$.

**Theorem 1.** *The price of the electricity future offering electricity in the time-period $[T, T + \Delta]$ at $t \leq T - \epsilon$, which we denote by $F(t, T, T + \Delta)$, computes according to:*

$$\delta \cdot F(t, T, T + \Delta) = \tilde{F}(t, T, T + \Delta)$$
$$+ \lambda \Delta \bar{Y} \left[ \Delta \left( \frac{1 - \varphi_\gamma(a)}{a} + \frac{1}{b} \right) + \frac{\varphi_\gamma(-b) e^{-b(T-t)}}{b^2} \left( e^{-b\Delta} - 1 \right) \right]$$
$$- \frac{D_t}{\kappa} \left( e^{-\kappa(T+\Delta)} - e^{-\kappa T} \right) + \kappa \int_{T}^{T+\Delta} \int_{t}^{u} e^{\kappa s} \theta(s)\, ds\, du, \qquad (4)$$

*where we denote the $\mathcal{F}_t$-measurable part of shot-noise component by*

$$\tilde{F}(t, T, T + \Delta) := \int_{T}^{T+\Delta} \sum_{\tau_i \leq t} h(u - \tau_i, \gamma_i, Y_i)\, du.$$

The term $\tilde{F}$ captures the part of the past shot-noise effects. In practice, if the market at $t$ is not in a extreme spike, $\tilde{F}$ can be safely neglected.

*Proof.* Following [18], the price of the future is given by an expectation under the risk-neutral martingale measure $\mathbb{Q}$. Hence[6],

$$\delta \cdot F(t, T, T + \Delta) = e_t\left( \int_{T}^{T+\Delta} S_u\, du \right) = e_t\left( \int_{T}^{T+\Delta} D_u\, du \right) + e_t\left( \int_{T}^{T+\Delta} J_u \right).$$

We first consider the expectation of the diffusive part and second the expectation of the shot-noise part. It is well-known[7] that (2) has the following explicit solution:

$$D_t = e^{-\kappa t} \left( D_0 + \kappa \int_{0}^{t} e^{\kappa s} \theta(s) ds \right) + \sigma \int_{0}^{t} e^{\kappa(s-t)} dB_s.$$

---

[6] We use the short notation $e_t(\cdot)$ for $e^{\mathbb{Q}}(\cdot | \mathcal{F}_t)$.
[7] For example, see [17].

Then, for $u > t$ we obtain $D_u = e^{-\kappa u}(D_t + \kappa \int_t^u e^{\kappa s}\theta(s)ds) + \sigma \int_t^u e^{\kappa(s-u)}dB_s$, such that

$$\int_T^{T+\Delta} e_t(D_u)\,du = -\frac{D_t}{\kappa}\left(e^{-\kappa(T+\Delta)} - e^{-\kappa T}\right) + \kappa \int_T^{T+\Delta}\int_t^u e^{\kappa s}\theta(s)\,ds\,du.$$

Second, consider the shot-noise part. Observe that

$$e_t\left(\int_T^{T+\Delta} J_u\,du\right) = \int_T^{T+\Delta} e_t\left(\sum_{\tau_i > t} h(u - \tau_i, \gamma_i, Y_i)\right)du$$

$$+ \int_T^{T+\Delta}\sum_{\tau_i \le t} h(u - \tau_i, \gamma_i, Y_i)\,du.$$

As a Poisson process has independent and stationary increments, the expectation on the r.h.s. computes to

$$e_t\left(\sum_{t < \tau_i \le u} h(u - \tau_i, \gamma_i, Y_i)\right) = e^{\mathbb{Q}}\left(\sum_{i=N_t+1}^{N_u} h(u - \tau_i, \gamma_i, Y_i)\right)$$

$$= e^{\mathbb{Q}}\left(\sum_{i=1}^{N_{u-t}} h(u - t - \tau_i, \gamma_i, Y_i)\right) = e(J_{u-t}).$$

This expectation can be computed using Lemma 1. We therefore compute $\bar{S}$:

$$\bar{S}(t) = e^{\mathbb{Q}}\left(e^{a[t(1-U_1)-\gamma_1]}\mathbb{1}_{\{t(1-U_1)\in[0,\gamma_1]\}} + e^{-b[t(1-U_1)-\gamma_1]}\mathbb{1}_{\{t(1-U_1)>\gamma_1\}}\right)$$

$$= \int_0^\infty\left[\int_0^{1-v/t} e^{-b[t(1-u)-v]}du + \int_{1-v/t}^1 e^{a[t(1-u)-v]}du\right]F_\gamma(dv),$$

where the distribution of $\gamma$ is denoted by $F_\gamma$. Computing the integrals we obtain that

$$\bar{S}(t) = \int_0^\infty\left[\frac{1}{bt}\left(1 - e^{-b(t-v)}\right) + \frac{1}{at}\left(1 - e^{-av}\right)\right]F_\gamma(dv)$$

$$= \frac{1}{bt}\left(1 - e^{-bt}\varphi_\gamma(-b)\right) + \frac{1}{at}\left(1 - \varphi_\gamma(a)\right).$$

Finally, we have to compute the following integrals of $\bar{S}$:

$$\int_{T-t}^{T-t+\Delta} u\bar{S}(u)\,du = \int_{T-t}^{T-t+\Delta}\left[\frac{1 - e^{-bu}\varphi_\gamma(-b)}{b} + \frac{1 - \varphi_\gamma(a)}{a}\right]du$$

$$= \Delta\left(\frac{1 - \varphi(a)}{a} + \frac{1}{b}\right) + \frac{\varphi_\gamma(-b)e^{-b(T-t)}}{b^2}\left(e^{-b\Delta} - 1\right).$$

Using Lemma 1 with the above expressions proves the theorem.

Coming back to the simple case in Example 1 where $\gamma$ was zero with probability $p_\gamma$ and $\tilde{\gamma}$ otherwise, we obtain a simple Laplace transform as $\varphi_\gamma(c) = p_\gamma + (1 - p_\gamma)\exp(-c\tilde{\gamma})$. Of course, there are many other possibilities where the Laplace transform is obtained in closed form (eg. Beta distribution, log-normal distribution or others).

*Example 2.* There are several interesting special cases or modifications of the above setting:

1. If $\theta(u) = \theta$, then the second line in (4) simplifies considerably to

$$\theta\Delta - \frac{D_t - \theta}{\kappa}\left(e^{-\kappa(T+\Delta-t)} - e^{-\kappa(T-t)}\right).$$

2. For incorporating seasonalities one frequently uses a mean-reversion level similar to $\theta(s) = \sin(\omega s)$. In this case we have that

$$\kappa\int_T^{T+\Delta}\int_t^u e^{\kappa s}\theta(s)\,ds\,du = \frac{\kappa}{(\kappa^2 + \omega^2)^2}\Big(\omega\Delta\cos(\omega t)e^{\kappa t}(\kappa^2 + \omega^2)$$
$$+ e^{\kappa(\Delta+T)}\left((\kappa^2 - \omega^2)\sin((T+\Delta)\omega) - 2\kappa\omega\cos(\omega(T+\Delta))\right)$$
$$- \kappa\Delta\sin(\omega t)e^{\kappa t}(\kappa^2 - \omega^2)$$
$$+ e^{\kappa T}\left(2\kappa\omega\cos(\omega T) + \sin(\omega T)(\omega^2 - \kappa^2)\right)\Big).$$

And hence we also obtain a closed-form expression for $\theta(s) = \omega_0 + \sin(\omega_1 s) + \sin(\omega_2 s)$.

3. The chosen Gaussian mean-reverting Diffusion may become negative. If the parameters are suitably chosen this probability might be small, but still positive. To overcome this difficulty one can use $S_t = \exp(D_t) + J_t$. It is also straightforward to compute the price of the future in this case as then

$$e_t(D_u) = \exp(D_t)e_t\left(\exp(D_t - D_u)\right)$$
$$= \exp\left(-\frac{D_t}{\kappa}\left(e^{-\kappa(T+\Delta)} - e^{-\kappa T}\right) + \kappa\int_T^{T+\Delta}\int_t^u e^{\kappa s}\theta(s)\,ds\,du\right.$$
$$\left.+ \frac{1}{2}\sigma^2\int_t^{T+\Delta}\left(\int_{T\vee s}^{T+\Delta} e^{\kappa(s-u)}\,du\right)^2 ds\right),$$

where the last line computes to

$$\frac{1}{2}\sigma^2\Bigg\{\frac{4e^{-\kappa(T+\Delta-t)} - 3 \quad e^{-2\kappa(T+\Delta-t)} + \left(e^{\Delta\kappa} - 1\right)^2\left(1 - e^{-2\kappa(T+\Delta-t)}\right)}{2\kappa^3}$$
$$+ \frac{T + \Delta - t}{\kappa^2}\Bigg\}.$$

# 3 Illustration

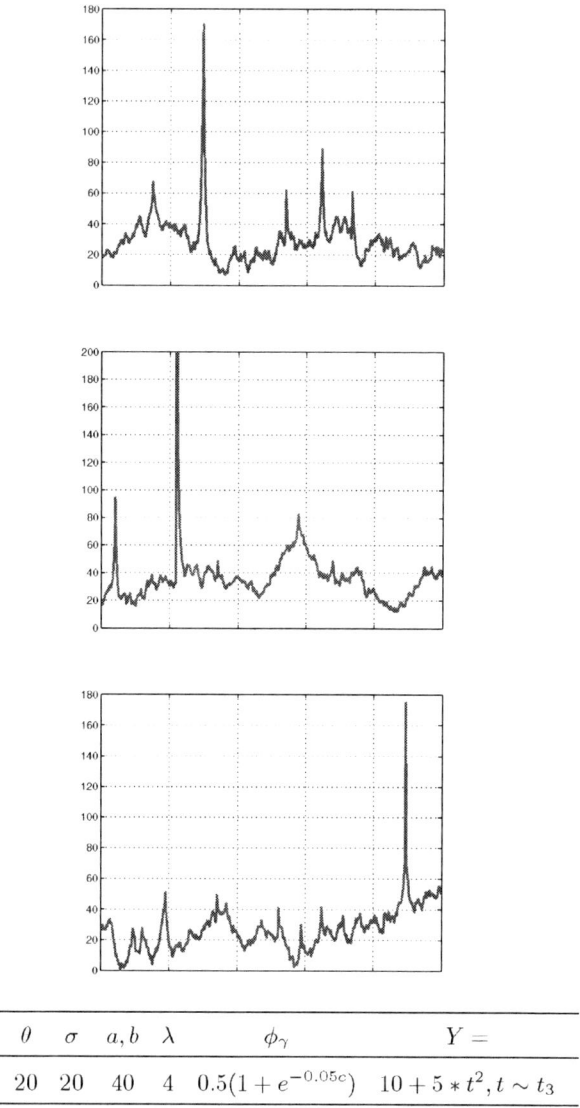

| $\kappa$ | $\theta$ | $\sigma$ | $a, b$ | $\lambda$ | $\phi_\gamma$ | $Y =$ |
|------|------|------|------|------|------|------|
| 0.5 | 20 | 20 | 40 | 4 | $0.5(1 + e^{-0.05c})$ | $10 + 5 * t^2, t \sim t_3$ |

**Fig. 2.** Simulations of the proposed shot-noise process over a horizon of 5 years. The parameters are as given above. The jump height $Y_i \sim 10 + 5 * \tilde{Y}_i^2$, where $\tilde{Y}_i$ are i.i.d. $t$-distributed with 3 degrees of freedom. Note that this specification does not include any seasonalities.

In this section we give some simulated paths, which illustrate the properties of the model. It should be noted that if the model is used for pricing, the specification under the risk-neutral measure matters. As used in Proposition 1, from a practical viewpoint it is reasonable that the model follows a Vasicek/shot-noise process under $\mathbb{P}$ as well as under $\mathbb{Q}$, of course with different parameters.

We assume constant $\theta$, i.e. no seasonality. The seasonalities have been discussed deeply in the literature, compare for example [12], [4] or [10]. As noted in [10], it might be profitable to choose a non-constant $\lambda$.

In Figure 2 we give several paths of the proposed model under the specifications in (1) and (2). It is clearly seen that the shot-noise model is mean-reverting (in this case to the constant level $\theta = 20$) and the spikes capture the empirically observed up-and-down shape.

In Figure 3 we give examples of computed futures price which illustrate the large variety of shapes which can be captured by the proposed model. Here we use $\theta(t) = \omega_0 + \sin(\omega_1 t)$ which in turn leads to the wavy structure of the futures curves. The left picture gives an example of a decreasing term structure. This is due to the different diffusion levels $D_0$. Note that the effect of $\tilde{F}$, thus the effect of past spikes declines rapidly due to the fast decay rate of the shot noise. Therefore the influence of a high spot electricity due to a spike on the futures curve is quite low, as it should be. The right plot shows an example of an increasing term structure. This is due to an increasing mean reversion level, such that the spot prices are expected to increase and therefore also the futures.

# 4 Estimation

One of the main points is of course estimation of the proposed model from historical data. This sections explores the use of the EM-algorithm for estimating shot-noise processes. The estimation of seasonalities is quite standard and we refer to [11] for further reading. It therefore remains to estimate the shot-noise as well as the diffusive part; a plot of the data after removal of seasonalities is given in Figure 4. We first give a short outline of the EM-algorithm in our setting, and provide the estimation results on daily data provided by the EEX[8]. Note that estimation always takes place under the real-world measure $\mathbb{P}$.

## 4.1 The EM-algorithm

Consider a pair of r.v. $X = (Y, Z)$, $Y \in \mathbb{R}^n$, $Z \in \mathbb{R}^m$. Think of $Y$ as observable quantities, and $Z$ of unobservable quantities. The aim is to estimate the distribution of $Y$ w.r.t. a parametric family $\{f_Y(\cdot; \phi) : \phi \in \Theta \subset \mathbb{R}^d\}$. However,

---

[8] EEX- European Energy Exchange, www.eex.de

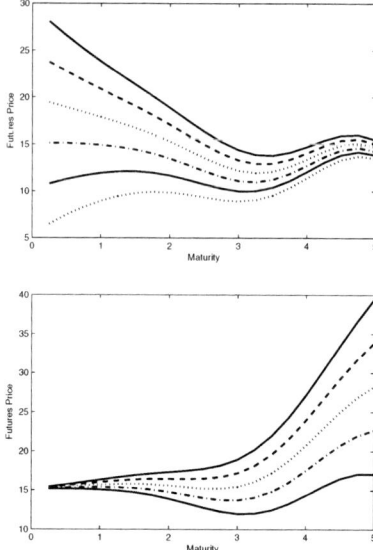

**Fig. 3.** Computed futures prices $F(0, T, \Delta)$ with $\Delta = 1$ `month` for the model as in Figure 2, but with seasonality of the type $\theta(t) = \omega_0 + \sin(\omega_1 t)$. Maturity $T$ varies from 0 to 5 `years`. <u>Left:</u> futures price for varying $D_0 = 5, 10, \ldots, 30$. <u>Right:</u> futures price for varying $\omega_0 = 0.02, \ldots, 0.1$.

the ML-estimate of $Y$ might not always be at hand, such that we need to make use of $Z$. The EM-algorithm maximizes the density of $X = (Y, Z)$ w.r.t. the distribution of $Z$ which is iteratively improved.

To this, let

$$L(\phi; \tilde{\phi}) := e_{\tilde{\phi}}\Big( \ln f_X(y, z; \phi) | Y = y \Big)$$
$$= \int \ln f_X(y, z; \phi) \, f_{Z|Y}(z|y; \tilde{\phi}) dz.$$

By Bayes' rule we are able to compute the conditional density of $Z$ given $Y$:

$$f_{Z|Y}(z|Y; \phi) \propto f_{Y|Z}(y|z; \phi) f_Z(z; \phi).$$

With this notation at hand we are able to state the EM-algorithm. Fix an initial value $\phi^0$. The iteration $\phi^k \to \phi^{k+1}$ consists of two steps:

**E-Step**   Compute $L(\phi; \phi^k)$
**M-Step**   Choose $\phi^{k+1}$ as maximizer of $L(\phi; \phi^k)$:

$$\phi^{k+1} := \arg\max_{\phi \in \Theta} L(\phi; \phi^k).$$

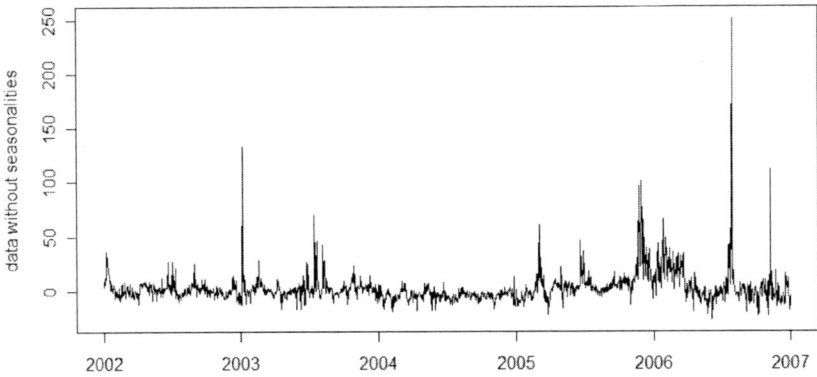

**Fig. 4.** The spot prices of energy (base load) after removal of seasonalities. Compare also Figure 1.

These steps are repeated until $|f_Y(\phi^{k+1}) - f_Y(\phi^k)| < \epsilon$.

Of course, the computation of $L(\phi; \phi^k)$ might be far from trivial depending on the considered model. For a large number of examples and applications we refer to [13]. The convergence of the EM-algorithm is proved in [21].

### 4.2 Application to the proposed model

The application of the EM-algorithm to the proposed model is done as follows. We consider Example 1 and are interested in estimating the parameter vector $\phi = (\kappa, \sigma, a, b, c, \lambda, p_\gamma)^\top$ and we assume that the distribution of $Y_1$ is described by a parameter $c$, i.e. $Y_1 \sim f_Y(\cdot; c)$. The observation consists of spot prices[9], for which we write $S = (S_1, \ldots, S_n)$. Meanwhile $S_i$ is a sum of a diffusive part and a shot-noise part, s.t. $S_i = D_i + J_i$. In the formulation of the EM-algorithm we therefore consider $X = (S, N, J)$, with $N_t = \sum_{\tau_i \leq t}$. Note that $S$ is observable and $N, J$ are not. Clearly, $D = S - J$.

We make the assumption that jumps occur directly at the considered time points and at a specific time point at most one jump occurs. This is reasonable if the chosen time grid is fine enough. Denote the time step by $\Delta$. To compute the likelihood function $L$ it is sufficient to have the common density of $S$, $N$ and $J$. Due to the dynamic nature of the processes we compute the density iteratively by

---

[9] Formally, we of course observe data on a certain time scale $t_1, \ldots, t_n$ such that the observations are $S_{t_1}, \ldots, S_{t_n}$, which we do not consider for expository purposes.

$$f_{X_n} = \prod_{i=1}^{n} f_{X_i|X_{i-1},\dots,X_1}(x_i|x_{i-1},\dots,x_1).$$

First, observe that the Euler discretisation of (2) immediately gives that

$$D_i|D_{i-1} \sim \mathcal{N}(D_{i-1}(1-\kappa\Delta), \sigma^2\Delta).$$

Second, as $N$ is a Poisson($\lambda$)-process, we have that

$$\mathbb{P}(N_i = N_{i-1}|N_{i-1}) = \exp(-\lambda\Delta).$$

As the third and last step we give the distribution of $J$ given $N$. Note that the process is piecewise deterministic. The process $J$ is not Markovian if $a \neq 0$. In the literature techniques for piecewise deterministic Markov processes have been applied to shot-noise processes of this type, compare [6]. We treat the two cases separately.

*Markovian case*

Assume that $a = 0$. Then $J$ is Markovian. Note that $J_i$ is a deterministic function of $J_{i-1}$ if no jump occurs, as in this case $J_i = J_{i-1}\exp(-b\Delta)$. Otherwise, if a jump occurred at time $i$, which is equivalent to $N_i > N_{i-1}$, then $J_i = J_{i-1}\exp(-b\Delta) + Y.$, where the $Y.$ are i.i.d. with density $f_Y$. We obtain

$$df_{J_i|J_{i-1}}(j_i) = \begin{cases} \delta_{\{j_i=j_{i-1}\exp(-b\Delta)\}}, & \text{if } N_i = N_{i-1} \\ f_Y\big(j_i - j_{i-1}\exp(-b\Delta)\big)dj_i, & \text{otherwise.} \end{cases}$$

*Non-Markovian case*

If $a$ is not zero, the case is more complicated. We just consider the case of Example 1, more general cases following similarly. Now we have to distinguish more cases. To begin with, note that there are two kinds of jumps. Jumps, where also $\gamma. = 0$ (which we call jumps of type 1) and jumps where $\gamma. = \tilde{\gamma}$ (called jumps of type 2). We additionally assume that $\tilde{\gamma}$ is sufficiently small such that we may neglect two jumps of type 2 in any interval of length up to $\tilde{\gamma}$. This leads to the following cases: first, if no jump of type 1 occurred at $i$ and the last jump of type 2 is before $i - \tilde{\gamma}$. Then $J_i = J_{i-1}\exp(-b\Delta)$. Second, if a jump of type 1 occurred at $i$, hence $J_i = J_{i-1}\exp(-b\Delta) + Y.$. Third, if a jump of type 2 occurred at $j \in \{i-\tilde{\gamma},\dots,i\}$. Then $J_i = J_j\exp(-b\Delta(i-j)) + Y.\exp(a\Delta(i-j))$. Summarizing we obtain that $df_{J_i|J_{i-1}}(j_i)$ equals

$$\begin{cases} f_Y\big(j_i - j_{i-1}\exp(-b\Delta)\big)dj_i, & \text{if } N_i > N_{i-1} \text{ and } \gamma_{N_i} = 0 \\ f_Y\Big(\big(j_i - j_{i-1}\exp(-b\Delta)\big)\exp(-a\Delta)\Big)dj_i & \text{if } N_i > N_{i-1} \text{ and } \gamma_{N_i} = \tilde{\gamma} \\ \delta_{\{j_i=j_j\exp(-b\Delta(i-j))+Y_{N_j}\exp(a\Delta(i-j))\}} & \text{if } N_i = N_j > N_{j-1}, \gamma_{N_j} = \tilde{\gamma} \\ \delta_{\{j_i=j_{i-1}\exp(-b\Delta)\}}, & \text{otherwise.} \end{cases}$$

With the above densities at hand the EM-algorithm is easily implemented. In the following section we apply the suggested method to electricity prices obtained from the EEX.

## 4.3 Estimation of the model on EEX data

We directly work on data where the seasonalities have been removed with standard methods, to illustrate the applicability of the method. A full statistical analysis and comparison with other models is beyond the scope of the article and will be pursued in future work ([15]).

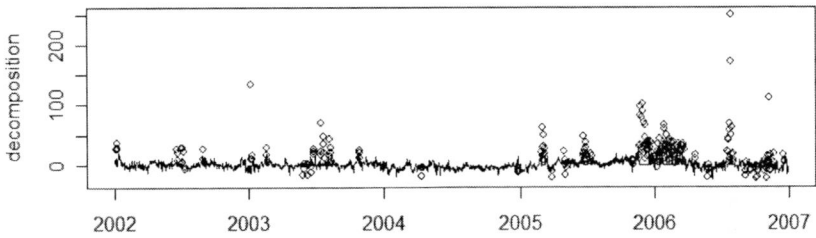

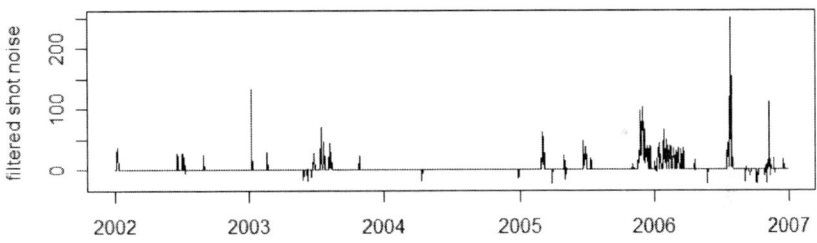

**Fig. 5.** Filtered shot-noise process from electricity data. The data consists of spot prices for base load from the European Energy Exchange, Leipzig.

| Parameter | $\kappa$ | $\sigma$ | $\sigma_Y$ | $\mu_Y$ | $\pi_1$ | $\pi_2$ | $\pi_3$ | $c$ |
|-----------|----------|----------|-----------|---------|---------|---------|---------|-----|
| Estimate | 0.2865 | 4.5762 | 60.34 | 17.4122 | 0.838 | 0.054 | 0.108 | 0.95 |

**Table 1.** Estimation results for the shot-noise model. See the text for details.

In Figure 5 the analyzed data is plotted as well as the filtered shot-noise parts. The graph on the top shows the data decomposed in the diffusive part (lines) as well as the shot-noise part (circles). The graph on the bottom shows the shot-noise part only. As there are negative as well as positive jumps in the time series we assume that the jumps $Y_i$ are normally distributed with mean $\mu_Y$ and standard deviation $\sigma_Y$.

For simplicity we consider the case with $p_\gamma = 0$ only. A further simplification speeds up the estimation process significantly: assuming that the decay rate $c$ is high enough (which is reasonable as the intention is to model extreme spikes by the shot-noise part) the effect of a jump is negligible after a small number of time steps. Then the diffusive and the jump part can be treated in one step distinguishing three cases: first, no jump occurs (exponential decay with rate $\kappa$); second, a jump occurred (exponential decay plus jump); third, a small time interval after the jump (exponential decay at rate $\kappa$ of the diffusive part plus exponential decay at rate $c$ of the jump part). Denote the probabilities to be in either case by $\pi_1, \pi_2$ and $\pi_3$, respectively. The estimation results are given in Table 1. $\hat{\pi}_1 = 0.838$, which corresponds to a jump intensity of 0.77, i.e. an average number of 13 jumps per year. In particular in the end of 2006 and the beginning of 2007 a large number of spikes were identified. The volatility of the diffusive part, $\sigma = 4.57$, shows the high variation in the data set. The standard deviation of the jumps, $\sigma_Y = 60.34$, is of course much higher, reflecting the extreme shocks captured by the shot-noise part. Finally, the decay rate of the jump part, $c = 0.95$, shows that the shot-noise part indeed drifts back very fast after occurring jumps.

Of course, the above analysis mainly suffices for an illustration of the concept and shows applicability of the proposed model as well as the estimation procedure. A deeper statistical analysis as well as a comparison to other models will be covered in [15].

# 5 Conclusion

This paper introduces a new model for spot electricity prices which easily captures the typical properties of electricity prices, namely seasonalities, extreme spikes and stochastic mean reversion. Moreover, the model allows for closed-form solutions of futures prices. Due to the flexibility of the model a large variety of shapes for the term structure of futures prices can be captured. It is shown how to use the EM-algorithm for statistical estimation of the model. The model is estimated using data from the European Energy Exchange.

## Acknowledgement

The author thanks Enrico Reiche for his excellent help in the statistical analysis.

# References

1. Altmann T., Schmidt T., Stute, W. (2006) A shot noise model for financial assets. Submitted

2. Benth F. E., Kallsen J., Meyer-Brandis T. (2006) A non-Gaussian Ornstein-Uhlenbeck process for electricity spot price modeling and derivatives pricing. Applied Mathematical Finance: Forthcoming.

3. Brémaud P. (1981) Point Processes and Queues. Springer Verlag, Berlin Heidelberg New York.

4. Cartea A., Figueroa, M. G. (2005) Pricing in electricity markets: a mean reverting jump diffusion model with seasonality. Applied Mathematical Finance: 12(4):313–335

5. Cox J. C., Ingersoll J. W., Ross, S. A. (1985) A theory of the term structure of interest rates. Econometrica 54:385–407

6. Dassios A., Jang J. (2003) Pricing of catastrophe reinsurance & derivatives using the cox process with shot noise intensity. Finance and Stochastics 7(1):73–95

7. Eberlein E., Stahl G. (2004) Both sides of the fence: a statistical and regulatory view of electricity risk. Energy & Power Risk Management 8(6):34-38

8. Filipović D. (2002) Separable term structures and the maximal degree problem. Mathematical Finance 12(4):341–349

9. Gaspar R. M., Schmidt, T. (2007) Shot-noise quadratic term structure models. Submitted

10. Geman H., Roncoroni, A. (2006) Understanding the fine structure of electricity prices. Journal of Business 79(3):1225–1261

11. Hylleberg S. (ed) (1992) Modelling Seasonality. Oxford University Press

12. Lucia J. J., Schwartz, E. S. (2002) Electricity prices and power derivatives: evidence from the nordic power exchange. Review of Derivatives Research 5:5–50

13. McLachlan J. G., Krishnan, T. (1997) The EM algorithm and extensions, John Wiley & Sons, New York

14. Protter P. (2004) Stochastic Integration and Differential Equations, 2nd edn. Springer Verlag, Berlin Heidelberg New York

15. Reiche E., Schmidt, T. (2007) A statistical analysis of models of electricity markets. Working paper

16. Schmidt T., Stute W. (2007) General shot-noise processes and the minimal martingale measure. Statistics & Probability Letters 77:1332–1338

17. Schmidt W. M. (1997) On a general class of one-factor models for the term structure of interest rates. Finance and Stochastics 1:3–24

18. Teichmann J. (2005) A note on nonaffine solutions of term structure equations with applications to power exchanges. Mathematical Finance 15(1):191–201

19. Vasiček O. (1977) An equilibrium characterization of the term structure. Journal of Financial Economics 5:177–188

20. Weron R. (2005) Heavy tails and electricity prices. The Deutsche Bundesbank's 2005 Annual Fall Conference (Eltville)

21. Wu C. F. J. (1983) On the convergence properties of the EM algorithm. The Annals of Statistics 11:95–103.

# Generalized Bayesian Nonlinear Quickest Detection Problems: On Markov Family of Sufficient Statistics

Albert N. Shiryaev*

Steklov Mathematical Institute of the Russian Academy of Sciences, Moscow, Russia. albertsh@mi.ras.ru.

**Summary.** We consider generalized Bayesian "nonlinear delay penalty" problems of the quickest detection of spontaneous appearing of "time-change" point $\theta \in [0, \infty]$ when the observable process changes its probability characteristics. For some classes of observable processes and penalty functions we describe the structure of the Markov family of "sufficient statistics" that gives a possibility to apply the methods of the general Markovian optimal stopping theory to solving of the quickest detection problems with a "nonlinear delay penalty".

## 1 Brownian motion model

To make the main ideas clearer let us consider first a *Brownian case*, where observable process $X = (X_t)_{t \geq 0}$ has the following structure:

$$X_t = \mu(t - \theta)^+ + \sigma B_t, \qquad t \geq 0, \tag{1}$$

or

$$X_t \begin{cases} \sigma B_t, & t \leq \theta, \\ \mu(t - \theta) + \sigma B_t, & t > \theta, \end{cases} \tag{2}$$

where $\mu \neq 0$ and $\sigma > 0$ are known constants and $\theta$ belongs to the set $[0, \infty]$. We interpret "time-change point" $\theta$ as a time when the "disorder" (the "spontaneous effect") has been appeared [1], [2].

In (1), (2) the process $B = (B_t)_{t \geq 0}$ is a standard Brownian motion defined on a filtered probability space $(\Omega, \mathcal{F}, (\mathcal{F}_t)_{t \geq 0}, \mathsf{P})$. Without loss of generality we may assume that $(\Omega, \mathcal{F})$ is the measurable space $(C, \mathcal{C})$ of continuous functions $\omega = (\omega_t)_{t > 0}$, $B_t(\omega) = \omega_t$, $\mathsf{P}$ is the Wiener measure, $\mathcal{F}_t = \mathcal{C}_t = \sigma\{\omega : \omega_s, s \leq t\}$. Assuming the canonical setting $(X_t(\omega) = \omega_t, t \geq 0)$, denote, for fixed $\theta$,

* Supported by the Organizing Committee of the Workshop on Mathematical Control Theory and Finance (Lisbon, April 10–14, 2007) and by Russian Foundation for Basic Research (grant 05-01-00944).

by $\mathsf{P}_\theta$ the distribution of the process $X$ described by (1). Then the system $(\Omega, \mathcal{F}, (\mathcal{F}_t)_{t\in[0,\infty)}, (\mathsf{P}_\theta)_{\theta\in[0,\infty]})$ is called a *probability-statistical experiment*.

If $\theta = \infty$, then the law $\mathsf{P}_\infty$ is the distribution of the process $X$ under assumption that the "disorder" never happened, i.e., $\mathsf{P}_\infty = \mathrm{Law}(\sigma B_t, \, t \geq 0)$. The law $\mathsf{P}_0$ is the distribution of the *$\sigma$-Brownian motion with (local) drift $\mu$*, i.e., $\mathsf{P}_0 = \mathrm{Law}(\mu t + \sigma B_t, \, t \geq 0)$.

Denote by $\tau = \tau(\omega)$ a finite stopping time with respect to the filtration $(\mathcal{F}_t)_{t\geq 0}$, where $\mathcal{F}_t = \sigma(w_s, \, s \leq t)$. For every $T > 0$ we consider the set

$$\mathfrak{M}_T = \{\tau : \; \mathsf{E}_\infty \tau \geq T\}, \tag{3}$$

where $\mathsf{E}_\infty \tau$ is the expected time of "sounding of the false alarm".

The generalized (linear) Bayesian problem as it was defined in [3] is, for given $T > 0$, to find a stopping time $\tau_T^*$, if it exists, such that

$$\int_0^\infty \mathsf{E}_\theta (\tau_T^* - \theta)^+ \, d\theta = \inf_{\tau \in \mathfrak{M}_T} \int_0^\infty \mathsf{E}_\theta (\tau - \theta)^+ \, d\theta. \tag{4}$$

This problem of the quickest detection was called in [3] *generalized Bayesian*, since in (4) parameter $\theta$ can be interpreted as a generalized random variable with the uniform distribution on $[0, \infty)$ (with respect to the Lebesgue measure on $[0, \infty)$).

*Remark* 1. In the paper [3, Remark 1.1] we explained how the generalized Bayesian problems can be obtained from Bayesian problems, where $\theta = \theta(\omega)$ is a random variable with exponential distribution $\mathsf{P}(\theta > t \,|\, \theta > 0) = e^{-\lambda t}$, $\mathsf{P}(\theta = 0) = \pi$, $\pi \in [0, 1)$, under the special limit transition $\lambda \to 0$, $\alpha \to 1$ such that $(1 - \alpha)/\lambda \to T$, where $\alpha$ is a probability of *false alarm*.

## 2 Linear-penalty case

For the Brownian model (1) we proved in [3, Lemma 2.1] that in the case of linear delay penalty $((\tau - \theta)^+$, see (4)) for any stopping time $\tau$

$$\int_0^\infty \mathsf{E}_\theta (\tau - \theta)^+ \, d\theta = \mathsf{E}_\infty \int_0^\tau \psi_t \, dt, \tag{5}$$

where the Markov process $\psi = (\psi_t)_{t\geq 0}$ satisfies the stochastic differential equation

$$d\psi_t = dt + \frac{\mu}{\sigma^2} \psi_t \, dX_t, \qquad \psi_0 = 0. \tag{6}$$

Hence the linear delay penalty problem (4) can be reduced to the following conditional optimal stopping problem for the Markov process $\psi = (\psi_t)_{t\geq 0}$: to find for every $T > 0$

$$\inf \mathsf{E}_\infty \int_0^\tau \psi_t \, dt, \tag{7}$$

where infimum is taken over the class of stopping times $\mathfrak{M}_T = \{\tau \colon \mathsf{E}_\infty \tau \geq T\}$.

From (5)–(7) we see that for the linear case the process $\psi = (\psi_t)_{t \geq 0}$ plays a role of the Markov "sufficient statistics". The main aim of the present paper is to describe the family of sufficient statistics, for some cases of nonlinear delay penalty $(G((\tau - \theta)^+))$, taking instead of (4) the following criteria: *to find a stopping time $\tau_T^*$, if it exists, such that*

$$\int_0^\infty \mathsf{E}_\theta G\big((\tau_T^* - \theta)^+\big)\, d\theta = \inf_{\tau \in \mathfrak{M}_T} \int_0^\infty \mathsf{E}_\theta G\big((\tau - \theta)^+\big)\, d\theta. \tag{8}$$

## 3 Nonlinear-penalty case

We assume here that penalty function $G = G(t)$, $t \geq 0$, in (8) has the representation

$$G(t) = \int_0^t g(s)\, ds, \tag{9}$$

where $g(s) \geq 0$ is a Lebesgue-measurable function with $g(s) = 0$ when $s < 0$.

For $\tau > \theta$

$$G(\tau - \theta) = \int_0^{\tau - \theta} g(s)\, ds = \int_0^\infty I(0 \leq s < \tau - \theta) g(s)\, ds$$
$$= \int_0^\infty I(u < \tau) g(u - \theta)\, du.$$

Therefore,

$$\mathsf{E}_\theta G\big((\tau - \theta)^+\big) = \mathsf{E}_\theta I(\tau > \theta) G(\tau - \theta)$$
$$= \mathsf{E}_\theta I(\tau > \theta) \int_\theta^\infty I(u < \tau) g(u - \theta)\, du$$
$$= \int_\theta^\infty g(u - \theta) \mathsf{E}_\theta I(u < \tau)\, du. \tag{10}$$

For $\theta \leq u$, taking into account that $\{u < \tau\} \in \mathcal{F}_u$, $\mathsf{P}_\theta \sim \mathsf{P}_u$, we get

$$\mathsf{E}_\theta I(u < \tau) = \mathsf{E}_u \frac{d\mathsf{P}_\theta}{d\mathsf{P}_u} I(u < \tau) = \mathsf{E}_u \frac{d(\mathsf{P}_\theta|\mathcal{F}_u)}{d(\mathsf{P}_u|\mathcal{F}_u)} I(u < \tau). \tag{11}$$

Let

$$L_t = \frac{d(\mathsf{P}_0|\mathcal{F}_t)}{d(\mathsf{P}_\infty|\mathcal{F}_t)}$$

be the likelihood (a Radon–Nikodým derivative) of the measure $\mathsf{P}_0|\mathcal{F}_t$ with respect to the measure $\mathsf{P}_\infty|\mathcal{F}_t$. (Notice that measures $\mathsf{P}_0$ and $\mathsf{P}_\infty$ are singular, $\mathsf{P}_0 \perp \mathsf{P}_\infty$, but for each $t \geq 0$ the measures $\mathsf{P}_0|\mathcal{F}_t$ and $\mathsf{P}_\infty|\mathcal{F}_t$ are equivalent, $\mathsf{P}_0|\mathcal{F}_t \sim \mathsf{P}_\infty|\mathcal{F}_t$.) It is well known (see, for example, [4], [5]) that for the Brownian model (1) ($\mathsf{P}_\infty$-a.s.)

$$L_t = \exp\left\{\frac{\mu}{\sigma^2}X_t - \frac{1}{2}\frac{\mu^2}{\sigma^2}t\right\}, \qquad t \geq 0, \tag{12}$$

and, by Itô's formula,

$$dL_t = \frac{\mu}{\sigma^2}L_t\,dX_t, \qquad t \geq 0, \quad L_0 = 1. \tag{13}$$

Since $\mathsf{P}_u(A) = \mathsf{P}_\infty(A)$ for all $A \in \mathcal{F}_u$, we get for $\theta \leq u$

$$\frac{d(\mathsf{P}_\theta|\mathcal{F}_u)}{d(\mathsf{P}_u|\mathcal{F}_u)} = \frac{d(\mathsf{P}_\theta|\mathcal{F}_u)}{d(\mathsf{P}_\infty|\mathcal{F}_u)} = \frac{\frac{d(\mathsf{P}_\theta|\mathcal{F}_u)}{d(\mathsf{P}_0|\mathcal{F}_u)}}{\frac{d(\mathsf{P}_\infty|\mathcal{F}_u)}{d(\mathsf{P}_0|\mathcal{F}_u)}} = \frac{L_u}{\frac{d(\mathsf{P}_0|\mathcal{F}_u)}{d(\mathsf{P}_\theta|\mathcal{F}_u)}}. \tag{14}$$

Note now that in the Brownian model (1) for $\theta \leq u$

$$\frac{d(\mathsf{P}_0|\mathcal{F}_u)}{d(\mathsf{P}_\theta|\mathcal{F}_u)} = \frac{d(\mathsf{P}_0|\mathcal{F}_\theta)}{d(\mathsf{P}_\theta|\mathcal{F}_\theta)} = L_\theta. \tag{15}$$

Hence from (11), (14), and (15) we get that for $\theta \leq u$

$$\mathsf{E}_\theta I(u < \tau) = \mathsf{E}_u \frac{L_u}{L_\theta}I(u < \tau) = \mathsf{E}_\infty \frac{L_u}{L_\theta}I(u < \tau). \tag{16}$$

From (10) and (16) we obtain the following representations:

$$\mathsf{E}_\theta G\big((\tau-\theta)^+\big) = \int_\theta^\infty g(u-\theta)\mathsf{E}_u \frac{L_u}{L_\theta}I(u < \tau)\,du$$

$$= \int_0^\infty g(u-\theta)\mathsf{E}_\infty \frac{L_u}{L_\theta}I(u < \tau)\,du$$

and

$$\int_0^\infty \mathsf{E}_\theta G\big((\tau-\theta)^+\big)\,d\theta = \int_0^\infty \left[\int_0^\infty g(u-\theta)\mathsf{E}_\infty \frac{L_u}{L_\theta}I(u < \tau)\,du\right]d\theta$$

$$= \mathsf{E}_\infty \int_0^\tau \left[\int_0^\infty g(u-\theta)\frac{L_u}{L_\theta}\,d\theta\right]du$$

$$= \mathsf{E}_\infty \int_0^\tau \left[\int_0^u g(u-\theta)\frac{L_u}{L_\theta}\,d\theta\right]du. \tag{17}$$

Introduce the notation

$$\Psi_u(g) = \int_0^u g(u-\theta)\frac{L_u}{L_\theta}\,d\theta, \qquad u \geq 0. \tag{18}$$

Then we see that

$$\inf_{\tau\in\mathfrak{M}_T} \int_0^\infty \mathsf{E}_\theta G\big((\tau-\theta)^+\big)\,d\theta = \inf_{\tau\in\mathfrak{M}_T} \mathsf{E}_\infty \int_0^\tau \Psi_u(g)\,du. \tag{19}$$

To say more about the structure of the process $(\Psi_u(g))_{u \geq 0}$ for the Brownian model, let us make some additional assumptions about function $g = g(t)$, $t \geq 0$.

We assume that

$$g(t) = \sum_{m=0}^{M} \sum_{n=0}^{N} c_{mn} e^{\lambda_m t} t^n, \tag{20}$$

where $\lambda_0 = 0$. (Cf. the representation for $g(t)$ in [6].)

Consider first the case $N = 0$:

$$g(t) = \sum_{m=0}^{M} c_{m0} e^{\lambda_m t} = c_{00} + \sum_{m=1}^{M} c_{m0} e^{\lambda_m t}. \tag{21}$$

From (18) for this case we get

$$\Psi_u(g) = \int_0^u \frac{L_u}{L_\theta} g(u - \theta) \, d\theta$$

$$= c_{00} \int_0^u \frac{L_u}{L_\theta} \, d\theta + \sum_{m=1}^{M} c_{m0} \int_0^u e^{\lambda_m (u-\theta)} \frac{L_u}{L_\theta} \, du. \tag{22}$$

Denote

$$\psi_u = \int_0^u \frac{L_u}{L_\theta} \, d\theta. \tag{23}$$

From (13) we find

$$d\psi_u = dL_u \cdot \int_0^u \frac{d\theta}{L_\theta} + du = \left( \frac{\mu}{\sigma^2} \int_0^u \frac{L_u}{L_\theta} \, d\theta \right) dX_u + du.$$

Therefore,

$$d\psi_u = du + \frac{\mu}{\sigma^2} \psi_u \, dX_u. \tag{24}$$

Similarly, for

$$\psi_u^{(m,0)} \stackrel{\text{def}}{=} \int_0^u e^{\lambda_m (u-\theta)} \frac{L_u}{L_\theta} \, du = \int_0^u \frac{L_u^{(m)}}{L_\theta^{(m)}} \, du, \quad \text{where} \quad L_u^{(m)} = e^{\lambda_m u} L_u,$$

we find that

$$d\psi_u^{(m,0)} = dL_u^{(m)} \int_0^u \frac{d\theta}{L_\theta^{(m)}} + du = \left( \lambda_m \, du + \frac{\mu}{\sigma^2} \, dX_u \right) \int_0^u \frac{L_u^{(m)}}{L_\theta^{(m)}} \, d\theta + du.$$

Hence

$$d\psi_u^{(m,0)} = \left( 1 + \lambda_m \psi_u^{(m,0)} \right) du + \frac{\mu}{\sigma^2} \psi_u^{(m,0)} \, dX_u. \tag{25}$$

Thus, if the function $g = g(t)$, $t \geq 0$, admits the representation (21), then the family of sufficient statistics

$$\left(\psi_u, \psi_u^{(1,0)}, \ldots, \psi_u^{(M,0)}\right)_{u \geq 0} \tag{26}$$

forms a Markov system with

$$\begin{cases} d\psi_u = du + \dfrac{\mu}{\sigma^2} \, \psi_u \, dX_u, \\ d\psi_u^{(m,0)} = \left(1 + \lambda_m \psi_u^{(m,0)}\right) du + \dfrac{\mu}{\sigma^2} \, \psi_u^{(m,0)} \, dX_u \end{cases} \tag{27}$$

and $\psi_0 = \psi_0^{(m,0)} = 0$, $m = 1, \ldots, M$.

Consider now the case $M = 0$:

$$g(t) = \sum_{n=0}^{N} c_{0n} t^n = c_{00} + \sum_{n=1}^{N} c_{0n} t^n. \tag{28}$$

If $1 \leq n \leq N$, we find for

$$\psi_u^{(0,n)} \overset{\text{def}}{=} \int_0^u \frac{L_u}{L_\theta} \, (u - \theta)^n \, d\theta$$

that

$$\begin{aligned} d\psi_u^{(0,n)} &= dL_u \cdot \int_0^u \frac{(u - \theta)^n}{L_\theta} d\theta + L_u \, d\left(\int_0^u \frac{(u - \theta)^n}{L_\theta} d\theta\right) \\ &= \frac{\mu}{\sigma^2} \, \psi_u^{(0,n)} \, dX_u + n \psi_u^{(0,n-1)} \, du, \end{aligned} \tag{29}$$

where $\psi_u^{(0,0)} = \psi_u$.

So, in the case (28) we have the following family of sufficient statistics:

$$\left(\psi_u, \psi_u^{(0,1)}, \ldots, \psi_u^{(0,N)}\right)_{u \geq 0},$$

with

$$\begin{cases} d\psi_u = du + \dfrac{\mu}{\sigma^2} \, \psi_u \, dX_u, \\ d\psi_u^{(0,n)} = n \psi_u^{(0,n-1)} \, du + \dfrac{\mu}{\sigma^2} \, \psi_u^{(0,n)} \, dX_u, \end{cases} \tag{30}$$

where $\psi_0 = \psi_0^{(0,n)} = 0$, $1 \leq n \leq N$, $\psi_u^{(0,0)} = \psi_u$.

This family forms a Markov system.

For general case (20) denote

$$\psi_u^{(m,n)} = \int_0^u \frac{L_u}{L_\theta} \, e^{\lambda_m u} (u - \theta)^n \, du. \tag{31}$$

Then

$$\psi_u^{(m,n)} = \int_0^u \frac{L_u^{(m)}}{L_\theta^{(m)}} \, (u - \theta)^n \, du$$

and for $1 \leq n \leq N$ and $1 \leq m \leq M$

$$d\psi_u^{(m,n)} = \left(\lambda_m + n \psi_u^{(m,n-1)}\right) du + \frac{\mu}{\sigma^2} \, \psi_u^{(m,n)} \, dX_u. \tag{32}$$

All these considerations lead to the following statement.

**Theorem 1.** *For the Brownian model* (1) *and nonlinear delay penalty function* $G(u) = \int_0^u g(t)\,dt$ *with* $g = g(t)$ *given by* (20) *the system*

$$\left(\psi_u, \; \psi_u^{(m,n)}, \; 0 \le m \le M, \; 0 \le n \le N\right)_{u \ge 0} \tag{33}$$

*forms the Markov family of sufficient statistics.*

*These statistics are diffusion processes satisfying the stochastic differential equations* (27), (30), (32). *The process* $\Psi_u(g) = \sum_{m=0}^{M} \sum_{n=0}^{N} c_{mn} \psi_u^{(m,n)}$.

## 4 Examples

(a) If $G(t) = t$, $t \ge 0$ (linear penalty), then there exists only one Markov sufficient statistics $(\psi_u)_{u \ge 0}$.

(b) If $G(t) = t^2$, $t \ge 0$, then $g(t) = 2t$ and

$$\Psi_u(g) = 2 \int_0^u \frac{L_u}{L_\theta} (u - \theta)\,d\theta = 2\psi_u^{(0,1)},$$

where

$$d\psi_u^{(0,1)} = \psi_u\,du + \frac{\mu}{\sigma^2}\,\psi_u^{(0,1)}\,dX_u.$$

Hence for the quadratic penalty function $G(t) = t^2$ we have a pair of sufficient statistics $(\psi_u, \psi_u^{(0,1)})_{u \ge 0}$ which form a diffusion Markov process.

(c) If $G(t) = (e^{\lambda t} - 1)/\lambda$, $t \ge 0$, $\lambda > 0$, then $g(t) = e^{\lambda t}$, $t \ge 0$. For this exponential delay penalty function $G(t)$, $t \ge 0$, we have one sufficient statistics $\psi_u^{(1,0)}$ with

$$d\psi_u^{(1,0)} = (1 + \lambda\psi_u^{(1,0)})\,du + \frac{\mu}{\sigma^2}\,\psi_u^{(1,0)}\,dX_u.$$

*Remark 2.* The *Bayesian* quickest detection problems with exponential delay penalty were considered in [7], [8] (for discrete and continuous time, respectively).

## 5 Some extensions

Let us analyze the considered Brownian model (1) from the standpoint of its properties used in the proof of Theorem.

First of all we see that the two "extreme" measures $\mathsf{P}_\infty$ and $\mathsf{P}_0$ play an essential role. These measures are the distributions of two processes $X^{(\infty)} = (X_t^{(\infty)})_{t \ge 0}$ and $X^{(0)} = (X_t^{(0)})_{t \ge 0}$ with

$$X_t^{(\infty)} = \sigma B_t$$

and

$$X_t^{(0)} = \mu t + \sigma B_t.$$

By analogy with these processes introduce for each $\theta \in (0, \infty)$ the process $X^{(\theta)} = (X_t^{(\theta)})_{t \geq 0}$ with

$$X_t^{(\theta)} \begin{cases} X_t^{(\infty)}, & t \leq \theta, \\ X_\theta^{(\infty)} + \left[X_t^{(0)} - X_\theta^{(0)}\right], & t > \theta. \end{cases} \tag{34}$$

Then we see that $X^{(\theta)}$ is exactly the process $X$ introduced in (1) under assumption that the "disorder" occurs at time $\theta$.

In our extensions of the Brownian model we make the assumption that, firstly, there are two (càdlàg) processes $X^{(\infty)} = (X_t^{(\infty)})_{t \geq 0}$ and $X^{(0)} = (X_t^{(0)})_{t \geq 0}$ interpreted as the observable processes when $\theta = \infty$ and $\theta = 0$, respectively, and, secondly, if a "disorder" occurs at time $\theta \in (0, \infty)$, then the observable process has the structure given in (34).

*Remark 3.* This is only one (among others) natural (constructive) model for the description of the structure of the observable process in the presence of "disorder". Sometimes it is reasonable to formulate the "disorder" problem as a problem of the change of some probabilistic characteristics of observable processes.

For the case of the Brownian model the likelihood process

$$L_t = \frac{d(\mathsf{P}_0 | \mathcal{F}_t)}{d(\mathsf{P}_\infty | \mathcal{F}_t)}, \qquad t \geq 0,$$

in whose terms the process $\Psi_u(g)$, $u \geq 0$, was expressed in (18), (22), has the following structure (see (13)):

$$dL_t = L_t \, dN_t, \tag{35}$$

where $N_t = (\mu/\sigma^2) \, dX_t \ (= (\mu/\sigma^2) \, dB_t$ with respect to the measure $\mathsf{P}_\infty)$.

For the general model (34), where $X^{(\infty)}$ and $X^{(0)}$ are càdlàg processes, it is natural to assume, instead of (35), that the process $L = (L_t)_{t \geq 0}$ satisfies the stochastic differential equation

$$dL_t = L_{t-} \, dN_t, \tag{36}$$

where $N = (N_t)_{t \geq 0}$ is a (local) martingale (with respect to the measure $\mathsf{P}_\infty$).

*Remark 4.* Generally speaking, the question about the possibility to have the representation (36) for the process $L = (L_t)_{t \geq 0}$ is rather complicated. In [5, Chap. III, §§ 5a–5c] one can find many results about structure of the process $L$ under different assumptions on measures $\mathsf{P}_0$ and $\mathsf{P}_\infty$. By Theorem 5.35 in [5, Chap. III] the representation (36) does hold if the (canonical) processes $X^{(\infty)}$ and $X^{(0)}$ have independent increments.

It is well known [5] that if $N = (N_t)_{t \geq 0}$ is a semimartingale, then (36) has (in the class of semimartingales) a unique solution which is given by the stochastic exponential:

$$L_t = L_0 \mathcal{E}(N)_t, \tag{37}$$

where

$$\mathcal{E}(N)_t = e^{N_t - \frac{1}{2}\langle N^c \rangle_t} \prod_{0 < s \le t} (1 + \Delta N_s) e^{-\Delta N_s} \tag{38}$$

and $\langle N^c \rangle$ is a quadratic characteristic of the continuous martingale part $N^c$ of $N$.

At the same time, if we start with the process $L = (L_t)_{t \ge 0}$, then a process $N = (N_t)_{t \ge 0}$ satisfying (37) can be obtained as a stochastic logarithm $\mathcal{L}og(L)$ of $L$:

$$N_t = \mathcal{L}og(L)_t, \tag{39}$$

where

$$\mathcal{L}og(L)_t = \int_0^t \frac{dL_s}{L_{s-}}. \tag{40}$$

If $L_0 = 1$, then for $\mathcal{L}og(L)_t$ we have the following representation [5]:

$$\mathcal{L}og(L)_t = \log L_t + \int_0^t \frac{d\langle L^c \rangle_s}{2L_{s-}^2} - \sum_{0 < s \le t} \left( \log \frac{L_s}{L_{s-}} + 1 - \frac{L_s}{L_{s-}} \right). \tag{41}$$

For example, suppose, that $X^{(\infty)}$ and $X^{(0)}$ are two Poisson processes with intensities $\lambda^{(\infty)}$ and $\lambda^{(0)}$. Then ($\mathsf{P}^{(\infty)}$-a.s.)

$$L_t = \exp\left\{ X_t \log \frac{\lambda^{(0)}}{\lambda^{(\infty)}} - t(\lambda^{(0)} - \lambda^{(\infty)}) \right\} \tag{42}$$

and

$$dL_t = \left( \frac{\lambda^{(0)}}{\lambda^{(\infty)}} - 1 \right) L_{t-}(dX_t - \lambda^{(\infty)} dt). \tag{43}$$

Therefore, by (40),

$$N_t = \left( \frac{\lambda^{(0)}}{\lambda^{(\infty)}} - 1 \right)(X_t - \lambda^{(\infty)}t). \tag{44}$$

Analysis of the proof of the identity

$$\int_0^\infty \mathsf{E}_\theta G\big((\tau - \theta)^+\big) \, d\theta = \mathsf{E}_\infty \int_0^\tau \Psi_u(g) \, du \tag{45}$$

presented above for the Brownian case (1) shows that this formula also hold, if in the model (34) the processes $X^{(\infty)}$ and $X^{(0)}$ have *stationary independent increments* (are PIIS in terminology of [5]) and are such that for each $t \ge 0$ the measures $\mathsf{P}_0 | \mathcal{F}_t$ and $\mathsf{P}_\infty | \mathcal{F}_t$ are equivalent. In this case we have also the representation (36), where $N = (N_t)_{t \ge 0}$ is a process with independent increments [5, Chap. III, Theorem 5.35].

Assume, in addition to the PIIS assumption for processes $X^{(\infty)}$ and $X^{(0)}$, that function $g(t)$ admits the representation (20). Then for

$$\psi_u = \int_0^u \frac{L_u}{L_\theta} \, d\theta$$

we get by Itô's formula that

$$d\psi_u = dL_u \cdot \int_0^u \frac{d\theta}{L_\theta} + du = L_{u-} dN_u \cdot \int_0^u \frac{d\theta}{L_\theta} + du$$
$$= \psi_{u-} dN_u + du. \tag{46}$$

Since $N = (N_u)_{u \geq 0}$ is the process with independent increments, we see that the stochastic differential equation

$$d\psi_u = du + \psi_{u-} dN_u \tag{47}$$

defines a Markov process.

Similarly, for statistics $\psi_u^{(m,n)}$ (see (31)) we get

$$d\psi_u^{(m,n)} = \left(\lambda_m + n\psi_u^{(m,n-1)}\right) du + \psi_{u-}^{(m,n)} dN_u. \tag{48}$$

Therefore, if in the model (34) *the processes $X^{(\infty)}$ and $X^{(0)}$ are processes with stationary independent increments such that $\mathsf{P}_0|\mathcal{F}_t \sim \mathsf{P}_\infty|\mathcal{F}_t$, $t \geq 0$, then the system of statistics (33) (with $\psi_u^{(m,n)} dN_u$ instead of $(\mu/\sigma^2)\psi_u^{(m,n)} dX_u$) forms a Markov family of sufficient statistics for optimal stopping problem*

$$\inf_{\tau \in \mathfrak{M}_T} \mathsf{E}_\infty \int_0^\tau \Psi_u(g) \, du.$$

# References

1. Shiryaev AN (1961) The detection of spontaneous effects. Soviet Math. Dokl., 2:740–743.
2. Shiryaev AN (1961) The problem of the most rapid detection of a disturbance in a statinary regime. Soviet Math. Dokl., 2:795–799.
3. Feinberg EA, Shiryaev AN (2006) Quickest detection of drift change for Brownian motion in generalized Bayesian ans minimax settings. Statist. Decisions, 24:1001–1025.
4. Liptser RS, Shiryaev AN (2001) Statistics of Random Processes. I: General Theory, 2nd ed.. Appl. Math. (New York), 5. Springer-Verlag, Berlin
5. Jacod J, Shiryaev AN (2003) Limit Theorems for Stochastic Processes, 2nd ed.. Grundlehren Math. Wiss., 288, Springer-Verlag, Berlin
6. Shiryaev AN (1964) On Markov sufficient statistics in non-additive Bayes problems of sequential analysis. Theory Probab. Appl., 9(4):604–618.
7. Poor HV (1998) Quickest detection with exponential penalty for delay.Ann. Statist., 26(6):2179–2205.
8. Beibel M (2000) A note on sequential detection with exponential penalty for the delay. Ann. Statist., 28(6):1696–1701.

# Necessary Optimality Condition for a Discrete Dead Oil Isotherm Optimal Control Problem

Moulay Rchid Sidi Ammi and Delfim F. M. Torres

Department of Mathematics, University of Aveiro,
3810-193 Aveiro, Portugal. `sidiammi@ua.pt`, `delfim@ua.pt`

**Summary.** We obtain necessary optimality conditions for a semi-discretized optimal control problem for the classical system of nonlinear partial differential equations modelling the water-oil (isothermal dead-oil model).

**Key words:** extraction of hydrocarbons; dead oil isotherm problem; optimality conditions.

## 1 Introduction

We study an optimal control problem in the discrete case whose control system is given by the following system of nonlinear partial differential equations,

$$
\begin{cases}
\partial_t u - \Delta \varphi(u) = div\,(g(u)\nabla p) & \text{in } Q_T = \Omega \times (0,T)\,, \\
\partial_t p - div\,(d(u)\nabla p) = f & \text{in } Q_T = \Omega \times (0,T)\,, \\
u|_{\partial\Omega} = 0\,, \quad u|_{t=0} = u_0\,, \\
p|_{\partial\Omega} = 0\,, \quad p|_{t=0} = p_0\,,
\end{cases}
\tag{1}
$$

which result from a well established model for oil engineering within the framework of the mechanics of a continuous medium [3]. The domain $\Omega$ is an open bounded set in $\mathbb{R}^2$ with a sufficiently smooth boundary. Further hypotheses on the data of the problem will be specified later.

At the time of the first run of a layer, the flow of the crude oil towards the surface is due to the energy stored in the gases under pressure in the natural hydraulic system. To mitigate the consecutive decline of production and the decomposition of the site, water injections are carried out, well before the normal exhaustion of the layer. The water is injected through wells with high pressure, by pumps specially drilled to this end. The pumps allow the displacement of the crude oil towards the wells of production. More precisely, the problem consists in seeking the admissible control parameters which minimize a certain objective functional. In our problem, the main goal is to distribute properly the wells in order to have the best extraction of the hydrocarbons. For this reason, we consider a cost functional containing different parameters arising in the process. To address the optimal control problem, we use the Lagrangian method to derive an optimality system: from the cost function we introduce a Lagrangian; then, we calculate the Gâteaux derivative of the Lagrangian with respect to its variables. This technique was used, in particular, by A. Masserey et al. for electromagnetic models of induction heating [1, 7], and by H.-C. Lee and T. Shilkin for the thermistor problem [5].

We consider the following cost functional:

$$
\begin{aligned}
J(u, p, f) = \frac{1}{2} \left\| u - U \right\|_{2, Q_T}^2 + \frac{1}{2} \left\| p - P \right\|_{2, Q_T}^2 \\
+ \frac{\beta_1}{2} \left\| f \right\|_{2q_0, Q_T}^{2q_0} + \frac{\beta_2}{2} \left\| \partial_t f \right\|_{2, Q_T}^2 .
\end{aligned}
\tag{2}
$$

The control parameters are the reduced saturation of oil $u$, the pressure $p$, and $f$. The coefficients $\beta_1 > 0$ and $\beta_2 \geq 0$ are two coefficients of penalization, and $q_0 > 1$. The first two terms in (2) allow to minimize the difference between the reduced saturation of oil $u$, the global pressure $p$ and the given data $U$ and $P$. The third and fourth terms are used to improve the quality of exploitation of the crude oil. We take $\beta_2 = 0$ just for the sake of simplicity. It is important to emphasize that our choice of the cost function is not unique. One can always add additional terms of penalization to take into account other properties which one may wish to control. Recently, we proved in [8] results of existence, uniqueness, and regularity of the optimal solutions to the problem of minimizing (2) subject to (1), using the theory of parabolic problems [4, 6]. Here, our goal is to obtain necessary optimality conditions which may be easily implemented on a computer. More precisely, we address the problem of obtaining necessary optimality conditions for the semi-discretized time problem.

In order to be able to solve problem (1)-(2) numerically, we use discretization of the problem in time by a method of finite differences. For a fixed real $N$, let $\tau = \frac{T}{N}$ be the step of a uniform partition of the interval $[0, T]$ and $t_n = n\tau$, $n = 1, \ldots, N$. We denote by $u^n$ an approximation of $u$. The discrete cost functional is then defined as follows:

$$
J(u^n, p^n, f^n) =
$$
$$
\frac{\tau}{2} \sum_{n=1}^{N} \int_{\Omega} \left\{ \left\| u^n - U \right\|_{2, \Omega}^2 + \left\| p^n - P \right\|_{2, \Omega}^2 + \beta_1 \left\| f^n \right\|_{2q_0, \Omega}^{2q_0} \right\} dx.
\tag{3}
$$

It is now possible to state our optimal control problem: find $(\bar{u}^n, \bar{p}^n, \bar{f}^n)$ which minimizes (3) among all functions $(u^n, p^n, f^n)$ satisfying

$$\begin{cases} \frac{u^{n+1}-u^n}{\tau} - \Delta\varphi(u^n) = div\,(g(u^n)\nabla p) & \text{in } \Omega\,, \\ \frac{p^{n+1}-p^n}{\tau} - div\,(d(u^n)\nabla p^n) = f^n & \text{in } \Omega\,, \\ u|_{\partial\Omega} = 0\,, \quad u|_{t=0} = u_0\,, \\ p|_{\partial\Omega} = 0\,, \quad p|_{t=0} = p_0\,. \end{cases} \tag{4}$$

The soughtafter necessary optimality conditions are proved in §3 under suitable hypotheses on the data of the problem.

## 2 Notation, hypotheses, and functional spaces

Our main objective is to obtain necessary conditions for a triple $\left(\bar{u}^n, \bar{p}^n, \bar{f}^n\right)$ to minimize (3) among all the functions $(u^n, p^n, f^n)$ verifying (4). In the sequel we assume that $\varphi$, $g$ and $d$ are real valued functions, respectively of class $C^3$, $C^2$ and $C^1$, satisfying:

(H1) $0 < c_1 \leq d(r)$, $\varphi(r) \leq c_2$; $|d'(r)|, |\varphi'(r)|, |\varphi''(r)| \leq c_3 \quad \forall r \in \mathbb{R}$.

(H2) $u_0,\ p_0 \in C^2\left(\bar{\Omega}\right)$, and $U, P \in L^2(\Omega)$, where $u_0,\ p_0,\ U,\ P : \Omega \to \mathbb{R}$, and $u_0|_{\partial\Omega} = p_0|_{\partial\Omega} = 0$.

We consider the following spaces:

$$W_p^1(\Omega) := \{u \in L^p(\Omega),\ \nabla u \in L^p(\Omega)\}\,,$$

endowed with the norm $\|u\|_{W_p^1(\Omega)} = \|u\|_{p,\Omega} + \|\nabla u\|_{p,\Omega}$;

$$W_p^2(\Omega) := \{u \in W_p^1(\Omega),\ \nabla^2 u \in L^p(\Omega)\}\,,$$

with the norm $\|u\|_{W_p^2(\Omega)} = \|u\|_{W_p^1(\Omega)} + \|\nabla^2 u\|_{p,\Omega}$; and the following notation:

$$W := \overset{\circ}{W}{}_{2q}^{2}(\Omega)\,;$$
$$\Upsilon := L^{2q}(\Omega)\,;$$
$$H := L^{2q}(\Omega) \times \overset{\circ}{W}{}_{2q}^{2-\frac{1}{q}}(\Omega)\,.$$

## 3 Main results

We define the following nonlinear operator corresponding to (4):

$$F : W \times W \times \Upsilon \longrightarrow H \times H$$
$$(u^n, p^n, f^n) \longrightarrow F(u^n, p^n, f^n)\,,$$

where

$$F\left(u^n, p^n, f^n\right) = \begin{pmatrix} \frac{u^{n+1}-u^n}{\tau} - \Delta\varphi(u^n) - div(g(u^n)\nabla p^n),\ \gamma_0 u^n - u_0 \\ \frac{u^{n+1}-u^n}{\tau} - div\,(d(u^n)\nabla p^n) - f^n,\ \gamma_0 p^n - p_0 \end{pmatrix}\,,$$

$\gamma_0$ being the trace operator $\gamma_0 u^n = u|_{t=0}$. Our hypotheses ensure that $F$ is well defined.

## 3.1 Gâteaux differentiability

**Theorem 1.** *In addition to the hypotheses (H1) and (H2), let us suppose that*

*(H3) $|\varphi'''| \leq c$.*

*Then, the operator $F$ is Gâteaux differentiable and for all $(e, w, h) \in W \times W \times \Upsilon$ its derivative is given by*

$$\delta F(u^n, p^n, f^n)(e, w, h) = \frac{d}{ds} F\left(u^n + se, p^n + sw, f^n + sh\right)|_{s=0}$$

$$= (\delta F_1, \delta F_2) = \begin{pmatrix} \xi_1, \xi_2 \\ \xi_3, \xi_4 \end{pmatrix},$$

$\xi_1 = e - div\left(\varphi'(u^n)\nabla e\right) - div\left(\varphi''(u^n)e\nabla u^n\right) - div\left(g(u^n)\nabla w\right)$
$-div\left(g'(u^n)e\nabla p^n\right)$, $\xi_2 = \gamma_0 e$, $\xi_3 = w - div\left(d(u^n)\nabla w\right) - div\left(d'(u^n)e\nabla p^n\right) - h$,
$\xi_4 = \gamma_0 w$. *Furthermore, for any optimal solution $\left(\bar{u}^n, \bar{p}^n, \bar{f}^n\right)$ of the problem of minimizing (3) among all the functions $(u^n, p^n, f^n)$ satisfying (4), the image of $\delta F\left(\bar{u}^n, \bar{p}^n, \bar{f}^n\right)$ is equal to $H \times H$.*

To prove Theorem 1 we make use of the following lemma.

**Lemma 1.** *The operator $\delta F(u^n, p^n, f^n) : W \times W \times \Upsilon \longrightarrow H \times H$ is linear and bounded.*

*Proof (Lemma 1).* For all $(e, w, h) \in W \times W \times \Upsilon$

$$\delta_{u^n} F_1(u^n, p^n, f^n)(e, w, h)$$
$$= e - div\left(\varphi'(u^n)\nabla e\right) - div\left(\varphi''(u^n)e\nabla u^n\right)$$
$$- div\left(g(u^n)\nabla w\right) - div\left(g'(u^n)e\nabla p^n\right)$$
$$= e - \varphi'(u^n)\triangle e - \varphi''(u^n)\nabla u^n.\nabla e - \varphi''(u^n)e\triangle u^n$$
$$- \varphi''(u^n)\nabla e.\nabla u^n - \varphi'''(u^n)e|\nabla u^n|^2 - g(u^n)\triangle w - g'(u^n)\nabla u^n.\nabla w$$
$$- g'(u^n)e\triangle p^n - g'(u^n)\nabla e.\nabla p^n - g''(u^n)e\nabla u^n.\nabla p^n,$$

where $\delta_{u^n} F$ is the Gâteaux derivative of $F$ with respect to $u^n$. Using our hypotheses we have

$$\|g''(u^n)e\nabla u^n.\nabla p^n\|_{2q,\Omega} \leq \|e\|_{\infty,\Omega}\|\nabla u^n.\nabla p^n\|_{2q,\Omega}$$
$$\leq \|e\|_{\infty,\Omega}\|\nabla u^n\|_{\frac{4q}{2-q},\Omega}\|\nabla p^n\|_{4,\Omega}$$
$$\leq c\|u^n\|_W\|p^n\|_W\|e\|_W.$$

Evaluating each term of $\delta_{u^n} F_1$, we obtain

$$\|\delta_{u^n} F_1(u^n, p^n, f^n)(e, w, h)\|_{2q,Q_T}$$
$$\leq c\left(\|u^n\|_W, \|p^n\|_W, \|f^n\|_\Upsilon\right)\left(\|e\|_W + \|w\|_W + \|h\|_\Upsilon\right). \quad (5)$$

In a similar way, we have for all $(e, w, h) \in W \times W \times \Upsilon$ that

$$
\begin{aligned}
\delta_{p^n} F_2(u^n, p^n, f^n)(e, w, h) &= w - div\,(d(u^n)\nabla w) - div\,(d'(u^n)e\nabla p^n) - h \\
&= w - d(u^n)\triangle w - d'(u^n)\nabla u^n.\nabla w - d'(u^n)e\triangle p^n \\
&\quad - d'(u^n)\nabla e.\nabla u^n - d'(u^n)e\nabla u^n.\nabla p^n - h\,,
\end{aligned}
$$

with $\delta_{p^n} F$ the Gâteaux derivative of $F$ with respect to $p^n$. Then, using again our hypotheses, we obtain that

$$
\begin{aligned}
\|\delta_{p^n} F_2(u^n, p^n, f^n)(e, w, h)\|_{2q,\Omega} &\leq \|w\|_{2q,\Omega} + \|\nabla w\|_{2q,\Omega} + c\|\triangle w\|_{2q,\Omega} \\
&\quad + c\|\nabla u^n.\nabla w\|_{2q,\Omega} + c\|e\triangle p^n\|_{2q,\Omega} \\
&\quad + c\|\nabla e.\nabla u^n\|_{2q,\Omega} + c\|e\nabla u^n.\nabla p^n\|_{2q,\Omega} + \|h\|_{2q,\Omega}\,. \quad (6)
\end{aligned}
$$

Applying similar arguments to all terms of (6), we then have

$$
\begin{aligned}
\|\delta_{p^n} F_2(u^n, p^n, f^n)&(e, w, h)\|_{2q,\Omega} \\
&\leq c\,(\|u^n\|_W, \|p^n\|_W, \|f^n\|_\Upsilon)\,(\|e\|_W + \|w\|_W + \|h\|_\Upsilon)\,. \quad (7)
\end{aligned}
$$

Consequently, by (5) and (7) we can write

$$
\begin{aligned}
\|\delta F(u^n, p^n, f^n)&(e, w, h)\|_{H \times H \times \Upsilon} \\
&\leq c\,(\|u^n\|_W, \|p^n\|_W, \|f^n\|_\Upsilon)\,(\|e\|_W + \|w\|_W + \|h\|_\Upsilon)\,.
\end{aligned}
$$

□

*Proof (Theorem 1).* In order to show that the image of $\delta F(\overline{u}, \overline{p}, \overline{f})$ is equal to $H \times H$, we need to prove that there exists $(e, w, h) \in W \times W \times \Upsilon$ such that

$$
\begin{aligned}
e - div\,(\varphi'(\overline{u^n})\nabla e) &- div\,(\varphi''(\overline{u^n})e\nabla\overline{u^n}) \\
&- div\,(g(\overline{u^n})\nabla w) - div\,(g'(\overline{u^n})e\nabla\overline{p^n}) = \alpha, \\
w - div\,(d(\overline{u^n})\nabla w) &- div\,(d'(\overline{u^n})e\nabla\overline{p^n}) - h = \beta, \qquad (8) \\
e|_{\partial\Omega} = 0\,, &\quad e|_{t=0} = b, \\
w|_{\partial\Omega} = 0\,, &\quad w|_{t=0} = a,
\end{aligned}
$$

for any $(\alpha, a)$ and $(\beta, b) \in H$. Writing the system (8) for $h = 0$ as

$$
\begin{aligned}
e - \varphi'(\overline{u^n})\triangle e &- 2\varphi''(\overline{u^n})\nabla\overline{u^n}.\nabla e - \varphi''(\overline{u^n})e\triangle\overline{u^n} - \varphi'''(\overline{u^n})e|\nabla\overline{u^n}|^2, \\
&- g(\overline{u^n})\triangle w - g'(\overline{u^n})\nabla\overline{u^n}.\nabla w - g'(\overline{u^n})e\triangle\overline{p^n} \\
&- g'(\overline{u^n})\nabla\overline{p^n}.\nabla e - g''(\overline{u^n})e\nabla\overline{u^n}.\nabla\overline{p^n} = \alpha, \\
w - d(\overline{u^n})\triangle w &- d'(\overline{u^n})\nabla\overline{u^n}.\nabla w - d'(\overline{u^n})e\triangle\overline{p^n} \qquad (9) \\
&- d'(\overline{u^n})\nabla\overline{u^n}.\nabla\overline{e} - d'(\overline{u^n})e\nabla\overline{u^n}.\nabla\overline{p^n} = \beta, \\
e|_{\partial\Omega} = 0\,, &\quad e|_{t=0} = b, \\
w|_{\partial\Omega} = 0\,, &\quad w|_{t=0} = a,
\end{aligned}
$$

it follows from the regularity of the optimal solution that $\varphi''(\overline{u^n})\triangle\overline{u^n}$, $\varphi'''(\overline{u^n})|\nabla\overline{u^n}|^2$, $g'(\overline{u^n})\triangle\overline{p^n}$, $g''(\overline{u^n})\nabla\overline{u^n}.\nabla\overline{p^n}$, $d'(\overline{u^n})\triangle\overline{p^n}$, and $d'(\overline{u^n})\nabla\overline{u^n}.\nabla\overline{p^n}$ belong to $L^{2q_0}(\Omega)$; $\varphi''(\overline{u^n})\nabla\overline{u^n}$, $g'(\overline{u^n})\nabla\overline{u^n}$, $g'(\overline{u^n})\nabla\overline{p^n}$, and $d'(\overline{u^n})\nabla\overline{u^n}$ belong to $L^{4q_0}(\Omega)$. This ensures, in view of the results of [4, 6], existence of a unique solution of the system (9). Hence, there exists a $(e, w, 0)$ verifying (8). We conclude that the image of $\delta F$ is equal to $H \times H$. $\square$

## 3.2 Necessary optimality condition

We consider the cost functional $J : W \times W \times \Upsilon \to \mathbb{R}$ (3) and the Lagrangian $\mathcal{L}$ defined by

$$\mathcal{L}\left(u^n, p^n, f^n, p_1, e_1, a, b\right) = J\left(u^n, p^n, f^n\right) + \left\langle F(u^n, p^n, f^n), \begin{pmatrix} p_1 & a \\ e_1 & b \end{pmatrix}\right\rangle,$$

where the bracket $\langle \cdot, \cdot \rangle$ denotes the duality between $H$ and $H'$.

**Theorem 2.** *Under hypotheses (H1)–(H3), if $\left(\overline{u^n}, \overline{p^n}, \overline{f^n}\right)$ is an optimal solution to the problem of minimizing (3) subject to (4), then there exist functions $\left(\overline{e_1}, \overline{p_1}\right) \in W_2^2(\Omega) \times W_2^2(\Omega)$ satisfying the following conditions:*

$$\overline{e_1} + div\left(\varphi'(\overline{u^n})\nabla e_1\right) - d'(\overline{u^n})\nabla\overline{p^n}.\nabla\overline{p_1} - \varphi''(\overline{u^n})\nabla\overline{u^n}.\nabla\overline{e_1}$$

$$-g'(\overline{u^n})\nabla\overline{p^n}.\nabla\overline{e_1} = \tau \sum_{n=1}^{N}(\overline{u^n} - U),$$

$$\overline{e_1}\big|_{\partial\Omega} = 0, \quad \overline{e_1}\big|_{t=T} = 0,$$

$$\overline{p_1} + div\left(d(\overline{u^n})\nabla\overline{p_1}\right) + div\left(g(\overline{u^n})\nabla\overline{e_1}\right) = \tau \sum_{n=1}^{N}\left(\overline{p^n} - P\right),$$

$$\overline{p_1}\big|_{\partial\Omega} = 0, \quad \overline{p_1}\big|_{t=T} = 0,$$

$$q_0\beta_1\tau \sum_{n=1}^{N}|\overline{f^n}|^{2q_0-2}\overline{f^n} = \overline{p_1}.$$

(10)

*Proof.* Let $\left(\overline{u^n}, \overline{p^n}, \overline{f^n}\right)$ be an optimal solution to the problem of minimizing (3) subject to (4). It is well known (cf. e.g. [2]) that there exist Lagrange multipliers $\left((\overline{p_1}, \overline{a}), (\overline{e_1}, \overline{b})\right) \in H' \times H'$ verifying

$$\delta_{(u^n, p^n, f^n)}\mathcal{L}\left(\overline{u^n}, \overline{p^n}, \overline{f^n}, \overline{p_1}, \overline{e_1}, \overline{a}, \overline{b}\right)(e, w, h) = 0 \quad \forall(e, w, h) \in W \times W \times \Upsilon,$$

with $\delta_{(u^n, p^n, f^n)}\mathcal{L}$ the Gâteaux derivative of $\mathcal{L}$ with respect to $(u^n, p^n, f^n)$. This leads to the following system:

$$\tau \sum_{n=1}^{N} \int_{\Omega} \left( (\overline{u^n} - U)e + (\overline{p^n} - P)w + q_0 \beta_1 |\overline{f^n}|^{2q_0-2} \overline{f^n} h \right) dx$$

$$- \int_{\Omega} \left( \left( e - div \left( \varphi'(\overline{u^n}) \nabla e \right) - div \left( \varphi''(\overline{u^n}) e \nabla \overline{u^n} \right) \right. \right.$$

$$\left. \left. - div \left( g(\overline{u^n}) \nabla w \right) - div \left( g'(\overline{u^n}) e \nabla \overline{p^n} \right) \right) \overline{e_1} \right) dx$$

$$- \int_{\Omega} \left( w - div \left( d(\overline{u^n}) \nabla w \right) - div \left( d'(\overline{u^n}) e \nabla \overline{p^n} \right) - h \right) \overline{p_1} \, dx$$

$$- \langle \gamma_0 e, \overline{a} \rangle + - \langle \gamma_0 w, \overline{b} \rangle = 0 \quad \forall (e, w, h) \in W \times W \times \Upsilon.$$

The above system is equivalent to the following one:

$$\int_{\Omega} \left( \tau \sum_{n=1}^{N} (\overline{u^n} - U)e - div \left( d'(\overline{u^n}) e \nabla \overline{p^n} \right) \overline{p_1} + e \, \overline{e_1} - div \left( \varphi'(\overline{u^n}) \nabla e \right) \overline{e_1} \right.$$

$$\left. - div \left( \varphi''(\overline{u^n}) e \nabla \overline{u^n} \right) \overline{e_1} - div \left( g'(\overline{u^n}) e \nabla \overline{p^n} \right) \overline{e_1} \right) dx$$

$$+ \int_{\Omega} \left( \tau \sum_{n=1}^{N} (\overline{p^n} - P)w + w \, \overline{p_1} - div \left( d(\overline{u^n}) \nabla w \right) \overline{p_1} - div \left( g(\overline{u^n}) \nabla w \right) \overline{e_1} \right) dx$$

$$+ \int_{\Omega} \left( q_0 \beta_1 \tau \sum_{n=1}^{N} |\overline{f^n}|^{2q_0-2} \overline{f^n} h - \overline{p_1} h \right) dx$$

$$+ \langle \gamma_0 e, \overline{a} \rangle + \langle \gamma_0 w, \overline{b} \rangle = 0 \quad \forall (e, w, h) \in W \times W \times \Upsilon. \tag{11}$$

In others words, we have

$$\int_{\Omega} \left( \tau \sum_{n=1}^{N} (\overline{u^n} - U) + d'(u) \nabla \overline{p^n} . \nabla \overline{p_1} - \overline{e_1} - div \left( \varphi'(\overline{u^n}) \nabla \overline{e_1} \right) \right.$$

$$\left. + \varphi''(\overline{u^n}) \nabla \overline{u^n} . \nabla \overline{e_1} + g'(u^n) \nabla \overline{p^n} . \nabla \overline{e_1} \right) e \, dx$$

$$+ \int_{\Omega} \left( \tau \sum_{n=1}^{N} (\overline{p^n} - P) + \overline{p_1} - div \left( d(\overline{u^n}) \nabla \overline{p_1} \right) - div \left( g(\overline{u^n}) \nabla \overline{e_1} \right) \right) w \, dx \tag{12}$$

$$+ \int_{\Omega} \left( q_0 \beta_1 \tau \sum_{n=1}^{N} |\overline{f^n}|^{2q_0-2} \overline{f^n} h - \overline{p_1} h \right) dx$$

$$+ \langle \gamma_0 e, \overline{a} \rangle + \langle \gamma_0 w, \overline{b} \rangle = 0 \quad \forall (e, w, h) \in W \times W \times \Upsilon.$$

Consider now the system

$$e_1 + div\left(\varphi'(\overline{u^n})\nabla e_1\right) - d'(\overline{u^n})\nabla\overline{p^n}.\nabla p_1 - \varphi''(\overline{u^n})\nabla\overline{u^n}.\nabla e_1$$

$$-g'(\overline{u^n})\nabla\overline{p^n}.\nabla e_1 = \tau\sum_{n=1}^{N}(\overline{u^n} - U),$$

$$p_1 + div\left(d(\overline{u^n})\nabla p_1\right) + div\left(g(\overline{u^n})\nabla e_1\right) = \tau\sum_{n=1}^{N}(\overline{p^n} - P), \tag{13}$$

$$e_1|_{\partial\Omega} = p_1|_{\partial\Omega} = 0, \quad e_1|_{t=T} = p_1|_{t=T} = 0,$$

with unknowns $(e_1, p_1)$ which is uniquely solvable in $W_2^2(\Omega) \times W_2^2(\Omega)$ by the theory of elliptic equations [4]. The problem of finding $(e, w) \in W \times W$ satisfying

$$e - div\left(\varphi'(\overline{u^n})\nabla e\right) - div\left(\varphi''(\overline{u^n})e\nabla\overline{u^n}\right) - div\left(g(\overline{u^n})\nabla w\right)$$
$$-div\left(g'(\overline{u^n})e\nabla\overline{p^n}\right) = sign(e_1 - \overline{e_1}),$$
$$w - div\left(d(\overline{u^n})\nabla w\right) - div\left(d'(\overline{u^n})e\nabla\overline{p^n}\right) = sign(p_1 - \overline{p_1}), \tag{14}$$
$$\gamma_0 e = \gamma_0 w = 0,$$

is also uniquely solvable on $W_{2q}^2(\Omega) \times W_{2q}^2(\Omega)$. Let us choose $h = 0$ in (12) and multiply (13) by $(e, w)$. Then, integrating by parts and making the difference with (12) we obtain:

$$\int_{\Omega}\Big(e - div\left(\varphi'(\overline{u^n})\nabla e\right) - div\left(\varphi''(\overline{u^n})e\nabla\overline{u^n}\right) - div\left(g(\overline{u^n})\nabla w\right)$$

$$-div\left(g'(\overline{u^n})e\nabla\overline{p^n}\right)\Big)(e_1 - \overline{e_1})\,dx \tag{15}$$

$$+ \int_{\Omega}\left(w - div\left(d(\overline{u^n})\nabla w\right) - div\left(d'(\overline{u^n})e\nabla\overline{p^n}\right)\right)(p_1 - \overline{p_1})\,dx$$

$$+\langle\gamma_0 e, \gamma_0\overline{e_1} - \overline{a}\rangle + \langle\gamma_0 w, \gamma_0\overline{p_1} - \overline{b}\rangle = 0 \quad \forall(e, w) \in W \times W.$$

Choosing $(e, w)$ in (15) as the solution of the system (14), we have

$$\int_{\Omega} sign(e_1 - \overline{e_1})(e_1 - \overline{e_1})\,dxdt + \int_{\Omega} sign(p_1 - \overline{p_1})(p_1 - \overline{p_1})\,dx = 0.$$

It follows that $e_1 = \overline{e_1}$ and $p_1 = \overline{p_1}$. Coming back to (15), we obtain $\gamma_0\overline{e_1} = \overline{a}$ and $\gamma_0\overline{p_1} = \overline{b}$. On the other hand, choosing $(e, w) = (0, 0)$ in (12), we get

$$\int_{\Omega}\left(\beta_1\tau\sum_{n=1}^{N}|\overline{f^n}|^{2q_0 - 2}\overline{f^n} - \overline{p_1}\right)h\,dx = 0, \forall h \in \Upsilon.$$

Then (10) follows, which concludes the proof of Theorem 2. $\square$

We claim that the results we obtain here are useful for numerical implementations. This is still under investigation and will be addressed in a forthcoming publication.

## Acknowledgments

The authors were supported by the *Portuguese Foundation for Science and Technology* (FCT) through the *Centre for Research on Optimization and Control* (CEOC) of the University of Aveiro, cofinanced by the European Community fund FEDER/POCI 2010. This work was developed under the post-doc project SFRH/BPD/20934/2004.

## References

1. Bodart O, Boureau AV, Touzani R (2001) Numerical investigation of optimal control of induction heating processes. Applied Mathematical Modelling, 25:697–712.
2. Fursikov AV (2000) Optimal control of distributed systems. Theory and applications, AMS, Providence, RI
3. Gagneux G, Madaune-Tort M (1996) Analyse mathématique de modèles non linéaires de l'ingénierie pétrolière. Mathématiques & Application 22, Springer-Verlag
4. Ladyzhenskaya OA, Solonnikov VA, Uraltseva NN (1967) Linear and quasi-linear equations of parabolic type, Transl. Math. Monogr. 23, AMS, Providence, RI
5. Lee H-C, Shilkin T (2005) Analysis of optimal control problems for the two-dimensional thermistor system, SIAM J. Control Optim., 44(1):268–282.
6. Lions J-L (1969) Quelques méthodes de résolution des problèmes aux limites non linéaires. Dunod, Paris
7. Masserey A, Rappaz J, Rozsnyo R, Swierkosz M (2005) Numerical integration of the three-dimensional Green kernel for an electromagnetic problem. J. Comput. Phys. 205(1):48–71.
8. Sidi Ammi MR, Torres DFM (2007) Existence and regularity of optimal solution for a dead oil isotherm problem. Applied Sciences (APPS), 9:5–12.

# Managing Operational Risk: Methodology and Prospects

Grigory Temnov*

Vienna University of Technology, Institute for Mathematical Methods in
Economics, Financial and Actuarial Mathematics,
Wiedner Hauptstrasse 8-10/105-1, A-1040 Vienna, Austria.
gtemnov@fam.tuwien.ac.at

**Summary.** In the present paper we describe a combined methodology used for
modelling and measuring operational risk. Our basic aim is to choose and adjust an
efficient combination of techniques in order to cover a range of problems associated
with OpRisk and justify our choice. We analyze each part of the methodology and
briefly overview some recent results, as well as the prospects of the future research.

## 1 Operational risk data: models for severity and frequency

### 1.1 Introduction. Motivation of the work

During latter decades, the problem of operational risk measurement has been
staying under a thorough attention of practitioners and theorists of actuarial
sciences. Lots of scientific papers devoted to operational risk appeared since
early nineties; quite a number of textbooks and monographs highlight the
whole spectrum of problems associated with OpRisk. Among recently published books, [23] can be mentioned as a detailed overview of methods for
OpRisk analysis with examples. Another detailed survey can be found in [7]
which is rather a practical guide to the management of OpRisk.

A variety of methods in the frame of the problem of OpRisk may appear
rather confusing for an analyst facing the problem in practice. That's why
the choice of the right combination of methods becomes very important. In
the present paper we briefly describe and justify the methodology which we
find effective and appropriate to cover the wide range of problems associated

---

\* This work has been done under the financial support by Christian Doppler Labor
http://www.prismalab.at.

The present paper complements and extends basic results included in the article by G. Temnov and R. Warnung (2008) "A Comparison of Loss Aggregation
Methods for Operational Risk", the Journal of Operational Risk, 3(1):3–23.

with OpRisk, from the basic capital allocation problem to the problems of the insurance of OpRisk. We illustrate the application of this methodology basing on realistic data and analyze the efficiency of the proposed algorithms.

Each part of the combined methodology that we propose was investigated by different authors, and our contributions can found in [14], [32], [27]. In the present work we refer to these results and briefly describe the main of them. Besides, some new problems are outlined and some brief outlook for the future work is made. Still, the main purpose of the present work is to propose a combined methodology for the OpRisk measurement which a practitioner could find useful and efficient.

## 1.2 Description of the data and points of investigation

Dealing with the OpRisk management, one is usually interested in the characteristics of the yearly aggregate loss distribution. Database of historical OpRisk losses occurred during some period normally serves as an information basis which the estimation should rely upon. The characteristics in interest are often called risk measures. Depending on the particular task, one may choose such common risk measures as Value-at-Risk or Expected Shortfall.

In the present work we deal with exactly such type of problems. The historical data available to us is divided by business lines (BL) and types of events (ET) which is a usual classification for OpRisk data. BL's are encoded as follows: corporate finance (BL 1), trading and sales (BL 2), retail banking (BL 3), commercial banking (BL 4), clearing (BL 5), agency services (BL 6), asset management (BL 7), retail brokerage (BL 8), and private banking (BL 9). Note that these notations can differ from one association for collecting OpRisk data to another.

According to this classification, each loss value corresponds to one particular BL and a particular ET. The division by BL's is important for the problem of the calculation of regulatory capital for OpRisk. Particularly, according to the Basel II recommendations, the Value-at-Risk estimators, defining the regulatory capital, should be done for every BL separately.

Concerning the division of the data by ET's, it becomes crucial dealing with such problems as insurance, when a financial institution wishes to insure losses corresponding to certain BL's and ET's (we shall say that such losses *belong to a particular BL-ET cell*).

Let us make some preliminary comments on the quality of the data before passing to the description of the methodology. Usually, OpRisk data is heavy tailed in loss severities, which means the essential role of rare but very severe losses, especially for some particular BL's. Our case is not an exclusion. Consequently, extreme value theory (EVT) is used in our investigation as an important and powerful tool for the analysis of heavy tailed data. We make some remarks on the application of EVT in our work in paragraph 1.3.

As usual for OpRisk, two historical databases of loss data are available: the *internal data* of our selected financial institution (in the following *the Bank*), and the *external data*. The latter contains the OpRisk data from a group of banks and could be an important contribution for the loss analysis.

As counting processes of occurrences, homogeneous and inhomogeneous Poisson processes are used, as well as negative binomial distributions arising from mixing the intensity of a Poisson process with a gamma distribution. However, the class of stochastic processes eligible as models for the process of occurrences is much wider and may include e.g. some models used normally in financial mathematics. We describe some prospects and outline the ideas for future work concerning the models for OpRisk frequency in paragraph 1.4.

In some cases, say for some particular BL's, internal data may be enough to produce a reliable fit for the parameters of severity and frequency distributions. Yet, if internal data is insufficient, then the external data can be used to estimate prior distributions of severity and frequency parameters. As a tool for a proper mixing of internal and external data, the *bayesian inference* is used in our investigation. It allows to obtain posterior distributions of the parameters with respect to the internal data. The application of the bayesian inference to our problem is described in paragraph 1.5.

Separate section is devoted to the analysis of methods for precise loss aggregation. Finally, we discuss how the whole methodology can be applied to other problems in interest, such as the insurance of OpRisk.

## 1.3 Applying extreme value theory

Extreme value theory, originated in the 1920's in the works of Fisher and Tippett [9], plays an important role in actuarial problems. It has been highlighted in series of books: [8], [21] to mention some. A number of papers on applications of extreme value theory particularly to OpRisk also appeared recently, see e.g. [22], [4] and others.

One of the main subjects of these investigations is the analysis of the parameters' estimation. It is no wonder that the parameters' estimators for heavy tailed distributions often lack precision. Furthermore, some certain difficulties arise in comparison and selection of models and distributions within the EVT. In particular, say two different distributions (even within one class) may both produce a satisfactory fit for the loss severity data. However, one may observe a multiple difference in resulting estimators for the VaR of the aggregate loss distributions depending on which of those distributions were chosen as a model for the single loss distributions.

Practically all authors who describe and contribute EVT, make a note that this theory must be applied very carefully, as it deals with extrapolating to the area of very large losses. The authors of [8], in their comprehensive review often address to the statement made by Richard Smith: "There is always going to be an element of doubt, as one is extrapolating into areas one doesn't know about". Here we present an example that illustrates the exclusive importance of choice of the appropriate model for the severity distribution and make a brief outlook for the work which we intend to do in this direction.

In order to obtain severity distributions we use the peaks-over-threshold ($POT$) method, see e.g. [8]. For modelling the distribution of exceedances over a certain threshold mainly two probability distributions were used: generalized Pareto distribution (GPD) and three-parameters Weibull distribution. The cdf for GPD is given by

$$G_{\xi,\mu,\beta}(x) = 1 - \left(1 + \xi\frac{x-\mu}{\beta}\right)^{-1/\xi}, \qquad \text{for } x > \mu, \qquad (1)$$

where $\mu \geq 0$ is the location parameter (in our modelling scheme it coincides with the selected threshold), $\beta$ is the scale parameter and $\xi$ is the shape parameter.

The cdf of the Weibull distribution, in its generalized form — with shape parameter $\xi > 0$, scale parameter $\beta > 0$ and location parameter $\mu \geq 0$ — can be written as

$$W_{\xi,\mu,\beta}(x) = 1 - e^{-((x-\mu)/\beta)^{\xi}}, \qquad \text{for } x > \mu. \qquad (2)$$

GPD plays a major role in EVT as it is often used as the limit distribution of scaled excesses over a high threshold. In the frame of our investigation, we notice that for certain business lines GPD overestimates the heavy–tailedness of the real loss distribution. For such BL's we compare GPD with Weibull distribution, see Figure 1.3.

For BL 2, the value of Kolmogorov-Smirnov test for the goodness of fit is 0.044 with the corresponding p-value 0.05 for the GPD with fitted parameters $(\xi, \mu, \beta) = (1.28, 20000, 21050)$, while for the fitted Weibull with parameters $(\xi, \mu, \beta) = (0.61, 10^5, 3.13 \cdot 10^6)$ this test gives 0.025 with the corresponding p-value equal to 0.028.

To illustrate the possible effect of using Pareto instead of Weibull with these parameters, one may calculate the difference between 0.999-quantiles of the corresponding single-loss distributions. In our case, the quantile for GPD is $113.8 \cdot 10^6$ which is 53% larger in relation to Weibull model, for which we have $74.4 \cdot 10^6$. For the aggregate loss distribution, the difference can be much more significant. Note that GPD with the shape parameter equal to 1.28 corresponds to the case of infinite expectation.

What could one do in order to reduce the risk of choosing the wrong distribution? A reasonable answer could be: introduce an additional backtesting of parameters' estimators corresponding to different models. We tend to contribute this problem by considering the so-called generalized moment fit, which could be used together with MLE and may serve as an additional indicator of goodness-of-fit. The work on this issue [19] is currently in progress.

## 1.4 Models for loss occurrences and trends testing

As already mentioned, we model counting processes with homogeneous and inhomogeneous Poisson processes, as well as negative binomial distributions.

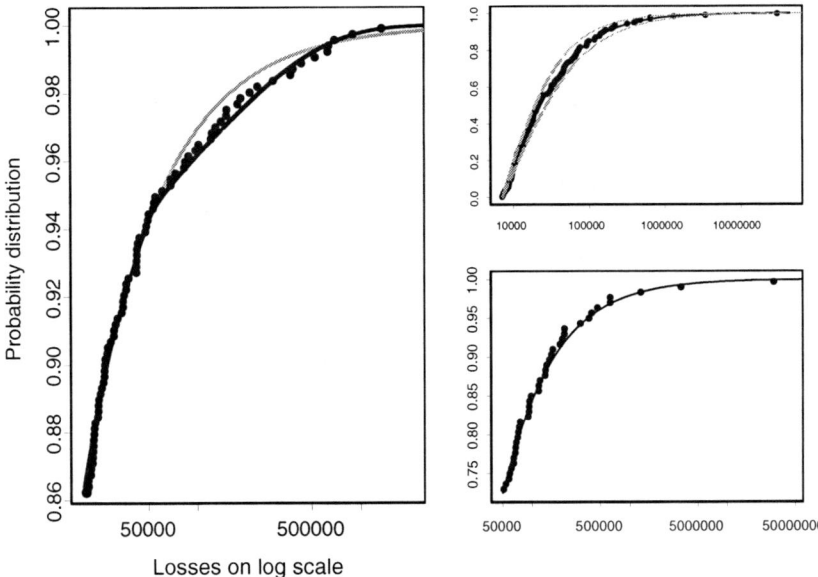

**Fig. 1.** Top right: BL 1 — GPD fit with error bounds (based on error bounds of severity parameters estimated via MLE), bottom right — tail of BL 1 fit, Left: BL2 — Comparison of GPD (grey) and Weibull (black) for the tail.

*Remark 1.* Working with the negative binomial distribution we use the following parameterization:

$$\mathbf{P}(N = k) = \binom{k + r - 1}{k} \left(\frac{1}{1 + b}\right)^r \left(\frac{b}{1 + b}\right)^k, \qquad (3)$$

with $r > 0$, $b > 0$. The expectation and the variance are equal to $\mathbf{E}(N) = rb$ and $\mathrm{Var}(N) = rb(b + 1)$.

Graphical diagnostics can be applied for checking the hypothesis that the number of losses follows a Poisson process. Specifically, if the counting process of losses follows a Poisson law with intensity $\lambda$, then the scaled inter-occurrence times

$$Z_i = \lambda \times (T_i - T_{i-1}), \qquad (4)$$

with $T_0 = 0$, should be unit exponentially distributed. This could be checked graphically with the corresponding QQ-plot. A scatterplot of the $Z_i$ also gives a notion about the evolution in time.

We find it useful to apply additional diagnostics in order to figure out how precisely the data follows the GPD law with estimated parameters, and also

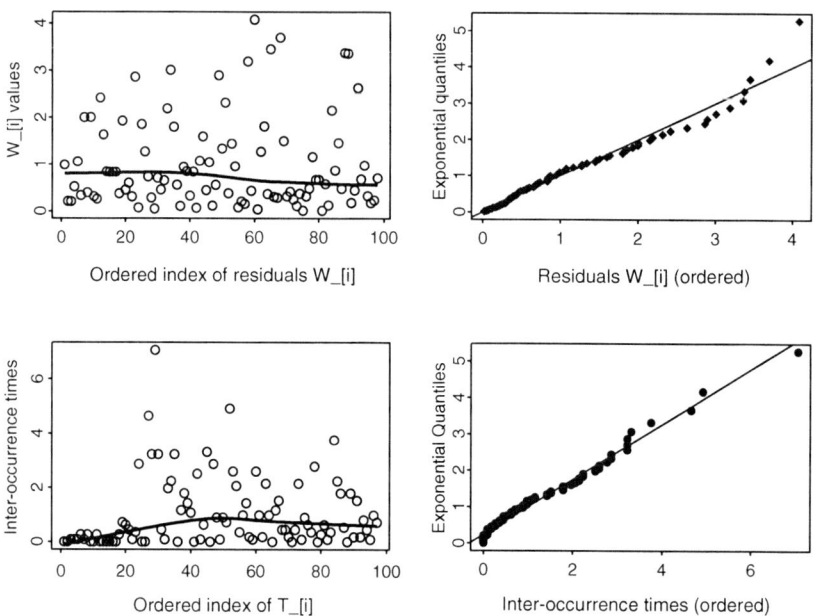

**Fig. 2.** Diagnostics for GPD-Poisson model, BL 3

to trace the possible evolution of the data in time within the selected time interval. Namely, residuals in the sense of [6]

$$W_i = \frac{1}{\xi} \log\left(1 + \xi \frac{X_i - \mu}{\beta}\right) \tag{5}$$

should be also i.i.d. unit exponentially distributed, if the hypothesis that the distribution of data $X_i^U$ exceeding the threshold $U$ follows GPD is right. This could also be checked using graphical diagnostics, if we construct a QQ-plot of the ordered data $W_i$ against exponential quantiles. Besides, the plot of the residuals $W_i$ against the order of occurrences, superimposed by a smooth curve fit (see Figure 1.3), allows to determine whether some evolution of the residuals $W_i$ in time exists (here we follow the methodology proposed in [22]).

Figure 1.3 shows some trend in the frequency of occurrences for BL 3. Note that, at the same time, ordered severity residuals don't show any evident trend, hence for BL 3 losses are becoming more frequent without consequent change of the severity. However, joint trends in the severity and frequency can be observed quite often in OpRisk data, and the modelling of joint trends is intensively discussed in the literature, see e.g. [22, 28].

If some trend in the frequency is detected, then we have to extrapolate it to the future time period over which the aggregate loss has to be considered and to estimate the expected frequency over this period. Hereafter we assume that the constant frequency parameter entering the formulae for the distribution of the aggregate losses is either the estimated constant or the predicted expected value.

We would like to give some references to the recent work and the work in progress dealing with trends in severity and joint trends. Particularly, [13] is devoted to the analysis of the impact of not taking the trend in severity into account. In [30], the problem of the parameters' estimation in the presence of joint trends is considered for the case when the threshold for the scaled losses is also varying in time.

*Alternative models*

As noticed, the analysis of the loss frequency processes often shows the change of the frequency in time, so that loss events become more or less frequent in average (see Figure 1.3). Observing trends in the frequency, one might be interested in finding the models which would be flexible enough to reflect both the general tendency in frequency change and the latest fluctuations.

In our opinion, one of the models deserving attention in the frame of this problem could be the so called "Nested-volatility model", introduced in [33], which incorporates short-memory, long-memory and jump dynamics in the volatility of asset returns. Though such kind of processes are used in application to finance, we find it interesting to apply this particular process to the risk problem. The Nested-volatility model can be represented as follows. It is assumed that the process of the arrival of occurrences (following the original notation, we shall call them "jumps" is this model) is a Poisson process with intensity changing from one occurrence to another as

$$\lambda_t = \lambda_0 + \rho \lambda_{t-1} + \eta \zeta(t). \tag{6}$$

So the current jump of the intensity depends on both its previous value and the intensity residual, with the intensity residual being defined as

$$\zeta(t) \equiv \mathbf{E}[N_{t-1} \mid I_{t-1}] - \lambda_{t-1}, \tag{7}$$

where $\mathbf{E}[N_{t-1} \mid I_{t-1}]$ is the ex-post expected number of jumps occurring from $t-2$ to $t-1$. In this model, the parameters $\rho$ and $\eta$ are to be estimated (e.g., by the maximum likelihood method).

More detailed consideration of the these models is beyond the scope of the present paper. However, implementations in practice and comparison to standard models, with respect to e.g. Akaike information criterion, show that such models can fit the OpRisk data quite well.

Obviously, the model (7) can be modified in order to fit the needs of OpRisk modelling better. Say, the homogeneous Poisson process can be replaced by an inhomogeneous one. Another way is to build a similar model on a base of negative binomial distribution. Besides, there is a possibility to combine the model like (7) with bayesian inference. We plan to devote a separate work to the frequency models in OpRisk.

## 1.5 Internal and external data: introduction to the bayesian methodology

The task of mixing internal and external data properly poses a very important and non-trivial problem. The use of external data becomes essential when there are not sufficient internal loss records to produce satisfactory distribution fits. At the same time, external data must not hide possible peculiarities of the internal data.

The problem of mixing the data from different sources has been addressed quite a lot in the literature, see e.g. [11], [10], [18]. Some authors found bayesian inference to be a preferable approach to this task. In particular, the works [31] and [18] deal with the application of bayesian approach to the task of combining internal loss data with expert opinions, as well as with external data. In [31] the common methodology of combining the bank data with expert opinions is described and some special cases including lognormal and classical Pareto distributions for loss severity are considered. Some important results concerning bayesian methodology for Pareto-type distributions were obtained in [16] and [25]. Here we make just a brief description of the bayesian inference (see e.g. [17] for more details) in application to the problem of mixing internal data with external data. Besides, we consider an example when the desired distributions can be calculated analytically (this example complements the results of [31]) and overview the methods for numerical modelling for all other cases.

Let us recall some assumptions and notations of the theory based on Bayes' theorem. The basic idea of this approach is to consider the parameter of the distribution as a random variable (if there are several parameters, then we deal with a random vector). Then the *prior distribution* $\pi(\theta)$ is defined as a probability distribution over the space of possible parameter values. Assume that a random variable $X$ has a distribution depending on the parameter vector $\theta$. Then the *model distribution* $f_{\mathbf{X}\,|\,\Theta}(\mathbf{x}|\theta)$ is the probability density for the observed sample given a particular value of the parameter (it is identical to the likelihood function). The *posterior distribution* denoted by $\pi_{\Theta\,|\,\mathbf{X}}(\theta\,|\,\mathbf{x})$ is the conditional probability distribution of the parameters given the observed data.

According to *Bayes' theorem* the posterior distribution of parameters is related to the prior distribution in the following way:

$$\pi_{\Theta\,|\,\mathbf{X}}(\theta\,|\,\mathbf{x}) \propto f_{\mathbf{X}\,|\,\Theta}(\mathbf{x}\,|\,\theta)\pi(\theta). \tag{8}$$

Let us consider how this theory can be applied to the problem of estimating severity distributions of operational losses. We should take into account both the internal data from the Bank and the external database, in which the loss data from many banks are accumulated. Values of parameters, estimated from the external data, correspond to the expectation under the prior distribution of the parameter vector while the variance of the prior distribution includes a measure of uncertainty of the estimate. Evaluating the model $f_{\mathbf{X}|\Theta}(\mathbf{x}|\theta)$ at the internal data points corresponding to the selected BL and plugging the result into (8), one is able to compute the posterior distribution of the parameters.

Applying this theory to the present problem, we have, first of all, to choose appropriate prior distributions. Under the assumption that the loss severity distribution is GPD and that the loss frequency follows a Poisson process, we need distributions for the shape and scale parameters of the GPD, as well as for the Poisson intensity of the occurrences.

Recall one of the key concepts of the bayesian theory: a prior distribution is said to be a **conjugate prior distribution** for a given model if the resulting posterior distribution is of the same type as the prior. Consider the Pareto distribution with the cdf given by

$$\widetilde{G}_{\xi,\beta}(y) = 1 - \left(1 + \frac{y}{\beta}\right)^{-1/\xi}, \qquad \text{for } y > 0. \tag{9}$$

*Remark 2.* Compare this to the GPD distribution (1). The difference between these two forms for Pareto distribution is important for us here, as they should be handled differently while applying bayesian inference. Often the Pareto distribution is used in the form $\widetilde{G}_{\xi,\beta}(y) = 1 - \left(\frac{y}{\beta}\right)^{-1/\xi}$, (for $y \geq \beta$). That is the form which was considered in [31].

Well known results (see e.g. [17]) tell us that if the model distribution is the Pareto distribution of the form (9) with the scale parameter $\beta$ equal to one, then the gamma distribution for the shape parameter $\xi$ is a conjugate prior. The same is true for the gamma distribution as prior for the Poisson intensity. In this case estimation of posterior parameters becomes rather simple. Unfortunately, for the scaled Pareto distribution with $\beta \neq 1$, as well as for the GPD given in (1), such a convenient choice for the prior distribution of $\xi$ does not exist.

In certain cases the model and the prior distribution can be chosen such that the joint distribution of the parameters (as a product of the likelihood function and the prior distribution) can be split into marginal distributions of each parameter. Then the posterior distribution of each parameter can be calculated directly. This is for example the case when the model distribution is the Pareto (9). If, moreover, the prior for the shape parameter is the gamma distribution then an explicit representation can be derived (see [25] for details):

$$\lambda^* = \frac{a + k}{b + T}, \tag{10}$$

$$\xi^* = \left[ \int \frac{c+k}{d + \sum_{i \leq k} \log(1 + X_i/\beta)} \tilde{p}(\beta) d\beta \right]^{-1}, \tag{11}$$

$$\beta^* = \int \beta \tilde{p}(\beta) d\beta, \tag{12}$$

where $k$ is the sample size, $T$ defines the period of the observation (expressed in the number of years), $a$ and $b$ are the parameters of the prior distribution of the intensity $\lambda$, which, in this model, has the gamma density $\gamma_{a,b}$. The parameters $c$ and $d$ come from the prior distribution of the shape parameter, where we assume that $1/\xi$ is distributed following $\gamma_{c,d}$. The posterior distribution of the scale parameter $\beta$, denoted by $\tilde{p}(\beta)$, is connected with the prior distribution $p(\beta)$ by the following expression

$$\tilde{p}(\beta) = \beta^{-k} \left( d + \sum_{i \leq k} \log(1 + Y_i/\beta) \right)^{-(c+k)} \cdot \exp\left( - \sum_{i \leq k} \log(1 + Y_i/\beta) \right) \cdot p(\beta).$$

In most other cases, however, no evident way to obtain an explicit expression for posterior parameters can be found. In these cases numerical methods such as Monte Carlo Markov chain (MCMC) have to be used to calculate marginal distributions from the joint distribution. We used MCMC for the calculation of posterior distributions for both GPD and Weibull model distributions.

Considering MCMC, both Gibbs sampler and Metropolis-Hastings algorithm (see e.g. [26]) are appropriate for the implementation of bayesian inference. Probably the most efficient for the tasks like this is the so-called *Random walk Metropolis-Hastings within Gibbs algorithm*. Its description and properties can also be found in [26].

Criticism about the bayesian approach often concerns the uncertainty of the choice of prior distributions. In situations like ours, when there is a separate database that serves for the prior parameters' estimation, it is natural to consider priors as random variables distributed according to the data that served for their estimation. When MLE is used for the estimation of priors, there are several well known methods to approximate the distribution of an MLE estimator. Particularly, one could apply the bootstrapping techniques or use the asymptotical distribution of the MLE estimator. However, each of these ways has its disadvantages. Besides, the distribution of the prior estimator should be obtained in a form convenient to link it with the numerical procedure of the bayesian estimation for posterior distributions. Altogether, it forms a complex of problems of parameter uncertainty in the frame of bayesian estimation within EVT. We briefly overview the existing results in Appendix, leaving some important problems for the future work.

In the present work, we used MLE estimator and its asymptotic variance (see Appendix) as the mean and the variance for the distribution of a corresponding prior. However, we didn't use normal distributions arising from the asymptotics of MLE as priors, but rather the distributions which are more convenient in the frame of the bayesian inference implementation. Though this is not strictly accurate approach and may lead to additional uncertainty, the loss of precision of the final result cannot be too crucial. As mentioned, we plan to address these matters in the future work in more details.

# 2 Estimation of the aggregate loss distribution and its characteristics

We pass to the brief description of the methods for loss aggregation which we used in our work. Besides the Monte Carlo technique and the method based on fast Fourier transform (FFT) which we deal with, the recursive methods are also quite effective techniques for the loss aggregation. We do not review the recursive methods here, as their detailed overview and comparison with FFT was made in [32].

The methodology described below is applied to separate BL's in order to calculate aggregate loss distribution for each stand-alone BL. However, if the number of internal loss events belonging to a particular cell of the BL-ET matrix is sufficient to produce satisfactory parameter adjustment, then we are also able to calculate the distribution of the aggregate loss for separate ET's within each BL. This issue will be mentioned once again in the frame of insurance.

We are interested in yearly aggregate loss distributions for given BL's, and in particular, the risk measure that we need to calculate for each BL is the Value-at-Risk:

$$\mathrm{VaR}_\alpha(S) = \inf\{\, s \in \mathrm{R} \,:\, \mathbf{P}(S > s) \leq 1 - \alpha \,\}.$$

The quantile level used is $\alpha = 99.9\%$, which corresponds to the usual capital charge under the Advanced Measurement Approach of the Basel II guidelines.

## 2.1 Monte Carlo simulations for aggregating operational risk

Monte Carlo simulation is widely used for operational risk aggregation. Due to its technical simplicity, it plays an important role in the problem of calculating aggregated risk and its results can be compared to the ones obtained by other methods. For each $\mathrm{BL}_i$ $i = 1, \ldots, m$ we proceed as follows:

- Simulate $n$ yearly losses, for short we denote them by $\widetilde{L_1}, \ldots, \widetilde{L_n}$, by the following procedure:
  - Simulate the number of loss occurrences in this BL i.e. a realization of $N_i$.
  - Subsequently simulate the corresponding loss sizes from the chosen severity distribution for each of these occurrences.

Dealing with heavy–tailed distributions, one usually needs quite a large number of simulations (normally, $\geq 10^5$ simulations for each BL).

- Put the obtained sample in increasing order to get the order statistics $\widetilde{L}_{1:n} \leq \cdots \leq \widetilde{L}_{n:n}$, where $\widetilde{L}_{1:n}$ denotes the smallest of the $n$ simulations and $\widetilde{L}_{n:n}$ the biggest simulated loss.
- The element at position $[\alpha n + 1]$ of the ordered sample, where $[\cdot]$ denotes rounding downwards, is an estimator of the quantile to the level $\alpha$ for the $BL_i$ — thus an estimator of VaR to the level $\alpha$ (e.g. choose $\alpha = 0.999$).

The estimation of high quantiles and thus also of the VaR by Monte Carlo techniques suffers from high variances. Monte Carlo techniques nevertheless provide a higher degree of flexibility when complicated structures appear, e.g. in the framework of insurance problems. However, FFT and recursive procedures, as deterministic methods, give in general more reliable results for comparably simple tasks such as calculating the VaR.

## 2.2 Fast Fourier transformation

Working with characteristic functions (for a fundamental overview see, e.g., [20]) instead of cdf's or densities allows to significantly simplify the calculation of compound distributions and therefore the calculation of the VaR. The application of characteristic functions (chf's) to the loss aggregation is discussed in the literature, e.g. in the presence of heavy tailed distributions (see, for example, [24]). While performing a series of convolutions of the severity distribution is clearly not feasible, under the assumption of independence between loss occurrences and loss sizes one can easily calculate the chf of the aggregate loss.

We shortly recall the derivation of the characteristic function of a compound sum of the form

$$S = \sum_{i=1}^{N} X_i \,, \tag{13}$$

where $N$ is the loss counting variable independent of the i.i.d. sequence $\{X_i\}$. We denote the cdf of the loss sizes by $F_X(x) := \mathbf{P}[X \leq x]$.

For a random variable $N$ taking only nonnegative integer values consider the probability generating function (pgf) $P_N(z) = \mathbf{E}[z^N] = \sum_{n=0}^{\infty} \mathbf{P}[N = n]z^n$ which is defined and analytic at least for $|z| \leq 1$. Considering the power series expansion of this function $P_N(z) = \sum_{n=0}^{\infty} p_n z^n$ one is able to retrieve the distribution $\mathbf{P}[N = n] = p_n$ for $n \geq 0$ by calculating the coefficients of $P_N(z)$.

Denoting the chf of the loss size distribution as

$$\widehat{f}_X(u) = \mathbf{E}[e^{iuX}] = \int_{-\infty}^{\infty} e^{iux} dF_X(x) \,, \quad \text{for } u \in \mathbf{R} \,,$$

and the chf of the distribution of the compound sum as $\widehat{f}_S(u)$, we have the well known representation

$$\widehat{f}_S(u) = P_N(\widehat{f}_X(u)). \tag{14}$$

For the occurrences following a Poisson distribution with intensity $\lambda$ formula (14) becomes

$$\widehat{f}_S(u) = \exp(\lambda(\widehat{f}_X(u) - 1))$$

and for the negative binomial distribution with parameters $r$ and $b$ the corresponding expression is

$$\widehat{f}_S(u) = \frac{1}{(1 + b - b\widehat{f}_X(u))^r}.$$

Hence, an obvious approach to compute the pdf of the aggregate loss for a fixed BL is to calculate the (14) and then to invert it numerically using the discrete Fourier transformation (DFT). Note that in many cases such as GPD the chf of loss sizes $\widehat{f}_{L_{i,1}}(u)$ can not be calculated explicitly, but it can be computed via DFT as well. An efficient algorithm is provided by the Fast Fourier transformation (FFT) which is available in most statistical or mathematical computer packages.

Thus, the scheme for calculating the density of the aggregate loss is the following:

- Choose an equidistant grid of points at which you want to approximate the loss density say $x_0 \leq \ldots \leq x_{n-1} := q_M$ where $q_M$ denotes some upper bound.
- Calculate the density of the severity distribution at these points $f_X(x_j)$ for $j = 1, \ldots, n$.
- From this sequence, compute the sequence $\widehat{f}_X$ for $l = 0, \ldots, n - 1$ using the FFT:

$$\widehat{f}_l = \sum_{k=0}^{n-1} e^{\frac{2\pi i}{n} kl} f(x_k). \tag{15}$$

- Plug the result into the expression for the chf of the aggregate loss (14) and perform the inverse Fourier transform in order to calculate an approximation of the density of the aggregate loss on this grid.

Using this calculated density we get an approximate quantile and thereby the Value-at-Risk.

*Remark 3.* Dealing with heavy–tailed distributions such as GPD, one faces the problem of properly choosing the right endpoint $q_M$, up to which one considers the loss size distributions. Obviously, fixing $n$ and choosing $q_M$ too high causes a coarser grid leading to an increase of the discretization error. On the other hand, $q_M$ should be chosen high enough in order to capture the desired quantile, e.g. in our case the aggregate loss cdf at $q_M$ must exceed 0.999.

*Aliasing reduction*

It is well known that FFT, especially applied to heavy tailed distributions, produces some typical errors, with so-called *aliasing error* being the most significant and dangerous for the final result.

Artificially adding zeros to the initial input sequence – so called *padding by zeros* – is, perhaps, the simplest tool for reducing the aliasing error. However, it is very time demanding. So, for the n-fold convolution one would have to use $\{\widetilde{f_k}\}$, $k := 0, \ldots, (nN - 1)$ with $\widetilde{f_k} = f_k$ for $k \leq N - 1$ and $\widetilde{f_k} = 0$ for $k > N - 1$ to eliminate the aliasing error completely.

In order to reduce the aliasing error significantly without a loss of the speed of calculations, another method was proposed in [27]. Specifically, one can apply a transformation which decreases the input values at the right-hand end of the grid and which can be easily inverted. Probably the simplest of such transformations is the exponential one. Obviously, if $f_\tau(x) := f(x)e^{-x/\tau}$ then

$$(f * g)(x) = e^{x/\tau}(f_\tau * g_\tau)(x).$$

The use of exponential window for aliasing reduction allows to increase the precision of the FFT significantly, as shown in [27]. We used this method in the present work as well.

# 3 An overview of the combined methodology and its results

In this chapter we present numerical results of the VaR calculation for all business lines. Since confidentiality does not allow us to present real values, absolute values of *all* results are fictitious, but the relative proportions are real.

## 3.1 Overview of results

Numerical results presented here concern at first hand Values-at-Risk for separate BL's, as due to the *Basel II* regulation rules [1], the basic recommendation for the OpRisk capital charge calculation is given by

$$C = \sum_{l=1}^{K} \text{VaR}_\alpha^l,$$

where $l = 1, \ldots, K$ denotes BL's, and $\text{VaR}_\alpha^l$ denotes the Value-at-Risk of the loss of the corresponding BL $l$. As already mentioned, the quantile level we need to capture is $\alpha = 99.9\%$.

*Remark 4.* Surely, simply summing up quantiles over the BL's to obtain the quantile of the total loss distribution might be too conservative, as it corresponds to the special case of perfect dependence between risks (see e.g. [2] for details). Accurate quantification of the dependencies between risks, including BL's, is a non-trivial problem. It has been discussed quite intensively in literature, see e.g. [21]. A contribution to this topic is also a subject for the future research.

Besides the numerical results concerning VaR, we also find it illustrative to present some graphical results for the computed aggregate loss distributions. On Figure 3.1, different shapes of the aggregate density of BL 1 and BL 7 are due to the difference in severity and frequency parameters: BL 7 is more heavy tailed in severity, but the frequency is higher for BL 1: the mean for observed loss frequency is 29.8, in contrast to 3.2 for BL7.

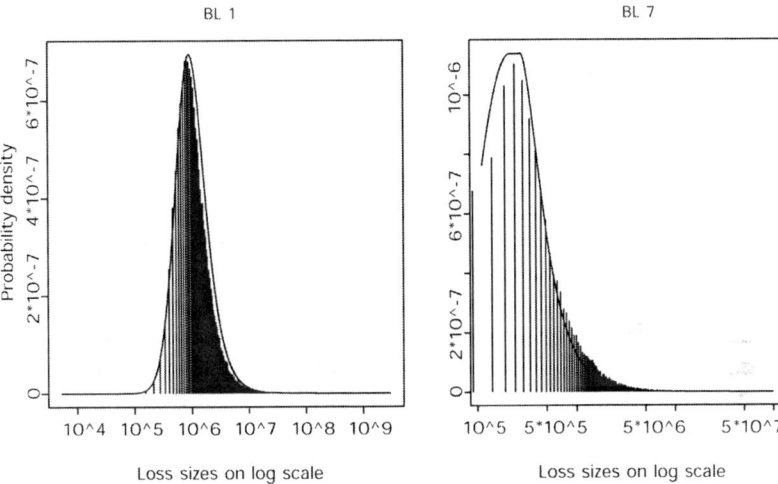

**Fig. 3.** Aggregate loss density plots for two selected BL's. Solid line — FFT results, histogram-type — Monte-Carlo.

In Table 1, we present results for Values-at-Risk, obtained with FFT and compare them with Monte Carlo results. Using the results of [27], we are able to estimate the absolute error of FFT results. To estimate approximate confidence intervals for Monte Carlo results, a well known method based on the sample order statistics and using Binomial distribution (see e.g. [12]) was used. This method allows to calculate approximate standard errors of Monte Carlo estimates and to estimate approximate error bounds at a selected level. In our analysis, we used the level 0.95 for confidence intervals.

As one observes from Table 1, Monte Carlo results for practically all BL's are far from the bounds of the true results, while FFT calculations have satisfactory precision.

*Remark 5.* The "true" results for the Value-at-Risk meant here are, of course, dependent on the estimated model parameters. To obtain a realistic picture of the absolute error bounds, one has to take into account confidence intervals for the estimated parameters. This complex problem is briefly outlined in Appendix.

| | Severity | Frequency[1] | MC | MC error bounds | FFT | FFT error bounds |
|---|---|---|---|---|---|---|
| **Line 1** GPD-NB | $\mu = 3500$ $\xi = 1.12$ $\beta = 7460$ | (33.07, 0.9) | 662 | $(646, 691)$ | 656.12 | $(655.5, 656.7)$ |
| **Line 2** W-P | $\mu = 5 \cdot 10^5$ $\xi = 0.61$ $\beta = 3.13 \cdot 10^6$ | 44.7 | 67.4 | $(65.9, 69.6)$ | 68.34 | $(68.0, 68.6)$ |
| **Lines 3,8,9** GPD-NB | $\mu = 7500$ $\xi = 0.71$ $\beta = 7375$ | (23.69, 2.9) | 33.5 | $(31.3, 36.2)$ | 32.33 | $(32.0, 32.7)$ |
| **Line 4** W-P | $\mu = 4 \cdot 10^5$ $\xi = 0.52$ $\beta = 1.38 \cdot 10^6$ | 7.1 | 26 | $(24.6, 28.8)$ | 27.3 | $(27.22, 27.29)$ |
| **Line 6** GPD-P | $\mu = 10^4$ $\xi = 1.15$ $\beta = 18100$ | 2.2 | 108.5 | $(95, 129)$ | 110.17 | $(109.93, 110.4)$ |
| **Line 7** GPD-P | $\mu = 10^4$ $\xi = 1.2$ $\beta = 15600$ | 3.2 | 212 | $(197, 234)$ | 209.47 | $(209.0, 209.9)$ |

**Table 1.** Fitted parameters and VaR for various business lines. The entries in the first column tell the models used for the severity, as well as for the frequency, where "W" stands for Weibull distribution, "P" for Poisson distribution and "NB" for negative binomial distribution

Business lines 3, 8 and 9 (retail banking, retail brokerage and private banking) are put together due to the advice by experts of the Bank who consider the mechanism of loss generation in these three BL's to be very similar. For BL's 2, 4, 6 and 7 external data were used to obtain prior estimates of the severity parameters first, and then bayesian inference was applied to adjust the parameters with respect to the internal data according to the bayesian

---

[1] For the negative binomial case we use the parametrization $(r, b)$ defined according to Remark 1.

methodology. In order to model losses below the thresholds which were used to fit the severity parameters, empirical distributions based on the internal data were used for all BL's (which is of special importance for BL 2 and BL 4 due to relatively high thresholds).

## 3.2 A note on the insurance of operational losses

We make a brief remark on how the described methodology for the capital charge calculation can be applied to the problem of the OpRisk insurance in its simple form.

As already mentioned, bayesian methodology allows us to make the parameters' adjustment with respect not only to separate BL's, but also w.r.t. selected ET's, i.e. for any cell of the BL-ET matrix, which contains enough data (so that the variance of the posterior parameter estimator would not be too large). While working with BL's only, the prior distribution of parameters is provided by the external data for this selected BL, and the internal data for this BL serves for the posterior parameters' estimation. If one needs to work also with separate ET's, then the estimated (posterior) parameters for the selected BL play a role of the prior estimators, and the internal losses within a particular cell of the BL-ET matrix are used in the bayesian inference to obtain posterior estimators.

In the frame of the insurance problem one is no longer that much interested in the high quantiles, but rather in such values as the mean of the distribution of aggregate insurance payments for particular cells of the BL-ET matrix, in order to estimate the fair insurance premium. Dealing with heavy tailed distributions, we can't work with such values as the mean. Instead, we were interested in calculating the median of the aggregate payments distributions, i.e. their 50%-quantiles. The calculation of the 50%-quantiles claims less computational efforts, for it is not necessary to calculate the aggregate distribution up to the very tail. Consequently, the single-loss distribution can be truncated at a lower level.

Parameters of the frequency of losses within a single BL and a single ET can also be calculated using the bayesian inference, whether we use the Poisson process or the model based on the negative binomial distribution. As soon as the parameters for both severity and frequency distributions are estimated, we then work with the usual model for the insurance payments $Y_i$

$$Y_i = \begin{cases} 0 & \text{if} \quad 0 < X_i < d \\ X_i - d & \text{if} \quad d \leq X_i < U \\ U - d & \text{if} \quad U \leq X_i < \infty \end{cases}, \tag{16}$$

where $X_i$ are the loss sizes, $d$ is the deductible level and $U$ is the upper limit.

The procedure of the calculation of aggregate insurance payments for the selected cell of the BL-ET matrix is analogous to the calculation of the aggregate for the single BL, described above. As in our investigation of the insurance problem the aim was the calculation of the median of the aggregate distribution, this resulting value is considered as the ideal value of the fair insurance premium and is used in the negotiations with an insurance company about the possible adjustment of the conditions of selected insurance policies.

## 3.3 Conclusions and outlook

In the present paper we highlighted several aspects of the methodology for modelling and measuring operational risk. We made some remarks on the use of EVT in the context of OpRisk and outlined some prospects concerning models for the frequency. Besides, a short overview of the bayesian inference as a tool for mixing internal and external operational loss data is made. In the frame of the problem of the loss aggregation we adapted the algorithm based of the FFT and found this method to be efficient and precise enough to our purpose. We have shown that, using this combined methodology for the risk modelling, a financial institution can cover quite a wide range of problems concerning operational risk, such as the calculation of regulatory capital and of the fair insurance premium for separate lines of business and types of events.

Quite a lot of opened questions mentioned in the paper form the field of the future research.

# A Error bounds estimation

Analysis of the impact of parameters' uncertainty of the aggregate loss characteristics is a very important problem. It has been discussed a lot in the literature, see e.g. [29], [3] for recent overviews.

*Confidence intervals according to asymptotic normality*

Recall that we used MLE estimation of parameters and combined it with bayesian inference for some BL's. In general, there are three common ways to estimate the errors in the situation like ours: to use the bootstrapping, the approach based on the asymptotic normality of MLE or to estimate the errors directly from the bayesian procedure. In the present work, we used mostly the method based on the asymptotic normality. The development of the technique for the accurate and strict estimation of parameters' uncertainty is also the field for the future research.

The problem of calculating approximate confidence intervals for MLE estimators is a classical problem of statistics. Estimating the variance of an MLE in the context of EVT is of particular interest, as the influence of the estimated parameter errors on the VaR is very strong for heavy–tailed distributions. The precision of MLE estimators for GPD was discussed, e.g., in [5, 17, 8, 15]. The well known way to calculate confidence intervals for MLE estimators using asymptotic normality (see [17] for details) allows to estimate the 0.95 confidence interval for the parameter $\theta$ as

$$\left(\widehat{\theta} - 1.96\sqrt{\mathrm{var}(\widehat{\theta})};\ \widehat{\theta} + 1.96\sqrt{\mathrm{var}(\widehat{\theta})}\right),\tag{17}$$

where $\widehat{\theta}$ is the corresponding maximum likelihood estimator and $\mathrm{var}(\widehat{\theta})$ denotes the variance which can be estimated by the inverse of the Fisher information matrix [17].

In cases when the bayesian methodology is used, one needs to take into account the errors associated with posterior distribution of the parameters. In [29], a simple approximation was proposed for this case. It was shown that the posterior distribution $\pi_{\Theta|\mathbf{X}}(\theta\,|\,\mathbf{x})$ which one calculates according to the relation (8) can be approximated by a multivariate normal distribution with covariance matrix calculated as the inverse of the Fisher information matrix with the elements $(\mathbf{I})_{ij} = -\partial^2 \ln \pi_{\Theta|\mathbf{X}}(\theta\,|\,\mathbf{x})/\partial\theta_i\partial\theta_j\big|_{\theta=\bar{\theta}}$, where $\bar{\theta}$ is the corresponding mean.

Applied to our problem, this method allows us to compute confidence intervals for the estimated parameters for each of the BL's in question. Then, in order to estimate the error bounds for corresponding Values-at-Risk, we can again apply one of MCMC algorithms. The results are presented in Table 2.

| Line No | Parameters (shape, scale) | Error bounds | VaR (via FFT) | lower bound | upper bound |
|---|---|---|---|---|---|
| Line 1 | $\xi = 1.12$ <br> $\beta = 7460$ | $(0.95\,,\ 1.29)$ <br> $(6326\,,\ 8594)$ | 656.12 | 115 | 3738 |
| Line 2 | $\xi = 0.61$ <br> $\beta = 3.13 \cdot 10^6$ | $(0.65,\ 0.57)$ <br> $(2.93\,,\ 3.13) \cdot 10^6$ | 68.34 | 54.2 | 88.2 |
| Lines 3,8,9 | $\xi = 0.71$ <br> $\beta = 7815$ | $(0.66\,,\ 0.76)$ <br> $(7268\,,\ 8362)$ | 32.33 | 19.2 | 55 |
| Line 4 | $\xi = 0.52$ <br> $\beta = 1.38 \cdot 10^6$ | $(0.58\,,\ 0.46)$ <br> $(1.21\,,\ 1.55) \cdot 10^6$ | 27.3 | 18 | 44 |
| Line 6 | $\xi = 1.15$ <br> $\beta = 18100$ | $(1.07\,,\ 1.23)$ <br> $(16833\,,\ 19140)$ | 110.17 | 59.6 | 203.8 |
| Line 7 | $\xi = 1.2$ <br> $\beta = 15600$ | $(1.1\,,\ 1.3)$ <br> $(14352\,,\ 16848)$ | 209.47 | 94 | 468 |

**Table 2.** VaR bounds from confidence intervals

*Remark 6.* The lower and upper bounds for VaR displayed in Table 2 are calculated considering the uncertainty of the severity parameters only. In order to determine strict bounds for VaR the uncertainty of the frequency parameters, as well as the error of the procedure of loss aggregation should also be taken into account. However, as shown above, the error of the loss aggregation via

FFT is relatively low in comparison to the error stemming from the parameter uncertainty. Concerning the uncertainty of the frequency parameters, one has to keep in mind that a future evolution is extrapolated.

## Acknowledgment

We would like to thank the anonymous referee for many constructive comments which have helped to improve the manuscript.

# References

1. Basel Committee of Banking Supervision (September 2001) Working paper on the regulatory treatment of operational risk. Available at the web site of the Bank for International Settlements http://www.bis.org/publ/bcbs118.htm
2. Böcker K, Klüppelberg C (2007) Modelling and measuring multivariate operational risk with Lévy copulas. Submitted for publication in PRMIA 2007 Enterprise Risk Management Symposium Award for Best Paper: New Frontiers in Risk Management Award. Available at http://www-m4.ma.tum.de/Papers/
3. Borowicz J, Norman J (2006) The effects of parameter uncertainty in extreme event frequency-severity model. In: Proceedings of 28th International Congress of Actuaries. Available at http://www.ica2006.com/Papiers/3093/3093.pdf.
4. Chavez-Demoulin V, Embrechts P,Nešlehová J (2006) Quantitative models for operational risk: extremes, dependence and aggregation, Journal of Banking and Finance 30(10):2635–2658.
5. Coles S (2001) An introduction to statistical modeling of extreme values. Springer Series in Statistics, Springer–Verlag London Ltd., London
6. Cox D Snell E (1968) A general definition of residuals (with discussion), Leuven Univercity Press, Leuven
7. Cruz M (2002) Modeling, measuring and hedging operational risk, Wiley Finance, John Wiley and Sons, Ltd., New York
8. Embrechts P, Klüppelberg C, Mikosch T (1997) Modelling extremal events for insurance and finance, Springer Verlag, Berlin
9. Fisher R, Tippett L (1928) Limiting forms of the frequency distribution of the largest or smallest number of the sample, Proc. Cambridge Philos. Soc. 24:180–190.
10. Frachot A, Roncalli T (2002) Internal data, external data and consortium data for operational risk measurement: How to pool data properly?. Crédit Lyonnais. Available at http://www.gloriamundi.org/
11. Frachot A, Roncalli T (2002) Mixing internal and external data for managing operational risk, Crédit Lyonnais. Available at http://www.gloriamundi.org/.
12. Glasserman P (2003) Monte Carlo Methods in Financial Engineering. Springer
13. Grandits P, Kainhofer R, Temnov G (2008) Analysis of a time-dependent model with the claims having Pareto distribution.
14. Grandits P, Temnov G (2007) A consistency result for the Pareto distribution in the presence of inflation. Available at http://www.fam.tuwien.ac.at/~gtemnov/

15. Han Z (2003) Actuarial modelling of extreme events using transformed generalized extreme value distributions and generalized Pareto distributions. The Ohio State University, Ohio, USA

16. Hesselager O (1993) A class of conjugate priors with applications to excess-of-loss reinsurance, Astin Bulletin 23:77–90.

17. Klugman S, Panjer H, Willmot G (2004) Loss models. From data to decisions (2nd edition). Wiley–Interscience, Hoboken, NJ, USA

18. Lambrigger D, Shevchenko P, Wüthrich M (2007) The quantification of operational risk using internal data, relevant external data and expert opinions, Journal of Operational Risk, 2(3):3–27.

19. Leitner J, Temnov G (2007) Estimation by fitting generalized moments. Preprint

20. Lukacs E (1960) Characteristic functions. Griffin and Co, London

21. McNeil A, Frey R, Embrechts P (2005) Quantitative Risk Management: Concepts, Techniques, and Tools. Princeton University Press

22. McNeil A, Saladin T (2000) Developing scenarios for future extreme losses using the POT method, Extremes and Integrated Risk Management. RISK books, London. Available at http://www.math.ethz.ch/mcneil/pub-list.html.

23. Panjer H (2006) Operational Risk. Modelling analytics. Wiley Series in Probability and Statistics. Wiley-Interscience, Hoboken, NJ

24. Rachev S, Mittnik S (2000) Stable paretian models in finance. Wiley, London–New York

25. Reiss R-D, Thomas M (1999) A new class of bayesian estimators in paretian excess-of-loss reinsurance, Astin Bulletin, 29(2):339–349.

26. Robert C, Casella G(2004) Monte Carlo Statistical Methods, 2nd Edition. Springer Texts in Statistics, New York

27. Schaller P, Temnov G (2007) Efficient and precise computation of convolutions: applying FFT to heavy tailed distributions, Submitted. Available at http://www.fam.tuwien.ac.at/~gtemnov/

28. Schmock U (1999) Estimating the value of the Wincat coupons of the Winterthur insurance convertible bond: a study of the model risk. ASTIN Bulletin, 29(1):101–163.

29. Shevchenko P (2007) Estimation of operational risk capital charge under parameter uncertainty, Preprint. To appear in the Journal of Operational Risk. Available at http://www.cmis.csiro.au/Pavel.Shevchenko/docs/OpRiskParameterUncertainty.htm

30. Shevchenko P, Temnov G (2008) Modelling operational risk using data reported above varying threshold. Preprint

31. Shevchenko P, Wüthrich M (2006) The structural modeling of operational risk via bayesian inference: combining loss data with expert opinions, Journal of Operational Risk. 1(3):3–26.

32. Temnov G, Warnung R (2008) A comparison of loss aggregation methods for operational risk. Journal of Operational risk. 3(1):3–23. Available at http://www.fam.tuwien.ac.at/~gtemnov/

33. Wang Y-H, Hsu C-C (2002) Short-memory, long-memory and jump dynamics in global financial markets, Preprint. Online-version (2002), available at http://www.ncu.edu.tw/fm/teacher/YHWang.htm

# Workshop on Mathematical Control Theory and Finance

Lisbon, 10-14 April 2007

## Scientific Committee

*Albert N. Shiryaev* (co-Chair), Steklov Mathematical Institute, Russia,

*Andrey V. Sarychev* (co-Chair), University of Florence, Italy,

*Andrei A. Agrachev* SISSA, Italy,

*Ole E. Barndorff-Nielsen*, University of Aarhus, Denmark,

*Jean-Paul Gauthier*, Université de Dijon, France,

*Terry Lyons*, University of Oxford, United Kingdom,

*Nizar Touzi*, École Polytechnique Paris, France.

## Organizing Committee

*Manuel Guerra* (Chair), ISEG/T.U.Lisbon,

*Maria do Rosário Grossinho*, ISEG/T.U.Lisbon,

*Fátima Silva Leite*, University of Coimbra,

*Eugénio Rocha*, University of Aveiro,

*Delfim Torres*, University of Aveiro.

# Tutorials

*Italo Capuzzo Dolcetta*, University of Rome "La Sapienza", Italy,

*Bronisław Jakubczyk*, Polish Academy of Science, Poland,

*Dmitry Kramkov*, Carnegie Melon University, USA,

*Albert N. Shiryaev*, Steklov Mathematical Institute, Russia,

*Nizar Touzi*, École Polytechnique Paris, France,

# Plenary Lectures

*Andrei A. Agrachev* SISSA, Italy,

*Ole E. Barndorff-Nielsen*, University of Aarhus, Denmark,

*Eugene A. Feinberg*, State University of New York at Stony Brook, USA,

*Jean-Paul Gauthier*, Université de Dijon, France,

*Terry Lyons*, University of Oxford, United Kingdom,

*Goran Peskir*, The University of Manchester, United Kingdom,

*Andrey V. Sarychev*, University of Florence, Italy,

*Albert N. Shiryaev*, Steklov Mathematical Institute, Russia,

*Xun-Yu Zhou*, Chinese University of Hong Kong, China.

Printed in the United States
132345LV00001B/1-66/P